Bacteria: Role in Health and Disease

Bacteria: Role in Health and Disease

Edited by Easton Scott

STATES
ACADEMIC PRESS
www.statesacademicpress.com

States Academic Press,
109 South 5th Street,
Brooklyn, NY 11249, USA

Visit us on the World Wide Web at:
www.statesacademicpress.com

ISBN: 978-1-63989-067-5

Cataloging-in-Publication Data

Bacteria : role in health and disease / edited by Easton Scott.
p. cm.
Includes bibliographical references and index.
ISBN 978-1-63989-067-5
1. Bacteria. 2. Bacterial diseases. 3. Bacteria--Therapeutic use. 4. Microorganisms--Therapeutic use. I. Scott, Easton.
QR74.8 .B33 2022
579.3--dc23

Table of Contents

Preface

This book aims to highlight the current researches and provides a platform to further the scope of innovations in this area. This book is a product of the combined efforts of many researchers and scientists, after going through thorough studies and analysis from different parts of the world. The objective of this book is to provide the readers with the latest information of the field.

Bacteria are the microscopic, prokaryotic, single-celled organisms that are a few micrometers in length and have different shapes. Bacteria are present in almost all the habitats including extreme habitats such as radioactive waste, acidic hot springs and the deep biosphere of the Earth's crust. Many species of bacteria are pathogenic and cause infectious diseases such as syphilis, leprosy, cholera, anthrax, bubonic plague, etc. The most common fatal diseases caused by bacteria are respiratory infections like tuberculosis. Despite numerous bacteria being pathogenic and causing diseases, there are some useful bacteria as well. For example, only bacteria possess the enzymes and genes required to synthesize vitamin B12 and provide it through the food chain. This book includes some of the vital pieces of work being conducted across the world, on various topics related to role of bacteria in health and disease. It presents researches and studies performed by experts across the globe. The extensive content of this book provides the readers with a thorough understanding of the subject.

I would like to express my sincere thanks to the authors for their dedicated efforts in the completion of this book. I acknowledge the efforts of the publisher for providing constant support. Lastly, I would like to thank my family for their support in all academic endeavors.

Editor

1

A Disulfide Bond in the Membrane Protein IgaA is Essential for Repression of the RcsCDB System

M. Graciela Pucciarelli[1,2,3†], Leticia Rodríguez[1†] and Francisco García-del Portillo[1]*

[1] Laboratorio de Patógenos Bacterianos Intracelulares, Departamento de Biotecnología Microbiana, Centro Nacional de Biotecnología-Consejo Superior de Investigaciones Científicas (CNB-CSIC), Madrid, Spain, [2] Departamento de Biología Molecular, Universidad Autónoma de Madrid, Madrid, Spain, [3] Centro de Biología Molecular Severo Ochoa-Consejo Superior de Investigaciones Científicas (CBMSO-CSIC), Madrid, Spain

Edited by:
Chew Chieng Yeo, Sultan Zainal Abidin University, Malaysia

Reviewed by:
Nadim Majdalani, National Institutes of Health (NIH), United States
Kevin D. Young, University of Arkansas for Medical Sciences, United States

***Correspondence:**
Francisco García-del Portillo
fgportillo@cnb.csic.es

† These authors have contributed equally to this work.

IgaA is an integral inner membrane protein that was discovered as repressor of the RcsCDB phosphorelay system in the intracellular pathogen *Salmonella enterica* serovar Typhimurium. The RcsCDB system, conserved in many members of the family *Enterobacteriaceae*, regulates expression of varied processes including motility, biofilm formation, virulence and response to envelope stress. IgaA is an essential protein to which, in response to envelope perturbation, the outer membrane lipoprotein RcsF has been proposed to bind in order to activate the RcsCDB phosphorelay. Envelope stress has also been reported to be sensed by a surface exposed domain of RcsF. These observations support a tight control of the RcsCDB system by RcsF and IgaA via mechanisms that, however, remain unknown. Interestingly, RcsF and IgaA have four conserved cysteine residues in loops exposed to the periplasmic space. Two non-consecutive disulfide bonds were shown to be required for RcsF function. Here, we report mutagenesis studies supporting the presence of one disulfide bond (C404-C425) in the major periplasmic loop of IgaA that is essential for repression of the RcsCDB phosphorelay. Our data therefore suggest that the redox state of the periplasm may be critical for the control of the RcsCDB system by its two upstream regulators, RcsF and IgaA.

Keywords: *Salmonella*, IgaA, periplasmic domain, cysteine, disulfide bond, RcsCDB

INTRODUCTION

The RcsCDB phosphorelay is a regulatory system conserved in most members of the family *Enterobacteriaceae* (Majdalani and Gottesman, 2007). A major role of this system is to monitor cell envelope stress, responding to alterations in outer membrane integrity and peptidoglycan structure (Farris et al., 2010; Evans et al., 2013; Konovalova et al., 2016). The tripartite RcsCDB system is atypical compared to most known phosphorelays, normally progressing from a sensor membrane protein to a cytosolic response regulator (Wolanin et al., 2002; Majdalani and Gottesman, 2005). In the RcsCDB system, the signal is transmitted from the sensor inner membrane protein RcsC to the intermediate membrane protein RcsD to end with phosphorylation of a conserved aspartate residue in the RcsB response regulator. The RcsCDB system controls expression of more than 40 genes involved in biofilm formation, synthesis of exopolysaccharide capsule, motility, and virulence among others (Hagiwara et al., 2003; Majdalani and Gottesman, 2007;

Mariscotti and Garcia-del Portillo, 2009; Howery et al., 2016). The genes of the RcsCDB regulon were initially classified in those regulated exclusively by RcsB and, a second group including those involved in exopolysaccharide synthesis, which are controlled by RcsB and the co-regulator RcsA (Dierksen and Trempy, 1996; Navasa et al., 2013). Recent studies in *Escherichia coli* demonstrate that RcsB can heterodimerize with other co-regulatory proteins (Pannen et al., 2016). RcsB has also been shown to have a notable conformational dynamism (Casino et al., 2017), which could explain its capacity for providing different responses depending the phosphorylation status and the type and intensity of the stimulus (Mariscotti and Garcia-del Portillo, 2009; Latasa et al., 2012).

The RcsCDB system displays a feature conserved in most other regulatory networks, regarding its rapid response to the stress signal followed by a progressive decrease in activity once the bacterium adapts to the new environmental conditions (Gao and Stock, 2017). Of interest, these regulatory systems are "prepared to act" as denoted by the presence of all components of the signaling cascade even in the absence of stimulus. Thus, isogenic mutants of *Salmonella enterica* serovar Typhimurium (*S.* Typhimurium) displaying differences in the expression of RcsB target genes produce similar relative levels of the RcsC, RcsD, and RcsB proteins (Dominguez-Bernal et al., 2004).

Two important regulatory elements acting upstream of the RcsCDB system are the outer membrane lipoprotein RcsF and the integral inner membrane protein IgaA. RcsF was first reported as a lipoprotein that transmits a stress signal to the inner membrane sensor RcsC following cell envelope perturbations (Majdalani et al., 2005). IgaA was discovered as an integral inner membrane protein that contributes to attenuate the growth rate of *S.* Typhimurium inside eukaryotic cells (Cano et al., 2001). Subsequent studies revealed that the mucoid phenotype exhibited by a mutant bearing a R188H mutation in IgaA was linked to over-activation of the RcsCDB phosphorelay (Cano et al., 2002; Dominguez-Bernal et al., 2004). IgaA is predicted to have four transmembrane domains with the R188 residue located in one of the cytosolic domains (Dominguez-Bernal et al., 2004). Unlike the wild type IgaA protein, produced at constant levels in actively growing and resting bacteria, the R188H variant is unstable in stationary phase (Dominguez-Bernal et al., 2004). Although the loss of IgaA can be supported in non-growing bacteria, genetic evidence obtained in *S.* Typhimurium and *E. coli* demonstrates that *igaA* is an essential gene (Cano et al., 2002; Cho et al., 2014). Of note, IgaA become dispensable if the RcsCBD system is genetically inactivated (Cano et al., 2002). Moreover, loss-of-function mutations in the RcsCDB system are selected at high rate when attempting to replace the wild-type *igaA* gene by a null allele (Mariscotti and Garcia-Del Portillo, 2008). Altogether, these observations reveal a critical function of IgaA as a dedicated repressor of the RcsCDB phosphorelay in actively growing bacteria. Transcriptomic analyses also pointed to a major role of IgaA in fine-tuning the RcsCDB phosphorelay (Mariscotti and Garcia-del Portillo, 2009).

A recent study has provided the first insights into the mechanism by which the upstream regulators, RcsF and IgaA, could control activity of the RcsCDB phosphorelay (Cho et al., 2014). These authors showed that in steady-state growth conditions, RcsF is exposed in the external face of the outer membrane via interaction with OmpA and BamA, the major component of the β-barrel assembly machinery. Following peptidoglycan stress, RcsF fails to interact with OmpA/BamA and, as a result, retained in the periplasmic space. In this condition RcsF binds to the major periplasmic domain of IgaA to activate the RcsCDB phosphorelay (Cho et al., 2014). Based on the previous functional data obtained with IgaA, the RcsF-IgaA interaction must therefore alleviate the repression that IgaA exerts on the RcsCDB system in non-stimulatory conditions. Envelope stress can also be directed sensed by the surface-exposed domain of RcsF when defects in lipopolysaccharide structure occur (Konovalova et al., 2016).

RcsF has four conserved cysteines that form disulfide bonds (Leverrier et al., 2011). The formation of these disulfide bonds in RcsF depends on DsbC, the main disulfide isomerase, which together with the disulfide oxidase DsbA, control the formation and correct configuration of disulfide bonds (Denoncin and Collet, 2013). Disulfide bridges can play a structural role, as stable bonds, or; alternatively, contribute to catalysis by forming reversible disulfide bonds in the catalytic site (Denoncin et al., 2013). This latter case is exemplified by periplasmic oxidoreductases such as DsbA and DsbC. Whether the disulfide bonds of RcsF play a role in the interaction with IgaA is unknown.

In this study, we investigated the presence of disulfide bonds in IgaA, which contains four conserved cysteine residues in its major periplasmic domain. Our results are consistent with the presence of a disulfide bond in IgaA that is important for its function as repressor of the RcsCDB phosphorelay.

MATERIALS AND METHODS

Bacterial Strains and Growth Conditions

The bacterial strains and plasmid used in this study are listed in Supplementary Table S1. Bacteria were cultured in Luria-Bertani (LB) broth at 37°C in shaking (150 rpm) conditions. To prepare material from mid-exponential cultures, the overnight culture was diluted 1:100 in fresh LB medium and collected at optical density (absorbance at 600 nm) of ~0.2–0.3. The remaining culture was incubated for additional 18 h to obtain stationary phase cultures. When required, the medium was supplemented with ampicillin (50 μg/ml), kanamycin (30 μg/ml), tetracycline (10 μg/ml), or chloramphenicol (10 μg/ml).

Mutagenesis of Periplasmic Cysteines

The mutagenesis was carried out with the Quick-change™ site-directed mutagenesis kit from Stratagene, following manufacture recommendations. The oligonucleotides used for these procedures, including those degenerated introducing the desired point mutations, are listed in Supplementary Table S2. A copy of the *S.* Typhimurium wild-type *igaA* gene was cloned in the pBAD18 vector (plasmid pNG1062, Supplementary Table S1) and used as template for the mutagenesis kit to obtain pNG1062-derivate plasmids containing the mutant alleles (C404S, C425S, C498, and C504S), which were cloned in

E. coli DH5α (Supplementary Table S1). The desired mutations were confirmed by sequencing. To generate the C404S–498S and C404S-C504S mutant alleles, a BlpI/BlpI fragment from the pLR1435 [pBAD18::*igaA*(C404S)] was used to replace the same region in pLR1438 [pBAD18::*igaA*(C498S)] and pLR1481 [pBAD18::*igaA*(C504S)] plasmids. The series of pNG1062-derivate plasmid (pBAD18 backbone) was transferred to the *S.* Typhimurium MD0835 strain [*igaA2*::KXX Δ(*apbE′-rcsC′*)], a mutant not producing IgaA and with an additional mutation in *rcsC* (Mariscotti and Garcia-Del Portillo, 2008). The production of the distinct IgaA variants with mutated cysteine residues was confirmed in the MD0835-derivate strains grown in LB-0.2% L-arabinose and subsequent analysis of total protein extracts by Western assay using anti-IgaA antibody (Cano et al., 2002).

Generation of *S.* Typhimurium Strains with *igaA* Mutant Alleles (C404S, C425S, C498S, C504S) Disposed in the Chromosome

To transfer the *igaA* mutant alleles to the chromosome, they were first moved from the pBAD cloning vector to the pCVD442 suicide vector (Donnenberg and Kaper, 1991), which has the counter selectable marker *sacB*, using *E. coli* DH5α as host strain (Supplementary Table S1). Since we did not initially know whether the *igaA* mutant alleles could support viability (*igaA* is an essential gene in a *rcsCDB*$^{+}$ background), the *igaA* mutant alleles were first moved to the chromosome of *S.* Typhimurium strain MD1446 (*igaA2*::KXX *zhf-6311*::Tn*10d*Tet *rcsC*::Mud*Q*). This was done by conjugation using as donor *E. coli* SM10λpir carrying the respective series of pCVD442 derivate plasmids. Loss of kanamycin resistance in the *S.* Typhimurium recipient strain was indicative of replacement of the *igaA* null allele (*igaA2*::KXX). The cysteine-defective *igaA* alleles were further transferred by P22 phage transduction to a clean wild-type background selecting by tetracycline resistance (flanking marker *zhf-6311*::Tn*10d*Tet). The *igaA* gene was PCR-amplified from all Tet^{R} transductants to confirm presence in the chromosome of the mutation in the codon of the corresponding cysteine residue. Whereas all single *igaA* mutants (C404S, C425S, C498S, and C504S) proved to support viability, no transductants were obtained attempting to transduce to a wild-type background the double mutant alleles C404S–C498S and C404S–C504S. This result was consistent with the viability test performed with the different *igaA* variants expressed from pBAD using different L-arabinose concentrations (see below, **Table 1**).

Western Blot Analyses

Preparation of protein extracts, electrophoresis and Western assay conditions using polyclonal rabbit anti-IgaA antibody were as described (Dominguez-Bernal et al., 2004).

AMS Alkylation Assays

These assays were performed using the 4′-acetamido-4′-maleimidylstilbene-2,2′-disulfonic acid (AMS) reagent, as described (Jurado et al., 2006).

β-galactosidase Assays

Levels of β-galactosidase derived from the *gmm*::*lacZ* transcriptional fusion were assayed as described by Miller, following the chloroform/SDS permeabilization procedure (Miller, 1972). For these assays, bacteria were grown in LB medium to mid-exponential phase ($OD_{600} \sim 0.2$–0.3).

Motility Assays

Motility of the different strains used was monitored by motility assays in soft agar plates, as described (Rosu et al., 2006).

Statistical Analysis

Data were analyzed by one-way ANOVA using Prism version 5.0 (Graph-Pad Software). Differences in values with $P < 0.05$ were considered significant.

RESULTS

IgaA Has Four Cysteines in the Periplasmic Domain Conserved in All Orthologs of Enteric Bacteria

Our previous studies in *S.* Typhimurium showed that IgaA is an inner membrane protein produced at relatively constant

TABLE 1 | Suppression of lethality associated to the *igaA*::Km null allele by ectopic expression of different IgaA variants.

% L-arabinose (w/v)	Number of Km^{R} transductants obtained with an *igaA*::Km P22 phage lysate							
	pBAD18 (empty vector)	***igaA*$^{+}$ pBAD18**	***igaA-C404S***	***igaA-C425S***	***igaA-C498S***	***igaA-C504S* pBAD18**	***igaA-C404S-C498S* pBAD18**	***igaA-C404S-C504S* pBAD18**
0	0	0	0	0	0	0	0	0
0.002	0	17/24	0	0	0	0	0	0
0.02	0	1250/1440	389/369*	437/384*	728/726*	384/390*	0	0
0.2	0	5500/6000	505/454*	393/360	716/720	408/412	484/500*	318/286*
2	0	10500/9000	687/612*	716/717	970/1138	600/350	543/672*	406/454*
5	0	3200/3400	850	618/716	768/815	730	540/672*	492/315*

These transductants were mucoid, denoting reduced repression of the RcsCDB system by the respective IgaA variant. No Km^{R} transductants bearing the igaA::Km null allele were obtained when using 2% glucose (w/v) instead of arabinose.

levels in all growing conditions (Dominguez-Bernal et al., 2004). Programs that predict transmembrane helix regions and protein topology (THMM, SOSUI, TMPred, PredictProtein) indicate that the 710 amino acid protein IgaA has five transmembrane segments, resulting in two cytosolic domains (residues 22–202, 247–336), one small periplasmic loop (221–225) and one major periplasmic domain (residues 358–652) (**Figure 1A**). Mutations in specific residues of the two cytosolic loops (R188H, T191P, G262R) as well as in the periplasmic domain (L514P, L643P) impact negatively the capacity of IgaA to repress the RcsCDB phosphorelay (Dominguez-Bernal et al., 2004). We further noted that the periplasmic domain of *S.* Typhimurium IgaA has four cysteines, (C404, C425, C498, C504) highly conserved in IgaA orthologs found in distinct genera of enteric bacteria (**Figure 1B**). Based on this observation, we assessed whether these periplasmic cysteines could form disulfide bonds and play an important role in function.

To determine the role played by the C404, C425, C498, and C504 residues, we generated isogenic *S.* Typhimurium mutants lacking each of these cysteines. To this aim, we first introduced in the chromosome the corresponding *igaA* point mutant allele using as recipient an *igaA*::KXX *rcsC* strain, to subsequently pass the allele to a wild-type (*rcsCDB*$^+$) genetic background by P22 phage transduction (see section "Material and Methods"). Importantly, none of the C404S, C425C, C498S, and C504S mutations affected protein stability in actively growing bacteria. Thus, in bacteria grown to exponential phase these IgaA variants were detected with similar levels than those of wild-type IgaA or the previously characterized R188H variant (**Figure 2A**) (Dominguez-Bernal et al., 2004). Nonetheless, all these four cysteine variants (C404S, C425S, C498S, and C504S) were unstable in stationary phase (**Figure 2A**). This phenomenon was reminiscent of that observed for other partially inactive variants such as R188H and L514P (Dominguez-Bernal et al., 2004). Therefore, we concluded that the elimination of any of the four conserved residues (C404, C425, C498, and C504) may lead to structural changes in IgaA that affect its stability when bacteria reach stationary phase. Considering the essentiality of IgaA linked to necessary repression of the RcsCDB system, all these cysteine variants were, however, expected to retain some partial function in growing bacteria.

Elimination of the Periplasmic Cysteines (C404, C425, C498, and C504) of IgaA Results in Distinct De-repression Levels of the RcsCDB System

To determine the capacity of the IgaA cysteine variants to repress the RcsCDB phosphorelay, we monitored several phenotypic traits associated to the activity of this regulatory system. We included production of colanic acid capsule, which is positively controlled by RcsCDB; and, flagella production, negatively regulated by the system. In a first series of experiments, we measured in actively growing (exponential) and resting (stationary phase) bacteria the expression levels of *gmm* (*wcaH*), a gene encoding GDP-mannose mannosyl hydrolase, an enzyme involved in colanic acid capsule synthesis. The results showed that the C404S

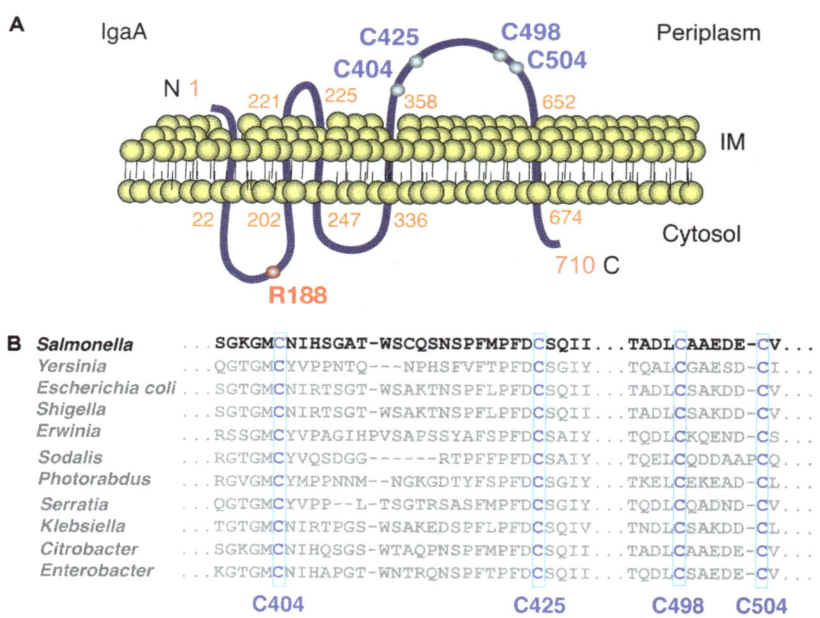

FIGURE 1 | The integral inner membrane protein IgaA has four conserved periplasmic cysteines. **(A)** Predicted topology of the *Salmonella* Typhimurium IgaA protein of 710 amino acids length. The periplasmic cysteines C404, C425, C498, and C504 as well as the cytosolic R188 residue important for IgaA function, are shown. **(B)** Alignment of the periplasmic domain of distinct IgaA orthologs from the indicated members of the *Enterobacteriaceae* family. Alignment was performed with the Clustal Omega tool.

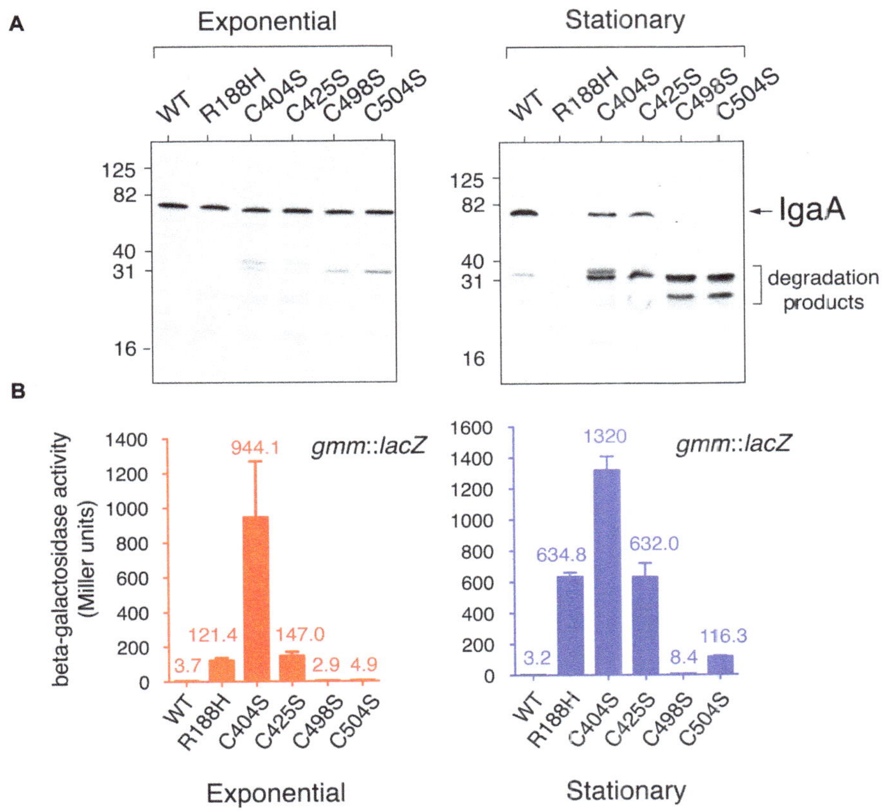

FIGURE 2 | *Salmonella* Typhimurium IgaA variants in conserved cysteines show distinct capacities to repress the RcsCDB system. **(A)** IgaA was detected in western assays using samples obtained from bacteria grown in LB medium at 37°C to mid-exponential (OD_{600} = 0.2) or stationary (OD_{600} = 2.0) phase. Note the degradation of the IgaA cysteine variants when bacteria reach the stationary phase. The previously characterized IgaA variant R188H (Dominguez-Bernal et al., 2004) was used as control of protein unstable in stationary phase. Numbers on the left indicate the position of molecular weight markers (in kDa). **(B)** Activity of the RcsCDB system monitored with a *gmm::lacZ* reporter fusion in bacteria growing to exponential or stationary phases. *gmm* encodes an enzyme involved in synthesis of the colanic acid capsule and is positively regulated by the RcsCDB system. Data are the average and standard deviation of three independent experiments. The averages are shown as numbers on top of the respective bars.

mutation caused partial de-repression of the RcsCDB system, even in actively growing bacteria when the protein remained stable (**Figures 2A,B**). Among the other mutants, a gradual variation in the level of RcsCDB activity was noted following the order C404S > C425S ~ R188H > C504S > C498S > wild-type (**Figure 2B**). Interestingly, the levels of the RcsCDB system in stationary phase inferred from the *gmm::lacZ* reporter fusion increased only slightly despite the partial degradation observed for some of the IgaA mutant proteins such as C404S, C425S, C498S, or C504S (**Figures 2A,B**). These data are consistent with a major role of IgaA in repressing the RcsCDB that is critical only during active growth. These data also indicated that C404 and C425 are cysteines more important for function in comparison to C498 and C504.

Additional phenotypic traits that were examined included the formation of mucoid colonies on plates (signal of capsule formation) and motility assays in soft agar plates. In agreement with the data obtained with the *gmm(wcaH)::lacZ* reporter fusion, the mutant producing IgaA-C404S was highly mucoid and non-motile (**Figures 3A,B**). For the rest of mutants, a gradual variation in RcsCDB activity was noted in the mucoidity and motility tests (**Figures 3A,B**), which in some cases were not completely matching the *gmm::lacZ* assays performed in liquid culture (**Figure 2B**). Thus, although the IgaA-C504S variant exhibited lower *gmm::lacZ* expression than R188H (**Figure 2B**), bacteria producing this C504S variant were slightly more mucoid on plates (**Figure 3A**). R188H, C425S, and C504S variants also displayed an intermediate phenotype in motility despite their variations in the *gmm::lacZ* assays or mucoidy on plates (**Figures 2B, 3A,B**). Such discrepancies in the different tests may be influenced by the different growth conditions used -liquid culture vs. solid agar media plates-. Interestingly, mutations in defined RcsB residues alter the phosphorylation status of this regulator with consequences in either mucoidy or motility, but not in both phenotypic traits (Casino et al., 2017). Some of the mutations described here in the periplasmic cysteines of IgaA could result in distinct RcsB~P/RcsB ratios, a hypothesis to be tested in future studies. Despite the minor phenotypic differences noted in IgaA for the R188H, C425S, and C504S variants; taken together, our data support that among the four periplasmic

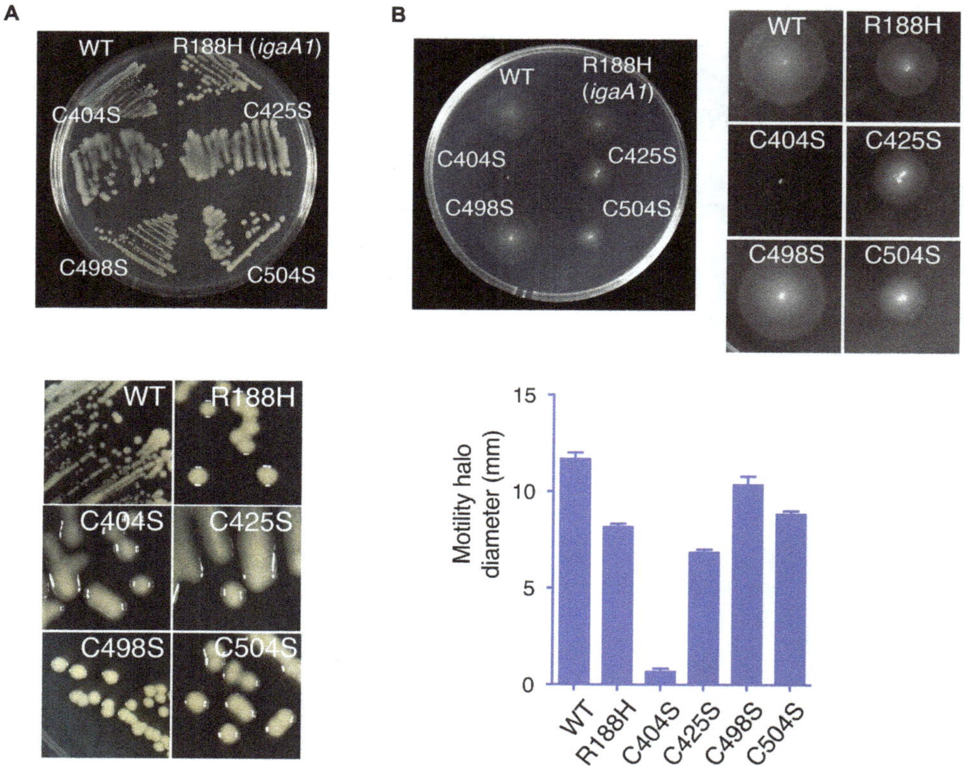

FIGURE 3 | The periplasmic cysteines C404 and C425 are essential for the negative regulation that IgaA imposes over the RcsCDB phosphorelay. **(A)** Mucoid phenotype of strains producing the C404S, C425S, C504S, and R188H variants. **(B)** Motility test in soft agar. Note that the high repression of motility in the strain producing the C404S variant. A graphic showing motility halo diameter (average and standard deviation from three independent assays) is also depicted.

cysteines analyzed, C404 and C425 are residues with a more critical role for repression of the RcsCDB phosphorelay.

IgaA Has One Disulfide Bond in the Periplasmic Domain

To elucidate whether the contribution of the periplasmic cysteines to IgaA function relies in the formation of disulfide bonds, we carried out alkylation experiments using the 4′-acetamido-4′-maleimidylstilbene-2,2′-disulfonic acid (AMS) reagent (Jurado et al., 2006; Denoncin et al., 2013). These experiments were performed in exponential phase (OD_{600} ~0.2–0.3), in which all IgaA variants with mutated cysteines are stable (see **Figure 2A**). AMS is a maleimide compound that binds to free thiol groups. Thus, it is possible to differentiate the presence of a disulfide bond if changes in electrophoretic mobility are detected in samples incubated with AMS and previously treated or not with a reducing agent such as di-thio-threitol (DTT). When S. Typhimurium wild type cells were incubated in solutions with or without DTT and further treated or not with AMS, we detected four IgaA forms with distinct electrophoretic mobility (**Figure 4A**). This result proved the presence of at least one disulfide bond in IgaA.

Besides the four periplasmic cysteines, IgaA of S. Typhimurium has four additional cysteines that map in the first and second transmembrane regions (C14, C219, respectively), the second cytosolic domain (C259) and, the short cytosolic domain encompassing the C-terminal end of the protein (C697). To discard the contribution of these cysteines to the mobility shift displayed by wild-type IgaA in the alkylation assays, we generated a variant lacking the periplasmic domain (ΔI358-W652). This variant, unlike the full-length protein, did not exhibit electrophoretic shift in cells exposed to DTT and subsequently to AMS (**Figure 4A**). Therefore, none of the non-periplasmic cysteines of IgaA contribute to the formation of disulfide bonds. Interestingly, the different electrophoretic forms of IgaA observed in the AMS alkylation assays in wild-type bacteria were also detected in a Δ*dsbA* mutant (**Figure 4A**). This result indicated that the disulfide bond present in IgaA can be formed in the absence of the major disulfide oxidase DsbA.

To define the configuration of the periplasmic disulfide bond(s) inferred in wild-type IgaA, we next performed AMS alkylation assays in isogenic strains lacking each of the four conserved periplasmic cysteines. These assays revealed an important contribution of C404 in the formation of a disulfide bond. Thus, unlike the C425S, C498S, and C504S variants, the C404S variant did not migrate differently in the presence/absence of DTT (**Figure 4B**). Interestingly, the four variants in the conserved periplasmic cysteines (C404, C425,

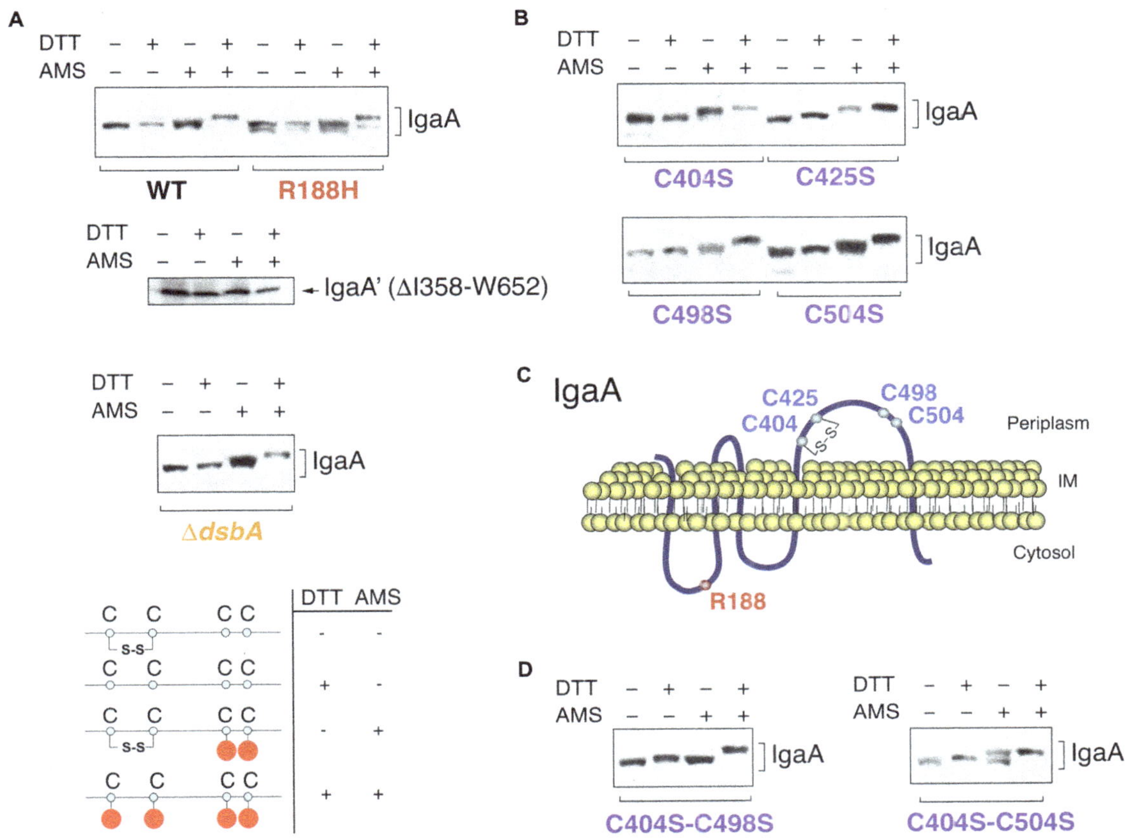

FIGURE 4 | The periplasmic domain of IgaA has a C404–C425 disulfide bond essential for function that can be formed independently of the major disulfide oxidoreductase DsbA. **(A)** Alkylation assays with the 4′-acetamido-4′-maleimidylstilbene-2,2′-disulfonic acid (AMS) reagent reveal electrophoretic shifts in wild-type IgaA but not in a variant lacking the periplasmic domain (DI358-W652). Note the co-existence of distinct forms in the IgaA-R188H mutant and the lack of effect of a Δ*dsbA* mutation in the presence of IgaA forms with distinct electrophoretic mobility. A scheme is shown denoting the changes expected for DTT/AMS incubations for an example of one disulfide bond. **(B)** AMS alkylation assays in the IgaA variants C404S, C425S, C498S, and C504S. Note that the lack of C404 results in no major changes in electrophoretic mobility in the samples treated or not with DTT. **(C)** Proposed configuration of the C404–C425 disulfide bond in the periplasmic domain of IgaA. **(D)** Detection of non-native disulfide bridges in the C404S–C498S and C404S–C504S IgaA variants. Note the similarity of electrophoretic mobility of the non-alkylated and alkylated samples not treated with DTT (see text for details).

C498, and C504) showed differences in mobility when comparing the samples further exposed to AMS (**Figure 4B**). This result led us to consider the presence of one "native" C404–C425 disulfide bond and contrasting outcomes when one or the other cysteine residue is mutated. In the absence of C404, we only see possible a non-native C498–C504 bond with minor consequences in electrophoretic migration as evidenced by the similar behavior of −/+ DTT samples (**Figure 4B**). In contrast, the lack of C425 might favor the formation of non-native C404–C498 or C404–C504 disulfide bonds, which could explain the clear electrophoretic shift observed in the −/+ DTT samples of the C425 mutant (**Figure 4B**). The more important role assigned to C404 and C425 in comparison to C498 and C504 agrees with the much higher RcsCDB activity registered when any of these two cysteines (C404 or C425) is lacking (**Figures 2B, 3A,B**).

Since the IgaA-R188H variant is only partially efficient in repressing the RcsCDB system (Cano et al., 2002; Dominguez-Bernal et al., 2004), we sought to determine whether this mutation could have effect on the C404–C425 disulfide bond. Unexpectedly, we detected molecules with distinct redox state that co-exist in this IgaA-R188H variant (**Figure 4A**). Thus, two forms with distinct electrophoretic mobility were clearly distinguishable in samples non-treated with DTT or the alkylating agent AMS (**Figure 4A**). Interestingly, DTT moved these two forms to a single one that was coincident to the fully reduced form of wild type IgaA (see **Figure 4A**). This result demonstrated that a mutation such as R188H mapping in a cytosolic loop can affect proper formation of a disulfide bond in the major periplasmic loop.

Taken together, these data are consistent with a model in which the function of IgaA as repressor of the RcsCDB system depends on the redox state of its periplasmic domain, which in its functional conformation may involve the formation of a C404–C425 disulfide bond (**Figure 4C**).

The Lack of the C404–C425 Disulfide Bond or the Formation of Alternative Non-native Disulfide Bonds Are Lethal in a RcsCDB$^+$ Background

To further support the essential role that the C404–C425 disulfide bond plays in repression of the RcsCDB phosphorelay, we generated IgaA double mutants. Our aim was to affect the native bridge (C404–C425) and to simultaneously impair alternative non-native disulfide bonds predicted by the AMS alkylation assays and involving either C498 or C504 (**Figure 4B**). Therefore, the new IgaA variants were C404S–C498S and C404S-C504S. The AMS alkylation assays revealed no difference in migration between the non-alkylated and alkylated samples not exposed to DTT for any of these two double mutants (**Figure 4D**). This result, much more evident in the case of the C404S–C498S variant, supported the absence of free periplasmic cysteines and, therefore, the presence a "non-native" disulfide bond (C425–C504). In the C404S-C504S variant, the alkylation assays revealed two bands in the case of the samples non-treated with DTT and exposed to AMS (**Figure 4D**), implying the co-existence of molecules with and without a non-native C425–C498 disulfide bond. This phenomenon resembled at some extent the co-existence of molecules with different redox state observed for the IgaA-R188H variant (**Figure 4A**).

We next assessed whether the non-native C425–C498 or C425–C504 bonds could provide functionality to the protein. To this aim, we analyzed the capacity of the different IgaA variants to suppress lethality associated to the presence of an *igaA*::Km null allele in an RcsCDB$^+$ genetic background (Cano et al., 2002; Mariscotti and Garcia-Del Portillo, 2008). The number of *igaA*::Km transductants obtained in strains bearing inducible pBAD18 vectors expressing the different IgaA variants was determined in the absence/presence of the inducer, L-arabinose. As a negative control, we used a strain with the pBAD18 empty vector, for which no Km^R transductants carrying the *igaA*:.km null allele were obtained regardless the absence/presence of inducer.

These assays showed that the C404S–C498S and C404S–C504S variants were not capable of suppressing the lethality associated to the *igaA*::km mutant when induced at 0.02% L-arabinose, a concentration sufficient to prevent lethality by any of the single mutants lacking each of the conserved cysteine residues (**Table 1**). Furthermore, we observed that when suppressing lethality at higher arabinose concentrations, all transductans were mucoid (**Table 1**). This result was indicative of the extremely limited function of these double C404S–C498S and C404S–C504S variants, even when produced at high levels. Therefore, the non-native C425-C498 and C425–C504 disulfide bonds are not optimal to provide function to IgaA.

DISCUSSION

In this study, we have examined the role in function of four conserved cysteines located in the periplasmic domain of the RcsCDB repressor IgaA. Our data prove the presence in the native protein of a periplasmic disulfide bond in the configuration C404-C425. The alkylation experiments suggest that a C404S mutation renders the protein unable to form any stable disulfide bond, a scenario slightly difference to that of the C425S mutation, in which non-native C404–C498 or C404–C504 bonds were inferred. This interpretation agrees with the high de-repression of the RcsCDB phosphorelay observed in *S.* Typhimurium strains expressing the IgaA-C404S variant. Importantly, our data discarded any compensatory role in function for the non-native disulfide bonds C404–C498 or C404-C504 that apparently occur in the C425S mutant. Therefore, not all disulfide bonds capable of forming in the periplasmic domain support equally IgaA function as RcsCDB repressor. A similar conclusion was reached for RcsF of *E. coli*, which has two not functionally equivalent disulfide bonds (Leverrier et al., 2011; Rogov et al., 2011), with one of them proposed to be more relevant for function (Leverrier et al., 2011).

The alkylation experiments performed with the Δ*dsbA* mutant discarded an absolute requirement of this disulfide oxidase for formation of the C404–C425 bond. Thus, different IgaA forms with distinct electrophoretic mobility were detected in this Δ*dsbA* mutant depending the presence/absence of DTT and/or AMS. This result opens the possibility of IgaA being recognized by alternative disulfide oxidases. Besides the pair DsbA/DsbB, *S.* Typhimurium encodes the paralogs DsbL and DsbI (Lin et al., 2009) and has an additional DsbA paralog, SrgA, encoded in the virulence plasmid (Bouwman et al., 2003). Interestingly, SsaC, a virulence-related protein related to the type III secretion system encoded in the *Salmonella* pathogenicity island 2 (SPI-2), was shown to be oxidized indistinctly by DsbA or SrgA (Miki et al., 2004). Further experiments are needed to confirm whether IgaA could be recognized by any of these alternative disulfide oxidases.

An unexpected finding of our study was the co-existence of IgaA molecules with different redox states. This situation, not observed for the wild-type IgaA, was evident for the R188H and the C404S–C504S variants (**Figures 4A,D**). Such behavior may reflect conformational plasticity in the periplasmic domain of IgaA, which may be altered at some degree when mutations in key residues important for function are introduced. Importantly, the observations with the R188H mutation proved that disrupting a key residue for function in the cytosolic side can have consequences in the structure of the periplasmic loop. This evidence points to signal transmission within the IgaA molecule from one to other side of the inner membrane.

Another aspect of interest found in the study was the instability of all IgaA variants lacking the conserved periplasmic cysteines when bacteria reached stationary phase. Proteolysis in stationary phase is not a common process. Some of the few cases known include YfgM and the formate dehydrogenase subunit FdoH, two substrates of the cytosolic FtsH protease (Westphal et al., 2012; Bittner et al., 2015). Based on our findings with the mutated IgaA variants, it is tempting to speculate on a protease that could monitor IgaA for proper folding in stationary phase.

Noteworthy, mass spectrometry analyses revealed RcsF and the protease DegP as partners interacting with the periplasmic domain of IgaA, (Cho et al., 2014). Whether DegP monitors correct folding of the IgaA periplasmic domain requires further experiments.

An important feature of all cysteine IgaA variants is that they were fully stable in actively growing bacteria (**Figure 2**), a condition in which de-repression of the RcsCDB system (measured by the *gmm::lacZ* fusion) was evident in some cases as those of C404S and C425S mutations (**Figure 2B**). This experimental evidence supports a mode of action of IgaA involving a C404–C425 disulfide bond, absolutely essential for repressing the RcsCDB system.

The gradation observed in the phenotypic assays involving mucoidity and motility and the discrepancy found for some of the IgaA variants tested (R188H, C425S, C504S), may reflect a functional link between IgaA and the conformational plasticity recently reported for RcsB (Casino et al., 2017). RcsB plasticity, which is directly linked to its phosphorylation status, could allow the bacteria to perceive many signals with intensities of the RcsCDB phosphorelay previously adjusted by IgaA. This idea of a correspondence between RcsB phosphorylation and IgaA function is now testable, for example by measuring phosphorylation status of RcsB in the different cysteine variants of IgaA reported here. We should also not discard other factors contributing to the role that IgaA has in shaping the RcsB regulon.

Another point of interest raised by our data accounts for the interaction recently reported for RcsF and IgaA as an event triggering the activation of the RcsCDB phosphorelay (Cho et al., 2014). Disulfide bonds are favored in oxidative environments as the periplasm (Denoncin and Collet, 2013; Goemans et al., 2014), and it is in this environment where the hypothetical RcsF-IgaA interaction takes place. Whether this interaction is influenced by the conserved cysteines of each partner is unknown and future studies directed to test the effect of cysteine mutations or the presence/absence of the C404–C425 disulfide bond could be therefore of much interest.

CONCLUSION

Our work provides evidence for the essential role played by one disulfide bond (C404–C425) of IgaA in its function as attenuator of the RcsCDB phosphorelay. Given the conservation of these two periplasmic cysteines in all IgaA orthologs known in enteric bacteria, the data reported here support an important structural role for this disulfide bond, probably facilitating an active conformation to the major periplasmic domain.

AUTHOR CONTRIBUTIONS

Experimental design, methodology and investigation: MP and LR. Conceptualization and supervision: FG-dP. Writing – original draft: MP, LR, and FG-dP. Writing – reviewing and editing: FG-dP.

ACKNOWLEDGMENT

We thank Juan J. Cestero (Centro Nacional de Biotecnología, CNB-CSIC) for the construction of the Δ*dsbA* mutant.

REFERENCES

Bittner, L. M., Westphal, K., and Narberhaus, F. (2015). Conditional proteolysis of the membrane protein Yfgm by the Ftsh protease depends on a novel n-terminal degron. *J. Biol. Chem.* 290, 19367–19378. doi: 10.1074/jbc.M115.648550

Bouwman, C. W., Kohli, M., Killoran, A., Touchie, G. A., Kadner, R. J., and Martin, N. L. (2003). Characterization of SrgA, a *Salmonella enterica* serovar typhimurium virulence plasmid-encoded paralogue of the disulfide oxidoreductase dsba, essential for biogenesis of plasmid-encoded fimbriae. *J. Bacteriol.* 185, 991–1000. doi: 10.1128/JB.185.3.991-1000.2003

Cano, D. A., Dominguez-Bernal, G., Tierrez, A., Garcia-Del Portillo, F., and Casadesus, J. (2002). Regulation of capsule synthesis and cell motility in *Salmonella enterica* by the essential gene igaA. *Genetics* 162, 1513–1523.

Cano, D. A., Martinez-Moya, M., Pucciarelli, M. G., Groisman, E. A., Casadesus, J., and Garcia-Del Portillo, F. (2001). *Salmonella enterica* serovar Typhimurium response involved in attenuation of pathogen intracellular proliferation. *Infect. Immun.* 69, 6463–6474. doi: 10.1128/IAI.69.10.6463-6474.2001

Casino, P., Miguel-Romero, L., Huesa, J., Garcia, P., Garcia-Del Portillo, F., and Marina, A. (2017). Conformational dynamism for DNA interaction in the *Salmonella* RcsB response regulator. *Nucleic Acids Res.* doi: 10.1093/nar/gkx1164 [Epub ahead of print].

Cho, S. H., Szewczyk, J., Pesavento, C., Zietek, M., Banzhaf, M., Roszczenko, P., et al. (2014). Detecting envelope stress by monitoring beta-barrel assembly. *Cell* 159, 1652–1664. doi: 10.1016/j.cell.2014.11.045

Denoncin, K., and Collet, J. F. (2013). Disulfide bond formation in the bacterial periplasm: major achievements and challenges ahead. *Antioxid. Redox Signal.* 19, 63–71. doi: 10.1089/ars.2012.4864

Denoncin, K., Nicolaes, V., Cho, S. H., Leverrier, P., and Collet, J. F. (2013). Protein disulfide bond formation in the periplasm: determination of the in vivo redox state of cysteine residues. *Methods Mol. Biol.* 966, 325–336. doi: 10.1007/978-1-62703-245-2_20

Dierksen, K. P., and Trempy, J. E. (1996). Identification of a second RcsA protein, a positive regulator of colanic acid capsular polysaccharide genes, in *Escherichia coli*. *J. Bacteriol.* 178, 5053–5056. doi: 10.1128/jb.178.16.5053-5056.1996

Dominguez-Bernal, G., Pucciarelli, M. G., Ramos-Morales, F., Garcia-Quintanilla, M., Cano, D. A., Casadesus, J., et al. (2004). Repression of the RcsC-YojN-RcsB phosphorelay by the IgaA protein is a requisite for *Salmonella* virulence. *Mol. Microbiol.* 53, 1437–1449. doi: 10.1111/j.1365-2958.2004.04213.x

Donnenberg, M. S., and Kaper, J. B. (1991). Construction of an eae deletion mutant of enteropathogenic *Escherichia coli* by using a positive-selection suicide vector. *Infect. Immun.* 59, 4310–4317.

Evans, K. L., Kannan, S., Li, G., De Pedro, M. A., and Young, K. D. (2013). Eliminating a set of four penicillin binding proteins triggers the Rcs phosphorelay and Cpx stress responses in *Escherichia coli*. *J. Bacteriol.* 195, 4415–4424. doi: 10.1128/JB.00596-13

Farris, C., Sanowar, S., Bader, M. W., Pfuetzner, R., and Miller, S. I. (2010). Antimicrobial peptides activate the Rcs regulon through the outer membrane lipoprotein RcsF. *J. Bacteriol.* 192, 4894–4903. doi: 10.1128/JB.00505-10

Gao, R., and Stock, A. M. (2017). Quantitative kinetic analyses of shutting off a two-component system. *mBio* 8:e00412-17. doi: 10.1128/mBio.00412-17

Goemans, C., Denoncin, K., and Collet, J. F. (2014). Folding mechanisms of periplasmic proteins. *Biochim. Biophys. Acta* 1843, 1517–1528. doi: 10.1016/j.bbamcr.2013.10.014

Hagiwara, D., Sugiura, M., Oshima, T., Mori, H., Aiba, H., Yamashino, T., et al. (2003). Genome-wide analyses revealing a signaling network of the RcsC-YojN-RcsB phosphorelay system in *Escherichia coli*. *J. Bacteriol.* 185, 5735–5746. doi: 10.1128/JB.185.19.5735-5746.2003

Howery, K. E., Clemmer, K. M., and Rather, P. N. (2016). The Rcs regulon in *Proteus mirabilis*: implications for motility, biofilm formation, and virulence. *Curr. Genet.* 62, 775–789. doi: 10.1007/s00294-016-0579-1

Jurado, P., De Lorenzo, V., and Fernandez, L. A. (2006). Thioredoxin fusions increase folding of single chain Fv antibodies in the cytoplasm of *Escherichia coli*: evidence that chaperone activity is the prime effect of thioredoxin. *J. Mol. Biol.* 357, 49–61. doi: 10.1016/j.jmb.2005.12.058

Konovalova, A., Mitchell, A. M., and Silhavy, T. J. (2016). A lipoprotein/beta-barrel complex monitors lipopolysaccharide integrity transducing information across the outer membrane. *Elife* 5:e15276. doi: 10.7554/eLife.15276

Latasa, C., Garcia, B., Echeverz, M., Toledo-Arana, A., Valle, J., Campoy, S., et al. (2012). *Salmonella* biofilm development depends on the phosphorylation status of RcsB. *J. Bacteriol.* 194, 3708–3722. doi: 10.1128/JB.00361-12

Leverrier, P., Declercq, J. P., Denoncin, K., Vertommen, D., Hiniker, A., Cho, S. H., et al. (2011). Crystal structure of the outer membrane protein RcsF, a new substrate for the periplasmic protein-disulfide isomerase DsbC. *J. Biol. Chem.* 286, 16734–16742. doi: 10.1074/jbc.M111.224865

Lin, D., Kim, B., and Slauch, J. M. (2009). DsbL and DsbI contribute to periplasmic disulfide bond formation in *Salmonella enterica* serovar Typhimurium. *Microbiology* 155, 4014–4024. doi: 10.1099/mic.0.032904-0

Majdalani, N., and Gottesman, S. (2005). The Rcs phosphorelay: a complex signal transduction system. *Annu. Rev. Microbiol.* 59, 379–405. doi: 10.1146/annurev.micro.59.050405.101230

Majdalani, N., and Gottesman, S. (2007). Genetic dissection of signaling through the Rcs phosphorelay. *Methods Enzymol.* 423, 349–362. doi: 10.1016/S0076-6879(07)23016-2

Majdalani, N., Heck, M., Stout, V., and Gottesman, S. (2005). Role of RcsF in signaling to the Rcs phosphorelay pathway in *Escherichia coli*. *J. Bacteriol.* 187, 6770–6778. doi: 10.1128/JB.187.19.6770-6778.2005

Mariscotti, J. F., and Garcia-Del Portillo, F. (2008). Instability of the *Salmonella* RcsCDB signalling system in the absence of the attenuator IgaA. *Microbiology* 154, 1372–1383. doi: 10.1099/mic.0.2007/015891-0

Mariscotti, J. F., and Garcia-del Portillo, F. (2009). Genome expression analyses revealing the modulation of the *Salmonella* Rcs regulon by the attenuator IgaA. *J. Bacteriol.* 191, 1855–1867. doi: 10.1128/JB.01604-08

Miki, T., Okada, N., and Danbara, H. (2004). Two periplasmic disulfide oxidoreductases, DsbA and SrgA, target outer membrane protein SpiA, a component of the *Salmonella* pathogenicity island 2 type III secretion system. *J. Biol. Chem.* 279, 34631–34642. doi: 10.1074/jbc.M402760200

Miller, J. H. (1972). *Experiments in Molecular Genetics*. Cold Spring Harbor, NY: Cold Spring Harbor Laboratory.

Navasa, N., Rodriguez-Aparicio, L., Ferrero, M. A., Monteagudo-Mera, A., and Martinez-Blanco, H. (2013). Polysialic and colanic acids metabolism in *Escherichia coli* K92 is regulated by RcsA and RcsB. *Biosci. Rep.* 33:e00038. doi: 10.1042/BSR20130018

Pannen, D., Fabisch, M., Gausling, L., and Schnetz, K. (2016). Interaction of the RcsB response regulator with auxiliary transcription regulators in *Escherichia coli*. *J. Biol. Chem.* 291, 2357–2370. doi: 10.1074/jbc.M115.696815

Rogov, V. V., Rogova, N. Y., Bernhard, F., Lohr, F., and Dotsch, V. (2011). A disulfide bridge network within the soluble periplasmic domain determines structure and function of the outer membrane protein RcsF. *J. Biol. Chem.* 286, 18775–18783. doi: 10.1074/jbc.M111.230185

Rosu, V., Chevance, F. F., Karlinsey, J. E., Hirano, T., and Hughes, K. T. (2006). Translation inhibition of the *Salmonella fliC* gene by the *fliC* 5′ untranslated region, *fliC* coding sequences, and FlgM. *J. Bacteriol.* 188, 4497–4507. doi: 10.1128/JB.01552-05

Westphal, K., Langklotz, S., Thomanek, N., and Narberhaus, F. (2012). A trapping approach reveals novel substrates and physiological functions of the essential protease FtsH in *Escherichia coli*. *J. Biol. Chem.* 287, 42962–42971. doi: 10.1074/jbc.M112.388470

Wolanin, P. M., Thomason, P. A., and Stock, J. B. (2002). Histidine protein kinases: key signal transducers outside the animal kingdom. *Genome Biol.* 3:REVIEWS3013.

2

The *Bacillus subtilis* Conjugative Plasmid pLS20 Encodes Two Ribbon-Helix-Helix Type Auxiliary Relaxosome Proteins that are Essential for Conjugation

Andrés Miguel-Arribas[1‡], Jian-An Hao[1,2‡], Juan R. Luque-Ortega[3‡], Gayetri Ramachandran[1†], Jorge Val-Calvo[1], César Gago-Córdoba[1], Daniel González-Álvarez[1], David Abia[1], Carlos Alfonso[3], Ling J. Wu[4] and Wilfried J. J. Meijer[1]*

[1] Department of Virology and Microbiology, Centro de Biología Molecular "Severo Ochoa" (CSIC-UAM), Instituto de Biología Molecular "Eladio Viñuela" (CSIC), Autonomous University of Madrid, Madrid, Spain, [2] The Institute of Seawater Desalination and Multipurpose Utilization (SOA), Tianjin, China, [3] Centro de Investigaciones Biológicas (CSIC), Madrid, Spain, [4] Centre for Bacterial Cell Biology, Institute for Cell and Molecular Biosciences, Newcastle University, Newcastle Upon Tyne, United Kingdom

Edited by:
Manuel Espinosa,
Centro de Investigaciones Biológicas (CSIC), Spain

Reviewed by:
Christopher Morton Thomas,
University of Birmingham,
United Kingdom
Elisabeth Grohmann,
Beuth University of Applied Sciences,
Germany
Fabián Lorenzo,
Universidad de La Laguna, Spain

***Correspondence:**
Wilfried J. J. Meijer
wmeijer@cbm.csic.es

Bacterial conjugation is the process by which a conjugative element (CE) is transferred horizontally from a donor to a recipient cell via a connecting pore. One of the first steps in the conjugation process is the formation of a nucleoprotein complex at the origin of transfer (*oriT*), where one of the components of the nucleoprotein complex, the relaxase, introduces a site- and strand specific nick to initiate the transfer of a single DNA strand into the recipient cell. In most cases, the nucleoprotein complex involves, besides the relaxase, one or more additional proteins, named auxiliary proteins, which are encoded by the CE and/or the host. The conjugative plasmid pLS20 replicates in the Gram-positive Firmicute bacterium *Bacillus subtilis*. We have recently identified the relaxase gene and the *oriT* of pLS20, which are separated by a region of almost 1 kb. Here we show that this region contains two auxiliary genes that we name $aux1_{LS20}$ and $aux2_{LS20}$, and which we show are essential for conjugation. Both $Aux1_{LS20}$ and $Aux2_{LS20}$ are predicted to contain a Ribbon-Helix-Helix DNA binding motif near their N-terminus. Analyses of the purified proteins show that $Aux1_{LS20}$ and $Aux2_{LS20}$ form tetramers and hexamers in solution, respectively, and that they both bind preferentially to $oriT_{LS20}$, although with different characteristics and specificities. *In silico* analyses revealed that genes encoding homologs of $Aux1_{LS20}$ and/or $Aux2_{LS20}$ are located upstream of almost 400 relaxase genes of the Rel_{LS20} family (MOB_L) of relaxases. Thus, $Aux1_{LS20}$ and $Aux2_{LS20}$ of pLS20 constitute the founding member of the first two families of auxiliary proteins described for CEs of Gram-positive origin.

Keywords: conjugation, relaxosome, auxiliary protein, DNA binding protein, Ribbon-Helix-Helix, antibiotic resistance, Firmicutes, horizontal gene transfer

INTRODUCTION

Bacteria exchange genetic material at gross scale, even between distantly related species, via different routes collectively called horizontal gene transfer (HGT) (for review see, Ochman et al., 2000; Frost et al., 2005; Thomas and Nielsen, 2005; Boto, 2010). Horizontal exchange of DNA provides bacteria instantly with a new set of gene(s) and hence is an important driver for the rapid adaptation and evolution of bacteria. Among the genes that are spread by HGT are those responsible for antibiotic resistance (AR), which poses a serious and increasingly worrisome economic and health problem at a global scale. Three main mechanisms are responsible for HGT: transformation through natural competence, transduction via bacterial phage, and conjugation (Ochman et al., 2000; Frost et al., 2005; Thomas and Nielsen, 2005). Of these, conjugation appears to be the route that is predominantly responsible for spreading AR genes (Mazel and Davies, 1999; Waters, 1999; Norman et al., 2009; Davies and Davies, 2010). Conjugation is the process by which a conjugative element (CE) is transferred from a donor cell to a recipient cell through a dedicated transportation pore connecting both cells. CEs contain all the genes required for processing the DNA, establishing contact with the recipient cell, those encoding the structural proteins of the connecting pore as well as those for transporting the DNA. CEs can be integrated in a bacterial chromosome or be present on plasmids, which are named integrative and conjugative elements (ICEs) and conjugative plasmids, respectively. Due to the enormous numbers and density of microbes and the constant replenishment of bacteria upon the intake of food and liquids, the intestinal gut of humans and animals is a niche that is particularly apt for emerging, pooling, and spreading AR (Sommer et al., 2009, 2010; Forsberg et al., 2012; Penders et al., 2013).

Conjugative elements are commonly present in Gram-positive (G+) and Gram-negative (G−) bacteria and the basic concepts of the transfer process are conserved (Alvarez-Martinez and Christie, 2009; De la Cruz et al., 2010; Smillie et al., 2010; Goessweiner-Mohr et al., 2013). However, whereas in most systems conjugation involves the transfer of a single DNA strand (see below), DNA is transferred in its double-stranded form during conjugation in G+ mycelial *Streptomyces* bacteria (Goessweiner-Mohr et al., 2013; Thoma and Muth, 2016), which is not further considered here. Conjugation starts with a process named mating pair formation (Mpf) in which a donor cell recognizes and interacts with a suitable recipient cell. Probably, this triggers the signal for processing the DNA of the CE and subsequent transfer of one of its strands, named T-strand, into the recipient cell. The sophisticated, multi-component pore connecting the donor and the recipient cell is named transferosome, which is a type IV secretion system (T4SS). The enzyme responsible for initiating the generation of the T-strand is a relaxase, a phosphodiesterase, that cleaves the DNA in a strand- and site-specific manner at a specific position called the *nic* site, which is located within the origin of transfer region (*oriT*). Relaxase-mediated cleavage generates a hydroxyl group at 3′ end of the *nic* site which functions as a primer for DNA elongation; i.e., the relaxase initiates a rolling-circle type of DNA replication (also named DNA transfer replication [Dtr]). Upon nicking, the relaxase remains covalently attached to the 5′-end of the nicked T-strand which is then transferred, together with the attached T-strand, into the recipient cell. In most cases the active site residue that becomes covalently attached to the T-strand concerns a tyrosine. However, very recently it has been shown that relaxases of the MOB_V family employ a histidine instead of a tyrosine residue to nick the DNA (Pluta et al., 2017). Due to its crucial role in conjugation, relaxases have attained considerable attention and several of them have been characterized in detail at the biochemical, functional and structural levels. In some cases, for instance ICE*Bs1* of *Bacillus subtilis* and the broad host range conjugative plasmid pIP501, the relaxase is the only protein that is required for processing the DNA (Kopec et al., 2005; Lee and Grossman, 2007; Grohmann et al., 2016). However, in the majority of cases additional protein(s), encoded either by the CE or the host, bind to the *oriT* and are involved in processing of the DNA. The nucleoprotein complex at *oriT* formed by the relaxase and additional proteins is called the relaxosome, and the additional proteins are named auxiliary or accessory proteins. Although their name may suggest that they play secondary role(s) in the processing reaction, most if not all of the auxiliary proteins studied so far have been shown to be essential for conjugation.

Most conjugation studies are based on CE present in G− bacteria, with knowledge on conjugation-related aspects in G+ bacteria lagging far behind. This is especially the case for auxiliary proteins (see Discussion). In our laboratory we study the conjugative plasmid pLS20 which was originally isolated from the Gram+ Firmicute bacterium *B. subtilis* natto IFO3335 (Tanaka et al., 1977). This strain is used for the fermentation of soybeans to produce "natto," a popular dish in South Asia, and hence it is conceivable that pLS20 or relatives play a role in the conjugation-mediated HGT in the gut of humans and animals. A derivative of pLS20 containing a chloramphenicol-resistance gene, pLS20cat, has been constructed (Itaya et al., 2006) and its sequence has been determined in our lab and in the lab of M. Itaya (Mitsuhiro Itaya, Keio University, Japan). All conjugation genes are located in one large operon spanning genes *28* till *74* according to our nomenclature (Singh et al., 2013). pLS20cat genes *25-27* are involved in regulating the expression of the conjugation genes (Singh et al., 2013; Ramachandran et al., 2014). Recently, we have identified and characterized the relaxase (gene *58*) and the *oriT* of pLS20cat, which we named Rel_{LS20} and $oriT_{LS20}$, respectively (Ramachandran et al., 2017). Contrary to many other plasmids, the relaxase gene and *oriT* are located within its large conjugation operon, and Rel_{LS20} turned out to be the founding member of a novel relaxase family containing >800 members.

Here, we addressed the question whether pLS20cat contains auxiliary relaxosome genes. We demonstrate that genes *56* and *57*, located in between the relaxase gene rel_{LS20} and $oriT_{LS20}$ are two auxiliary genes that are essential for conjugation and denominated them as $aux1_{LS20}$ and $aux2_{LS20}$, respectively. Both gene products were purified and biochemical analyses showed that one of them formed tetramers and the other hexamers in solution. We also show that the proteins bind to distinct DNA motifs present in $oriT_{LS20}$. *In silico* analyses revealed that a large fraction of the relaxase genes coding for the MOB_L

family of relaxases are preceded by genes encoding homologs of $Aux1_{LS20}$ and/or $Aux2_{LS20}$. The findings obtained for $Aux1_{LS20}$ and $Aux2_{LS20}$ are placed in perspective with other auxiliary proteins of CE present in G+ and G− organisms.

MATERIALS AND METHODS

Bacterial Strains, Plasmids, Media and Oligonucleotides

Escherichia coli and *B. subtilis* strains were grown in Luria-Bertani (LB) liquid medium or on 1.5% LB agar plates. When appropriate, media were supplemented with the following antibiotics: ampicillin (100 μg/ml), erythromycin (1 and 150 μg/ml in *B. subtilis* and *E. coli*, respectively), chloramphenicol (5 μg/ml), spectinomycin (100 μg/ml), and kanamycin (10 and 30 μg/ml in *B. subtilis* and *E. coli*, respectively). *B. subtilis* strains used were isogenic with *B. subtilis* strain 168 and are listed in Supplementary Table S1. Plasmids and oligonucleotides used are listed in Supplementary Tables S2, S3, respectively. All oligonucleotides were purchased from Isogen Life Science, Netherlands.

Transformation

Escherichia coli cells were transformed using standard methods (Sambrook et al., 1989). Preparation of competent *B. subtilis* cells and transformation were carried as described before (Bron et al., 1989). Transformants were selected on LB agar plates with appropriate antibiotics. pLS20cat encodes a protein, Rok_{LS20}, that inhibits the development of competence by repressing *comK*, the key transcriptional activator of competence genes (Singh et al., 2012). Therefore, to manipulate genes on pLS20cat we prepared competent cells of a pLS20cat-harboring strain that contains a chromosomal P_{xyl}-*comK* fusion (PKS56) using a standard protocol (Singh et al., 2012).

Construction of Plasmids and Strains

The correctness of sequences of all cloned PCR fragments was confirmed by sequence analysis. Amplification by PCR of pLS20cat regions was performed using as template total DNA isolated from pLS20cat harboring strain PKS11. Details regarding the construction of integration vectors based on plasmids pDR110 (*amyE* integration vector with IPTG-inducible P_{spank} promoter) or pAX01 (*lacA* integration vector with xylose-inducible P_{xyl} promoter) are given in Supplementary Table S2. In summary, gene *56* was cloned under the control of the P_{xyl} promoter or the P_{spank} promoter. In addition, genes *56-57-58*, genes *57-58*, or gene *58* were cloned behind the P_{spank} promoter. Plasmid DNA of the constructed pAXO1 and pDR110 derivatives was isolated from *E. coli* cells and then used to transform competent *B. subtilis* cells. Double-crossover integration into the chromosome was checked by PCR in the case of the pAXO1-derivatives. When pDR110 derivatives were used to transform competent *B. subtilis* cells, double cross over integration was tested by the loss of amylase activity. The pLS20cat genes 58 (rel_{LS20}), 57 ($aux1_{LS20}$) and 56 ($aux2_{LS20}$) were cloned in the *E. coli* expression vector pET28b+ to generate fusion genes containing a C-terminal $his_{(6)}$ extension. Details regarding these cloning strategies are given in Supplementary Table S2. The resulting derivatives of pET28b+ were constructed using *E. coli* strain XL1-Blue. Once verified its correctness, the plasmids were transformed into *E. coli* strain BL21(DE3).

Conjugation Assays

Conjugation was carried out in liquid medium as described previously (Singh et al., 2013). The effect of ectopic expression of a given gene placed under the control of the inducible P_{spank} and/or P_{xyl} promoter on conjugation was studied by adding the inducer (1 mM IPTG, 1% xylose) to prewarmed LB medium used to dilute overnight cultures of the donor cells.

Analytical Ultracentrifugation Experiments

Sedimentation velocity (SV), sedimentation equilibrium (SE), and dynamic light scattering (DLS) assays and processing of the data, including estimations of molar masses of the relaxosome proteins from the hydrodynamic measurements, were carried out using the same conditions to those used before in the analysis of Rel_{LS20} (Ramachandran et al., 2017).

Over Expression and Purification of Recombinant Rel_{LS20}, $Aux1_{LS20}$, and $Aux2_{LS20}$ Containing a C-Terminal $His_{(6)}$ Tag

Recombinant versions of Rel_{LS20}, $Aux1_{LS20}$, and $Aux2_{LS20}$ were expressed and purified using similar protocols. In brief, *E. coli* BL21(DE3) cells containing plasmid pAND83 ($rel_{LS20}His_{(6)}$), or pHJA56 ($aux1_{LS20}His_{(6)}$), or pHJA57 ($aux2_{LS20}His_{(6)}$) were inoculated in fresh LB media complemented with 30 μg/ml kanamicin and grown at 37°C with shaking (200 rpm). At an OD_{600} of about 0.6, IPTG was added to a final concentration of 1 mM to induce the recombinant protein and growth was continued for 2 h. Next, cells were collected by centrifugation and processed as described before (Singh et al., 2012). The nickel-column purified proteins (>95% pure) were finally dialysed against buffer B (20 mM Tris-HCl pH 8.0, 1 mM EDTA, 500 mM NaCl, 10 mM $MgCl_2$, 7 mM β-mercaptoethanol, 50% v/v glycerol) and stored in aliquots at −80°C. Bradford assay and OD_{280} determination were used to determine the protein concentrations.

Gel Retardation Assays

Gel retardation assays were essentially carried out as described before (Singh et al., 2012). Thus, different DNA fragments were amplified by PCR using pLS20cat as template. The resulting PCR fragments were purified and 170 ng of DNA [200 or 362 bp] (with or without 220 ng of control DNA [176 bp]) were incubated on ice in binding buffer [20 mM Tris HCl pH 8, 1 mM EDTA, 5 mM $MgCl_2$, 0.5 mM DTT, 100 mM KCl, 10% (v/v) glycerol, 0.05 mg ml^{-1} BSA] without or with purified $Aux1_{LS20}$ or $Aux2_{LS20}$ to a fixed final concentration of 90 nM (Supplementary Figure S3) or using twofold increasing concentrations ranging from 0.09 to 5.76 μM (**Figure 3**) in a total volume of 16 μl. The negative control, corresponding to bp numbers 63,774–63,950 of accession number NC_015148.1, has an AT-content that is very similar to the AT content of the *oriT* fragment (61.4 vs. 61.1%). This DNA corresponds to sequences located inside a gene (gene *24*), lowering the possibility that it harbors particular features for recruiting a transcriptional regulator or other DNA binding protein. In addition, it is predicted to lack a static bend. After careful mixing, samples were incubated for 20 min at 30°C, placed back on ice for 10 min, then loaded onto 2% agarose gel in 0.5XTBE. Electrophoresis was carried out in 0.5XTBE at 50 V at 4°C. Finally, the gel was stained with ethidium bromide, destained in 0.5XTBE and photographed with UV illumination.

In Silico Analyses

Identification of Mob_L Members

Rel_{LS20} was used as a query sequence to execute a psi-blast (version 2.6.1+) search against the NCBI nr protein database (July, 2017), allowing up to 10 rounds of reiteration with an *e*-value threshold of 1e-15 (Altschul et al., 1997, 2005; Schaffer et al., 2001) producing 1445 hits. The program "USEARCH" (version v10.0.240_i86linux32) was then used to identify and remove redundant sequences showing 100% identity (Edgar, 2010), resulting in 1249 unique hits showing high similarity to Rel_{LS20}.

Identification of Putative Auxiliary Proteins

Protein sequences of $Aux1_{LS20}$ and $Aux2_{LS20}$ were used as query against the NCBI nr protein database (July 2017) using psi-blast (version 2.6.1+), with an e-value threshold of 1e-6 and 1e-7, respectively, until no new hits were retrieved. The sequence identifiers obtained from psi-blast, were crossed with the sequence identifiers preceding the MOB_L family relaxase members, obtained from the nucleotide entries from they were translated.

Prediction of Secondary Structure for $Aux1_{LS20}$ and $Aux2_{LS20}$ Homologs

Corresponding sequences were submitted to the RaptorX property web server (Wang et al., 2016) and predictions for β-strands and α-helices along the sequences were plotted with "R"[1]) (R Core Team, 2017).

RESULTS

Identification of Putative Relaxosome Genes of pLS20cat by *in Silico* Analysis

Recently, we have shown that pLS20cat gene *58* is essential for conjugation and that it encodes the relaxase, which we named Rel_{LS20} (Ramachandran et al., 2017). In these studies we also identified the *nic* site of Rel_{LS20} and delineated the functional *oriT*, named $oriT_{LS20}$, to a region of 362 bp. Remarkably, $oriT_{LS20}$ and rel_{LS20} are separated by a region of 865 bp, which has been annotated to contain two relatively small putative genes, designated genes *56* and *57* (Singh et al., 2013, see **Figure 1** for a schematic view of this region). Often, but not always, conjugative plasmid-located relaxase genes are accompanied by small auxiliary relaxosome genes that generally are located upstream of the relaxase gene. This prompted us to investigate whether genes *56* and *57* might encode auxiliary relaxosome genes of pLS20cat. *In silico* analyses of pLS20cat genes *56* and *57* show that, firstly, rel_{LS20} is translationally coupled to the preceding gene *57* [i.e., the stop (TAA) and start codon (ATG) of genes *57* and rel_{LS20}, respectively, overlap; see **Figure 1**], and only a small intergenic region of 183 bp separates gene *57* from its preceding gene *56*. Second, gene *56* and *57* are both small genes (79 and 147 codons, respectively). And third, the proteins encoded by these genes are both putative DNA binding proteins predicted to contain a Ribbon-Helix-Helix (RHH) motif in their N-terminal regions. An overview of the secondary structure prediction of both proteins and their homology with CopG, a paradigm of RHH DNA binding protein (Gomis-Ruth et al., 1998; Del Solar et al., 2002), is shown in Supplementary Figure S1. This figure shows that both $Aux1_{LS20}$ and $Aux2_{LS20}$ contain several lysine and arginine residues near the end of their predicted helix 1 and beginning of helix 2. The corresponding region in known RHH structures has been shown to be close to the phosphate backbone of the DNA (for example see, Schildbach et al., 1999). In summary, *in silico* analyses suggested that the two small genes *56* and *57* preceding the relaxase gene rel_{LS20} may encode auxiliary relaxosome proteins.

pLS20cat Genes *56* and *57* Are Essential for Conjugation

Previously, we engineered a derivative of pLS20cat, pLS20Δ56-58, in which the putative genes *56-57* together with the relaxase gene rel_{LS20} (gene *58*) have been deleted, and demonstrated that this plasmid was deficient in conjugation. Conjugation of pLS20Δ56-58 was restored when all three genes (*56-58*), were ectopically expressed from the IPTG-inducible P_{spank} promoter at the chromosomal *amyE* locus, but not in the absence of gene *58*, showing that Rel_{LS20} was essential for conjugation (Ramachandran et al., 2017). We used a similar approach to study whether genes *56* and/or *57* were essential for conjugation. Thus, we constructed strain GR153, which harbors pLS20Δ56-58 and also contains rel_{LS20} (gene *58*), but not *56* and *57*, under the control of the P_{spank} promoter at the *amyE* locus. We then employed this strain as donor to determine the conjugation efficiencies using a standard protocol (see Materials and Methods). Strains PKS11, GR149 and GR150 were included as controls. As shown in **Table 1**, the efficiency of conjugation observed for the wild type plasmid pLS20cat was in the range of 10^{-3}, which is similar to values reported previously under similar conditions (Singh et al., 2013; Ramachandran et al., 2014, 2017). As reported before (Ramachandran et al., 2017), conjugation was observed for pLS20Δ56-58 only when genes *56-58* were expressed from the chromosome (**Table 1**, strain GR149 and GR150). Importantly, no transconjugants were obtained when strain GR153 (*amyE*::P_{spank}-rel_{LS20}, pLS20Δ56-58) was used as donor in conjugation experiments, regardless of whether they were grown in the presence or absence of IPTG. These results

[1]https://www.R-project.org/

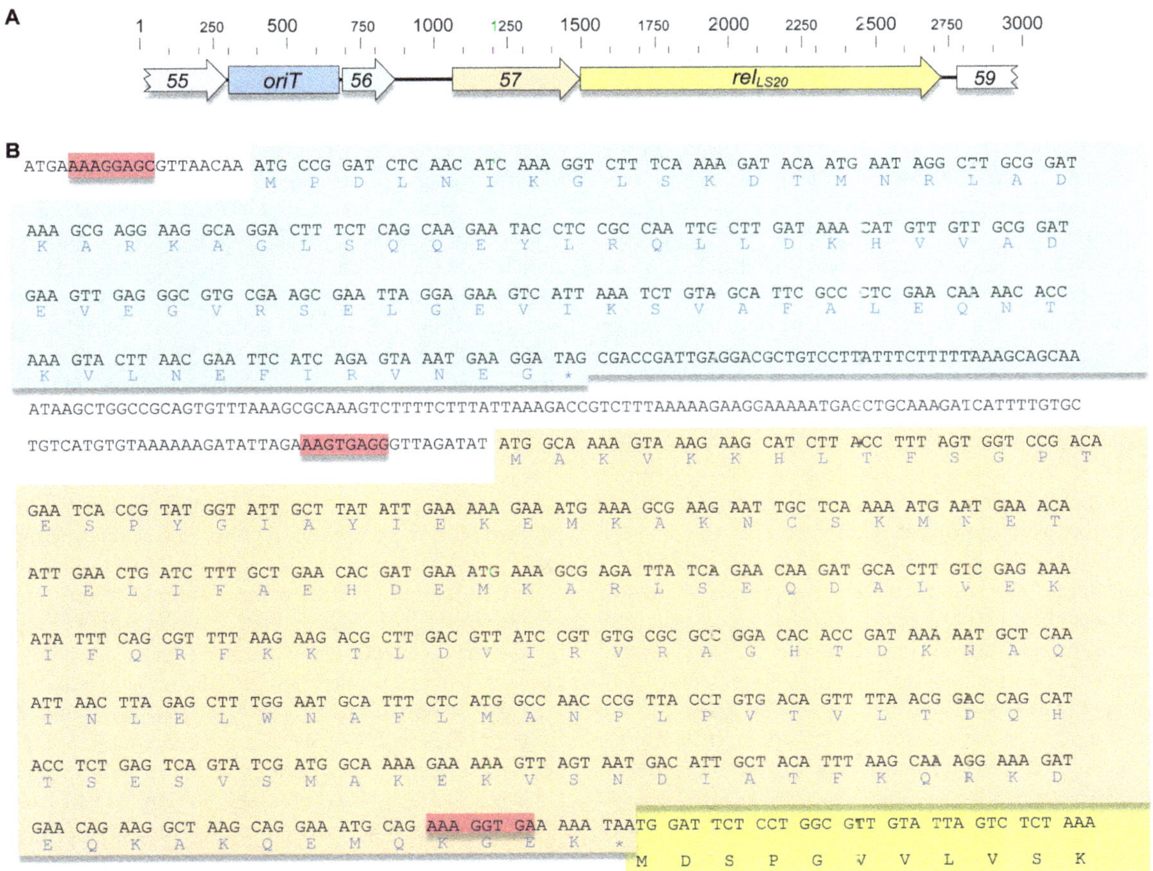

FIGURE 1 | The pLS20cat relaxosome region. **(A)** Genetic organization of the pLS20cat genes *55–59*. The genes, their sizes and orientation of transcription are indicated with arrows. Genes *55* and *59* are colored gray. Genes *56*, *57*, and *58* (*rel*$_{LS20}$) are colored green, orange, and yellow, respectively. The same color code is used in "**B**," as well as in **Figure 4** (see below). The 362 bp *oriT*$_{LS20}$ region is indicated with a blue box labeled *oriT*. Base pair numbering is given on the top. **(B)** DNA sequence of genes *56* and *57* and their deduced protein sequences. Stop codons are indicated with an asterisk and likely R bosomal Binding sites (RBS) are highlighted with a red box. Note that genes *57* and *rel*$_{LS20}$ are translationally coupled. Only the first 11 codons of the *rel*$_{LS20}$ gene are given.

TABLE 1 | pLS20cat genes *56* and *57* are required for conjugation.

Strain	Genotype	Plasmid	Inductor	Conjugation efficiency*
PKS11	168	pLS20cat	−	7.9×10^{-3}
GR149	168	pLS20Δ56-58	−	$<10^{-7}$
GR150	168, *amyE*::P$_{spank}$ *56-58*	pLS20Δ56-58	−	$<10^{-7}$
			+	3.2×10^{-3}
GR153	168, *amyE*::P$_{spank}$ *58*	pLS20Δ56-58	−	$<10^{-7}$
			+	$<10^{-7}$
GR197	168, *amyE*::P$_{spank}$ *57-58*	pLS20Δ56-58	−	$<10^{-7}$
			+	$<10^{-7}$
GR200	168, *amyE*::P$_{spank}$ *58, lacA*::P$_{xyl}$-*56*	pLS20Δ56-58	−	$<10^{-7}$
			+	$<10^{-7}$
GR225	168, *amyE*::P$_{spank}$ 57-*58, lacA*::P$_{xyl}$-*56*	pLS20Δ56-58	−	$<10^{-7}$
			+	5.8×10^{-4}

**Conjugation efficiencies were calculated as transconjugants/donor, and correspond to the mean value of at least three independent experiments. When indicated, the inducer IPTG was added at a final concentration of 1 mM in the case of strains GR150 and GR197. In the case of GR200 and GR225, the final concentrations of the inducers was 1 mM (IPTG) and 1% (xylose).*

showed that pLS20cat gene *56* and/or *57* are necessary for conjugation.

We next tested whether only one or both genes were required for conjugation. For this, we constructed the pLS20Δ56-58-harboring strains GR197 and GR200 in which rel_{LS20} together with either gene *57* (strain GR197) or gene *56* (strain GR200) could be induced from the bacterial genome. When used as donor, no transconjugants were obtained for each strain regardless whether they were grown in the absence or presence of the inductor(s) (see **Table 1**), demonstrating that both genes are essential for conjugation.

In the above conjugation experiments, one or a combination of genes *56*, *57*, rel_{LS20} was complemented by expressing them from the IPTG-inducible P_{spank} promoter for all the strains except for strain GR200. In this strain rel_{LS20} is controlled by P_{spank} at the *amyE* locus and gene *56* by the xylose-inducible P_{xyl} promoter at the *lacA* locus. To rule out the possibility that transconjugants were not obtained for donor strain GR200 because the genes were expressed from different promoters at a different locus, we constructed strain GR225 in which gene *56* was placed under the control of the P_{xyl} promoter at *lacA*, and genes *57* and *58* under the control of the P_{spank} promoter at *amyE*. Transconjugants were obtained for this strain when cells were grown in the presence of both inducers (**Table 1**), demonstrating that the gene products expressed from the two different promoters and chromosomal loci were all functional. These results demonstrate therefore that besides rel_{LS20} genes *56* and *57* are also required for conjugation. Taking into account these results, together with the structural organization of these genes with respect to rel_{LS20} and $oriT_{LS20}$, the *in silico* analyses presented above, and additional evidence presented below, we conclude that pLS20cat gene *56* and *57* encode auxiliary relaxosome proteins which we name $Aux1_{LS20}$ and $Aux2_{LS20}$, respectively.

In Vitro Analysis of the Relaxosome Proteins $Aux1_{LS20}$ and $Aux2_{LS20}$, and Rel_{LS20}

Oligomerization State Determined by Analytical Ultracentrifugation and DLS Techniques

To characterize the auxiliary relaxosome proteins *in vitro*, we purified $Aux1_{LS20}$ (Mw 10,601 Da) and $Aux2_{LS20}$ (Mw 18,605 Da) from *E. coli*, each fused to a $His_{(6)}$ tag at its C-terminus. We first determined the oligomerization state of the proteins, and also investigated putative interactions among them and with Rel_{LS20}, using two complementary analytical ultracentrifugation approaches, i.e., SV and SE (**Figures 2A–D**), together with DLS experiments using the same experimental conditions.

Sedimentation profiles obtained by SV assays showed $Aux1_{LS20}$ as a single species with an experimental sedimentation coefficient of 2.5 S ($s_{20,w}$ = 2.9 S) compatible with a moderately elongated tetrameric form of the protein ($f/f_0 = 1.5$) (**Figure 2A**). Subsequent analysis of $Aux1_{LS20}$ gave a *D*-value of $52.5 \pm 0.3\ \mu m^2/s$. The obtained *S*- and *D*-values, once introduced in the Svedberg equation, yielded an apparent molar mass of 46,290 Da. SE data, best-fit analysis to single species model gave an average molecular mass of 42,200 Da ± 300 Da, confirming that $Aux1_{LS20}$ is a tetramer in solution (**Figure 2B**).

In the case of $Aux2_{LS20}$, analysis of the sedimenting boundaries showed a sedimentation profile with a main peak corresponding to 90.0% of the total proteins at 4.4 S ($s_{20,w} = 5.1$ S), together with a second peak at 3.3 S ($s_{20,w}$ = 3.8 S) encompassing 7% of the sample (**Figure 2C**). The *S*-value of the main peak is compatible with the theoretical behavior of a spherical $Aux2_{LS20}$ tetramer (f/f_0 = 1.2), as well as with a moderately elongated hexamer (f/f_0 = 1.6). DLS analysis of $Aux2_{LS20}$ yielded a *D* of $38.2 \pm 1.0\ \mu m^2/s$, which combined with the obtained *S*-value of 4.4 in the Svedberg formula resulted in an apparent molar mass of 113,400 Da that is very close to the molecular mass of $Aux2_{LS20}$ hexamers (111,630 Da). SE experiments were decisive for establishing the oligomerization state of $Aux2_{LS20}$, as the best fit of the SE data gave an average molecular mass of 111,300 ± 1,200 Da, unequivocally demonstrating that $Aux2_{LS20}$ forms hexamers in solution (**Figure 2D**). In summary, the outcome of three complementary experimental approaches showed that $Aux1_{LS20}$ and $Aux2_{LS20}$ form tetramer and hexamers in solution, respectively.

Previously, we determined that purified Rel_{LS20} forms monomers in solution (Ramachandran et al., 2017). To study possible interactions between the relaxosome proteins in solution we used combinations of $Aux1_{LS20}$, $Aux2_{LS20}$ and Rel_{LS20} and subjected these to SV experiments (Supplementary Figure S2). No additional peaks with increased *S*-values reflecting new protein hetero-complexes were obtained in any of the combinations tested implying that the relaxosome proteins of pLS20cat do not interact in solution, at least not under the conditions tested.

$Aux1_{LS20}$ and $Aux2_{LS20}$ Bind Specifically to $oriT_{LS20}$

Electrophoretic Mobility Shift Assays (EMSA) were performed to study the DNA binding properties of $Aux1_{LS20}$ and $Aux2_{LS20}$. The results presented in **Figure 3** show that both auxiliary proteins bound DNA, and that both bound preferentially to $oriT_{LS20}$. Nevertheless, there were distinct differences in binding characteristics between the two proteins. The addition of $Aux1_{LS20}$ resulted in the appearance of only one retarded species of $oriT_{LS20}$, and even at the highest concentration tested $Aux1_{LS20}$ did not bind to the negative control DNA (**Figure 3**, left panel). One retarded $oriT_{LS20}$ species was also observed for $Aux2_{LS20}$ at low concentrations. However, higher $Aux2_{LS20}$ concentrations resulted in the appearance of additional shifted species of $oriT_{LS20}$. In addition, at higher concentrations $Aux2_{LS20}$ bound also to the negative control DNA, and at the highest concentration tested a smear of retarded species was observed (**Figure 3**, right panel). These results show that both proteins bind preferentially to $oriT_{LS20}$, but $Aux1_{LS20}$ appears to bind $oriT_{LS20}$ with a higher specificity than $Aux2_{LS20}$.

To delineate further the binding sites of $Aux1_{LS20}$ and $Aux2_{LS20}$ we generated thirteen overlapping 200 bp DNA fragments (F21–F33) covering the $oriT_{LS20}$ region with a sliding window of 25 bp, and used them in EMSA. The results presented in Supplementary Figure S3 show that $Aux1_{LS20}$ bound to fragments F22-F29, which share the 25 bp sequence 5′-CAAATAAATCTGGTACCACGAAAAA-3′ located in the 5′

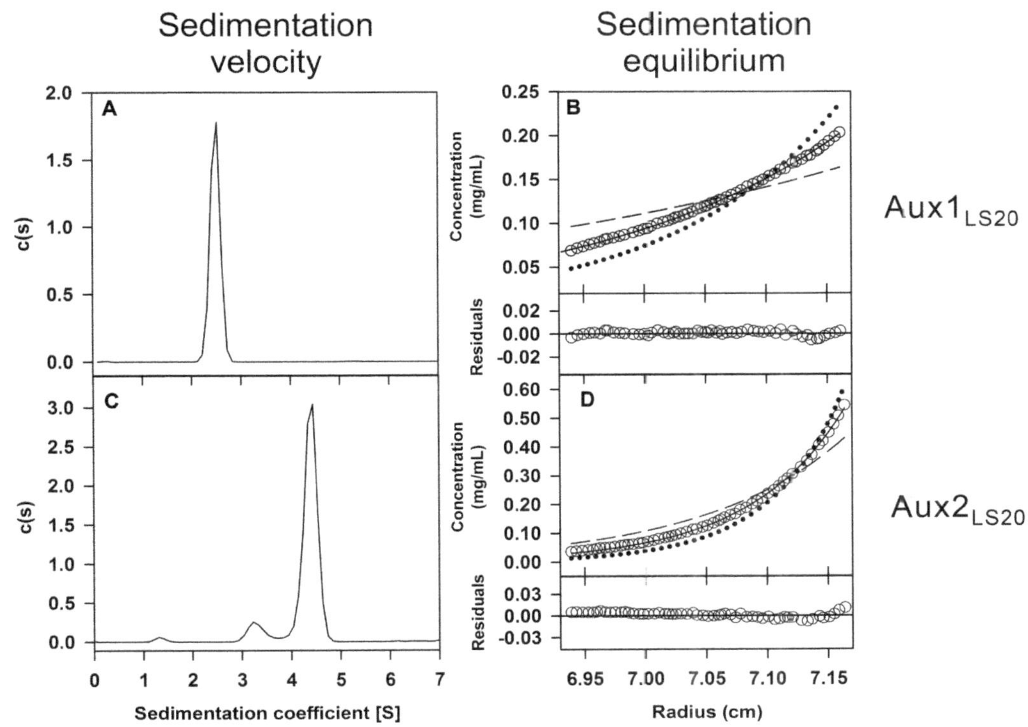

FIGURE 2 | Determination of the oligomerization state of the relaxosome proteins Aux1$_{LS20}$ and Aux2$_{LS20}$. Purified proteins in solution at 12 μM were studied by sedimentation velocity (SV) and sedimentation equilibrium (SE). Plots **(A,C)** are representations of the sedimentation coefficient distribution c(s) profiles obtained by SV and correspond to Aux1$_{LS20}$ and Aux2$_{LS20}$, respectively. Concentration gradients obtained by SE assays: **(B)** Data obtained with Aux1$_{LS20}$ (empty circles) and best-fit analysis assuming a protein tetramer (black line), dimer (dashed line) and hexamer (dotted line) species model; **(D)** Data collected with Aux2$_{LS20}$ (empty circles) and best-fitting assuming a protein hexamer (black line), tetramer (dashed line) and octamer (dotted line) species model. Lower part in the SE plots represents the difference between experimental data and estimated values for the best fit to a single species model (residuals).

half of *oriT*$_{LS20}$. This sequence contains the inverted repeat 5′-TGGTACCA-3′, which could be the binding site of Aux1$_{LS20}$. In the case of Aux2$_{LS20}$ retarded species of *oriT*$_{LS20}$ with strong and weak intensity were observed for fragments F21–F25 and F26–F28, respectively. No shifts were observed for fragments F29–F33 at the protein concentration used. This shows that Aux2$_{LS20}$ binds the 5′ half region of *oriT*$_{LS20}$ upstream of Aux1$_{LS20}$. The sequence motif 5′-TGTGCAT-3′ is present three times in a directed repeated orientation in the 5′ half of *oriT*$_{LS20}$. While fragments F21–F25 each contain the three 5′-TGTGCAT-3′ motifs, fragment F26 contains only two, and the motif is present only once on fragments F27 and F28. This suggests that the motif 5′-TGTGCAT-3′ may be the preferred binding site for Aux2$_{LS20}$. It is worth mentioning that two of the 5′-TGTGCAT-3′ motifs are embedded within a larger motif (5′-TTTATGTGCATT-3′).

Over 400 Members of the MOB$_L$ Family of Relaxase Genes Contain Upstream Genes Encoding Homologs of Aux1$_{LS20}$ and/or Aux2$_{LS20}$

Previously, we reported that the pLS20cat-encoded Rel$_{LS20}$ constitutes the founding member of a novel, large family of relaxases that we named MOB$_L$, which contained 817 members that were almost exclusively encoded in bacteria belonging to the phylum Firmicutes (Ramachandran et al., 2017). We wanted to know whether other MOB$_L$ relaxase genes were also preceded by genes encoding putative homologs of Aux1$_{LS20}$ and/or Aux2$_{LS20}$. To study this we first determined the current number of MOB$_L$ relaxase genes, applying the same method as that used in our previous study; i.e., we performed a psi-blastp search of the NCBI nr database using Rel$_{LS20}$ as a query. After removing redundant sequences this search now resulted in 1,453 hits that showed high similarity with Rel$_{LS20}$ (threshold value $P = 1e\text{-}15$). Next, the corresponding DNA accession number of each identified MOB$_L$ relaxase was retrieved, which was subsequently used to generate a database that contains the accession number of each MOB$_L$ member together with that of the protein encoded by the gene located upstream and downstream of the relaxase gene. We then performed the same procedures for Aux1$_{LS20}$ and Aux2$_{LS20}$; i.e., we identified proteins sharing a high level of similarity with Aux1$_{LS20}$ and Aux2$_{LS20}$ and generated databases that contained these accession numbers together with those of the proteins encoded by the flanking genes. Finally, the three databases were crossed to identify those MOB$_L$ members that are preceded by a gene encoding a putative homolog of Aux1$_{LS20}$ and/or Aux2$_{LS20}$. This approach revealed 387 MOB$_L$

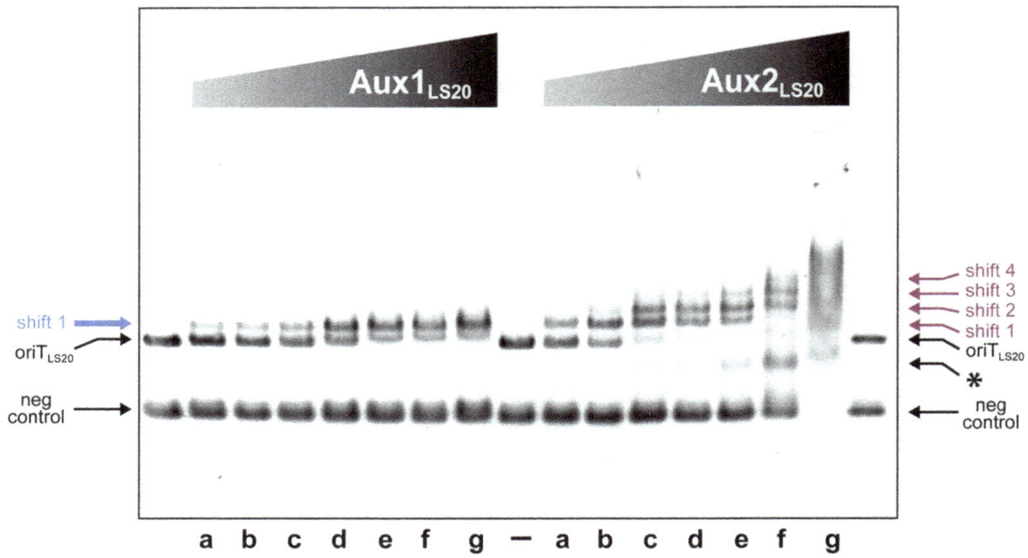

FIGURE 3 | Auxiliary protein Aux1$_{LS20}$ and Aux2$_{LS20}$ are DNA binding proteins that preferentially bind to *oriT*$_{LS20}$. DNA binding activity of Aux1$_{LS20}$ and Aux2$_{LS20}$ was analyzed by EMSA. Each lane contained two DNA fragments; one corresponding to *oriT*$_{LS20}$ (OriT$_{LS20}$, 362 bp, 170 ng), and a control fragment (neg control, 176 bp, 220 ng, corresponding to pLS20 positions 63,774–63,950 of accession number NC_015148.1). –, loaded without protein. Increasing concentrations of Aux1$_{LS20}$ or Aux2$_{LS20}$ were prepared using twofold dilution steps, and ranged from 0.09 (lanes "a") to 5.76 μM (lanes "g"). Positions of unbound control DNA (neg control) and *oriT*$_{LS20}$ (oriT$_{LS20}$) are indicated. The single shifted *oriT*$_{LS20}$ species observed for Aux1$_{LS20}$ is indicated on the left with a blue arrow. The different shifted *oriT*$_{LS20}$ species observed for Aux2$_{LS20}$ are indicated on the right with purple arrows, and the shifted species of the control DNA fragment is indicated with an asterisk.

relaxase genes that were preceded by a gene encoding a putative Aux2$_{LS20}$ homolog; and of these 87 contained an additional Aux1$_{LS20}$ homolog encoding gene upstream. Without exception, the identified MOB$_L$ relaxase genes having upstream gene(s) encoding putative homologs of Aux1$_{LS20}$ and/or Aux2$_{LS20}$ are all present in bacteria belonging to the phylum Firmicutes. Although stringent settings were used to identify proteins sharing high similarity with Aux1$_{LS20}$ or Aux2$_{LS20}$, this does not automatically imply that the identified proteins will contain a Ribbon-Helix-Helix motif in their N-terminal region, which is a characteristic feature of both Aux1$_{LS20}$ and Aux2$_{LS20}$ (see above). We therefore carried out secondary structure prediction for all the putative Aux1$_{LS20}$ and Aux2$_{LS20}$ homologs identified (see Materials and Methods). The results of these analyses, which are presented in Supplementary Table S4, show that 86 of the 87 (98.9%), and 384 of the 387 (99.2%) putative homologs of Aux1$_{LS20}$ and Aux2$_{LS20}$, respectively, contain a typical Ribbon-Helix-Helix signature in their N-terminal region, and thereby support the view that they are auxiliary proteins of the corresponding relaxase. In summary, these analyses provide compelling evidence that almost 400 MOB$_L$ relaxase genes are preceded by a gene encoding an Aux2$_{LS20}$ homolog, and that in 87 of these cases this putative auxiliary gene is preceded by another auxiliary gene encoding an Aux1$_{LS20}$ homolog. Consequently, pLS20 encoded Aux1$_{LS20}$ and Aux2$_{LS20}$ are the founding members of two families of Ribbon-Helix-Helix type auxiliary proteins that are encoded by Firmicutes bacteria.

DISCUSSION

In this study we have demonstrated that the pLS20cat genes *56* (*aux1*$_{LS20}$) and *57* (*aux2*$_{LS20}$) encode the auxiliary relaxosome proteins of pLS20cat. Combined with our previously published results (Ramachandran et al., 2017), we have identified the relaxosome module of pLS20cat that includes *oriT*$_{LS20}$ and the downstream genes *aux1*$_{LS20}$, *aux2*$_{LS20}$, and *rel*$_{LS20}$. This module is embedded within the large conjugation operon of pLS20cat (Singh et al., 2013). In addition, we have provided strong evidence that Aux1$_{LS20}$ and Aux2$_{LS20}$ constitute the founding member of corresponding families of Ribbon-Helix-Helix type auxiliary proteins whose genes precede a large fraction of the MOB$_L$ type relaxase genes. Thereby, our results provide a better understanding of the relaxosome components present on Gram+ mobile elements in general and particularly those belonging to the phylum Firmicutes.

The results presented here, together with those obtained previously (Ramachandran et al., 2017), show that *aux1*$_{LS20}$ and *aux2*$_{LS20}$ encode trans-acting proteins that are essential for conjugation. We also showed that Aux1$_{LS20}$ and Aux2$_{LS20}$ form tetramers and hexamers in solution, respectively, and we detected no interaction between the three pLS20 relaxosome proteins under the conditions tested. We cannot exclude the possibility that they interact when they form a nucleoprotein complex at *oriT*$_{LS20}$. Aux1$_{LS20}$ bound with high specificity to a region of 25 bp located about 100 bp upstream of the *nic* site that contains the inverted repeated sequence 5′-TGGTACCA-3′.

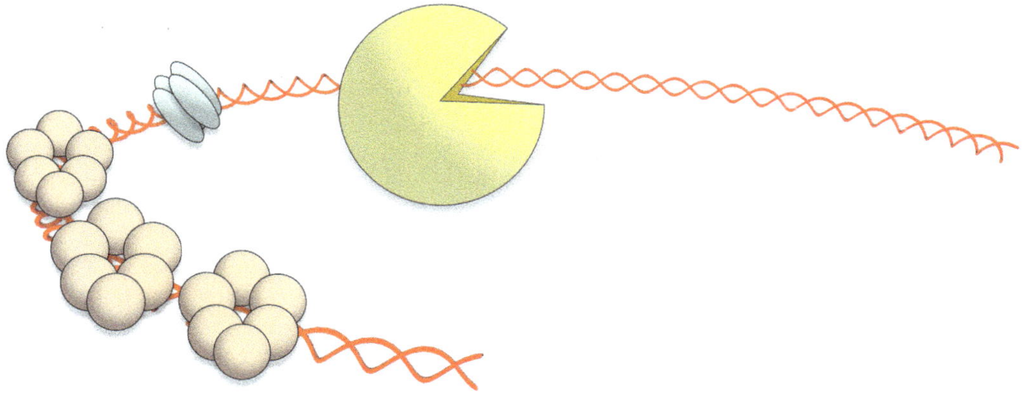

FIGURE 4 | Schematic model of the nucleoprotein complex formed at $oriT_{LS20}$. DNA is shown as a red double helix. The tetrameric $Aux1_{LS20}$ unit, the hexameric $Aux2_{LS20}$ units and the monomeric Rel_{LS20} are depicted as green, orange, and yellow cartoons, respectively. It is well possible that binding of one or more of the auxiliary proteins to $oriT_{LS20}$ alters the topology of the DNA. This is not taken into account in this simplified model.

The preferred binding site of $Aux2_{LS20}$ resulted to be a 140 bp fragment located at the 5′ half of $oriT_{LS20}$ and that contains three times the sequence 5′-TGTGCAT-3′. In our previous study (Ramachandran et al., 2017), we showed that a derivative of $oriT_{LS20}$ that includes the *nic* site and the binding site for $Aux1_{LS20}$, but lacks the 5′-located 100 bp containing two of the three 5′-TGTGCAT-3′ motifs was not functional *in vivo*. The topology of DNA can have a large effect on the binding characteristics of DNA binding proteins and which in turn may affect their function (Gimenes et al., 2008; Fogg et al., 2012). The *oriT* regions of several conjugative plasmids contain an intrinsic bend that is thought to be important for optimal binding and functionality of the relaxosome proteins (for review see, De la Cruz et al., 2010). We have demonstrated that the $oriT_{LS20}$ region is also intrinsically bent, and that the bend is located in the 5′ half of $oriT_{LS20}$ (Ramachandran et al., 2017), which we show here corresponds to the region where $Aux1_{LS20}$ and $Aux2_{LS20}$ (preferentially) bind. When we combine the results obtained here and in our previous study a picture emerges that is schematically presented in **Figure 4**. $Aux1_{LS20}$ and $Aux2_{LS20}$ bind to the left half of $oriT_{LS20}$ that is intrinsically bent and we envisage that the formation of this nucleoprotein complex contributes to optimal functioning of Rel_{LS20}. In other systems, auxiliary proteins have been described to stimulate relaxase-mediated nicking at *oriT* by recruiting the relaxase to *oriT*, probably by facilitating the relaxase to access the *nic* site, and/or by acting as molecular wedges to melt double-stranded DNA (reviewed in, Alvarez-Martinez and Christie, 2009). Thus, it is conceivable that the auxiliary proteins of pLS20 fulfill similar function(s).

Most of our knowledge on auxiliary proteins is related to those encoded by conjugative plasmids replicating in G− bacteria; in particular, the auxiliary proteins of F and related plasmids have been studied in detail at the functional, biochemical and structural levels (for review see, Alvarez-Martinez and Christie, 2009; De la Cruz et al., 2010; Wong et al., 2012). Upon binding, TraY and TraM of plasmid F bent the DNA and therewith play important roles in organizing the relaxosome complex at *oriT* and influencing the nicking reaction of the relaxase. In addition, they both play a role in gene expression by regulating the activity of their own promoters. TraM also has a key role in delivering the relaxosome to the conjugative pore by interacting with its cognate T4CP (Wong et al., 2011; Peng et al., 2014). Future studies are needed to determine whether the auxiliary proteins of pLS20 fulfill similar functions to those of F, although it is doubtful that $Aux1_{LS20}$ and $Aux2_{LS20}$ play a role in gene regulation due to the different genetic organization. In the case of F, the monocistronic *traM* gene is located directly downstream of its *oriT*. *TraM* is followed by another monocistronic gene, *traJ*, which in turn is followed by a large multicistronic operon in which *traY* is the first gene (Zatyka and Thomas, 1998). In the case of pLS20, though, the relaxosome genes are embedded within the large conjugation operon and are under the control of the main conjugation promoter P_c that is located almost 26 kbp upstream of $aux1_{LS20}$ (Singh et al., 2013; Ramachandran et al., 2014). At present, we cannot fully exclude the possibility that the relaxosome genes of pLS20cat are controlled by an additional promoter that is regulated by $Aux1_{LS20}$ or $Aux2_{LS20}$. RNAseq data showed, however, that repression of the main conjugation promoter results in silencing of the relaxosome genes, as well as other genes in the conjugation operon of pLS20cat (Singh et al., 2013).

Far less is known about auxiliary proteins encoded by conjugative plasmids of Gram+ origin. The monomeric Helix-Turn-Helix protein TraN of the *Enterococcus faecalis* conjugative plasmid pIP501 binds to its *oriT* region, which suggested that it might be an auxiliary protein of pIP501. However, recent results revealed that *traN* is not essential for conjugation, and it is now believed that it may be a repressor of conjugation by regulating either the expression of the conjugation operon or activity of the relaxase TraA (Goessweiner-Mohr et al., 2014; Grohmann et al., 2016).

The auxiliary proteins PcfF encoded by the *E.nterococcus faecalis* plasmid pCF10, and LtrF of *Lactococcus lactis* plasmid pRS01 share 47% sequence identity. As far as we know, these are the only auxiliary proteins encoded by conjugative plasmids of Gram+ origin that have been studied in some detail (Chen et al., 2007, 2008). The *pcfF* and *ltrF* genes are essential for conjugation and purified PcfF and LtrF bind their cognate *oriT*s. Moreover, evidence supports a model in which PcfF recruits the relaxase PcfG to *oriT*, and that PcfF, probably in conjunction with the relaxase PcfG, interacts with its cognate T4CP and hence plays an important role in delivering the relaxosome to the conjugative pore.

Several auxiliary proteins of conjugative plasmids of Gram− origin are described to contain a RHH motif. These include, TraY and TraM of F plasmid, TrwA of R388, VirC2 of *Agrobacterium tumefaciens*, NikA of R64, TraJ of RP4, MobC of RSF1010, MbeC of ColE1, MobC of RA3 (Bowie and Sauer, 1990; Zhang and Meyer, 1997; Moncalian and De la Cruz, 2004; Ragonese et al., 2007; Yoshida et al., 2008; Lu et al., 2009; Varsaki et al., 2009; Godziszewska et al., 2016). For some of them structure-based mutational analyses have demonstrated the importance of the RHH motif in *oriT* binding as well as relaxase recruitment (Yoshida et al., 2008; Lu et al., 2009). Interestingly, Aux1$_{LS20}$ and Aux2$_{LS20}$ are also predicted to contain an RHH DNA-binding domain in their N-terminal region (Supplementary Figure S1). In addition, our *in silico* analyses predict that the auxiliary PcfF and LtrF proteins of Gram+ *E. faecalis* pCF10 and *L. lactis* pRS01 plasmids, respectively, also contain an RHH motif in their N-terminal region (our unpublished results). The presence of a likely RHH motif in Aux1$_{LS20}$ and Aux2$_{LS20}$ is therefore in line with the conclusion that they are auxiliary proteins. More importantly, the observation that the auxiliary proteins encoded by plasmids pLS20, pRS01, and pCF10, replicating in Gram+ bacteria, all contain a predicted RHH motif indicates that this is a conserved motif in auxiliary proteins encoded by CEs of both Gram− and Gram+ origin, and suggests that auxiliary proteins share a common ancestor. We have made use of this feature, combined with the genetic organization, to identify putative auxiliary genes located upstream of the MOB$_L$ type relaxase genes that encode homologs of Aux1$_{LS20}$ and Aux2$_{LS20}$. This strategy resulted in the identification of about 400 and 90 genes encoding homologs of Aux2$_{LS20}$ and Aux1$_{LS20}$, respectively; 99.2% (Aux2$_{LS20}$) and 98.9% (Aux1$_{LS20}$) of these homologs were predicted to contain a Ribbon-Helix-Helix motif in their N-terminal region. These results reinforce therefore the view that an N-terminal Ribbon-Helix-Helix DNA binding motif is a characteristic feature of auxiliary relaxosome proteins. In addition, these data showed that Aux1$_{LS20}$ and Aux2$_{LS20}$ are the founding members of two families of auxiliary proteins whose genes are genetically linked to a MOB$_L$ type relaxase gene. In summary, we have demonstrated that pLS20cat genes *56* (*aux1*$_{LS20}$) and *57* (*aux2*$_{LS20}$) encode the auxiliary proteins of pLS20 that are essential for conjugation, and that they form the founding members of families of auxiliary relaxosome proteins that are encoded in Firmicutes bacteria.

AUTHOR CONTRIBUTIONS

All authors listed have made substantial, direct experimental and/or intellectual contribution to the work. AM-A, J-AH, GR, CG-C, DG-A, and JV-C generated all plasmids and strains, purified proteins and executed all the experiments except the ultracentrifugation studies, which were performed by JL-O and CA. DA performed *in silico* analyses contributed to the general design and analyses of the results. LW and WM designed the experimental plan and were principally responsible for analyzing the results and writing the paper. WM supervised AM-A, J-AH, GR, CG-C, DG-A and JV-C.

ACKNOWLEDGMENTS

We thank Jose Belio for help with preparing the Figures, and Margarita Salas and Jeff Errington for their support on our work. We also want to acknowledge helpful discussion with other lab members, and particularly want to mention Praveen K. Singh.

REFERENCES

Altschul, S. F., Madden, T. L., Schaffer, A. A., Zhang, J., Zhang, Z., Miller, W., et al. (1997). Gapped BLAST and PSI-BLAST: a new generation of protein database search programs. *Nucleic Acids Res.* 25, 3389–3402.

Altschul, S. F., Wootton, J. C., Gertz, E. M., Agarwala, R., Morgulis, A., Schäffer, A. A., et al. (2005). Protein database searches using compositionally adjusted substitution matrices. *FEBS J.* 272, 5101–5109. doi: 10.1111/j.1742-4658.2005.04945.x

Alvarez-Martinez, C. E., and Christie, P. J. (2009). Biological diversity of prokaryotic type IV secretion systems. *Microbiol. Mol. Biol. Rev.* 73, 775–808. doi: 10.1128/MMBR.00023-09

Boto, L. (2010). Horizontal gene transfer in evolution: facts and challenges. *Proc. Biol. Sci.* 277, 819–827. doi: 10.1098/rspb.2009.1679

Bowie, J. U., and Sauer, R. T. (1990). TraY proteins of F and related episomes are members of the Arc and Mnt repressor family. *J. Mol. Biol.* 211, 5–6.

Bron, S., Peijnenburg, A., Peeters, B. P. H., Haima, P., and Venema, G. (1989). "Cloning and plasmid (in)stability in *Bacillus subtilis*," in *Genetic Transformation and Expression*, eds L. O. Butler, C. R. Harwood, and B. E. B. Moseley (Andover: Intercept Ltd.), 205–219.

Chen, Y., Staddon, J. H., and Dunny, G. M. (2007). Specificity determinants of conjugative DNA processing in the *Enterococcus faecalis* plasmid pCF10 and the Lactococcus lactis plasmid pRS01. *Mol. Microbiol.* 63, 1549–1564.

Chen, Y., Zhang, X., Manias, D., Yeo, H. J., Dunny, G. M., and Christie, P. J. (2008). *Enterococcus faecalis* PcfC, a spatially localized substrate receptor for type IV secretion of the pCF10 transfer intermediate. *J. Bacteriol.* 190, 3632–3645. doi: 10.1128/JB.01999-07

Davies, J., and Davies, D. (2010). Origins and evolution of antibiotic resistance. *Microbiol. Mol. Biol. Rev.* 74, 417–433. doi: 10.1128/MMBR.00016-10

De la Cruz, F., Frost, L. S., Meyer, R. J., and Zechner, E. L. (2010). Conjugative DNA metabolism in Gram-negative bacteria. *FEMS Microbiol. Rev.* 34, 18–40. doi: 10.1111/j.1574-6976.2009.00195.x

Del Solar, G., Hernandez-Arriaga, A. M., Gomis-Ruth, F. X., Coll, M., and Espinosa, M. (2002). A genetically economical family of plasmid-encoded transcriptional repressors involved in control of plasmid copy number. *J. Bacteriol.* 184, 4943–4951.

Edgar, R. C. (2010). Search and clustering orders of magnitude faster than BLAST. *Bioinformatics* 26, 2460–2461. doi: 10.1093/bioinformatics/btq461

Fogg, J. M., Randall, G. L., Pettitt, B. M., Sumners de, W. L., Harris, S. A., and Zechiedrich, L. (2012). Bullied no more: when and how DNA shoves proteins around. *Q. Rev. Biophys.* 45, 257–299. doi: 10.1017/S0033583512000054

Forsberg, K. J., Reyes, A., Wang, B., Selleck, E. M., Sommer, M. O., and Dantas, G. (2012). The shared antibiotic resistome of soil bacteria and human pathogens. *Science* 337, 1107–1111. doi: 10.1126/science.1220761

Frost, L. S., Leplae, R., Summers, A. O., and Toussaint, A. (2005). Mobile genetic elements: the agents of open source evolution. *Nat. Rev. Micobiol.* 3, 722–732.

Gimenes, F., Takeda, K. I., Fiorini, A., Gouveia, F. S., and Fernandez, M. A. (2008). Intrinsically bent DNA in replication origins and gene promoters. *Genet. Mol. Res.* 7, 549–558.

Godziszewska, J., Moncalian, G., Cabezas, M., Bartosik, A. A., De la Cruz, F., and Jagura-Burdzy, G. (2016). Concerted action of NIC relaxase and auxiliary protein MobC in RA3 plasmid conjugation. *Mol. Microbiol.* 101, 439–456. doi: 10.1111/mmi.13401

Goessweiner-Mohr, N., Arends, K., Keller, W., and Grohmann, E. (2013). Conjugative type IV secretion systems in Gram-positive bacteria. *Plasmid* 70, 289–302. doi: 10.1016/j.plasmid.2013.09.005

Goessweiner-Mohr, N., Eder, M., Hofer, G., Fercher, C., Arends, K., Birner-Gruenberger, R., et al. (2014). Structure of the double-stranded DNA-binding type IV secretion protein TraN from *Enterococcus*. *Acta Crystallogr. D. Biol. Crystallogr.* 70, 2376–2389. doi: 10.1107/S1399004714014187

Gomis-Ruth, F. X., Sola, M., Acebo, P., Parraga, A., Guasch, A., Eritja, R., et al. (1998). The structure of plasmid-encoded transcriptional repressor CopG unliganded and bound to its operator. *EMBO J.* 17, 7404–7415.

Grohmann, E., Goessweiner-Mohr, N., and Brantl, S. (2016). DNA-binding proteins regulating pIP501 transfer and replication. *Front. Mol. Biosci.* 3:42. doi: 10.3389/fmolb.2016.00042

Itaya, M., Sakaya, N., Matsunaga, S., Fujita, K., and Kaneko, S. (2006). Conjugational transfer kinetics of pLS20 between *Bacillus subtilis* in liquid medium. *Biosci. Biotechnol. Biochem.* 70, 740–742.

Kopec, J., Bergmann, A., Fritz, G., Grohmann, E., and Keller, W. (2005). TraA and its N-terminal relaxase domain of the Gram-positive plasmid pIP501 show specific *oriT* binding and behave as dimers in solution. *Biochem. J.* 387, 401–409.

Lee, C. A., and Grossman, A. D. (2007). Identification of the origin of transfer (oriT) and DNA relaxase required for conjugation of the integrative and conjugative element ICEBs1 of *Bacillus subtilis*. *J. Bacteriol.* 189, 7254–7261.

Lu, J., den Dulk-Ras, A., Hooykaas, P. J., and Glover, J. N. (2009). *Agrobacterium tumefaciens* VirC2 enhances T-DNA transfer and virulence through its C-terminal ribbon-helix-helix DNA-binding fold. *Proc. Natl. Acad. Sci. U.S.A.* 106, 9643–9648. doi: 10.1073/pnas.0812199106

Mazel, D., and Davies, J. (1999). Antibiotic resistance in microbes. *Cell Mol. Life Sci.* 56, 742–754.

Moncalian, G., and De la Cruz, F. (2004). DNA binding properties of protein TrwA, a possible structural variant of the Arc repressor superfamily. *Biochim. Biophys. Acta* 1701, 15–23.

Norman, A., Hansen, L. H., and Sorensen, S. J. (2009). Conjugative plasmids: vessels of the communal gene pool. *Philos. Trans. R. Soc. Lond B Biol. Sci.* 364, 2275–2289. doi: 10.1098/rstb.2009.0037

Ochman, H., Lawrence, J. G., and Groisman, E. A. (2000). Lateral gene transfer and the nature of bacterial innovation. *Nature* 405, 299–304.

Penders, J., Stobberingh, E. E., Savelkoul, P. H., and Wolffs, P. F. (2013). The human microbiome as a reservoir of antimicrobial resistance. *Front. Microbiol.* 4:87. doi: 10.3389/fmicb.2013.00087

Peng, Y., Lu, J., Wong, J. J., Edwards, R. A., Frost, L. S., and Mark Glover, J. N. (2014). Mechanistic basis of plasmid-specific DNA binding of the F plasmid regulatory protein, TraM. *J. Mol. Biol.* 426, 3783–3795. doi: 10.1016/j.jmb.2014.09.018

Pluta, R., Boer, D. R., Lorenzo-Diaz, F., Russi, S., Gomez, H., Fernandez-Lopez, C., et al. (2017). Structural basis of a histidine-DNA nicking/joining mechanism for gene transfer and promiscuous spread of antibiotic resistance. *Proc. Natl. Acad. Sci. U.S.A.* 114, E6526–E6535. doi: 10.1073/pnas.1702971114

R Core Team (2017). *R: A Language, and Environment for Statistical Computing.* Vienna: R Foundation for Statistical Computing.

Ragonese, H., Haisch, D., Villareal, E., Choi, J. H., and Matson, S. W. (2007). The F plasmid-encoded TraM protein stimulates relaxosome-mediated cleavage at oriT through an interaction with TraI *Mol. Microbiol.* 63, 1173–1184.

Ramachandran, G., Miguel-Arribas, A., Abia, D., Singh, P. K., Crespo, I., Gago-Cordoba, C., et al. (2017). Discovery of a new family of relaxases in Firmicutes bacteria. *PLOS Genet.* 13:e1006586. doi: 10.1371/journal.pgen.1006586

Ramachandran, G., Singh, P. K., Luque-Ortega, J. R., Yuste, L., Alfonso, C., Rojo, F., et al. (2014). A complex genetic switch involving overlapping divergent promoters and DNA looping regulates expression of conjugation genes of a gram-positive plasmid. *PLOS Genet.* 10:e1004733. doi: 10.1371/journal.pgen.1004733

Sambrook, J., Fritsch, E. F., and Maniatis, T. (1989). *Molecular Cloning: A Laboratory Manual.* Cold Spring Harbor, NY: Cold Spring Harbor Laboratory Press.

Schäffer, A. A., Aravind, L., Madden, T. L., Shavirin, S., Spouge, J. L., Wolf, Y. I., et al. (2001). Improving the accuracy of PSI-BLAST protein database searches with composition-based statistics and other refinements. *Nucleic Acids Res.* 29, 2994–3005.

Schildbach, J. F., Karzai, A. W., Raumann, B. E., and Sauer, R. T. (1999). Origins of DNA-binding specificity: role of protein contacts with the DNA backbone. *Proc. Natl. Acad. Sci. U.S.A.* 96, 811–817.

Singh, P. K., Ramachandran, G., Duran-Alcalde, L., Alonso, C., Wu, L. J., and Meijer, W. J. (2012). Inhibition of *Bacillus subtilis* natural competence by a native, conjugative plasmid-encoded comK repressor protein. *Environ. Microbiol.* 14, 2812–2825. doi: 10.1111/j.1462-2920.2012.02819.x

Singh, P. K., Ramachandran, G., Ramos-Ruiz, R., Peiro-Pastor, R., Abia, D., Wu, L. J., et al. (2013). Mobility of the native *Bacillus subtilis* conjugative plasmid pLS20 is regulated by intercellular signaling. *PLOS Genet.* 9:e1003892. doi: 10.1371/journal.pgen.1003892

Smillie, C., Garcillán-Barcia, M. P., Francia, M. V., Rocha, E. P. C., and De la Cruz, F. (2010). Mobility of plasmids. *Microbiol. Mol. Biol. Rev.* 74, 434–452.

Sommer, M. O., Church, G. M., and Dantas, G. (2010). The human microbiome harbors a diverse reservoir of antibiotic resistance genes. *Virulence* 1, 299–303. doi: 10.4161/viru.1.4.12010

Sommer, M. O., Dantas, G., and Church, G. M. (2009). Functional characterization of the antibiotic resistance reservoir in the human microflora. *Science* 325, 1128–1131. doi: 10.1126/science.1176950

Tanaka, T., Kuroda, M., and Sakaguchi, K. (1977). Isolation and characterization of four plasmids from *Bacillus subtilis*. *J. Bacteriol.* 129, 1487–1494.

Thoma, L., and Muth, G. (2016). Conjugative DNA-transfer in *Streptomyces*, a mycelial organism. *Plasmid* 87–88, 1–9. doi: 10.1016/j.plasmid.2016.09.004

Thomas, C. M., and Nielsen, K. M. (2005). Mechanisms of, and barriers to, horizontal gene transfer between bacteria. *Nat. Rev. Microbiol.* 3, 711–721.

Varsaki, A., Moncalian, G., Garcillan-Barcia, M. P., Drainas, C., and De la Cruz, F. (2009). Analysis of ColE1 MbeC unveils an extended ribbon-helix-helix family of nicking accessory proteins. *J. Bacteriol.* 191, 1446–1455. doi: 10.1128/JB.01342-08

Wang, S., Li, W., Liu, S., and Xu, J. (2016). RaptorX-Property: a web server for protein structure property prediction. *Nucleic Acids Res.* 44, W430–W435. doi: 10.1093/nar/gkw306

Waters, V. L. (1999). Conjugative transfer in the dissemination of beta-lactam and aminoglycoside resistance. *Front. Biosci.* 4:D433–D456.

Wong, J. J., Lu, J., Edwards, R. A., Frost, L. S., and Glover, J. N. (2011). Structural basis of cooperative DNA recognition by the plasmid conjugation factor, TraM. *Nucleic Acids Res.* 39, 6775–6788. doi: 10.1093/nar/gkr296

Wong, J. J., Lu, J., and Glover, J. N. (2012). Relaxosome function and conjugation regulation in F-like plasmids - a structural biology perspective. *Mol. Microbiol.* 85, 602–617. doi: 10.1111/j.1365-2958.2012.08131.x

Yoshida, H., Furuya, N., Lin, Y. J., Guntert, P., Komano, T., and Kainosho, M. (2008). Structural basis of the role of the NikA ribbon-helix-helix domain in initiating bacterial conjugation. *J. Mol. Biol.* 384, 690–701. doi: 10.1016/j.jmb.2008.09.067

Zatyka, M., and Thomas, C. M. (1998). Control of genes for conjugative transfer of plasmids and other mobile elements. *FEMS Microbiol. Rev.* 21, 291–319.

Zhang, S., and Meyer, R. (1997). The relaxosome protein MobC promotes conjugal plasmid mobilization by extending DNA strand separation to the nick site at the origin of transfer. *Mol. Microbiol.* 25, 509–516.

3

Successful Establishment of Plasmids R1 and pMV158 in a New Host Requires the Relief of the Transcriptional Repression of their Essential *rep* Genes

José Á. Ruiz-Masó, Luis M. Luengo, Inmaculada Moreno-Córdoba, Ramón Díaz-Orejas and Gloria del Solar*

Molecular Microbiology and Infection Biology Department, Centro de Investigaciones Biológicas, Consejo Superior de Investigaciones Científicas, Madrid, Spain

Edited by:
Tatiana Venkova,
Fox Chase Cancer Center, United States

Reviewed by:
Matxalen Llosa,
University of Cantabria, Spain
Elisabeth Grohmann,
Beuth University of Applied Sciences, Germany
Antonio Juárez,
University of Barcelona, Spain

****Correspondence:***
Gloria del Solar
gdelsolar@cib.csic.es

Although differing in size, encoded traits, host range, and replication mechanism, both narrow-host-range theta-type conjugative enterobacterial plasmid R1 and promiscuous rolling-circle-type mobilizable streptococcal plasmid pMV158 encode a transcriptional repressor protein, namely CopB in R1 and CopG in pMV158, involved in replication control. The gene encoding CopB or CopG is cotranscribed with a downstream gene that encodes the replication initiator Rep protein of the corresponding plasmid. However, whereas CopG is an auto-repressor that inhibits transcription of the entire *copG-repB* operon, CopB is expressed constitutively and represses a second, downstream promoter that directs transcription of *repA*. As a consequence of the distinct regulatory pathways implied by CopB and CopG, these repressor proteins play a different role in control of plasmid replication during the steady state: while CopB has an auxiliary role by keeping repressed the regulated promoter whenever the plasmid copy number is above a low threshold, CopG plays a primary role by acting coordinately with RNAII. Here, we have studied the role of the regulatory circuit mediated by these transcriptional repressors during the establishment of these two plasmids in a new host cell, and found that excess Cop repressor molecules in the recipient cell result in a severe decrease in the frequency and/or the velocity of appearance of transformant colonies for the cognate plasmid but not for unrelated plasmids. Using the pMV158 replicon as a model system, together with highly sensitive real-time qPCR and inverse PCR methods, we have also analyzed the effect of CopG on the kinetics of repopulation of the plasmid in *Streptococcus pneumoniae*. We show that, whereas in the absence of CopG pMV158 repopulation occurs mainly during the first 45 min following plasmid transfer, the presence of the transcriptional repressor in the recipient cell severely impairs the replicon repopulation and makes the plasmid replicate at approximately the same rate as the chromosome at any time after transformation, which results in maximal plasmid loss rate in the absence of selection. Overall, these findings indicate that unrepressed activity of the Cop-regulated promoter is crucial for the successful colonization of the recipient bacterial cells by the plasmid.

Keywords: plasmid repopulation, establishment phase replication, R1 replicon, pMV158 replicon, Cop transcriptional repressors, plasmid replication rate

INTRODUCTION

Plasmids specify replication control systems that enable them to maintain a characteristic steady-state concentration (copy number) in their host cell. These regulatory systems are *trans*-acting and can sense and correct stochastic up and down fluctuations of the plasmid copy number in individual cells. By adjusting the replication initiation rate in response to changes in the intracellular plasmid concentration, control systems manage to keep the steady-state condition, where every plasmid copy replicates, on average, once per cell generation (Nordström, 1990). Steady-state control of plasmid replication has been analyzed at a deep level and a variety of different regulatory circuits have been mechanistically characterized (del Solar and Espinosa, 2000; Das et al., 2005; Nordström, 2006). Antisense RNA (asRNA)-mediated control of plasmid replication is widely spread in theta- and rolling circle-replicating plasmids from Gram-positive and Gram-negative bacteria (del Solar and Espinosa, 2000; Brantl, 2014). The small regulatory RNAs involved in control of plasmid replication and copy number are *bona fide* (i.e., *cis*-encoded) asRNAs, as they are encoded on the DNA strand opposite to an RNA essential for replication initiation, namely a pre-primer or an mRNA for the replication initiator protein (Rep). These asRNAs base-pair to their target (sense) RNA to inhibit its function and/or the completion of its synthesis through a variety of mechanisms, including inhibition of primer maturation, transcription attenuation, prevention of formation of a translation activator RNA pseudoknot, and inhibition of translation of either *rep* or a leader-peptide reading frame to which *rep* is translationally coupled (del Solar and Espinosa, 2000; Brantl, 2014). Most frequently, asRNAs controlling plasmid replication are metabolically unstable, *trans*-acting inhibitory elements, whose synthesis is directed by unregulated and strong promoters. These features enable them to sense and correct rapidly up and down fluctuations of the plasmid copy number in individual cells (del Solar and Espinosa, 2000; Wagner et al., 2002; Brantl, 2014).

Although the asRNA is the sole replication control element in some plasmids (pT181 family, IncB/IncIα family, ColE2), it is accompanied, in others, by a regulatory protein that acts either as a transcriptional repressor (Cop proteins in R1, Inc18, and pMV158 families) or as an RNA-binding protein (Rom/Rop protein in ColE1-like plasmids) (del Solar and Espinosa, 2000; Brantl, 2014). Rom/Rop and CopB of ColE1- and R1-like replicons, respectively, have been largely considered as mere auxiliary elements because of their rather secondary role in the steady-state plasmid replication control, when the activity of these proteins is almost saturating (Nordström et al., 1984; Rosenfeld and Grover, 1993; Atlung et al., 1999; Summers, 2009). In contrast, efficient replication control of plasmids of the Inc18 and pMV158 families requires the coordinated participation of the asRNA and of the transcriptional repressor Cop protein, both elements playing a primary regulatory role (del Solar and Espinosa, 2000; Brantl, 2014).

Plasmid R1, originally isolated from *Salmonella enterica* serovar Paratyphi, is a low-copy-number, multiresistance, conjugative plasmid of the IncFII incompatibility group. It has a narrow host-range restricted to the *Enterobacteriaceae* family. The R1 elements and circuits involved in steady-state plasmid replication control and maintenance have been studied in great detail (Nordström et al., 1984; Olsson et al., 2004; Nordström, 2006). In addition to the origin of replication (*oriR1*), the R1 basic replicon includes the *repA* gene for the replication initiator protein and the two replication control genes, *copB* and *copA*, encoding, respectively, the transcriptional repressor CopB protein and the CopA asRNA (**Figure 1**). The RepA protein is rate limiting for initiation of replication. The essential *repA* gene is transcribed from two promoters, namely P_{copB} and P_{repA}. The upstream P_{copB} promoter directs constitutive transcription of *copB*, *tap* (the leader peptide reading frame) and *repA*, which is translationally coupled to *tap*. Transcription from the downstream CopB-regulated P_{repA} promoter gives rise to the shorter bicistronic *tap-repA* mRNA (**Figure 1**). When unrepressed, the P_{repA} promoter is about twice as strong as P_{copB}, although under normal conditions during the steady-state plasmid replication, P_{repA} is almost totally (90%) switched off by CopB-mediated repression (Olsson et al., 2004). Since CopB acts as a tetramer (Riise and Molin, 1986), the activity of P_{repA} is likely to be strongly dependent on plasmid concentration. It has been shown that the presence of extra copies of *copB in trans*, which further reduces the already low activity of P_{repA}, increases 7-fold the rate of loss of a Par^{+} derivative of the R1 basic replicon (Olsson et al., 2004). The steady-state activity of P_{repA} is thought to stabilize the plasmid inheritance both by speeding up R1 replication in cells with very few plasmid copies (thus decreasing the frequency of these cells) and by slightly increasing the average plasmid concentration. CopB also plays a main role in the coupling between the *kis-kid* auxiliary maintenance system and the basic replicon of the plasmid (López-Villarejo et al., 2015). The switch of this coupling is the antitoxin Kis, whose levels decrease in cells with lower-than-average plasmid copy number. Decrease of Kis concentration activates the Kid toxin, which is an RNase with two efficient target sites in the intergenic region of the *copB-repA* mRNA (Pimentel et al., 2005). Cleavage

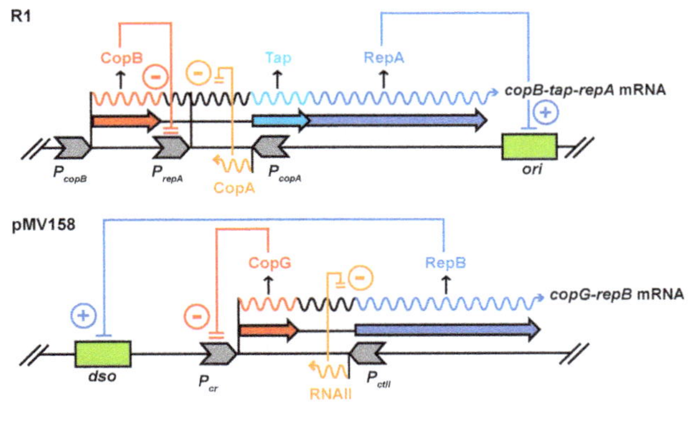

FIGURE 1 | Schematic comparison of the mechanisms controlling expression of the essential *rep* gene in R1 and pMV158. In R1, *repA* is transcribed from P_{copB} and P_{repA} promoters, and CopB represses transcription from the stronger P_{repA} promoter. Gene *copA*, which encodes an asRNA, overlaps the intergenic region of the *copB-tap-repA* operon. CopA exerts an indirect control on *repA* translation by inhibiting synthesis of Tap protein, to which RepA synthesis is coupled. In pMV158, the *copG-repB* operon is transcribed from the P_{cr} promoter, which is negatively regulated by transcriptional repressor CopG. asRNAII, transcribed from P_{ctII} promoter, binds to its complementary sequence in the *copG-repB* mRNA, thus inhibiting translation of the replication initiator *repB* gene.

at these sites reduces the CopB levels, which leads to activation of P_{repA} and subsequent increase in plasmid replication efficiency. Hence, the *kis-kid* system coupled to the CopB-mediated loop functions as a safety device when the plasmid copy number is very low.

Promiscuous plasmid pMV158 was originally isolated from a clinical strain of *Streptococcus agalactiae* (Burdett, 1980) and subsequently transferred to a large number of bacterial genera and species. The detailed analysis of its replicon has made pMV158 the prototype of a vast family of rolling circle-replicating plasmids (Ruiz-Masó et al., 2015; Boer et al., 2016). The pMV158 basic replicon includes a compact region containing the double-strand origin (*dso*) (Ruiz-Masó et al., 2007) as well as the genes that encode the replication initiator protein (RepB) (Ruiz-Masó et al., 2004; Boer et al., 2009) and the two replication control elements (transcriptional repressor CopG and asRNA RNAII) (del Solar et al., 1995; Gomis-Rüth et al., 1998; Hernández-Arriaga et al., 2009; López-Aguilar et al., 2015) (**Figure 1**). Like the *rep* gene product of R1, RepB of plasmid pMV158 is the rate-limiting factor for the replication initiation process. Unlike R1, the streptococcal plasmid expresses the essential *rep* gene from a single promoter (P_{cr}), which directs cotranscription of the *copG-repB* operon and is subjected to CopG-mediated regulation (del Solar et al., 1990) (**Figure 1**). In the unrepressed state, P_{cr} seems to be even stronger than promoter P_{ctII} that directs synthesis of countertranscript RNAII, a situation that contrasts with the plasmid replication control systems based exclusively on asRNA, where the transcription rate of the essential RNA is constant but rather low compared with that of the asRNA (del Solar and Espinosa, 2000). Unsuccessful repression of promoter P_{cr} in pMV158 derivatives encoding a defective CopG repressor leads to a 5-fold increase in the plasmid copy number (del Solar et al., 1990, 1995). On the other hand, the presence *in trans* of high dosages of the autoregulated *copG* gene has been shown to decrease by ~35% the steady-state copy number of the pMV158 replicon in *S. pneumoniae* (del Solar et al., 1995).

Despite the quite deep current knowledge about the involvement of the Cop transcriptional repressors in the control of the steady-state plasmid replication, very little has been reported so far on their role in the establishment phase replication. The Cop regulatory elements have been proposed to play an important role during plasmid establishment in a new bacterium based on the fact that the Cop-regulated promoter, when unrepressed, determines high transcription rates of the essential *rep* gene (del Solar et al., 1990, 1995; del Solar and Espinosa, 2000; Olsson et al., 2004; Brantl, 2014). Yet, only non-published results have been invoked in a few articles to suggest the involvement of the Cop regulatory loops of pMV158 and R1 in the establishment phase replication of these plasmids (Nordström and Nordström, 1985; del Solar and Espinosa, 2000; Olsson et al., 2004).

In this work, we have analyzed the effect of the Cop proteins of R1 and pMV158 on the establishment of these plasmids. To this end, we have electrotransferred a mini-R1 derivative (pKN1562) to *Escherichia coli* and *Salmonella* Typhimurium (*S*. Typhimurium hereafter) cells that either contain or lack a compatible recombinant plasmid encoding CopB. Similarly, we have transferred pLS1 (a *mob*$^-$ derivative of pMV158, Lacks et al., 1986) to naturally-competent pneumococcal cells and to electrocompetent cells of *Staphylococcus aureus* either containing or lacking a compatible recombinant plasmid encoding CopG. We show that, irrespective of the system employed, the presence of these proteins in the recipient cell selectively impairs the establishment of the cognate plasmid, resulting in a decrease in the frequency of total or early transformant colonies. By using the pneumococcal host as a model system, we also show that repopulation of the pMV158 replicon is almost abolished when autoregulated *copG* is supplied *in trans* at a high gene dosage.

MATERIALS AND METHODS

Bacterial Strains and Plasmids

Bacterial strains employed, and their uses for this work, are summarized in **Table 1**. Plasmid constructions used throughout this study, as well as their relevant features, are listed in **Table 2**. *S. pneumoniae* 708 was the host for pLS1, pLS1*cop7*, pC194, pCGA3, pCGA3n, pCGA30, and pAMβ1 plasmids. Pneumococcal cells were grown at 37°C in AGCH medium (Lacks et al., 1986) supplemented with 0.3% sucrose and 0.2% yeast extract. *S. aureus* RN4220 was the host for pLS1, pT181*cop608*, pC194, pCGA3, pCGA3n, and pCGA30 plasmids. Staphylococcal cells were grown at 37°C in brain heart infusion medium (BHI, Difco). *E. coli* C600 and *S. Typhimurium* SL1344 were the hosts for pUC18-*copB*, pKN1562, and pACYC184 plasmids; cells were grown at 30°C in Lysogeny broth (LB) medium.

TABLE 1 | Strains used in this study.

Bacteria		
Bacteria	**Characteristics**	**Source**
Escherichia coli C600	K12 derivate	Appleyard, 1954
Salmonella Typhimurium SL1344	Wild-type strain	Hoiseth and Stocker, 1981 NC_016810.1
Staphylococcus aureus RN4220	Restriction-deficient mutant of strain 8325-4	Kreiswirth et al., 1983
Streptococcus pneumoniae 708	*end*-1 *exo*-1 *trt*-1 *hex*-4 *mal*M594	Lacks and Greenberg, 1977

Plasmidic and Genomic DNA Preparations

Plasmidic DNA (pDNA) content from pneumococcal transformants was analyzed by preparing total DNA crude extracts as described (del Solar et al., 1987). These DNA preparations were also used to estimate the relative plasmid copy number from the ratio between the intensities of the plasmid and chromosome DNA bands quantified for the plasmid of interest relative to a plasmid control whose copy number has been precisely determined, after correcting for the difference in size of both plasmids (del Solar et al., 1995). Plasmids pLS1, pLS1*cop7*, pC194, pCGA3, pCGA3n, and pCGA30 were isolated from *S. pneumoniae* 708 and purified by two consecutive CsCl/ethidium bromide density gradient centrifugations, as described (Lacks et al., 1986). Plasmid pAMβ1 was isolated from *S. pneumoniae* 708 and purified by alkaline lysis as described (Stassi et al., 1981). Plasmid pT181*cop608*, isolated from *S. aureus* RN4220, and plasmids pKN1562, pUC18-*copB*, and pACYC184, isolated from *E. coli* C600, were purified using a Jetstar Plasmid Midiprep Kit (Genomed). pDNA content from staphylococcal transformants was analyzed by the same alkaline lysis method used for *S. pneumoniae*. In both midipreps and alkaline lysis procedures lysostaphin (50 μg/ml) was added to the cell resuspension buffer in order to facilitate staphylococcal cells lysis.

Genomic DNA (gDNA) used as template for real-time quantitative PCR (qPCR) and inverse PCR (iPCR) was isolated from pneumococcal cultures in exponential growth phase, which was determined by measurement of optical density at 650 nm. The DNA was extracted from cells of *S. pneumoniae* 708 with different plasmid content by using the Wizard® Genomic DNA Purification Kit (Promega) optimized for *S. pneumoniae*. Cells resuspended in 50 mM EDTA were incubated with 0.04% of deoxycholate and 0.1 mg/ml of Proteinase K for 10 min at 37°C. Next, and before proceeding with the lysis step, the cellular suspension was quickly frozen on a mixture of dry ice and ethanol and stored at −80°C. With this method, aliquots taken at different time intervals were processed simultaneously from the lysis step. Moreover, 0.05 μg/ml of glycogen (molecular biology grade) was added in the isopropanol precipitation step to facilitate gDNA recovery from diluted samples. Contrarily to the samples where no glycogen was added, the gDNA yield of the samples treated with glycogen was found to be nearly proportional to the total amount of lysed cells. Concentration of the gDNA was determined with a Qubit fluorometer by using the Qubit HS dsDNA Assay Kit (Molecular Probes).

Purified gDNA was digested with EcoRI, a restriction enzyme that linearizes the pLS1 DNA but leaves intact the plasmidic and chromosomal amplicons (i.e., the DNA segments to be amplified in the qPCR assays). This method has been developed to obtain accurate qPCR-based copy number results for plasmids (Providenti et al., 2006).

Calculation of the Experimental Plasmid Loss Rate

The experimental loss rate (L_{ex}) of pLS1 and pLS1*cop7* in newly transformed pneumococcal cells was calculated from the equation (Gerdes et al., 1985):

$$T/T_0 = (1 - L_{ex})^n, \tag{1}$$

where T_0 and T are, respectively, the fractions of transformants *ab initio* and after n generations. This equation can be converted into a linear function by taking logarithms,

$$\log(T/T_0) = \log(1 - L_{ex})n, \tag{2}$$

where log(1-L_{ex}) is the slope of the linear regression fit in the plot of the experimental values of log(T/T_0) against the number of cell generations (n).

Transformation of Bacterial Species with Plasmid DNA

Transformation of *E. coli* C600 and *S.* Typhimurium SL1344 cells was performed by electroporation essentially as described (Dower et al., 1988). Competent cultures of *E. coli* C600 and *S.* Typhimurium SL1344 and those of the same strains harboring pUC18-*copB* as the resident plasmid were transformed with 0.2 μg of plasmid DNA of pKN1562 or pACYC184. Transformants were grown on LB-agar plates with antibiotic selection according to the resistance carried by the plasmids: 50 μg/ml of kanamycin (Km) for pKN1562, 50 μg/ml of ampicillin (Amp) for pUC18-*copB*, or 20 μg/ml of chloramphenicol (Cm) for pACYC184. Competent cells of *S. aureus* RN4220 were prepared and transformed by electroporation following the procedure depicted in Augustin and Götz (1990). Competent staphylococcal cultures (50 μl) harboring pC194, pCGA3, pCGA3n, or pCGA30 as the resident plasmid were transformed with 0.5 μg of DNA of the donor plasmid (pLS1 or pT181). After allowing for phenotypic expression (60 min), all cultures were treated for 30 min with 0.5 μg/ml of tetracycline (Tc), a sub-inhibitory concentration of the antibiotic that allows induction of the pT181 *tet* gene. Cells transformed with pLS1 or pT181 were selected on BHI-agar plates containing 5 μg/ml of Tc, for selection of the entering plasmid, and 3 μg/ml of Cm, for resident plasmid selection.

Competent cells of *S. pneumoniae* 708 were prepared and transformed as described (López et al., 1982). Three independent lots of naturally competent cells were prepared from each of the four different strains harboring pC194, pCGA3, pCGA3n, or pCGA30 as the resident plasmid. Since the development of pneumococcal competence is influenced by many factors (Attaiech et al., 2015), including the exact composition of the

TABLE 2 | Plasmids used in this study.

Plasmids				
Plasmids	**Characteristics**	**Antibiotic marker**	**Copy number**	**Source**
pACYC184	Plasmid vector containing the minimal p15A replicon	Cm^R Tc^R	≈15[a]	Chang and Cohen, 1978
pAMβ1	Conjugative plasmid from *Enterococcus faecalis* DS5	Em^R	≈7[c]	Leblanc and Lee, 1984
pCGA3	*copG* cloned in the high copy number vector pC194	Cm^R	≈200[e]	del Solar et al., 1995
pCGA3n	*copG* cloned in the medium copy number vector pC194n	Cm^R	≈25[e]	del Solar et al., 1995
pCGA30	pCGA3 derivative with a 48-bp deletion that includes the−10 sequence of the P_{cr} promoter, and the *SD* sequence and the two first codons of *copG*.	Cm^R	≈200[e]	del Solar et al., 1993
pC194	Rolling circle replicating plasmid from *S. aureus*	Cm^R	≈200[e]	del Solar et al., 1995
pKN1562	Wild-type mini-R1 plasmid carrying the wt kis-kid system	Km^R	1–2	Molin et al., 1979
pLS1	Non-mobilizable pMV158 derivative plasmid (Δ*mobM*, Δ*ssoU*)	Tc^R	46.0 ± 5.0[c,d]	Stassi et al., 1981
pLS1*cop7*	pLS1 copy-up mutant	Tc^R	182.1 ± 15.5[c,d]	del Solar et al., 1990
pT181-*cop608*	pT181-based high copy number mutant plasmid	Tc^R	800–1,000[b]	Rasooly et al., 1994
pUC18- *copB*	*copB* cloned in the high copy number vector pUC18	Amp^R	≈500[a]	López-Villarejo et al., 2015

Plasmid copy number (PCN) determined in E. coli C600[a], S. aureus[b], and S. pneumoniae[c], [d]PCN determined by RT-qPCR in this work (± standard error), [e]Approximate PCN value estimated by comparative quantitation of total DNA extracted from pneumococcal strains containing the target plasmid or the control pLS1 plasmid; PCN of the latter has been determined by RT-qPCR.

semi-defined AGCH culture medium, the 12 competent cultures were each tested for their level of competence by transformation with chromosomal DNA from a strain able to grow in maltose. Although the transformation efficiency for maltose utilization varied significantly from lot to lot of competent cells of the same strain, the level of competence for chromosomal transformation of the competent cells prepared in parallel was, consistently, 1.5 to 2-, 2 to 3-, and 4 to 5-fold higher for the strains containing pCGA30, pCGA3n, and pCGA3, respectively, than for the strain harboring pC194. The reason for the apparent increase in the natural competence of *S. pneumoniae* when the DNA of gene *copG* is present has not been investigated yet. Cultures (1 ml) of competent pneumococcal cells were transformed with 0.25 μg of DNA of the donor plasmid (pLS1 or pAMβ1). After allowing for phenotypic expression (70 min), cultures were induced with 0.5 μg/ml of Cm for 20 min. Transformants were selected using agar plates containing 1 μg/ml of Tc for pLS1 or 1 μg/ml of erythromycin for pAMβ1. In these plates, selection for the resident plasmid (3 μg/ml of Cm) was maintained. Since pneumococci grow best when protected from air, the basal AGCH-agar layer containing the cells was overlaid with AGCH-agar medium.

Repopulation Kinetics Assays

For repopulation kinetics experiments, cultures of competent pneumococcal cells harboring pCGA3 and pCGA30 were subjected to a modified version of the transformation procedure described in López et al. (1982) that yielded 10-fold higher competence levels. Competent cultures (OD_{650} = 0.3) were diluted 1/20 in 10 ml of AGCH medium supplemented with 0.3% of sucrose, 0.001% of $CaCl_2$ and a sub-inhibitory concentration of Cm (0.5 μg/ml), in order to keep the induced expression of the *cat* gene. The cells were cultured at 37°C to an OD_{650} of 0.3, and the cultures were cooled to 30°C for 15 min. Then, the cells were transformed with 2 μg of DNA of the pMV158 derivative (pLS1 or pLS1*cop7*) by incubation for 30 more min at the same temperature. To stop the transformation process, pancreatic DNase I was added to a final concentration of 2 μg/ml, and the incubation at 30°C was prolonged for 20 more min. Next, the cultures were diluted 1/10 in pre-warmed (37°C) AGCH medium supplemented with 0.3% sucrose, 0.2% yeast extract and 0.5 μg/ml of Cm, and incubated at 37°C up to 150 min. Immediately after dilution, and at the indicated time intervals, 10 and 0.1-ml aliquots of the cultures were withdrawn and used, respectively, to extract the gDNA and to determine the number of total viable cells (c.f.u./ml) and the fraction of transformants. Transformants were selected in three-layered AGCH-medium agar plates containing 3 μg/ml of Cm and 1 μg/ml of Tc. Cells were deposited in the basal layer that was overlaid with a second layer of AGCH-agar medium. The plates were then incubated at 37°C for 2 h before antibiotics for selection were included in the third layer and spread across the rest of the plate by diffusion.

No-transformation control experiments were also performed to ensure that plasmid DNA amplified by real-time qPCR arose from the transformed cells and was not contaminant DNA that escaped from the DNase I digestion. For this purpose, we followed the same transformation procedure as described above but adding simultaneously 2 μg of pLS1*cop7* DNA and DNase I (2 μg/ml), in order to avoid transformation. Aliquots of 10 ml of the no-transformed cultures were taken immediately after dilution and after 30 min of incubation at 37°C. These cell aliquots were processed as described to obtain gDNA. In all cases, before proceeding with the gDNA isolation protocol, cells were washed with 10 ml of 1X PBS (phosphate-buffered saline).

The total number of cell generations (n) was calculated according to the following equation:

$$n = log(V/V_0)/log2, \quad (3)$$

where V is the number of viable cells (c.f.u.) at any of the times analyzed and V_0 is the initial value of viable cells.

A similar expression was used to calculate the number of gDNA duplications (D_C) at a given time interval (*ti-tj*):

$$D_{C_{i-j}} = log(gDNA_j/gDNA_i)/log2, \quad (4)$$

where $gDNA_j$ and $gDNA_i$ are the amounts (ng) of gDNA obtained (after precipitation in the presence of glycogen, see above) at the times *tj* and *ti*, respectively.

Determination of the Copy Number of a Specific pMV158 Amplicon Relative to a Chromosomal Amplicon in Transformed Pneumococcal Cells by qPCR

Two primer sets specific to the PcrA helicase single-copy reference gene (*pcrA*) of *S. pneumoniae R6* (Hoskins et al., 2001) and to the tetracycline resistance TetL protein gene (*tetL*) of pMV158 were designed. Oligonucleotide primers sets (**Table 3**) were designed with Primer3 v0.4.0 (Koressaar and Remm, 2007; Untergasser et al., 2012) based on the pLS1 sequence (NC_010096.1) and on the *S. pneumoniae* R6 (NC_003098.1) *pcrA* sequence. Criteria used during primer design were that primers had predicted Tm of ~59°C and that they generated amplicons ~140 bp in length.

qPCRs were conducted in a total volume of 20 μl using a LightCycler® 96 real-time detection system (Roche) and the FastStart Essential DNA Green Master (Roche), as per manufacturer's recommendations. Decimally diluted EcoRI-digested gDNA preparations (14, 1.4, 0.14 ng per reaction) were analyzed using 0.5 μM (final concentration) of the specific forward and reverse primers of either primer-pair used (**Table 3**). To prepare the reactions and minimize pipetting errors 2 μl of template DNA were added to individual qPCRs. Thermal cycling conditions were as follows: initial denaturation at 95°C for 5 min, followed by 40 cycles of 95°C for 10 s (denaturation), 59°C for 30 s (primer annealing), and 72°C for 20 s (elongation). A melting curve analysis of the PCR products, with a temperature gradient of 0.1°C/s from 59 to 95°C, was performed to confirm the purity and specificity of the PCR products. Two independent qPCR trials were conducted for each template source. In each trial, triplicate samples of the three different amounts of template were analyzed. Control samples without template DNA were also analyzed.

Relative copy number (*CN*) of the pMV158 amplicon was calculated using equation:

$$CN = \left(1 + E_{pcrA}\right)^{Ct_{pcrA}} / (1 + E_{tetL})^{Ct_{tetL}}, \quad (5)$$

where E_{pcrA} and E_{tetL} are, respectively, the PCR amplification efficiencies of the chromosomal and plasmid amplicons, and Ct_{pcrA} and Ct_{tetL} are the mean threshold cycle values obtained for the corresponding amplicons. A CN value was calculated for each of the three template concentrations analyzed, and the mean and standard deviation of the six values (two independent trials with three different template concentrations each) were determined.

E values of target (E_{tetL}) and reference (E_{pcrA}) sequences were empirically calculated for each qPCR trial. For that purpose, mean *Ct* values were plotted against the logarithm of the amount of total DNA template in the assay. From the slope of the curve generated by linear regression of the plotted points, the PCR amplification efficiency was determined according to the equation:

$$E = 10^{-1/slope} - 1, \quad (6)$$

Although the *E* values for both amplicons were higher than 0.9, we have chosen Equation (5) to calculate the relative copy number of the plasmid amplicon as it allows taking into account the slight differences between E_{target} and $E_{reference}$ that we have observed.

Determination of the Relative Amount of Circular Plasmid DNA in Transformed Pneumococcal Cells by iPCR

The plasmidic DNA (pDNA) present in the gDNA isolated from the transformed pneumococcal cells was used as template to perform an inverse PCR protocol with a primer set of divergent oligonucleotides. iPCR was performed using the Phusion High Fidelity (HF) (Thermo Scientific) DNA polymerase. Amplification reactions (20 μl) contained 0.7 ng of gDNA and 0.5 μM of the specific forward and reverse primers (**Table 3**). Thermal cycling conditions comprised 25 cycles (98°C for 10 s, 59.5°C for 30 s, and 72°C for 1 min and 25 s) plus a final extension step of 10 min at 72°C. The amplification reaction yielded a linear dsDNA fragment corresponding to almost the entire pLS1*cop7* plasmid. The products of iPCR were analyzed on 0.8% agarose gels, stained with GelRed (Biotium), and quantified with the aid of a Gel Doc (BIO-RAD) system. At least three gels with DNA products obtained in each of three independent iPCR assays were analyzed. *In vitro* DNA amplification in these iPCR assays was based on equation:

$$P = P_0(1 + E)^C, \quad (7)$$

where P and P_0 are, respectively, the amount of amplified linear pDNA product and the initial amount of template pDNA in the gDNA used for the amplification reaction; E is the amplification efficiency, and C is the number of cycles. Irrespective of the gDNA concentration used, the ratio between P and P_0 is kept constant for a given C provided there is no exhaustion of the primers and dNTPs required for DNA synthesis (and hence E is kept constant). We then confirmed that the employed iPCR conditions fulfilled this requirement for gDNA concentrations ranging from half to twice that used for the analysis of the kinetics of plasmid repopulation.

On the other hand, we have defined the relative plasmid amplification occurring in the transformants during the time interval t_i-t_j as the ratio between the relative numbers of plasmid molecules (with respect to the total gDNA) at times t_j and t_i, which can be estimated from the intensity of the bands corresponding to the linear pDNA products obtained in the iPCR assays using the same amount of gDNA extracted at different

TABLE 3 | Oligonucleotides used in this study.

Target	Accession No.	Sequence (5′-3′)[a]	Length (nt)	Primer position	Product size (bp)
Sequence of primers used for real-time qPCR					
pcrA	NC_003098.1	F: GAGTTGGTTGAGTCCGTCCT	20	980570–980589	144
		R: TGTCACATCCGTGGTGTCAT	20	980713–980694	
tetL	NC_010096.1	F: TGCGAGTACAAACTGGGTGA	20	1879–1898	146
		R: ACCCAATTACCGACCCGAAA	20	2024–2005	
Sequence of primers used for iPCR					
del-5	NC_010096.1	F: GTTTGAGGCTCGTCAAATC	19	514–532	4403
dso2	NC_010096.1	R: CAGCTCTAAGGCTAAAGGCG	20	508–489	

[a] *F and R indicate forward and reverse primers, respectively.*

times after transformation. Considering that, in the time interval t_i to t_j, the plasmid replication rate (i.e., the ratio of pDNA duplications, D_P, to gDNA duplications, D_C,) has a value of R, the relative plasmid amplification in this interval is given by the following equation:

$$Pl_{t_j}/Pl_{t_i} = 2^{D_P}/2^{D_C} = 2^{(RD_c - D_c)} = 2^{D_c(R-1)}, \quad (8)$$

where Pl_{t_j} and Pl_{t_i} are the relative intensities of the amplified pDNA products at times t_j and t_i, respectively. Equation (8) can be converted into a linear function by taking logarithms:

$$log(Pl_{t_j}/Pl_{t_i}) = D_C\,(R-1)\,log2, \quad (9)$$

therefore, the R value in the time interval t_i to t_j was calculated from the equation:

$$R = \left(log(Pl_{t_j}/Pl_{t_i})/D_c log2\right) + 1 \quad (10)$$

Statistical Analysis

ANOVA was run to determine whether experimental Q ratios differed among groups of staphylococcal strains (p-values < 0.05 were considered significant).

RESULTS

The Presence of the Cop Repressor Protein of R1 or pMV158 in the Recipient Cell Decreases the Frequency and/or the Velocity of Appearance of Colonies Transformed with the Cognate Plasmid

To know whether an initially unrepressed transcription of the essential *rep* gene is required for successful establishment of plasmids R1 and pMV158 in a new cell, we compared the efficiencies with which recipient cells that contain or lack the Cop transcriptional repressor of either plasmid were transformed in parallel with plasmids harboring the cognate replicon or an unrelated replicon. It is worth noting that the incoming plasmids carry their own *cop* genes, and hence these assays aim to analyze the importance of the recipient cells having the Cop protein already synthesized upon plasmid entrance.

The existence of a specific effect of R1 CopB on the establishment of the cognate plasmid was tested by transforming either plasmid-free or pUC18-*copB*-carrying *E. coli* C600 and *S.* Typhimurium SL1344 (WT) cells with DNAs of plasmids pKN1562 (a mini-R1 derivative) or pACYC184 (harboring the R1-unrelated p15A replicon), both of which are compatible with the pUC18 replicon. Plasmid pUC18-*copB* provides *in trans* a very high dosage of the *copB* gene cloned under control of its own constitutive promoter. When *S.* Typhimurium cells were transformed with pACYC184, similar transformation efficiencies were obtained irrespective of the presence of CopB in the recipient bacteria (**Figure 2A**). Also, the frequency of transformation of *S.* Typhimurium cells lacking CopB with pKN1562 was basically the same as with the p15A-derivative plasmid (**Figure 2A**). Contrarily, the presence of CopB most severely impaired the efficiency of transformation with the cognate plasmid containing the R1-replicon, as no transformant colonies appeared within 24 h of incubation of the plates (**Figure 2A**). The drastic and specific effect of CopB on the establishment of the R1 replicon was also observed in *E. coli*, where the presence of resident pUC18-*copB* reduced by more than two orders of magnitude the number of colonies transformed with pKN162, without affecting the efficiency of transformation with pACYC184 (**Figure 2B**).

With respect to the pMV158 system, its CopG repressor protein was also shown to significantly impair the establishment of the plasmid, although a differential effect was observed between *S. pneumoniae* and *S. aureus* (**Figure 3**). In these assays, resident plasmids pC194, pCGA30, pCGA3n, and pCGA3 provided no *copG*, inactive *copG* and medium and high dosages of the autoregulated active *copG* gene, respectively (see **Table 2**). The specific effect of CopG on the establishment of pMV158 in a new cell was analyzed by comparing the efficiency with which recipient strains containing each of these resident plasmids were transformed in parallel with the pMV158-derivative plasmid (pLS1) and with a pMV158-unrelated plasmid (pAMβ1 in *S. pneumoniae* and pT181-*cop608* in *S. aureus*).

In the pneumococcal host, the plasmids whose establishment was to be analyzed were introduced by natural transformation, a horizontal gene transfer mechanism that requires the development of a transient physiological property named competence. Many different factors have been shown to affect

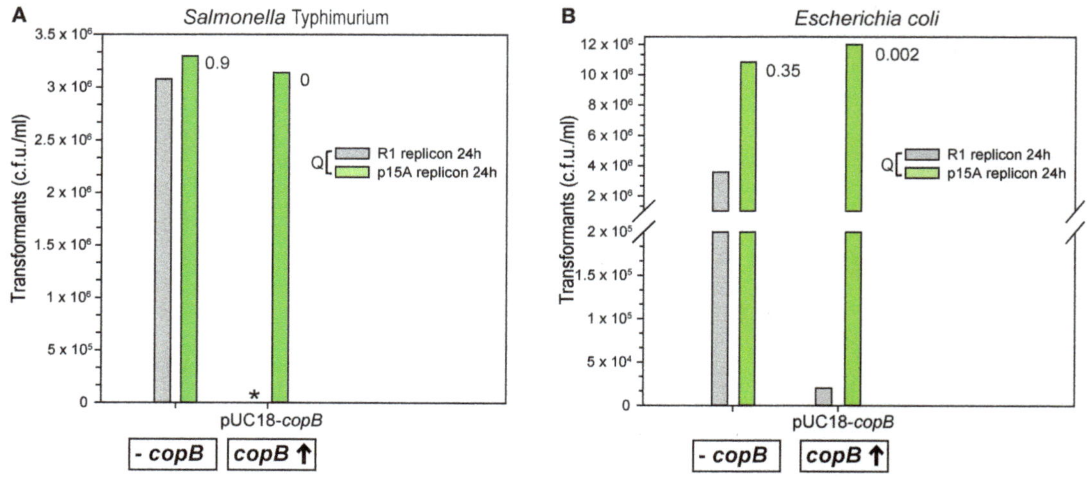

FIGURE 2 | The presence of excess CopB in the recipient cell dramatically and specifically decreases the efficiency of transformation with the R1 replicon. The vertical bar graphs show the number of transformant colonies per ml that appeared after transforming either plasmid-free or pUC18-*copB*-carrying *S.* Typhimurium **(A)** and *E. coli* **(B)** cells with DNAs of plasmids pKN1562 (a mini-R1 derivative) or pACYC184 (harboring the R1-unrelated p15A replicon), both of which are compatible with the pUC18 replicon. The same volumes were plated for all transformed cultures; by plating this volume, 500–1,000 p15A-transformant colonies were counted. Transformant colonies were counted after incubation for 24 h at 30°C. The asterisk in **(A)** indicates the absence of transformants after transforming *S.* Typhimurium SL1344 carrying pUC18-*copB* with pKN1562. The ratio (Q) between the number of transformants per ml obtained with pKN1562 and that obtained with pACYC184 is indicated in the graphs on the right of the corresponding vertical bars.

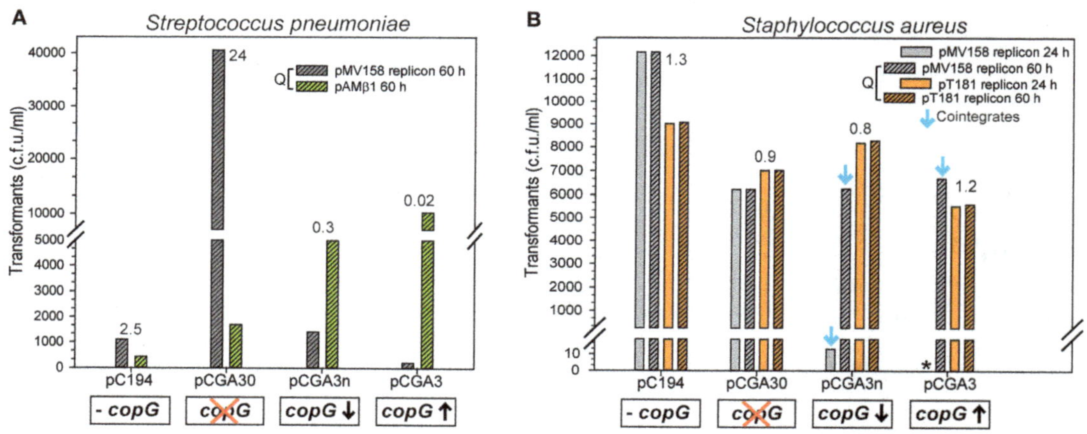

FIGURE 3 | The presence of CopG in the recipient cell decreases the frequency or the velocity of appearance of transformants for the pMV158 replicon in a selective and dosage-dependent manner. **(A)** The vertical bar graph shows the frequency of transformants after transforming *S. pneumoniae* harboring different plasmids with pLS1 (pMV158 replicon) or pAMβ1 (pMV158-unrelated replicon). The resident plasmids provided no *copG* (pC194), inactive *copG* (pCGA30), and medium and high dosages of active *copG* gene (pCAG3n and pCGA3, respectively). The ratio (Q) between the frequency of transformants colonies obtained with pLS1 and that obtained with pAMβ1, counted after 60 h of incubation at 37°C, is indicated on the top of the corresponding vertical bars. **(B)** *S. aureus* cells, harboring the same set of plasmids as described in **(A)**, were transformed with plasmids containing the replicon of either pMV158 or pT181. Vertical bars represent the frequency of transformants colonies counted after 24 h or 60 h of incubation at 37°C. The down facing blue arrow symbol indicates that the pMV158 replicon is only present as a plasmid cointegrate in the transformants. The ratio (Q) between the frequency of transformants colonies obtained with pLS1 and that obtained with pT181*cop608*, counted after 60 h of incubation at 37°C, is indicated on the top of the corresponding vertical bars. The asterisk in **(B)** indicates the absence of transformant colonies appeared within 24 h of incubation. The data presented in this figure summarize the results obtained in typical transformation experiments of *S. pneumoniae* and *S. aureus*. Two additional transformation experiments were performed for each species and the results with respect to the inhibitory effect of CopG on the transformation with the pMV158 replicon were similar to those shown here.

competence (Attaiech et al., 2015), and we actually observed quite different chromosomal transformation frequencies in the various strains used. Namely, the strain harboring pCGA3 showed the highest transformation frequency, followed by the strain containing pCGA3n, next that harboring pCGA30, and finally the strain with the pC194 vector exhibited the lowest competence level (see Material and Methods). The same qualitative trends were observed when analyzing the efficiencies of transformation of the various strains with plasmid pAMβ1 (whose replicon is not repressed by CopG), although in this case quantitatively larger differences were observed among them (**Figure 3A**). In order to normalize the frequencies of transformation with the pMV158 derivative with respect to the level of competence for plasmid transfer, the ratio (Q) between the transformation efficiencies with the pMV158 derivative and with pAMβ1 was used as parameter. This allowed us to evaluate the specific effect of CopG on the establishment of its cognate replicon. Since the pAMβ1 transformants grew slowly, total transformant colonies were only counted after 60 h of incubation at 37°C, regardless of which plasmid was transferred into the recipient cells. When the void pC194 vector was the resident plasmid, pLS1 yielded 2.5-fold more transformants than pAMβ1. This ratio increased to 24 when the recipient cells contained high-copy-number plasmid pCGA30 providing an inactive *copG* gene. The observed increase in the transformation efficiency is most likely due to facilitation of plasmid transformation by the existence of homology between the plasmid and the genome of the recipient cell, which in this case arises from the presence of a *copG* fragment in both the resident and the entering plasmid. The phenomenon of facilitation has only been reported to occur in natural transformation systems where donor DNA enters the competent cells in a linear ssDNA form (López et al., 1982). In spite of the potential facilitation of the transfer of the pMV158 derivative due to the presence of the entire *copG* gene in the recipient cells, the ratio of pLS1 to pAMβ1 transformant colonies was reduced to 0.3 and to 0.02 when the resident plasmid provided, respectively, medium (pCGA3n) and high (pCGA3) dosages of active *copG* (**Figure 3A**). Two other independent lots of competent cells of the various strains were transformed in parallel with pLS1 and pAMβ1, and similar Q ratios were obtained in both cases. These results show that the presence of CopG in the recipient cell severely and specifically impairs the success of the establishment of the cognate pMV158 replicon by decreasing the frequency of transformation, as we have also shown to be the case with CopB of the R1 system in both *E. coli* and *S.* Typhimurium (**Figure 2**).

In the staphylococcal host (**Figure 3B**), slightly different electrotransformation efficiencies were observed for the various strains, although all pT181-*cop608* transformant colonies appeared within the first 24 h of incubation irrespective of the presence or absence of CopG in the recipient cells. This was also the case for the pLS1 transformant colonies provided that the recipient strains lacked CopG (i.e., when the resident plasmids were pC194 or pCGA30). In contrast, virtually all pLS1 transformant colonies of the recipient strains carrying medium or high dosages of active *copG* could only be detected after 24 h of incubation, and thus they were counted at 60 h after plating. Despite the delay caused by CopG in the growth of the staphylococcal cells transformed with the pMV158 replicon, the ratio of pLS1 to pT181-*cop608* transformant colonies after 60-h incubation was close to 1, regardless of the plasmid resident in the recipient strain. Two more *S. aureus* transformation experiments were performed using the frozen stocks of electrocompetent cells. Although in these experiments the efficiencies of transformation with either plasmid were lower than when freshly-prepared cells were used, the Q ratios remained near constant. The mean Q value for all the strains was 1.1. Analysis of variance (ANOVA) of all experimental Q ratios indicated that there were no significant differences among groups, i.e., the same final frequency of pLS1 transformants was obtained irrespective of the presence of CopG in the recipient cell. Hence, in *S. aureus* CopG seems to specifically impair transformation with the pMV158 replicon by decreasing the velocity of growth, but not the final frequency, of the transformants (**Figure 3B**).

To investigate the basis of the differential effect of CopG on the establishment of the pMV158 replicon in *S. pneumoniae* and *S. aureus*, we analyzed the plasmid content of various transformants of either species grown under selective pressure for both the resident and the incoming plasmid. This analysis was facilitated because the medium or high steady-state copy number of both resident and newly-acquired plasmids in these bacteria allows us to visualize the plasmid bands after electrophoretic separation and staining of DNA minipreps. In *S. pneumoniae*, total crude DNA extracts showed the presence of the expected resident and newly-acquired plasmids in all transformant clones (**Figure 4A**). When the co-resident plasmid provided no active *copG* (pC194 and pCGA30) or medium dosages of the active gene (pCGA3n), the steady-state copy number of pLS1 in the corresponding transformant clones remained the same as in the homoplasmid situation, whereas a ~35% decrease in the pLS1 copy number was observed if this plasmid coexisted with pCGA3, which provides high dosages of the active *copG* gene. A similar effect of the different dosages of *copG* supplied *in trans* on the steady-state copy number of the pMV158 replicon has been previously reported in transformant clones arising from the reverse transformations (i.e., when pneumococcal cells carrying pLS1 were transformed with the various pC194 derivatives) (del Solar et al., 1995). On the other hand, plasmid DNA minipreps of staphylococcal transformant clones (**Figure 4B**) only revealed the presence of the two expected plasmids when the resident plasmid was pC194 (no *copG*) or pCGA30 (inactive *copG*). In the latter case, however, an additional slight DNA band appeared that, according to restriction analysis, corresponded to a cointegrate generated by homologous recombination between the resident and the newly-acquired plasmids through the pMV158 DNA region cloned in the resident plasmid (not shown). When the resident plasmid carried an active *copG* gene (pCGA3 or pCGA3n), no separate pLS1 plasmid could be observed and the pMV158 replicon was only present as cointegrate (**Figure 4B**). It is worth noting that, unlike total DNA extracts from *S. pneumoniae*, the method used to extract the plasmid DNA from *S. aureus* gave random yields and thus could not be employed to estimate the plasmid copy number. Since CopG selectively inhibits the pMV158 replicon, generation of cointegrates with the resident plasmid allows incoming pLS1 to escape from replication inhibition and yet to reach the concentration required to render the host cell resistant to the antibiotic (Tc) with which transformants are selected. According to the results obtained, the exclusive presence of pLS1 as cointegrate occurs in all staphylococcal transformants where an active *copG* gene is provided by the resident plasmid (**Figures 3B, 4B**), whereas cointegrate formation does not seem to be the strategy used by the equivalent pneumococcal transformants (**Figures 3A, 4A**), even though cointegration between the incoming and resident plasmids can also occur in this bacterium (Figure S1). This distinct behavior might arise from differences in the frequency with which this Campbell-like recombination takes place in these two bacteria. Cointegrate formation seems to be ultimately

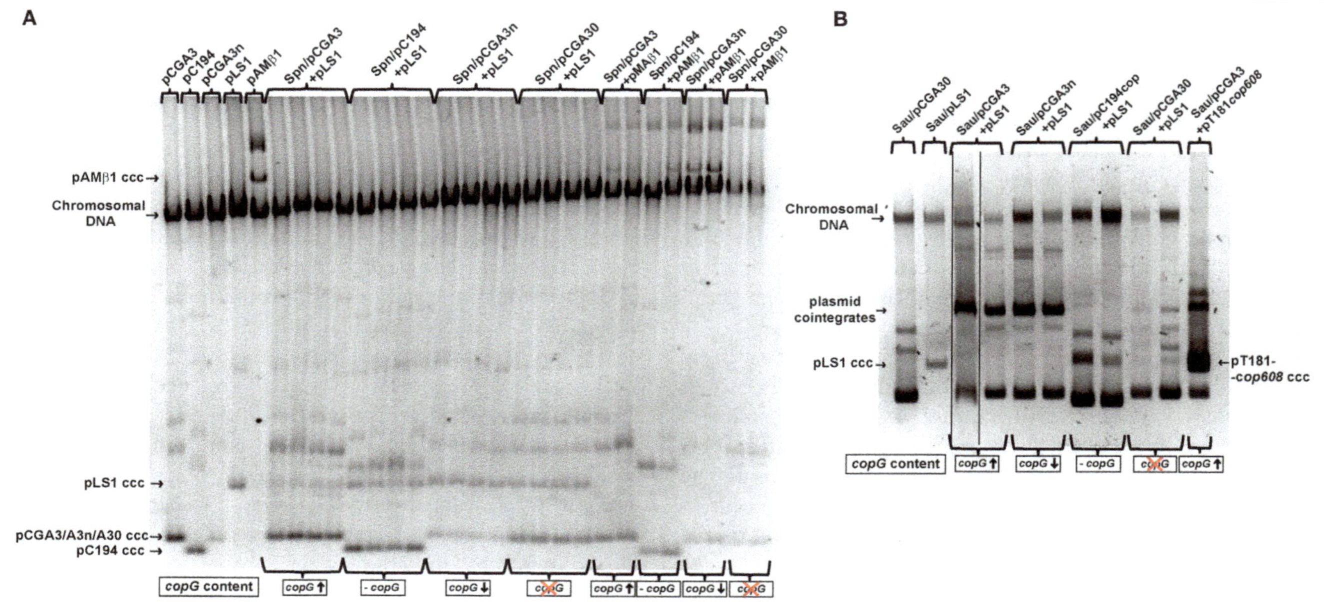

FIGURE 4 | Analysis of the plasmid DNA content of the indicated transformant clones grown under selective pressure for both the resident and the incoming plasmid. **(A)** Pneumococcal cells harboring pCAG3, pC194, pCGA3n, or pCGA30 were transformed with pLS1 and pAMβ1. The gel shows the total DNA content of several transformant clones. Four transformant clones were analyzed for each transformation with pLS1 and only two clones in the case of each transformation with pAMβ1. The combination of recipient strain and incoming plasmid is indicated on the top of the gel. The *copG* gene content of the resident plasmid in the recipient strain is indicated below the gel image. Homoplasmid strains harboring pCGA3, pC194, pCGA3n, pLS1, or pAMβ1 were used as controls. Supercoiled monomeric forms of the five plasmids are indicated in the gel; monomeric forms of pCGA3, pCGA3n, and pCGA30 have a similar electrophoretic mobility. **(B)** Staphylococcal cells harboring pCAG3, pCGA3n, pCGA30, or pC194 were transformed with pLS1 and pT181*cop608*. The gel shows the plasmid DNA content of several transformant clones. Two transformant clones were analyzed for each transformation experiment with pLS1, and one clone resulting from the transformation of *S. aureus*/pCGA3 with pT181*cop608* was also analyzed. The combination of recipient strain and incoming plasmid is indicated on the top of the gel. As in **(A)**, the *copG* content of the recipient strain in indicated below the gel image. Homoplasmid strains harboring pCGA3 or pLS1 were used as controls. Supercoiled monomeric forms of pLS1 and pT181*cop608* are indicated in the gel.

responsible for the delayed appearance of the staphylococcal transformant colonies.

To further analyze the role of the Cop proteins in the kinetics of plasmid repopulation, we chose the pMV158 replicon and its pneumococcal host as a model system because of a number of reasons. Importantly, compared to R1, the pMV158 replicon has a higher copy number in both the staphylococcal and the streptococcal host, and hence the amplitude of the plasmid amplification during the establishment phase replication is expected to be also greater, thus increasing the accuracy of the analysis. Moreover, the pneumococcal host of pMV158 was selected because we have a deep knowledge of it and we have set up a higher-frequency transformation protocol. And last but not least, no cointegrates that could mask repopulation of the pMV158 replicon have been observed in this system.

The Presence of the pMV158 CopG Repressor Protein in the Recipient Cell Results in Segregational Instability of the Incoming pMV158 Replicon

As a first approach to study the effect of CopG-mediated transcriptional repression of the essential *repB* gene on the success and velocity of repopulation of the pMV158 replicon, we tested the stability of inheritance of newly-acquired pLS1 in the transformant population of pneumococcal cells carrying null, medium, or high dosage of active *copG* (**Figure 5**). To this end, we analyzed the change in the numbers of total viable cells and transformants, as well as in the fraction of transformants, at various times after transformation. From these data, the experimental rate of loss of newly-acquired pLS1 from the transformants was calculated (see Material and Methods). The 3-layer *S. pneumoniae* plating method used in these assays (see Material and Methods) allowed the isolation of agar-embedded transformant c.f.u., so that plasmids could repopulate and express their antibiotic resistance determinant before selective pressure was applied.

When CopG-free recipient cells were employed (**Figures 5A,B**), no loss of pLS1 from the transformants could be inferred, although the fraction of transformants appeared to display a slight decreasing trend. This could be due to the pLS1 burden on the host, which for the steady-state plasmid concentration has been shown to cause an 8–9% increase in the cell doubling time (Hernández-Arriaga et al., 2012), so that the transformants slowly become overgrown by non-transformants.

When the plasmid resident in the recipient cell provided medium dosage of the active *copG* gene (**Figure 5C**), pLS1 showed unstable inheritance during division of the

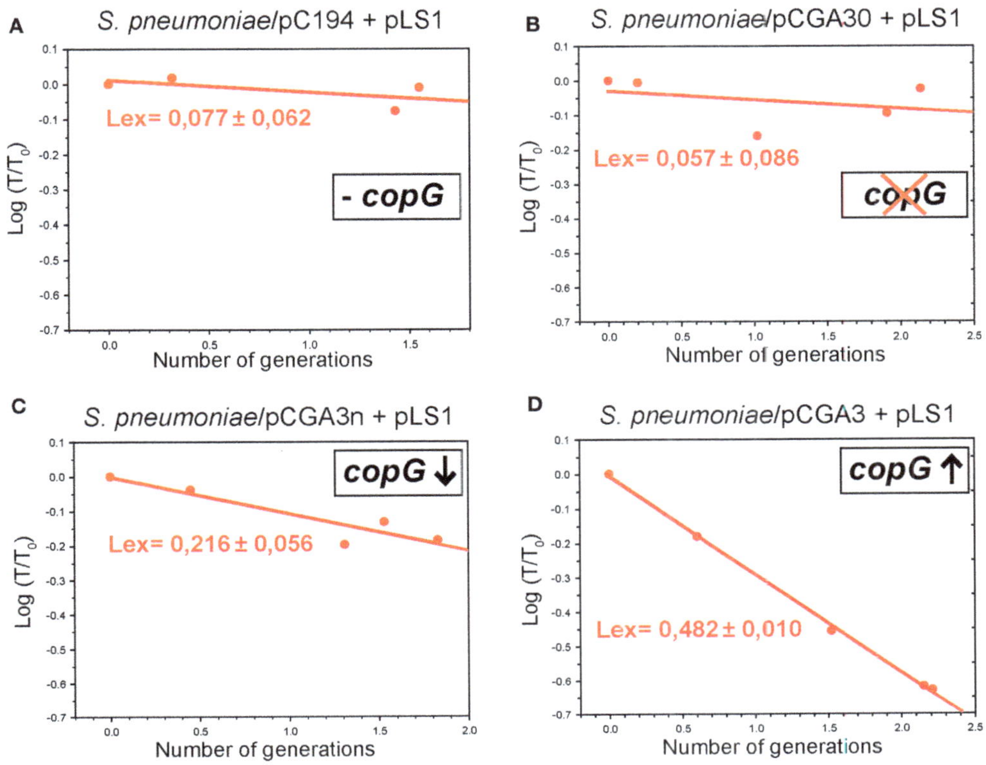

FIGURE 5 | The presence of CopG in the recipient cell unstabilizes the inheritance of the incoming pMV158 replicon in a dosage-cependent manner. The graphs represent the changes in the fraction of transformants retaining the newly-acquired pMV158 replicon (pLS1) when growing for several generations in the absence of selective pressure for the plasmid. The recipient pneumococcal strains lacked *copG* **(A)**, had a truncated version of *copG* **(B)**, or harbored a medium **(C)** or high **(D)** dosage of the active *copG* gene. The experimental loss rate (L_{ex}) of pLS1 was calculated from the slope of the linear regression model of the plot of the experimental values according to Equation (2) (red circles and lines). T_0 and T are, respectively, the fractions of transformants *ab initio* and after n generations.

transformants, with a quite high loss rate per cell and generation (~0.2).

A near-maximal loss rate (~0.5) was determined for newly-acquired pLS1 when the recipient cells contained plasmid pCGA3, which provides high dosages of active *copG* gene (**Figure 5D**). This maximum segregational instability implies that most frequently pLS1 is inherited by only one of the two daughter cells resulting from division of the transformants. It should be mentioned that, in the pneumococcal host, pLS1 has been shown to be segregationally stable during the steady state, both in the homoplasmid situation (del Solar et al., 1987) and in the presence of recombinant plasmid pCGA3 (del Solar et al., 1995). Hence, plasmid loss in the transformants that harbor extra copies of the active *copG* gene can certainly be ascribed to failures in the establishment phase replication of the pMV158 derivative.

The CopG Repressor Protein Impairs pMV158 Repopulation in the Transformants by Decreasing the Plasmid Replication Rate

As a further step toward the characterization of the role of CopG during the establishment phase replication of the pMV158 replicon, we have analyzed the kinetics of repopulation of plasmid pLS1*cop7*, a copy-up derivative of pLS1, in pneumococcal cells carrying high dosages of either active or inactive *copG* gene. Compared to pLS1, pLS1*cop7* has a single-point mutation in the *copG* gene, thus encoding a defective CopG protein that leads to a 5-fold increase in the plasmid copy number (del Solar et al., 1990, 1995). The use of pLS1*cop7* in these assays ensures that the effects observed when the resident plasmid carries an active *copG* gene arise from the CopG protein present in the recipient cells and not from that encoded by the incoming plasmid, and also allows determining the replicon repopulation kinetics in the absence of any CopG. Moreover, the use of this copy-up derivative of pMV158 was expected to increase the amplitude of the replicative amplification in the case that repopulation occurred.

Plasmid stability assays showed that pLS1*cop7* was inherited rather stably (with no plasmid loss being inferred) when the recipient cells lacked CopG (**Figure 6B**). In fact, the fraction of transformants was kept almost constant along several generations of cell growth in the absence of selection for the incoming plasmid (**Table 4**). In contrast, pLS1*cop7* was lost from the CopG-containing transformants at about the maximum possible rate (0.5; **Figure 6A**), as can be inferred from the transformant fraction being inversely proportional to the number of total cells (**Table 4**). The results of the segregational stability of newly-acquired pLS1*cop7* coincided with those obtained with incoming pLS1 (**Figure 5**), thus showing that the amount of CopG provided by the plasmid resident in the recipient cells suffices to severely impair repopulation of the pMV158 replicon.

The issue of the impact of CopG on the kinetics of pLS1*cop7* repopulation was first addressed by qPCR assays aimed at quantifying the variation of the number of copies of the incoming plasmid relative to the chromosome during the growth of the total bacterial population in the absence of Tc (**Figures 6C,D**). When the recipient cells contained resident pCGA30 and hence no functional CopG was provided, an abrupt decline (from ~1 to ~0.04) in the relative copy number of the plasmid amplicon within the total population was observed during the first 30 min of bacterial growth. This decrease was followed by a slower increase in the number of copies of the plasmid amplicon until a value of ~0.32 was reached after 150 min (**Table 5B** and **Figure 6D**). On its turn, a smaller decrease (from ~0.07 to ~0.01) followed by a rather constant relative copy number of the plasmid amplicon was observed when the recipient cells provided CopG (**Table 5A** and **Figure 6C**). This transient high concentration of the plasmid specific amplicon that decayed very rapidly in the absence of cell division was most unlikely to correspond to intact plasmid molecules, and could rather reflect the features of the mechanism of natural transformation in *S. pneumoniae*. It is worth mentioning that donor plasmid DNA enters the pneumococcal cell as ssDNA segments of both strands, and that two or more fragments of the opposite strands must anneal through overlapping regions at their ends to generate a circular plasmid form with partial dsDNA regions. This is followed by DNA synthesis to reconstruct the intact plasmid molecule. It should also be noted that, in order to have a high frequency of transformation, we added 300–500 plasmid DNA molecules per bacterial cell. Thus, a number of ssDNA molecules that have entered the cell will harbor the plasmid amplicon to be amplified in the qPCR assay, although most of them will never render a reconstructed plasmid molecule and will be degraded by the cellular nucleases instead. The actual content in intact plasmid molecules was then analyzed by iPCR employing gDNA and two divergent primers that specifically annealed to DNA sequences within the pMV158 replicon (**Figure 6G**). Immediately after transformation with pLS1*cop7*, while the relative amount of the qPCR-detected pMV158 amplicon in the CopG-free recipient cells was maximal ($t = 0$; **Figure 6D**), iPCR amplification rendered the faintest band of specific full-length plasmid DNA (**Figure 6F**). A similar result was observed when the CopG-containing cells were transformed with pLS1*cop7*, although in this case the initial amount of the pMV158 amplicon was much lower (**Figures 6C,E**). The observed discrepancy between the initial relative amounts of pMV158 amplicon and intact pLS1*cop7* DNA molecules led us to conclude that, in fact, most of the amplicons to be amplified in the qPCR assays were not carried on reconstituted plasmids but on smaller DNA fragments. Hence, the relative copy number of the donor plasmid shortly after entrance is better evaluated from the iPCR assays. At longer times after transformation (30 min and further), when the DNA fragments carrying the amplicon would have declined, an almost perfect match was observed between the intracellular amplifications of the pLS1*cop7* amplicon (evaluated by qPCR) and of the intact plasmid molecules (evaluated by iPCR). This match was observed irrespective of whether the transformants carried or lacked CopG (**Tables 5A,B**; see also the Discussion). This coincidence allowed us to determine the rate of pLS1*cop7* replication (R, defined as the ratio of plasmid to gDNA duplications, and calculated according to Equation 10) along the time, based on the iPCR data of the plasmid amplification which, unlike the qPCR data, were unaffected by the presence of DNA fragments carrying the pMV158 amplicon.

Although varied slightly along the time, and even reached a value of ~2 in the interval between 15 and 30 min after transformation, the replication rate (R) of pLS1*cop7* in the transformants carrying CopG fluctuated around 1 (**Table 5A** and **Figure 6C**). An R value of ~1, which was also inferred from the ratio between pDNA and gDNA total duplications ($R = 1.1$; **Table 5A**), implies that the plasmid replicated at the same average velocity as the gDNA during the time interval analyzed and, hence, that it failed to repopulate.

When no CopG was present, an overall R value of ~2.8 (the ratio of pDNA to gDNA total duplications; **Table 5B**) was found for incoming pLS1*cop7* during the entire 150-min interval analyzed, although the plasmid replication rate varied substantially along the time (**Figure 6D** and **Table 5B**). Repopulation of pLS1*cop7* mainly occurred during the first 45 min after completion of transformation, with a peak in the plasmid replication rate in the interval between 15 and 30 min (**Figure 6D** and **Table 5B**). Afterwards, the plasmid replication rate decreased asymptotically to 1, indicating that some repopulation still occurred at these longer times. There seemed to be also a small increase in the duplication rate of the pMV158 replicon when bacterial growth slowed down (in the 120–150 min interval), which would be also reflected in an increase of the plasmid copy number (**Figure 6D** and **Table 5B**). Whether this could indicate an increase of the pMV158 copy number during the stationary growth phase of its host is a potential matter for future investigation.

DISCUSSION

In this work we have analyzed for the first time the role played by the Cop regulatory loops of R1 and pMV158 in plasmid establishment. The establishment phase replication, which amplifies the plasmid from an initially low concentration to the steady-state copy number, is a crucial process in the biology of naturally transferable plasmids that may importantly affect the success of their colonization and spreading. Nevertheless, plasmid replication during the establishment phase has been much less studied than the steady-state replication. In a pioneer work by Highlander and Novick (1987) the kinetics of repopulation of various pT181 derivatives that carried or lacked a functional asRNA control system were analyzed in *S. aureus* by determining the replication rates and copy numbers of the plasmids after radioactive *in vivo* labeling of total gDNA. A bit

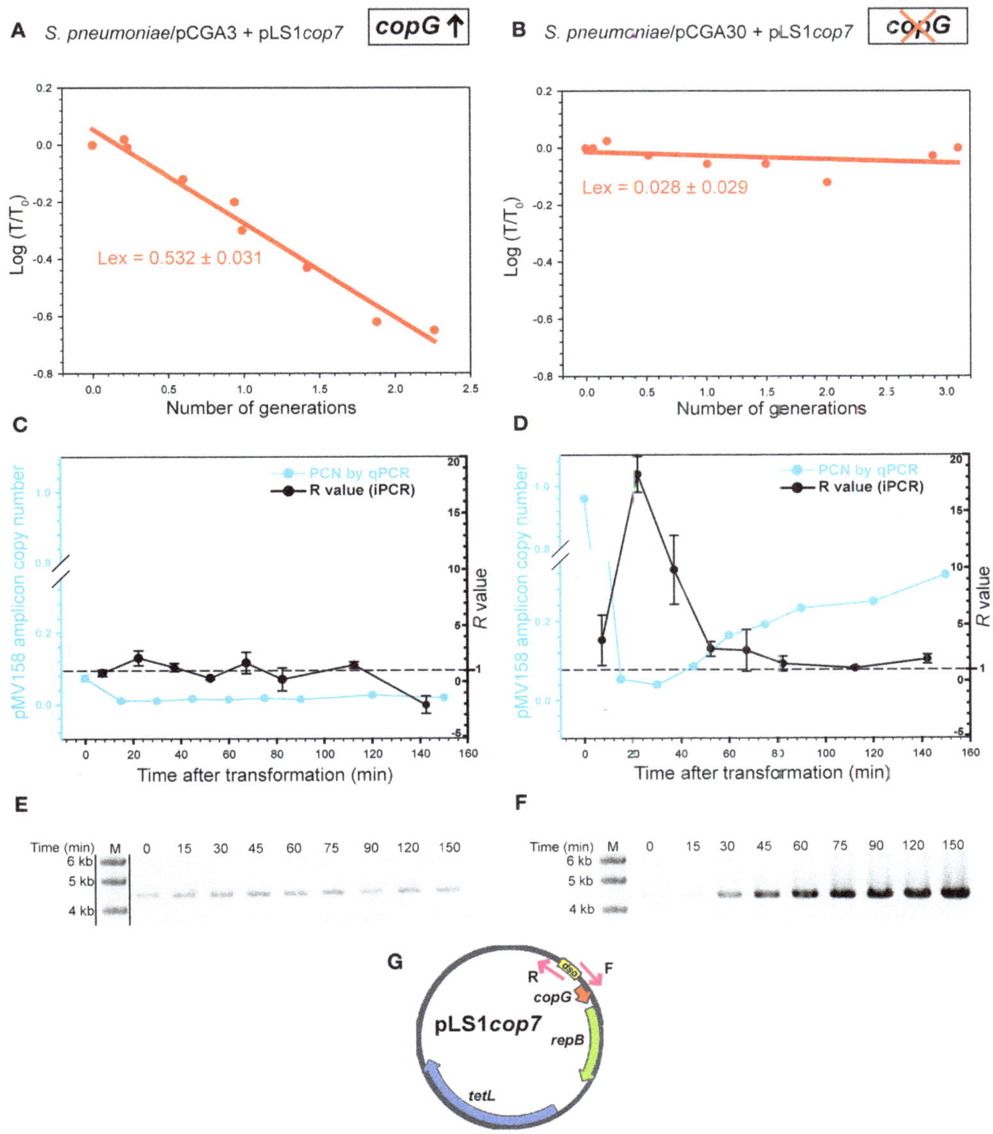

FIGURE 6 | The presence of CopG in the recipient cell impairs repopulation of the pMV158 replicon by decreasing the plasmid replication rate. Changes in the fraction of transformants retaining newly-acquired pLS1*cop7* (a copy-up pMV158 derivative) were analyzed in pneumococcal strains that either harbor a high dosage of active *copG* **(A)** or a truncated version of *copG* **(B)**. The experimental loss rate (L_{ex}) of pLS1*cop7* was calculated from the slope of the linear regression model of the plot of the experimental values according to Equation (2) (red circles and lines). T_0 and T are, respectively, the fractions of transformants *ab initio* and after *n* generations. The graphs in **(C,D)** show the impact of CopG on the kinetics of pLS1*cop7* repopulation. A qPCR approach was used to calculate the variation in the copy number of a specific amplicon of the incoming pLS1*cop7* plasmid relative to the chromosome during the growth of the total bacterial population containing pCGA3 **(C)** or pCGA30 **(D)** as resident plasmid for 150 min after transformation (left y-axis; blue circle and lines). The pLS1*cop7* replication rate (*R* value), defined as the ratio of plasmid to gDNA duplications, was calculated at different time intervals following transformation of pneumococcal cells harboring pCGA3 **(C)** or pCGA30 **(D)**. Determination of *R* was based on the iPCR data of the *in vivo* plasmid amplification (black circles and lines) and its value (right y-axis) was calculated according to Equation (10). Discontinuous horizontal line in graphs of **(C,D)** denotes an *R* value of 1, which characterizes the steady-state plasmid replication. The mean (symbols) and standard deviation (error bars) of all the experimental points in the graphs of **(C,D)** are displayed. Panels **(E,F)** show the iPCR analysis of the gDNA samples obtained at the indicated times after transformation of pneumococcal cells carrying pCGA3 and pCGA30, respectively, with pLS1*cop7*. iPCR assays were carried out by using a pair of divergent primers specific for the pMV158 replicon (**Table 3** and **G**) and the Phusion DNA polymerase. Lane M, DNA molecular weight standard (NZYDNA ladder III; NZYTECH). Note that lanes M are the same in **(E,F)** because, in fact, both images of these panels arise from the same gel. Dividing lines in **(E)** indicate grouping of different parts of the same gel. The original image of the gel used for **(E,F)** composition is shown in Figure S2. A schematic representation of pLS1*cop7* displaying the plasmid regions complementary to the divergent primers is shown in **(G)**. Genes *copG*, *repB*, and *tetL*, as well as the *dso* region, are indicated.

TABLE 4 | Total number of cells, number of generations, and % of transformants at different times after transformation during the growth of the indicated strains in the absence of selective pressure.

Time after transformation (min)	*S. pneumoniae*/pCGA3 + pLS1*cop7*			*S. pneumoniae*/pCGA30 + pLS1*cop7*		
	Total number of cell generations[a]	Total cell count (c.f.u./ml)	% of transformants	Total number of cell generations[a]	Total cell count (c.f.u./ml)	% of transformants
0	0	3.48×10^7	0.89	0	2.32×10^7	0.16
15	0.01	3.52×10^7	0.94	0.07	2.48×10^7	0.16
30	0.23	4.11×10^7	0.87	0.17	2.62×10^7	0.17
45	0.60	5.27×10^7	0.66	0.52	3.32×10^7	0.15
60	0.94	6.72×10^7	0.56	1.01	4.77×10^7	0.14
75	0.99	6.94×10^7	0.45	1.50	6.59×10^7	0.14
90	1.42	9.32×10^7	0.33	2.01	9.37×10^7	0.12
120	1.88	12.85×10^7	0.21	2.89	17.20×10^7	0.15
150	2.26	16.80×10^7	0.19	3.10	19.95×10^7	0.16

[a] *Total number of cell generations was calculated according to Equation (3).*

later, the establishment phase replication of ColE1 was studied by using Southern blot to determine the number of phasmids containing the plasmid replicon per *E. coli* cell as a function of time after infection (Merlin and Polisky, 1993). It was found that a phasmid containing an up mutation in the RNA II primer promoter replicated at a 15-fold faster rate than the wild type, thus early highlighting the importance of rapid synthesis of the essential RNA II in ColE1 plasmid establishment. So far, however, the scarce characterized examples of repopulation have missed out the analysis of the effect on plasmid establishment of transcriptional repressors that either exert an auxiliary role or act synergistically with an antisense RNA in controlling the steady-state replication. As a first approach to address this analysis, we have investigated whether the presence of the R1 or pMV158 Cop repressor protein in the recipient cell affects the frequency and velocity of appearance of the plasmid transformant colonies. Moreover, by taking pMV158 and its pneumococcal host as a model system, we have developed a new approach for evaluating the kinetics of plasmid repopulation that is based on the estimation of the plasmid loss rate in transformants and on the use of non-radioactive highly sensitive qPCR and iPCR methods.

The results of the frequencies of transformation of different bacterial species with the R1 or the pMV158 replicons show that, when supplied in the recipient cell, CopB from R1 and CopG from pMV158 severely and selectively impair the establishment of their cognate plasmids. Actually, a dramatic decrease in the efficiency of transformation of *E. coli* or *S.* Typhimurium with the R1 replicon, and of *S. pneumoniae* with the pMV158 replicon can be observed when the cognate Cop repressor is present in the recipient cells (**Figures 2, 3A**). On its turn, in *S. aureus* cointegration with the resident plasmid allows the incoming pMV158 derivative to overcome the CopG-mediated inhibition of its replicon, so that only a delay in the appearance of the transformant colonies, but not a decrease in the final frequency of them, is observed in this bacterium. As seen for the pMV158 replicon/pneumococcal host system, inhibition of plasmid establishment requires the presence of an active *copG* gene and depends on the dosage of this gene (**Figure 3A**). Since the R1 and pMV158 Cop proteins repress transcription of the respective plasmid *rep* gene from a strong promoter, the requirement of fully unrepressed expression of the essential *rep* genes for the successful establishment of these replicons can be inferred from the results of the transformation experiments performed in this work. The coincidence between the impairing effects of preexisting CopB and CopG on the efficiency of establishment of their cognate replicon in a new host leads us to think that the Cop repressor-mediated blockage of plasmid repopulation observed in the pMV158/pneumococcal host system can be extrapolated to R1 entering its *Enterobacteriaceae* host.

As shown in **Figure 7**, unsuccessful repopulation of the incoming plasmid may lead to unstable inheritance of the plasmid during division of the transformed cells. The rates at which the pMV158 derivatives (wild-type or copy-up mutant plasmids) are lost during the culture of pneumococcal transformants carrying or lacking a high dosage of autoregulated active *copG* gene match quite well the two extreme theoretical cases of failure or immediate plasmid repopulation, respectively (**Figure 7**). In fact, a loss rate close to the maximum value (0.5) was found in the first case whereas no significant loss was observed in the second one (**Figures 5A,B,D, 6A,B**). When a medium dosage of autoregulated *copG* gene is provided by the recipient cell, an intermediate rate of loss of the pMV158 replicon was observed (**Figure 5C**). Actually, these results indicate that the presence of a functional *copG* gene in the recipient cell causes a dosage-dependent impairment in the pMV158-replicon repopulation that leads to the unstable inheritance of the underpopulated plasmid.

The kinetics of repopulation of the pMV158 replicon, and the effect of the CopG repressor protein on it, have been studied by qPCR and iPCR using total gDNA, prepared at different times after transformation, as template. Irrespective of whether the

TABLE 5 | Kinetics of genomic and plasmid DNA replication during plasmid establishment.

S. pneumoniae/pCGA3 + pLS1*cop7*								
Time after transformation (min)	**Total number of cell generations[a]**	**Copy number of the plasmid amplicon (qPCR)**	**Time interval**	**Amplification of plasmid amplicon[b,d] (qPCR)**	**Amplification of plasmid molecules[c,d] (iPCR)**	**R[e]**	**gDNA duplications**	**pDNA duplications**
A								
0	0	0.074 ± 0.009						
15	0.01	0.011 ± 0.002	0–15	0.2	1.0	0.8 ± 0.3	0.21	0.17
30	0.23	0.011 ± 0.001	15–30	1.0	1.3	2.1 ± 0.7	0.28	0.60
45	0.60	0.016 ± 0.001	30–45	1.4	1.1	1.3 ± 0.4	0.60	0.80
60	0.94	0.014 ± 0.001	45–60	0.9	0.9	0.3 ± 0.1	0.35	0.12
75	0.99	0.018 ± 0.001	60–75	1.3	1.0	1.7 ± 1.0	0.16	0.27
90	1.42	0.015 ± 0.001	75–90	0.8	0.9	0.2 ± 1.0	0.41	0.09
120	1.88	0.026 ± 0.001	90–120	1.7	1.5	1.5 ± 0.3	0.96	1.43
150	2.26	0.018 ± 0.001	120–150	0.7	0.8	−2.1 ± 0.8	0.06	−0.10
							Total = 3.03	Total = 3.36

S. pneumoniae/pCGA30 + pLS1*cop7*								
Time after transformation (min)	**Total number of cell generations[a]**	**Copy number of the plasmid amplicon (qPCR)**	**Time interval**	**Amplification of plasmid amplicon[b,d] (qPCR)**	**Amplification of plasmid molecules[c,d] (iPCR)**	**R[e]**	**gDNA duplications**	**pDNA duplications**
B								
0	0	0.960 ± 0.010						
15	0.07	0.054 ± 0.020	0–15	0.1	1.4	3.0 ± 1.7	0.24	0.72
30	0.17	0.040 ± 0.001	15–30	0.7	5.1	14.1 ± 1.2	0.18	2.54
45	0.52	0.085 ± 0.013	30–45	2.1	2.3	7.7 ± 2.3	0.18	1.39
60	1.01	0.164 ± 0.021	45–60	1.9	1.7	2.4 ± 0.5	0.54	1.30
75	1.50	0.191 ± 0.025	60–75	1.2	1.2	2.3 ± 1.4	0.20	0.46
90	2.01	0.232 ± 0.023	75–90	1.2	1.1	1.4 ± 0.5	0.34	0.48
120	2.89	0.249 ± 0.018	90–120	1.1	1.1	1.1 ± 0.1	1.10	1.22
150	3.10	0.315 ± 0.025	120–150	1.3	1.2	1.7 ± 0.3	0.38	0.65
							Total = 3.16	Total = 8.76

[a] Total number of cell generations was calculated according to Equation (3). [b,c] The factor by which the [b] plasmid amplicon or the [c] plasmid molecules copy number is increased in the indicated time interval. [d] Amplification factor for a certain time interval can be calculated as the product of the amplification factors of the reference time intervals. [e] R value was calculated according to Equation (10).

pneumococcal cells harbor or not the *copG* gene, replication of the gDNA (consisting mainly of chromosomal DNA) appears to begin earlier than cellular division after the 30 to 37°C shift that follows the transformation step (**Table 5**). The estimated number of total gDNA duplications in the 150-min interval is about three for both kinds of recipient cells, a value that almost equals the number of total cell generations in the case of the recipient cells containing the inactive *copG* gene, whereas it is somewhat higher than the value obtained when the recipient cells contain a functional *copG* gene (**Table 5**).

In the absence of CopG, the highest repopulation rate occurs during the first 45 min after the entrance of the plasmid DNA in the cell, with a peak in the interval between 15 and 30 min (**Figure 6D** and **Table 5B**). According to the results obtained by qPCR (**Table 5B**), and taking also into account the fraction of transformants (**Table 4**), the average copy number of pLS1*cop7* in the transformant cells is estimated to be about 200 per chromosome equivalent at 150 min after plasmid transfer. This value coincides with the reported steady-state copy number of the copy-up pMV158 derivative (**Table 2**), indicating that repopulation has been accomplished by this time. Moreover, no overshoot of the steady-state copy number was observed and instead this value was asymptotically approached (**Figures 6D,F**). Overall, about nine duplications of the pMV158 derivative took place within the transformants harboring an inactive *copG* gene in the 150-min time interval during which the gDNA duplicated around three times (**Table 5B**). This represents a 50- to 60-fold relative amplification of the plasmid DNA (see Equation 8),

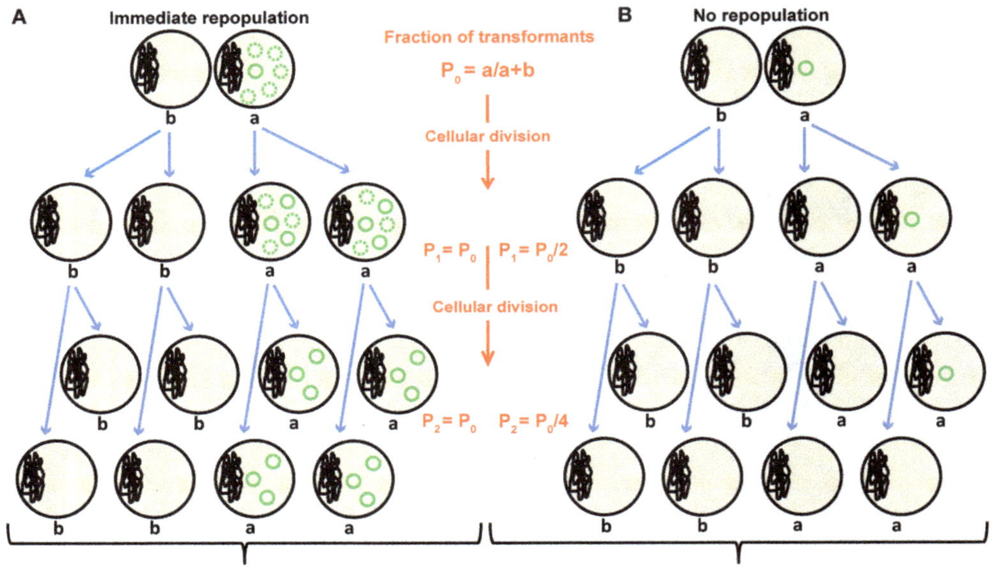

FIGURE 7 | Effect of the velocity of plasmid repopulation on the segregational stability of an incoming plasmid. Two extreme theoretical cases of immediate **(A)** or null **(B)** plasmid repopulation are displayed. These models assume that the plasmid does not burden the host cell. We defined "P_n" as the fraction of transformants after *n* generations; "a" is the initial number of transformants, and "b" is the number of non-transformed cells within the total bacterial population. In the case that immediate repopulation occurs **(A)**, the steady-state plasmid copy number is reached before division of the transformants, and the plasmid is stably inherited to both daughter cells. Dotted circles in **(A)** represent the number of new plasmid molecules generated in each cell generation. If no repopulation occurs **(B)**, the single, unreplicated plasmid copy is inherited by only one of the two daughter cells resulting from division of the transformants, giving rise to maximal segregational instability.

which can also be determined from the amplification factor of the plasmid copy number (**Table 5B**).

In the presence of CopG, the relative average plasmid copy number in the total bacterial population is kept near constant, as shown by the results from both the qPCR and the iPCR assays (**Figure 6C,E** and **Table 5A**). This is also consistent with an overall plasmid replication rate (R) of 1.1 along the entire 150-min time period after transformation (**Table 5A**), which means that, on average, the incoming pMV158 replicon underwent about the same number of duplications as the gDNA. An R value of around 1, which in fact characterizes the steady-state replication of any plasmid, demonstrates the unsuccessful repopulation of the pMV158 replicon when CopG is present in the recipient cell. However, based on the maximum possible value of plasmid loss rate that is observed (**Figure 6A**), a total absence of plasmid replication ($R \sim 0$ at any time interval) should have been expected instead (**Figure 7B**). We could envisage two potential explanations for this. One possibility is that in most cells the plasmid replicates at approximately the same rate as the chromosome, thus giving rise to cells containing at least two plasmid copies at division; if so, the observed loss rate would require that the two sister plasmid molecules tend to segregate together into the same daughter cell. An alternative explanation is that plasmid replication manages to evade inhibition by CopG in a small fraction of the transformants so that repopulation to the steady-state plasmid copy number is achieved only in these cells, and hence the maximum value of plasmid loss rate (0.5) would not be significantly altered. Be that as it may, according to the qPCR assays the relative average plasmid copy number in the total bacterial population increases ~1.65-fold during the time interval between 15 min (when the plasmid DNA fragments are considered to be negligible) and 150 min after transformation (**Table 5A**). When corrected for the fraction of transformants (**Table 4**), the average plasmid copy number per chromosome in the sub-population of transformed cells is shown to increase from ~1.2 to ~9.5 between 15 min and 150 min after transformation, although at the latter time the plasmid copies might not be evenly distributed among the transformants, as discussed above. Moreover, once the transformant clones are selected for the incoming plasmid, pLS1*cop7* seems to be "forced" to repopulate until reaching the steady-state relative copy number (~33) that is observed in the heteroplasmid situation (Figure S1).

It should be noted that we have obtained similar negative effects on plasmid establishment due to the presence of Cop proteins in the recipient cells when repopulation followed either of two distinct horizontal transfer mechanisms, namely a natural transformation or an electrically-induced transfer. Although the R1 and pMV158 derivatives used in this work did not allow the analysis of plasmid repopulation following conjugative transfer, the significance of fully unrepressed transcription of their essential *rep* genes can most likely be extrapolated to any process implying the establishment phase replication. Establishment phase repopulation is a pivotal step within the process of horizontal plasmid spread by which a unique (or very few) intact dsDNA molecule of the donor plasmid is amplified to the steady-state characteristic copy number. Depending on the transfer mechanism employed, one full-length circular dsDNA plasmid copy can enter directly the cell (electroporation), can be reconstructed from overlapping ssDNA fragments of both strands (pneumococcal natural transformation), or can be generated by synthesis of the DNA strand complementary to the one that was transferred by conjugation. Nevertheless, plasmid repopulation (and its Cop-mediated inhibition) only depends on the replicon function and is independent of the mechanism used for plasmid transfer. In this sense, it is worth noting that similar replication kinetics were observed for the repopulation of pT181 derivatives in *S. aureus* irrespective of whether the initial low plasmid copy number was achieved by shutoff of the replication of thermosensitive mutants or by introduction of the plasmid DNA into the bacterial cells through high-frequency (50%) transduction (Highlander and Novick, 1987).

Carrying an emergency mechanism that enables strong expression of the essential *rep* gene allows rapid and successful repopulation, which can benefit especially those plasmids whose lifestyle includes colonization of new hosts. This is in fact the case, among others, of conjugative or mobilizable plasmids R1, pIP501, and pMV185, all of which harbor a strong promoter directing transcription of the *rep* gene that is subjected to repression by a plasmid-encoded Cop protein. Although a crucial role in plasmid establishment has been proposed for the Cop regulatory loops of these plasmids, no empirical demonstration of it had been reported so far (Nordström and Nordström, 1985; del Solar et al., 1990; del Solar and Espinosa, 2000; Olsson et al., 2004; Brantl, 2014). We think that the requisite, observed in the present work, of fully unrepressed *rep* transcription for the successful establishment of R1 and pMV158 can be extended to other plasmids encoding a similar Cop regulatory loop.

Understanding the role of the Cop regulatory loop that switches on/off transcription of the essential *rep* gene may help the design and development of new strategies to control spreading of undesirable plasmids among bacterial populations or to prevent transfer of a specific plasmid to a potential host in mating experiments with multiple plasmids.

AUTHOR CONTRIBUTIONS

RD-O conceived the idea of bringing together the independent observations obtained with the R1 and pMV158 plasmid systems to prepare a joining article about the role of Cop repressors on plasmid establishment. RD-O and IM-C designed and performed the transformation experiments with the R1 replicon. LL performed the transformation experiments with the pMV158 replicon. JR-M carried out the experiments of the kinetics of pMV158 repopulation and the analysis of plasmid stability in the transformants, wrote the Material and Methods section and also prepared most of the figures. GdS designed the experimental approach and the formulation of the kinetics of repopulation and plasmid stability analyses, participated in the experiments with the pMV158 system and wrote most of the manuscript. All the authors discussed the results and corrected the entire manuscript.

ACKNOWLEDGMENTS

Thanks are due to members of our labs for helpful discussions, and particularly to Lidia de Tapia for her contribution to the *Salmonella* transformation experiments and to Sergio Barata, Javier Nicolas Garay Novillo and Diego García de la Morena for critical reading of the manuscript.

REFERENCES

Appleyard, R. K. (1954). Segregation of new lysogenic types during growth of a doubly lysogenic strain derived from *Escherichia coli* K12. *Genetics* 39, 440–452.

Atlung, T., Christensen, B. B., and Hansen, F. G. (1999). Role of the rom protein in copy number control of plasmid pBR322 at different growth rates in *Escherichia coli* K-12. *Plasmid* 41, 110–119. doi: 10.1006/plas.1998.1386

Attaiech, L., Minnen, A., Kjos, M., Gruber, S., and Veening, J. W. (2015). The ParB-*parS* chromosome segregation system modulates competence development in *Streptococcus pneumoniae*. *MBio* 6, e00662-15. doi: 10.1128/mBio.00662-15

Augustin, J., and Götz, F. (1990). Transformation of *Staphylococcus epidermidis* and other staphylococcal species with plasmid DNA by electroporation. *FEMS Microbiol. Lett.* 54, 203–207. doi: 10.1111/j.1574-6968.1990.tb03997.x

Boer, D. R., Ruiz-Masó, J. A., Rueda, M., Pethoukov, M., Svergun, D. I., Orozco, M., et al. (2016). Conformational plasticity of RepB, the replication initiator protein of promiscuous streptococcal plasmid pMV158. *Sci. Rep.* 6:20915. doi: 10.1038/srep20915

Boer, D. R., Ruiz-Masó, J. A., López-Blanco, J. R., Blanco, A. G., Vives-Llàcer, M., Chacón, P., et al. (2009). Plasmid replication initiator RepB forms a hexamer reminiscent of ring helicases and has mobile nuclease domains. *EMBO J.* 28, 1666–1678. doi: 10.1038/emboj.2009.125

Brantl, S. (2014). Plasmid replication control by antisense RNAs. *Microbiol. Spectr.* 2:PLAS-0001-2013. doi: 10.1128/microbiolspec.PLAS-0001-2013

Burdett, V. (1980). Identification of tetracycline-resistant R-plasmids in *Streptococcus agalactiae* (group B). *Antimicrob. Agents Chemother.* 18, 753–760. doi: 10.1128/AAC.18.5.753

Chang, A. C., and Cohen, S. N. (1978). Construction and characterization of amplifiable multicopy DNA cloning vehicles derived from the P15A cryptic miniplasmid. *J. Bacteriol.* 134 1141–1156.

Das, N., Valjavec-Gratian, M., Basuray, A. N., Fekete, R. A., Papp, P. P., Paulsson, J., et al. (2005). Multiple homeostatic mechanisms in the control of P1 plasmid replication. *Proc. Natl. Acad. Sci. U.S.A.* 102, 2856–2861. doi: 10.1073/pnas.0409790102

del Solar, G., Acebo, P., and Espinosa, M. (1995). Replication control of plasmid pLS1: efficient regulation of plasmid copy number is exerted by the combined action of two plasmid components, CopG and RNA II. *Mol. Microbiol.* 18, 913–924. doi: 10.1111/j.1365-2958.1995.18050913.x

del Solar, G., and Espinosa, M. (2000). Plasmid copy number control: an ever-growing story. *Mol. Microbiol.* 37, 492–500. doi: 10.1046/j.1365-2958.2000.02005.x

del Solar, G., Moscoso, M., and Espinosa, M. (1993). *In vivo* definition of the functional origin of replication (*ori*(+)) of the promiscuous plasmid pLS1. *Mol. Gen. Genet.* 237, 65–72. doi: 10.1007/BF00282785

del Solar, G. H., Pérez-Martín, J., and Espinosa, M. (1990). Plasmid pLS1-encoded RepA protein regulates transcription from *repAB* promoter by binding to a DNA sequence containing a 13-base pair symmetric element. *J. Biol. Chem.* 265, 12569–12575.

del Solar, G. H., Puyet, A., and Espinosa, M. (1987). Initiation signals for the conversion of single stranded to double stranded DNA forms in the streptococcal plasmid pLS1. *Nucleic Acids Res.* 15, 5561–5580. doi: 10.1093/nar/15.14.5561

Dower, W. J., Miller, J. F., and Ragsdale, C. W. (1988). High efficiency transformation of *E. coli* by high voltage electroporation. *Nucleic Acids Res.* 16, 6127–6145. doi: 10.1093/nar/16.13.6127

Gerdes, K., Larsen, J. E., and Molin, S. (1985). Stable inheritance of plasmid R1 requires two different loci. *J. Bacteriol.* 161, 292–298.

Gomis-Rüth, F. X., Solà, M., Acebo, P., Párraga, A., Guasch, A., Eritja, R., et al. (1998). The structure of plasmid-encoded transcriptional repressor CopG unliganded and bound to its operator. *EMBO J.* 17, 7404–7415. doi: 10.1093/emboj/17.24.7404

Hernández-Arriaga, A. M., Espinosa, M., and del Solar, G. (2012). Fitness of the pMV158 replicon in *Streptococcus pneumoniae*. *Plasmid* 67, 162–166. doi: 10.1016/j.plasmid.2012.01.005

Hernández-Arriaga, A. M., Rubio-Lepe, T. S., Espinosa, M., and del Solar, G. (2009). Repressor CopG prevents access of RNA polymerase to promoter and actively dissociates open complexes. *Nucleic Acids Res.* 37, 4799–4811. doi: 10.1093/nar/gkp503

Highlander, S. K., and Novick, R. P. (1987). Plasmid repopulation kinetics in *Staphylococcus aureus*. *Plasmid* 17, 210–221. doi: 10.1016/0147-619X(87)90029-1

Hoskins, J., Alborn, W. E. Jr., Arnold, J., Blaszczak, L. C., Burgett, S., DeHoff, B. S., et al. (2001). Genome of the bacterium *Streptococcus pneumoniae* strain R6. *J. Bacteriol.* 183, 5709–5717. doi: 10.1128/JB.183.19.5709-5717.2001

Hoiseth, S. K., and Stocker, B. A. (1981). Aromatic-dependent *Salmonella typhimurium* are non-virulent and effective as live vaccines. *Nature* 291, 238–239. doi: 10.1038/291238a0

Koressaar, T., and Remm, M. (2007). Enhancements and modifications of primer design program Primer3. *Bioinformatics* 23, 1289–1291. doi: 10.1093/bioinformatics/btm091

Kreiswirth, B. N., Lofdahl, S., Betley, M. J., O'Reilly, M., Schlievert, P. M., Bergdoll, M. S., et al. (1983). The toxic shock syndrome exotoxin structural gene is not detectably transmitted by a prophage. *Nature* 305, 709–712. doi: 10.1038/305709a0

Lacks, S. A., and Greenberg, B. (1977). Complementary specificity of restriction endonucleases of *Diplococcus pneumoniae* with respect to DNA methylation. *J. Mol. Biol.* 114, 153–168. doi: 10.1016/0022-2836(77)90289-3

Lacks, S. A., López, P., Greenberg, B., and Espinosa, M. (1986). Identification and analysis of genes for tetracycline resistance and replication functions in the broad-host-range plasmid pLS1. *J. Mol. Biol.* 192, 753–765. doi: 10.1016/0022-2836(86)90026-4

Leblanc, D. J., and Lee, L. N. (1984). Physical and genetic analyses of streptococcal plasmid pAM beta 1 and cloning of its replication region. *J. Bacteriol.* 157, 445–453.

López, P., Espinosa, M., Stassi, D. L., and Lacks, S. A. (1982). Facilitation of plasmid transfer in *Streptococcus pneumoniae* by chromosomal homology. *J. Bacteriol.* 150, 692–701.

López-Aguilar, C., Romero-López, C., Espinosa, M., Berzal-Herranz, A., and del Solar, G. (2015). The 5′-tail of antisense RNAII of pMV158 plays a critical role in binding to the target mRNA and in translation inhibition of *repB*. *Front. Genet.* 6:225. doi: 10.3389/fgene.2015.00225

López-Villarejo, J., Lobato-Márquez, D., and Díaz-Orejas, R. (2015). Coupling between the basic replicon and the *kis-kid* maintenance system of plasmid R1: modulation by Kis antitoxin levels and involvement in control of plasmid replication. *Toxins* 7, 478–492. doi: 10.3390/toxins7020478

Merlin, S., and Polisky, B. (1993). Analysis of establishment phase replication of the plasmid ColE1. *J. Mol. Biol.* 230, 137–150. doi: 10.1006/jmbi.1993.1131

Molin, S., Stougaard, P., Uhlin, B. E., Gustafsson, P., and Nordstrom, K. (1979). Clustering of genes involved in replication, copy number control, incompatibility, and stable maintenance of the resistance plasmid R1drd-19. *J. Bacteriol.* 138, 70–79.

Nordström, K. (1990). Control of plasmid replication - how do DNA iterons set the replication frequency? *Cell* 63, 1121–1124. doi: 10.1016/0092-8674(90)90405-4

Nordström, K. (2006). Plasmid R1-Replication and its control. *Plasmid* 55, 1–26. doi: 10.1016/j.plasmid.2005.07.002

Nordström, K., Molin, S., and Light, J. (1984). Control of replication of bacterial plasmids: genetics, molecular biology, and physiology of the plasmid R1 system. *Plasmid* 12, 71–90. doi: 10.1016/0147-619X(84)90054-4

Nordström, M., and Nordström, K. (1985). Control of replication of FII plasmids: comparison of the basic replicons and of the *copB* systems of plasmids R100 and R1. *Plasmid* 13, 81–87. doi: 10.1016/0147-619X(85)90060-5

Olsson, J. A., Paulsson, J., and Nordström, K. (2004). Effect of the CopB auxiliary replication control system on stability of maintenance of Par+ plasmid R1. *J. Bacteriol.* 186, 207–211. doi: 10.1128/JB.186.1.207-211.2004

Pimentel, B., Madine, M. A., and de la Cueva-Méndez, G. (2005). Kid cleaves specific mRNAs at UUACU sites to rescue the copy number of plasmid R1. *EMBO J.* 24, 3459–3469. doi: 10.1038/sj.emboj.7600815

Providenti, M. A., O'Brien, J. M., Ewing, R. J., Paterson, E. S., and Smith, M. L. (2006). The copy-number of plasmids and other genetic elements can be determined by SYBR-Green-based quantitative real-time PCR. *J. Microbiol. Methods* 65, 476–487. doi: 10.1016/j.mimet.2005.09.007

Rasooly, A., Wang, P. Z., and Novick, R. P. (1994). Replication-specific conversion of the *Staphylococcus aureus* pT181 initiator protein from an active homodimer to an inactive heterodimer. *EMBO J.* 13, 5245–5251.

Riise, E., and Molin, S. (1986). Purification and characterization of the CopB replication control protein, and precise mapping of its target site in the R1 plasmid. *Plasmid* 15, 163–171. doi: 10.1016/0147-619X(86)90034-X

Rosenfeld, R., and Grover, N. B. (1993). Control of mini-R1 plasmid replication: a computer simulation. *Plasmid* 29, 94–116. doi: 10.1006/plas.1993.1012

Ruiz-Masó, J. A., López-Zumel, C., Menéndez, M., Espinosa, M., and del Solar, G. (2004). Structural features of the initiator of replication protein RepB encoded by the promiscuous plasmid pMV158. *Biochim. Biophys. Acta Proteins Proteomics* 1696, 113–119. doi: 10.1016/j.bbapap.2003.09.010

Ruiz-Masó, J. A., Lurz, R., Espinosa, M., and del Solar, G. (2007). Interactions between the RepB initiator protein of plasmid pMV158 and two distant DNA regions within the origin of replication. *Nucleic Acids Res.* 35, 1230–1244. doi: 10.1093/nar/gkl1099

Ruiz-Masó, J., Machón, C., Bordanaba-Ruiseco, L., Espinosa, M., Coll, M., and del Solar, G. (2015). Plasmid rolling-circle replication. *Microbiol Spectr.* 3:PLAS-0035-2014. doi: 10.1128/microbiolspec.PLAS-0035-2014

Stassi, D. L., López, P., Espinosa, M., and Lacks, S. A. (1981). Cloning of chromosomal genes in *Streptococcus pneumoniae*. *Proc. Natl. Acad. Sci. U.S.A.* 78, 7028–7032. doi: 10.1073/pnas.78.11.7028

Summers, D. K. (2009). *The Biology of Plasmids*. Oxford, UK: Blackwell Publishing Ltd. doi: 10.1002/9781444313741

Untergasser, A., Cutcutache, I., Koressaar, T., Ye, J., Faircloth, B. C., Remm, M., et al. (2012). Primer3–new capabilities and interfaces. *Nucleic Acids Res.* 40:e115. doi: 10.1093/nar/gks596

Wagner, E. G., Altuvia, S., and Romby, P. (2002). Antisense RNAs in bacteria and their genetic elements. *Adv. Genet.* 46, 361–398. doi: 10.1016/S0065-2660(02)46013-0

4

Bad Phages in Good Bacteria: Role of the Mysterious *orf63* of λ and Shiga Toxin-Converting $\Phi 24_B$ Bacteriophages

Aleksandra Dydecka [1†], Sylwia Bloch [1†], Ali Rizvi [2], Shaili Perez [2], Bozena Nejman-Falenczyk [1], Gracja Topka [1], Tomasz Gasior [3], Agnieszka Necel [1], Grzegorz Wegrzyn [1], Logan W. Donaldson [2] and Alicja Wegrzyn [3]*

[1] Department of Molecular Biology, Faculty of Biology, University of Gdansk, Gdansk, Poland, [2] Department of Biology, York University, Toronto, ON, Canada, [3] Institute of Biochemistry and Biophysics, Polish Academy of Sciences, Warsaw, Poland

Edited by:
Manuel Espinosa,
Centro de Investigaciones Biológicas (CSIC), Spain

Reviewed by:
Radoslaw Pluta,
International Institute of Molecular and Cell Biology in Warsaw (IIMCB), Poland
Ramon Diaz Orejas,
Consejo Superior de Investigaciones Científicas (CSIC), Spain
Dhruba Chattoraj,
National Institutes of Health, United States

***Correspondence:**
Alicja Wegrzyn
alicja.wegrzyn@biol.ug.edu.pl

[†] These authors have contributed equally to this work.

Lambdoid bacteriophages form a group of viruses that shares a common schema of genome organization and lifecycle. Some of them can play crucial roles in creating the pathogenic profiles of *Escherichia coli* strains. For example, Shiga toxin-producing *E. coli* (STEC) acquired *stx* genes, encoding Shiga toxins, via lambdoid prophages (Stx phages). The results obtained so far present the evidence for the relation between the *exo-xis* region of the phage genome and lambdoid phage development, however molecular mechanisms of activities of the *exo-xis* genes' products are still unknown. In view of this, we decided to determine the influence of the uncharacterized open reading frame *orf63* of the *exo-xis* region on lambdoid phages development using recombinant prophages, λ and Stx phage $\Phi 24_B$. We have demonstrated that *orf63* codes for a folded protein, thus, it is a functional gene. NMR spectroscopy and analytical gel filtration were used to extend this observation further. From backbone chemical shifts, Orf63 is oligomeric in solution, likely a trimer and consistent with its small size (63 aa.), is comprised of two helices, likely intertwined to form the oligomer. We observed that the deletion of phage *orf63* does not impair the intracellular lambdoid phage lytic development, however delays the time and decreases the efficiency of prophage induction and in consequence results in increased survival of *E. coli* during phage lytic development. Additionally, the deletion of phage *orf63* negatively influences expression of the major phage genes and open reading frames from the *exo-xis* region during prophage induction with hydrogen peroxide. We conclude, that lambdoid phage *orf63* may have specific functions in the regulation of lambdoid phages development, especially at the stage of the lysis vs. lysogenization decision. Besides, *orf63* probably participates in the regulation of the level of expression of essential phage genes and open reading frames from the *exo-xis* region during prophage induction.

Keywords: Shiga toxin-producing *Escherichia coli* (STEC), lambdoid bacteriophages, lytic development, *exo-xis* region, open reading frames

INTRODUCTION

The significance of Shiga toxin-producing *E. coli* (STEC) as a public health problem was first recognized in 1982 during an investigation of an outbreak of hemorrhagic colitis associated with consumption of contaminated hamburgers (Riley et al., 1983). Since then, STEC strains have been implicated in many outbreaks of diarrhea world-wide. Quite recently (2011), the Shiga toxin-producing *E. coli* serotype O104:H4 was responsible for a serious epidemic outbreak in Germany (Bloch et al., 2012; Muniesa et al., 2012). STEC pathogens can cause serious food poisoning with bloody diarrhea in humans (Nataro and Kaper, 1998). Their main virulence factors are Shiga toxins, encoded by *stx* genes located in genomes of bacteriophages which occur in bacteria as prophages (Mizutani et al., 1999). These bacteriophages are called Shiga toxin-converting or Stx, for short, and belong to the lambdoid family of phages (Schmidt, 2001). All phages within this group indicate high similarities in the lifecycle and genomic organization to bacteriophage λ, the most reviewed member of this family (Wegrzyn et al., 2012). In the prophage state, most of phage genes, including *stx* genes, are not transcribed due to inhibition caused by the phage cI repressor. As a consequence, Shiga toxins are not produced under such conditions. Expression of *stx* as well as other phage genes occurs effectively only after prophage induction. In most cases, this process requires activation of the RecA-dependent bacterial S.O.S. response which is provoked by factors causing appearance of single-stranded DNA fragments. Activated RecA protein stimulates cleavage of the S.O.S. regulon repressor, the LexA protein, and the cI phage repressor. Prophage induction and subsequent phage lytic development lead to production of progeny phage particles and Shiga toxins, followed by their release from the lysed cell (Licznerska et al., 2016b). In the regulation of the lysis-vs. -lysogenization decision after infection of the host cell by a bacteriophage, both phage- and host-encoded proteins play important roles (for a review, see Wegrzyn et al., 2012). Among environmental factors influencing the decision, the crucial are temperature, nutrients availability and multiplicity of infection (m.o.i.). Lytic growth is supported by high temperature, rich medium and high m.o.i., while low temperature, starvation and low m.o.i. favor lysogenization. At the molecular level, the major players supporting lytic and lysogenic pathways are Cro and cI proteins, respectively. They are transcriptional regulators, and Cro represses expression of the *c*I gene, whereas cI downregulates transcription from two major "lytic" promoters (pL and pR, which provide mRNAs for *cro* and other "lytic" genes, encoding proteins involved in all processes during production of phage progeny) while stimulating its own expression by activation of the pM promoter. Thus, the result of the competition between Cro and cI is crucial for choosing one of the alternative developmental pathways. Since shortly after infection no cI protein is present, another transcription regulator, the cII protein (whose gene is transcribed from pR), is a key player in this game. This protein activates the second promoter for *c*I expression, pE. Therefore, cII activity decides on the Cro or cI predominance. In fact, cII is a subject of various regulatory mechanisms acting in response to different environmental conditions, including those playing major roles in the lysis-vs. -lysogenization decision (see Wegrzyn et al., 2012, for details).

An evolutionarily conserved region of lambdoid bacteriophage genome, located between *exo* and *xis* genes (so called "the *exo-xis* region"), contains several genes and open reading frames (**Figure 1**). Quite surprisingly, until recently, the role of this region in bacteriophage development was almost completely unknown. Recent studies indicated that overexpression of genes from the *exo-xis* region' impaired lysogenization of *E. coli* by bacteriophage λ (Loś et al., 2008b) and enhanced induction of prophages λ and $\Phi 24_B$ (one of Shiga toxin-converting phages) (Bloch et al., 2013). The Ea8.5 protein, encoded by a gene located in the *exo*-xis region, contains a fused homeodomain/zinc-finger fold (Kwan et al., 2013), which suggest a regulatory role for this protein. Interestingly, prophage induction with mitomycin C or hydrogen peroxide caused different expression patterns of genes from the *exo-xis* region; such differences were observed in both phages, λ and $\Phi 24_B$ (Bloch et al., 2014). Moreover, phages with deletions in the *exo-xis* region responded to the oxidative stress in a different manner relative to wild-type phages (Licznerska et al., 2016a). Therefore, it is important to determine structures and functions of particular proteins encoded in the *exo-xis* region.

In this work, we have focused on *orf63*. In our preliminary experiments with mutants in particular genes and ORFs from the *exo-xis* region, deletion of *orf63* gave one of the strongest effects (Licznerska et al., 2016a). Moreover, the transcription factor YqhC has been recognized as a potential partner for interaction with Orf63 in yeast two-hybrid study of phage-host interactions (Blasche et al., 2013). Therefore, we decided to investigate structure and functions of *orf63* and its product, the Ofr63 protein, in more detail.

MATERIALS AND METHODS

The *orf63* Gene Expression and Protein Purification

A codon-optimized *orf63* gene (NCBI ID: 2703507) for high level expression *E. coli* was synthesized by ATUM (Menlo Park, CA) and supplied in plasmid pD441-NH for direct transformation of a BL21 host strain (Novagen). Amino terminal 6xHis and Flag (DYKDDDDK) tags were included to facilitate affinity purification and detection. Milligram quantities of isotopically labeled 6xHis-Flag-Orf63 for NMR spectroscopy were obtained from a 1.0 L fermentation in a minimal medium containing 1 g $^{15}NH_4Cl$, 3 g of ^{13}C-glucose, and 1 g of ^{15}N-^{13}C Celtone algal extract (CIL; Cambridge, MA). The cell pellet was dissolved in T300 buffer (20 mM Tris-HCl, 300 mM NaCl, 0.05% NaN_3) and lysed by French press and sonication. The Orf63 protein purified from the bacterial soluble fraction by Nickel-NTA affinity chromatography (Qiagen) that included a 10 mM imidazole wash step and a 20 mM EDTA elution step, all in T300 buffer. A subsequent gel filtration chromatography step (Sephacryl-100, HiLoad 16/60; GE Life Sciences) was employed to further purify the Orf63 protein and exchange it into NMR spectroscopy buffer (5 mM Tris-HCl, 0.15 M NaCl, 0.05% NaN_3).

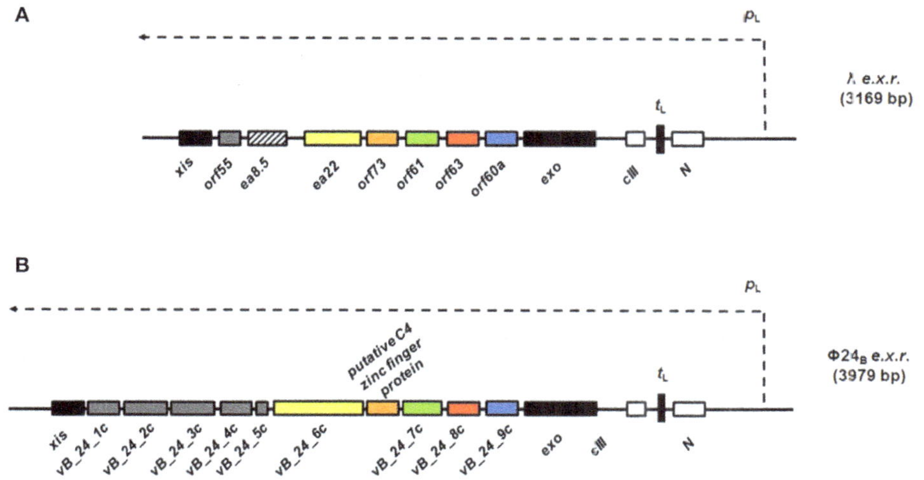

FIGURE 1 | Maps of genes and open reading frames (ORFs) from the region located between *exo* and *xis* genes (black rectangles), the *exo-xis* region or *e.x.r.*, of lambdoid bacteriophages: λ **(A)** and $\Phi 24_B$ **(B)**. In the case of phage λ **(A)**, the *exo-xis* region consists of two recognized genes: *ea22* and *ea8.5*, and five additional ORFs, named: *orf60a*, *orf63*, *orf61*, *orf73*, and *orf55*, which expression is under control of the p_L promoter (thin dashed arrow). Comparatively, the *exo-xis* region of phage $\Phi 24_B$ **(B)** contains additional ORFs (gray rectangles), but there is no homolog of the *ea8.5* gene of phage λ **(A)**, rectangle with black stripes). Note, that some ORFs from the *exo-xis* region of phage $\Phi 24_B$ **(B)**: *vb_24B_9c* (blue rectangles), *vb_24B_8c* (red rectangles), *vb_24B_7c* (green rectangles), *putative C4 zinc finger protein* (orange rectangles), and *vb_24B_6c* (yellow rectangles) are homologs of phage λ *orf60a*, *orf63*, *orf61*, *orf73*, *ea22* **(B)**, respectively. In spite of the differences in composition of both λ and $\Phi 24_B$ *exo-xis* regions, attention needs to paid to highly conserved sequences of the *orf60a-orf73* regions among lambdoid bacteriophages (≥70% nucleotide and amino acid sequence identity) (Bloch et al., 2013). The regulatory genes: *N* and cIII are marked as white rectangles and t_L terminator is indicated as black vertical rectangle.

Gel Filtration Assay to Estimate Apparent Molecular Weight

Chromatograms of eleven proteins on the same gel filtration column used to purify Orf63 were compiled to produce a standard curve describing the relationship between retention volume and molecular weight. Specific proteins used for the standard curve included the HACS1 SH3 domain (10.6 kDa), the AIDA1 PTB domain (22.0 kDa), the monomeric and dimeric states CASKIN2 SAM domain tandem (20.2 / 40.5 kDa), the CASKIN2 SAM1 domain (10.3 kDa), the SHP2 adaptor SH2 domain (14.1 kDa), the AIDA1 SAM domain tandem (16.4 kDa), two deletion mutants of the La RRM domain (14.6 / 15.9 kDa), the Crk2 adaptor SH2 domain (15.5 kDa), dimeric glutathione S-transferase (52 kDa), and calmodulin (18.8 kDa).

NMR Spectroscopy

A 0.5 mM sample of ^{13}C,^{15}N-labeled Orf63 was prepared for NMR spectroscopy in NMR buffer supplemented with 10% D_2O. All experiments were performed at 310 K using a Bruker Avance 700 MHz NMR spectrometer equipped with a cryogenically cooled 5 mm probe at the York University Life Sciences Building Central Facility. Backbone (HN, N, CA, CB, C') assignments were achieved using a set of conventional triple resonance experiments (HNCA, HNCACB, CBCAcoNH, HNCO, HNcaCO) incorporating sparse sampling for the optimum sensitivity and resolution. Datasets were processed with NMRpipe (Delaglio et al., 1995) and istHMS (Hyberts et al., 2012) and interpreted with CCPN Analysis (Skinner et al., 2015).

Bacteria, Bacteriophages and Plasmids

The *E. coli* strains, bacteriophages and plasmids used in *in vivo* work are presented in **Table 1**. Work with these strains was approved by the Ministry of Environment (decision no. 189/2016). The *E. coli* lysogens were obtained using the following lambdoid phages: λ, λΔ*orf63*, $\Phi 24_B$ or $\Phi 24_B$Δ*orf63* (Bloch et al., 2013; Licznerska et al., 2016b). In the first step, phage lysates were prepared. Bacterial cultures were grown at 37°C to $A_{600} = 0.1$. Then, mitomycin C (Sigma—Aldrich) was added to all flasks to a final concentration of 1 μg/ml. The incubation with shaking was continued for about 12 h. To obtain lysates, bacterial debris were centrifuged (2,000 × g for 10 min at 4°C) and supernatants were filtered through the 0.22-μm-pore-size filters (Sigma—Aldrich). In the next step, the lysogenization procedure was carried out. Briefly, *E. coli* strain C600 was cultivated at 37°C to $A_{600} = 0.2$. Then, 4 ml of bacterial culture was centrifuged (2,000 × g for 5 min at RT), the pellet was washed with TCM buffer (10 mM Tris-HCl, 10 mM $MgSO_4$, 10 mM $CaCl_2$, pH 7.2; Sigma—Aldrich) and suspended in LB medium (Sigma—Aldrich) supplemented with $MgSO_4$ (phages λ and λΔ*orf63*) or with $MgSO_4$ and $CaCl_2$ (phages $\Phi 24_B$ and $\Phi 24_B$Δ*orf63*) to a final concentration of 10 mM. Bacteriophages were added to the suspensions to m.o.i. of 10. Following incubation at 37°C, the mixtures were spread on LB agar plates. After overnight incubation at 37°C, bacterial colonies were tested for the presence of prophages by using UV irradiation (this procedure is described in detail in the next section). For construction of the plasmid pSB_orf63_λ, nucleotide sequence of *orf63* from phage λ was amplified by PCR with primers: Fλorf63_EcoRI (5'GGA GAA TTC GGC TGT ATG CAC AAA

TABLE 1 | Bacterial strains, bacteriophages and plasmids used for *in vivo* experiments.

E. coli strains, bacteriophages or plasmids	Relevant genotype or description	References
***E. coli* LABORATORY STRAINS**		
MG1655	F– λ– *ilvG rfb-50 rph-1*	Jensen, 1993
MG1655 (λ)	MG1655 bearing λ prophage	Bloch et al., 2013
MG1655 (λΔ*orf63*)	MG1655 bearing λ prophage with deletion of *orf63*	Licznerska et al., 2016a
MG1655 ($\Phi 24_B$)	MG1655 bearing $\Phi 24_B$ prophage	Bloch et al., 2013
MG1655 ($\Phi 24_B \Delta$*orf63*)	MG1655 bearing $\Phi 24_B$ prophage with deletion of *vb_24B_8c*, the homolog of λ*orf63*	Licznerska et al., 2016a
C600	F– *tonA21 thi-1 thr-1 leuB6 lacY1 glnV44 rfbC1 fhuA1* λ^-	Appleyard, 1954
C600 (λ)	C600 bearing λ prophage	This study, by lysogenization
C600 (λΔ*orf63*)	C600 bearing λ prophage with deletion of *orf63*	This study, by lysogenization
C600 ($\Phi 24_B$)	C600 bearing $\Phi 24_B$ prophage	This study, by lysogenization
C600 ($\Phi 24_B \Delta$*orf63*)	C600 bearing $\Phi 24_B$ prophage with deletion of *vb_24B_8c*, the homolog of λ*orf63*	This study, by lysogenization
BACTERIOPHAGES		
λ	carries a frameshift mutation relative to Ur-lambda	Hendrix and Duda, 1992
λΔ*orf63*	$\Phi 24_B$ phage with deletion of *orf63*	Licznerska et al., 2016a
$\Phi 24_B$	λ*stx2::cat*	Allison, 2003
$\Phi 24_B \Delta$*orf63*	$\Phi 24_B$ phage with deletion of *vb_24B_8c*, the homolog of λ*orf63*	Licznerska et al., 2016a
PLASMIDS		
pUC18	*ori* pMB1 (pBR322 derivative), *bla*, Amp^R	Thermo fisher scientific
pSB_orf63_λ	as pUC18 but bearing the *orf63* from phage λ	This study
pSB_orf63_$\Phi 24_B$	as pUC18 but bearing the *orf63* from phage $\Phi 24_B$	This study

GC) and Rλorf63_BamHI (5′ GAG GAT CCT GCA TTC CGT GGT TGT C), and phage DNA as a template, which was isolated by using MasterPure™ Complete DNA and RNA Purification Kit (Epicenter). Then, the λ*orf63* was ligated with fragment of plasmid pUC18 (insert and vector were digested with EcoRI and BamHI restrictions endonucleases; Thermo Scientific), bearing an ampicillin resistance gene and sequence of p_{lac} promoter. The plasmid pSB_orf63_$\Phi 24_B$ was constructed according to similar procedure. To amplify a DNA fragment containing *vb_24B_8c* sequence (the homolog of λ*orf63*) by PCR method, two primers: F$\Phi 24_B$orf63_EcoRI (5′GGA GAA TTC GGC TGT ATG CAC AAA GC) and R$\Phi 24_B$orf63_BamHI (5′GTA GGA TCC TTG TCA TGC CGG GTC) were used. Next, plasmid pUC18 and insert were cut with EcoRI and BamHI enzymes and ligate by the T4 DNA ligase (Thermo Scientific). The construction of pUC18 derivatives was confirmed by DNA sequencing (Genomed).

Media and Growth Conditions

All *in vivo* experiments were performed in LB liquid medium (Sigma—Aldrich) supplemented with 10 mM $MgSO_4$ (phage λ or phage λΔ*orf63*) or with 10 mM $MgSO_4$ and 10 mM $CaCl_2$ (phage $\Phi 24_B$ or phage $\Phi 24_B \Delta$*orf63*), and with 50 μg/ml ampicillin (if necessary) (Sigma—Aldrich). To stimulate Orf63 protein production from the recombinant pUC18 derivatives, overnight bacterial cultures were diluted 1:100 in fresh LB medium and treated with IPTG (A&A Biotechnology) to a final concentration of 1 mM. Then, host bacteria were grown in aeration condition, achieved by shaking, at 30°C to $A_{600} = 0.1$ or 0.2 (the optical density of bacterial cells was dependent on the experimental conditions described in the following chapters).

Double Overlay Plaque Assay

Bacteriophage titration was performed on the standard Petri dishes (Alchem) filled with 25 ml of LB agar (1.5% agar; Sigma—Aldrich), according to a procedure described by Sambrook and Russell (2001), with some modification. The top layer was prepared by mixing 2 ml of LB agar (0.7% agar; Sigma—Aldrich) with 1 ml of the overnight bacterial cell culture. To obtain visible plaques formed by Stx phages, the bottom agar was supplemented with sublethal concentration of chloramphenicol (Sigma—Aldrich). This antibiotic was effective in increasing of size of plaques of phage $\Phi 24_B$ and its derivative, which possessed in genomes chloramphenicol resistance gene (**Table 1**). As described previously (Łoś et al., 2008a), the *cm* gene expression, especially after phage infection of *E. coli* bacteria, may have the positive influence on cellular productivity by decreasing the inhibitory effects of the antibiotic on protein synthesis. To determine the number of phages per ml of suspension (PFU/ml), serial 10-fold dilutions were prepared in TM buffer (10 mM Tris–HCl, 10 mM $MgSO_4$; pH 7.2). Then, appropriate volume of each dilution of phage lysate was spotted onto double agar layer. The plates were incubated at 37°C overnight, plaques were counted, and the phage titer was calculated.

One-Step Growth Experiments in Phage-Infected Bacteria

To investigate the intracellular lytic development of lambdoid phages the one-step-growth experiment was prepared using the method described by Wegrzyn et al. (1995), with a minor modification (Bloch et al., 2014; Nejman-Falenczyk et al., 2015). Host bacteria were grown in LB medium at 30°C to $A_{600} = 0.2$. In the next step, 10 ml of a bacterial culture was centrifuged (2,000 × g for 10 min at 4°C). The pellet was suspended in 1 ml of LB medium supplemented with 3 mM sodium azide (Sigma—Aldrich). Bacteriophages were added to *E. coli* cells to m.o.i. of 0.05. After 10 min incubation at 30°C, unadsorbed phages were removed by three times washing in LB medium with 3 mM sodium azide (2,000 × g for 10 min at 4°C). Then, 25 μl of the suspension was added to 25 ml of LB medium prewarmed to 30°C (time 0) and cultivated in an incubator shaker. The number of infection centers were determined at times: 5, 10, 15 min after infection by mixing 0.1 ml of the sample with 0.9 ml of an overnight culture of appropriate indicator bacteria and 2 ml of top agar. Next, the mixture was poured onto LB agar plate (phages λ and λΔ*orf63*) or LB agar plate with 2.5 μg/ml chloramphenicol (phage $\Phi 24_B$ and $\Phi 24_B\Delta$*orf63*). Samples taken at later times were treated with chloroform (POCH), shaken vigorously and cleared by centrifugation (2,000 × g for 5 min at RT). The phage lysate was diluted in TM buffer and titrated under permissive condition. Plates were incubated at 37°C overnight. The number of viruses released from each infected cell (burst size) was calculated as a ratio of phage titer to the titer of infection centers.

Prophage Induction with Hydrogen Peroxide

Bacteria lysogenic for lambdoid phages were grown in LB medium at 30°C to $A_{600} = 0.1$. Next, the culture was divided into two aliquots. One of them was treated with 1 mM hydrogen peroxide (Sigma—Aldrich) to provoke the prophage induction. The second one was a control without an induction agent. The cultivation was continued at 30°C. At indicated times samples were harvested, mixed with chloroform and vortexed for 1 min. The suspension was centrifuged for 5 min in a microfuge at RT. The supernatant was diluted in TM buffer and 2.5 μl of each serial dilution was dropped onto a freshly prepared double-layer LB agar in plastic Petri dishes. Plates were incubated at 37°C overnight. The relative phage titer was estimated by subtracting the phage titer determined in non-induced cultures from the phage titer estimated in induced cultures.

Survival of Host Bacteria after Bacteriophage Infection

To estimate the percentage of surviving cells after bacteriophage infection the procedure created by Sambrook and Russell (2001) was used, with a minor modification. A bacterial culture was grown at 30°C to $A_{600} = 0.2$. Then, 4 ml of the sample was centrifuged (2,000 × g for 10 min at 4°C). The supernatant was discarded and the pellet was washed with 0.85% sodium chloride (POCH) (2,000 × g for 10 min at 4°C). Finally, the bacterial pellet was suspended in 1 ml of LB medium supplemented with $MgSO_4$ (phage λ and phage λΔ*orf63*) or with $MgSO_4$ and $CaCl_2$ (phage $\Phi 24_B$ and $\Phi 24_B\Delta$*orf63*) to a final concentration of 10 mM. The suspension was incubated for 30 min at 30°C and then phage particles were added to m.o.i. of 1, 5, 10. The mixture was kept for 15 min (phage λ and phage λΔ*orf63*) or 30 min (phage $\Phi 24_B$ and $\Phi 24_B\Delta$*orf63*) at 30°C. In the next step, serial dilutions in 0.85% sodium chloride were prepared and 40 μl of each dilution was spread on LB agar plates. After overnight incubation at 37°C, percentage of surviving *E. coli* bacteria was calculated relative to bacterial culture in which TM buffer was added instead of phage particles.

Efficiency of Prophage Formation after Bacterial Virus Infection

Efficiency of lysogenization was estimated according to Arber et al. (1983) and Wegrzyn et al. (1992), with some modification. Host bacteria were cultured at 30°C to $A_{600} = 0.2$. Next, 1 ml of the sample was centrifuged (2,000 × g for 10 min at 4°C). Bacterial culture was washed with TCM buffer twice, and then pellet was suspended in the same buffer. Bacteriophages were added to bacterial cells to m.o.i. of 1, 5, 10. The mixture was incubated at 30°C. Then, serial dilutions were prepared and 20 μl of each suspension was spread on LB agar plates prior to overnight incubation at 37°C. The next day, 96 colonies were passaged in each well of a 96-well plate with 200 μl of LB medium and shaken at 37°C to $A_{600} = 0.1$. To estimate a percent of lysogens among survivors, bacterial cultures were treated with UV light at 50 J/m^2 (the dose used routinely for lambdoid prophage induction) and incubated at 37°C for 2 h. Following induction, putative lysogens were mixed with chloroform, centrifuged (2,000 × g for 10 min at 4°C) and the water phase was spotted onto a double-layer LB agar (phage λ and phage λΔ*orf63*) or a double-layer LB agar supplemented with chloramphenicol to a final concentration of 2.5 μg/ml (phage $\Phi 24_B$ and $\Phi 24_B\Delta$*orf63*). Efficiency of lysogenization was calculated as a percent of lysogens relative to all tested bacterial cells. Lysogens were also infected with the same phage to check their resistance to superinfection, as described previously (Wegrzyn et al., 1992).

Prophage Induction and Extraction of RNA

Induction of tested prophages was provoked in lysogenic bacteria by addition of hydrogen peroxide to a final concentration of 1 mM. At the appropriate time, 10^9 bacterial cells were harvested, treated with 10 mM sodium azide and deep frozen in liquid nitrogen (this procedure was necessary to inhibit the growth of host bacteria). Total RNA from all samples were isolated with the High Pure RNA Isolation Kit (Roche Applied Science). To remove DNA from RNA preparations the TURBO DNA-free™ Kit (Life Technologies) was used. The quality and quantity of total isolated RNA were analyzed by a NanoDrop spectrophotometer and agarose gel electrophoresis. The contamination of DNA from RNA samples was also tested by routine PCR and qRT-PCR.

cDNA Synthesis from an RNA Template

To synthesize cDNA from an RNA template, the Transcriptor Reverse Transcriptase and random hexamer primers (Roche Applied Science) were used, according to the protocol supplied from the provider. 1.25 μg of the total RNA was taken for each reaction. Finally, mixture was diluted 10-fold and tested in qRT-PCR.

TABLE 2 | Primers used for RT-Qpcr.

Primer name	Sequence (5′→ 3′)
pF_λ_ea8.5	GGGCAAGTATCGTTTCCACC
pR_ λ_ea8.5	GCAATGTGCGAGAAATGACTG
pF_λ_ea22	GCAGTTCCAGCACAATCGAT
pR_ λ_ea22	AATGCATGACGACTGGGGAT
pF_λ_orf73	CACTTCGAACCTCTCTGTTTACTG
pR_ λ_orf73	CAGGGTTGTCGGACTTGTG
pF_λ_orf61	TTAGCCTGACGGGCAATG
pR_ λ_orf61	CCGACATGGGACTTGTTCA
pF_λ_orf60a	GCATACAGCCCCTCGTTTAT
pR_ λ_orf60a	CCGAAATCCACTGAAAGCAC
pF_λ_cIII	ATTCTTTGGGACTCCTGGCTG
pR_ λ_cIII	GTAAATTACGTGACGGATGGAAAC
pF_λ_N	CTCGTGATTTCGGTTTGCGA
pR_ λ_N	AAGCAGCAAATCCCCTGTTG
pF_λ_cro	ATGCGGAAGAGGTAAAGCCC
pR_ λ_cro	TGGAATGTGTAAGAGCGGGG
pF_λ_cII	TCGCAATGCTTGGAACTGAGA
pR_ λ_cII	CCCTCTTCCACCTGCTGATC
pF_λ_Q	TTCTGCGGTAAGCACGAAC
pR_ λ_Q	TGCATCAGATAGTTGATAGCCTTT
pF_λ_R	ATCGACCGTTGCAGCAATA
pR_ λ_R	GCTCGAACTGACCATAACCAG
pF_Φ24B_ea22	TCAGCAACATGGCATTCACT
pR_ Φ24B_ea22	GGTTGGGAAGCTGAGAGTTG
pF_Φ24B_orf73	CGAACCTCTCTGTTTACTGATAAGC
pR_ Φ24B_orf73	TTCAGGGTTGTCGGACTTGT
pF_Φ24B_orf61	TTAGCCTGACGGGCAATG
pR_ Φ24B_orf61	CCGACATGGGACTTGTTCA
pF_Φ24B_orf60a	CATACAGCCCCTCGTTTAT
pR_ Φ24B_orf60a	CCGAAATCCACTGAAAGCAC
pF_Φ24B_cIII	ATTCTTTGGGACTCCTGGCTG
pR_Φ24B_cIII	GTAAATTACGTGACGGATGGAAAC
pF_Φ24B_N	AGGCGTTTCGTGAGTACCTT
pR_ Φ24B_N	TTACACCGCCCTACTCTAAGC
pF_Φ24B_cro	CGAAGGCTTGTGGAGTTAGC
pR_ Φ24B_cro	GTCTTAGGGAGGAAGCCGTT
pF_Φ24B_cII	TGATCGCGCAGAAACTGATTTAC
pR_Φ24B_cII	GACAGCCAATCATCTTTGCCA
pF_Φ24B_O	AAGCGAGTTTGCCACGAT
pR_Φ24B_O	GAACCCGAACTGCTTACCG
pF_Φ24B_Q	GGGAGTGAGGCTTGAGATGG
pR_ Φ24B_Q	TACAGAGGTTCTCCCTCCCG
pF_Φ24B_R	GGGTGGATGGTAAGCCTGT
pR_ Φ24B_R	TAACCCGGTCGCATTTTTC
pF_E.coli_16SrRNA	CCTTACGACCAGGGCTACAC
pR_ E.coli_16SrRNA	TTATGAGGTCCGCTTGCTCTC

qRT-PCR Assay and Data Analysis

The pattern of genes expression after prophage induction was performed by using the LightCycler® 480 Real-Time PCR System (Roche Applied Science), LightCycler® 480 SYBR Green I Master (Roche Applied Science) and cDNA samples. Transcription rates of genes of lambdoid bacteriophages were compared in parallel to the *16S rRNA* housekeeping gene (according to a procedure described by Strauch et al. (2008), which expression was stable during prophage induction provoked by hydrogen peroxide. All primers were created by Primer3web version 4.0.0 and are listed in **Table 2**. Each reaction mixture consisted of: 2x SYBR Green I Master Mix, 6.25 ng/μl cDNA and 200 nM specific primers. qRT-PCR amplifications were performed for 55 cycles. To confirm the specificity of primers, melting curve for each product was analyzed. The relative changes in gene expressions were determined by E-Method and calculated by the following formula: Normalized relative ratio $= E_t^{\text{CT (target) calibrator} - \text{CT (target) sample}} / E_r^{\text{CT (reference) calibrator} - \text{CT (reference) sample}}$, where E_t is the PCR efficiency of target and E_r means the PCR efficiency of reference. The sample before the addition of the inductor (the time point "zero") was a calibrator. The raw run data for tested lambdoid phages were transferred using the "LC480 Conversion: conversion of raw LC480 data" software and then, PCR efficiency for each gene was calculated by LinRegPCR program, which was successfully used previously (Bloch et al., 2014, 2015; Nejman-Falenczyk et al., 2015; Licznerska et al., 2016a).

Statistical Analysis

Each experiment was repeated three times and variation among replicates was presented as the error bars indicating the standard deviation (SD). All data comparisons were made by using Student's *t*-test. Significant differences were marked by asterisks when $P < 0.05$ (∗) or $P < 0.01$ (∗∗).

RESULTS

The Oligomeric State of Orf63

Samples from four independent preparations of 6xHis-Flag tagged Orf63 (15 aa. tag + 63 aa. protein) eluted as one peak on a preparative gel filtration column with an average retention volume of 59.5 mL corresponding to an apparent molecular weight of ~26 kDa (**Figure 2A**). Since affinity-tagged Orf63 is only 9 kDa, the gel filtration results suggest that Orf63 is oligomeric with a trimer as the most plausible configuration. This estimate is most accurate if Orf63 has the characteristics of a globular protein to match the standards used.

The *orf63* Gene Encodes a Folded Protein

Consistent with the observation that Orf63 is oligomeric in solution, NMR spectra of Orf63 at 298 K (25°C) suffered from considerable resonance line broadening that was characteristic of proteins > 20 kDa in overall molecular weight. Consequently, a higher temperature of 310 K (37°C) was chosen for all NMR studies to increase the tumbling time of the protein that, in turn, improves the sensitivity of triple resonance experiments. In **Figure 2B**, a ^{1}H-^{15}N HSQC spectrum is presented. The amide resonances in this two-dimensional spectrum are disperse indicating that the protein is folded. The combined analysis of several triple resonance (^{1}H, ^{13}C, ^{15}N) spectra lead to the determination of backbone (HN,

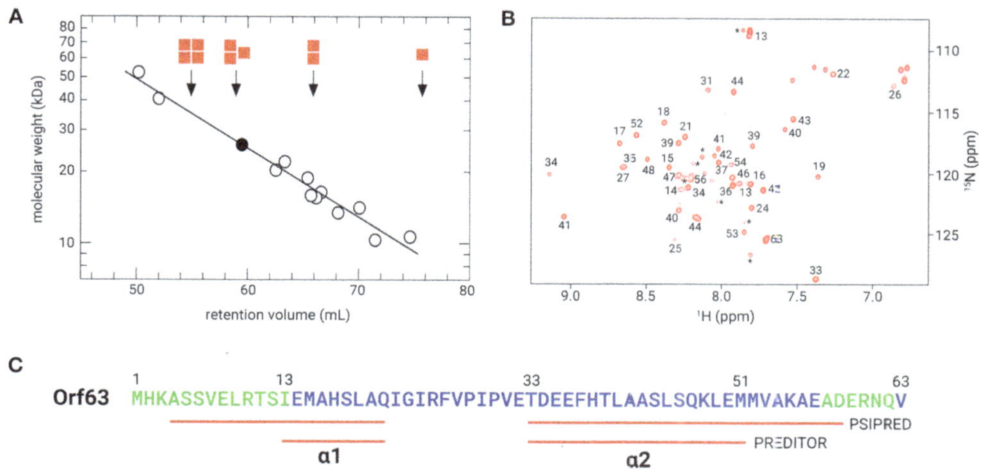

FIGURE 2 | Orf63 structural insights. (A) The retention volumes for a set of proteins ranging from 10–67 kDa were plotted against molecular weight to produce the log-linear relationship shown (open circles). The K_{av} value observed for Orf63 corresponds to an apparent molecular weight of ~28 kDa (closed circle). For reference, red squares denote the expected retention volumes of monomeric, dimeric, trimeric, and tetrameric species of the 9 kDa affinity tagged Orf63 protein. **(B)** A representative ^{1}H-^{15}N HSQC spectrum of Orf63 at 37°C demonstrating the characteristic resonance dispersion of a folded protein. Unassigned residues are labeled with an asterisk. Lines denote the amino resonances of three asparagine and glutamine side chains. **(C)** A conventional strategy was used to assign the backbone (HN, N, CA, CB, C') chemical shifts of Orf63. The extent of these assignments is colored blue on the sequence. The chemical shift data were then used as input to PREDITOR (Berjanskii et al., 2006) which identified two helices labeled α1 and α2. Using sequence data alone as input to PSIPRED (McGuffin et al., 2000), longer helices are predicted.

N, CA, CB, C') chemical shift assignments for residues 14–52 of Orf63. Resonances for the amino terminal affinity tags and from residue 53 onwards to the carboxy-terminus were either not observed or unassignable. Thus, the NMR data suggest that the folded region of Orf63 includes from residues 14–52.

Structural Characteristics of Orf63

Several statistical methods are available to predict secondary structure from backbone chemical shift data with a high degree of accuracy. As shown in **Figure 2C**, two helices are predicted (α1: 13–21; α2: 33–50). The secondary structure determined from chemical shift data is consistent with the secondary structure of Orf63 predicted from sequence information alone, although the helical boundaries are different.

The Sequence of Putative *orf63* Products is Conserved among Lambdoid Bacteriophages

Since experiments shown in **Figure 2** indicated that *orf63* encodes a protein, we have tested similarity of the putative proteins encoded by *orf63* of different lambdoid bacteriophages. Thus, scores of pairwise alignments of the predicted amino acid sequences of *orf63* from six such phages have been calculated. As demonstrated in **Table 3**, all these putative proteins are similar to each other. This indicate that the high similarity is kept at the protein level of Orf63 of lambdoid bacteriophages.

TABLE 3 | Scores of pairwise alignments of the predicted amino acid sequences of *orf63* from six analyzed lambdoid phages: λ phage (NC_001416), Φ24$_B$ phage (HM208303), 933W phage (NC_000924), VT2 Sakai phage (AP000422), Stx1 converting phage (NC_004913), and Stx2 converting phage II (NC_004914).

	λ	Φ24$_B$	933W	VT2 Sakai	Stx1	Stx2_II
λ		86	86	87	87	87
Φ24$_B$			100	98	98	98
933W				98	98	98
VT2 Sakai					100	100
Stx1						100
Stx2_II						

Pairwise scores are simply the number of identities between the two sequences, divided by the length of the alignment, and represented as a percentage. The multiple sequence alignment was performed using the ClustalW algorithm.

Efficiency of Lysogenization and Prophage Induction in the Absence of *orf63*

Since previous studies suggested that genes from the *exo-xis* region might be involved in the regulation of bacteriophage development (Bloch et al., 2013, 2014; Licznerska et al., 2016a), we have tested two crucial controlled steps in the lambdoid phage life cycle, the lysis-vs.-lysogenization decision, and prophage induction. We found that lysogenization efficiency was significantly increased in bacteriophages λ and Φ24$_B$ devoid of *orf63* (**Figures 3A,B**, respectively) though this phenomenon was more pronounced in λ (the effects were seen at all tested m.o.i.) (**Figure 3A**) than in Φ24$_B$ (significant effects observed only at m.o.i. = 10) (**Figure 3B**). Also, survival rates of bacterial cells (i.e., cells lysogenized and not infected) in populations

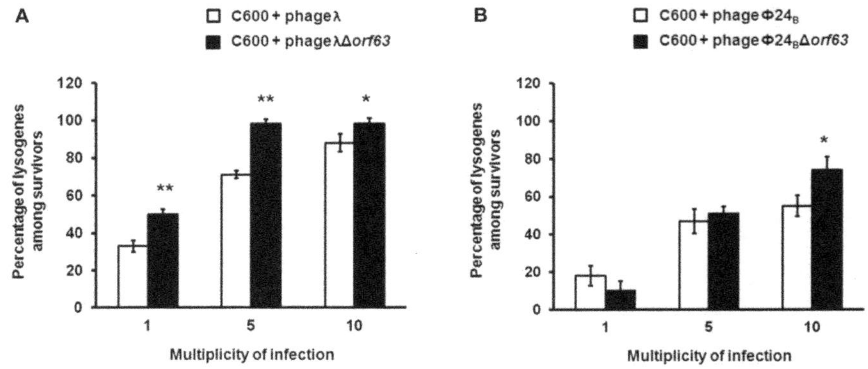

FIGURE 3 | Efficiency of lysogenization of *E. coli* C600 strain with lambdoid bacteriophages: λ and $\Phi 24_B$ (□ in **(A,B)**, respectively) or their deletion mutants λΔ*orf63* and $\Phi 24_B\Delta$*orf63* (■ in **A,B**, respectively). Results are presented as mean values ±*SD* from three independent experiments. Statistical analysis (*t* test) was performed for results from each m.o.i. (multiplicity of infection) between wild type phage and its deletion mutant. Significant differences are marked by asterisks $P < 0.05$ (*) or $P < 0.01$ (**).

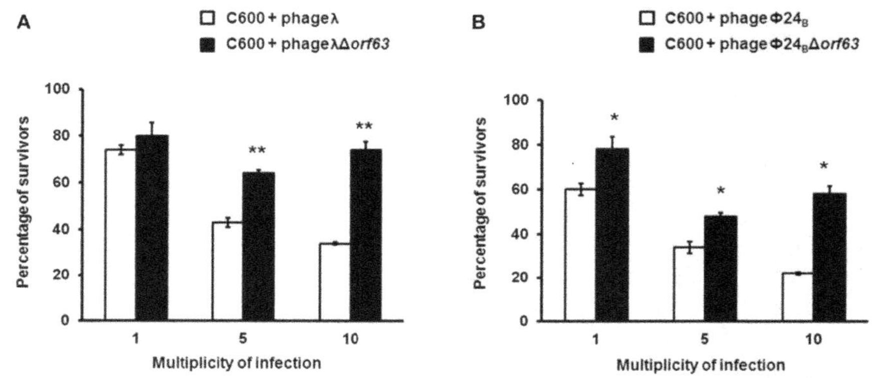

FIGURE 4 | Survival (%) of the wild-type strain *E. coli* C600, after infection with lambdoid bacteriophages: λ and $\Phi 24_B$ (□ in **(A,B)**, respectively) or their deletion mutants λΔ*orf63* and $\Phi 24_B\Delta$*orf63* (■ in **A,B**, respectively). Mean values from three independent experiments ±*SD* are shown. Statistical analysis were performed for each m.o.i. by *t* test. The significance of differences between fractions of bacterial cells surviving the infection with λ and λΔ*orf63* as well as Φ24B and Φ24BΔ*orf63* are observed and marked by asterisks $P < 0.05$ (*) or $P < 0.01$ (**).

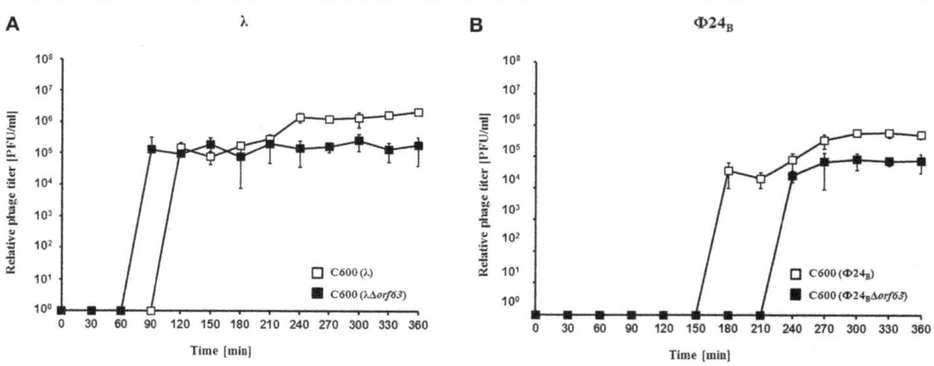

FIGURE 5 | Development of λ and λΔ*orf63* **(A)** or $\Phi 24_B$ and $\Phi 24_B\Delta$*orf63* **(B)** bacteriophages after prophage induction with hydrogen peroxide at 30 °C. *E. coli* C600 bacteria were either lysogenic with wild-type phages λ and $\Phi 24_B$ (□ in **A,B**, respectively) or their deletion mutants λΔ*orf63* and $\Phi 24_B\Delta$*orf63* (■ in **(A,B**, respectively). Phage lytic development was initiated by addition of H_2O_2 to final concentration of 1 mM at time 0. The presented results are mean values from three independent experiments with error bars indicating *SD* (note that in the most cases, the bars are smaller than sizes of symbols). Results are shown as PFU (plaque forming units) per one ml.

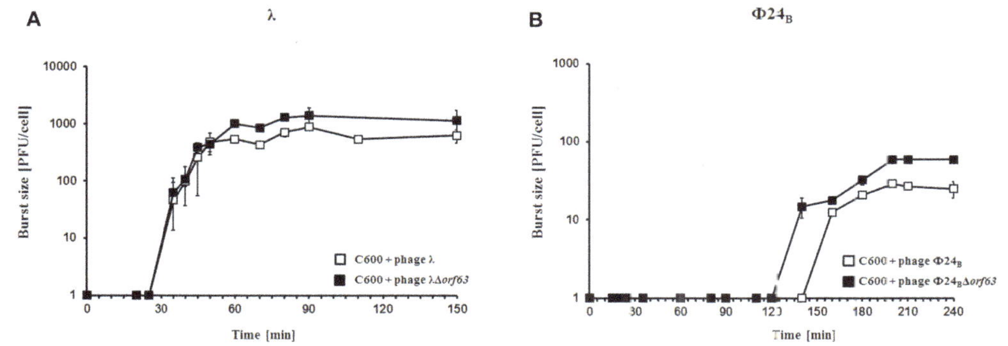

FIGURE 6 | Development of λ and λΔ*orf63* **(A)** or $\Phi24_B$ and $\Phi24_B\Delta$*orf63* **(B)** bacteriophages following phage infection of *E. coli* bacteria. Host *E. coli* strains were infected with wild-type phages λ and $\Phi24_B$ (□ in **A,B**, respectively) or their deletion mutants λΔ*orf63* and $\Phi24_B\Delta$*orf63* (■ in **A,B**, respectively) at time 0. The presented results are mean values ±*SD* from three independent experiments. Results are shown as PFU (plaque forming units) per cell.

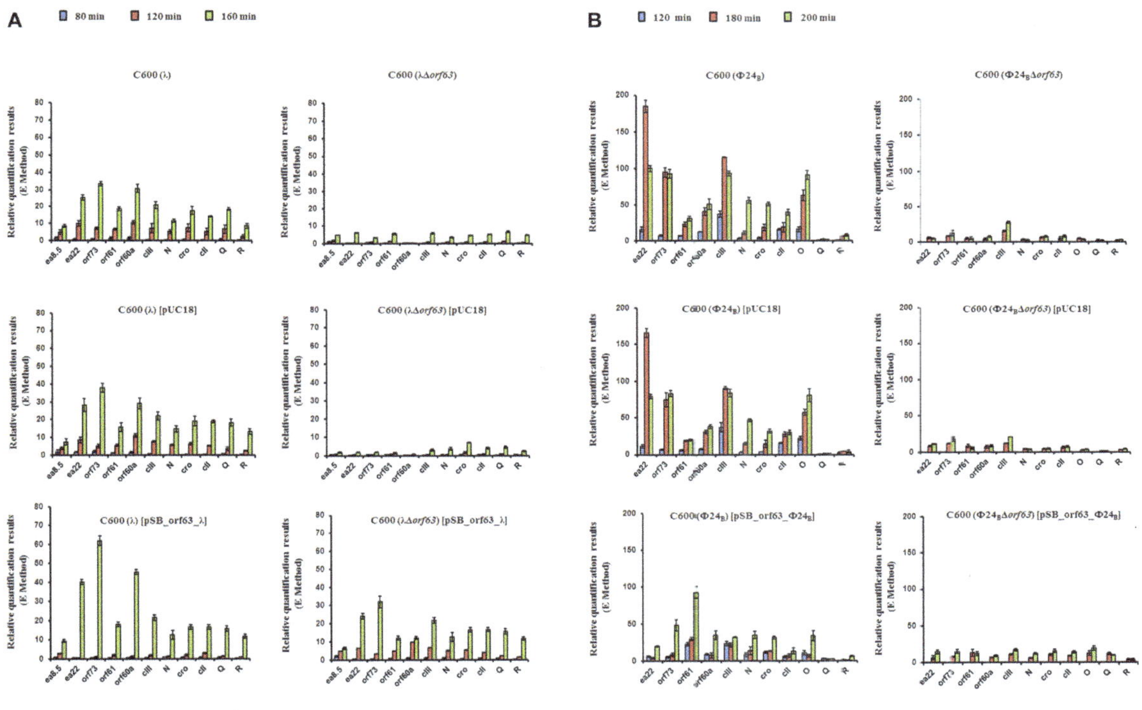

FIGURE 7 | Levels of transcripts of indicated genes of wild type phages λ and $\Phi24_B$ and their deletion mutants in the absence or presence of control plasmid [pUC18] or plasmid bearing the deleted *orf63* [pSB_orf63], assessed by quantitative reverse RT-PCR analysis, aftser prophage induction with 1 mM H_2O_2 in *E. coli* C600 host at 30°C. Levels of transcripts corresponding to particular genes were determined at following times after induction: 80 (blue), 120 (red) or 160 (green) minutes in case of phages λ and λΔ*orf63* **(A)**, and 120 (blue), 180 (red) or 200 (green) minutes in case of phages $\Phi24_B$ and $\Phi24_B\Delta$*orf63* **(B)**. Results are presented as mean values from three independent experiments with error bars indicating *SD*.

infected with Δ*orf63* or $\Phi24_B$ Δ*orf63* were higher than those in experiments with wild-type λ or $\Phi24_B$ (**Figures 4A,B**, respectively), supporting the conclusion that lysogenization is more effective for the mutant, indeed.

To test efficiency of prophage induction, we have estimated the number of phages appearing after prophage induction with hydrogen peroxide (one of natural prophage inducers occurring in human intestine, the common habitat of *E. coli*). Deletion of *orf63* caused a lower phage titer after prophage induction for both λ and $\Phi24_B$ (**Figures 5A,B**, respectively). However, when measured kinetics of phage development following infection of *E. coli* cells at low m.o.i. (0.05), we found that phages λ and $\Phi24_B$ devoid of *orf63* gave even more progeny per infected cell than their wild-type counterparts (**Figures 6A,B**, respectively). Therefore, combining results of experiments presented in **Figures 5**, **6**, one can conclude that deletion of *orf63* influences efficiency of prophage induction in both λ and $\Phi24_B$. Since formation of progeny phages is definitely not impaired in the

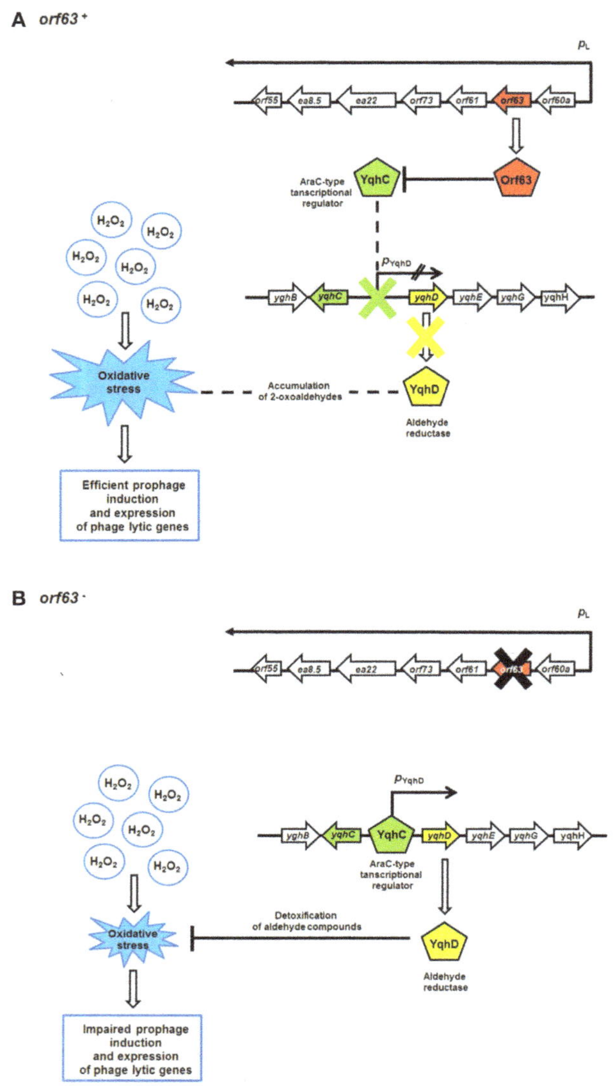

FIGURE 8 | A model for Orf63-mediated modulation of bacteriophage development. Mechanisms operating in the presence **(A)** or absence **(B)** of the orf63 function are shown. Transcripts are shown as black arrows, with arrowheads indicating directionality of transcription. The lack of the activity of the p_{YghD} promoter is marked as two black inclined lines. Phage and *E. coli* (host) proteins are presented as pentagons. Stimulations of processes or protein production/activity are shown by open arrows. Negative regulations of processes or protein functions are represented by blunt-ended lines.

absence of *orf63* when lytic development starts after infection, we suggest that lower phage titer after prophage induction indicates lower efficiency of this process in phages devoid of this gene.

Deletion of *Orf63* Influences Expression of Genes of λ and Φ24$_B$ Phages

Since experiments described above indicated that *orf63* function is involved in the regulation of lysogenization and prophage induction in both λ and Φ24$_B$, we aimed to measure expression of selected bacteriophage genes in *E. coli* cells after hydrogen peroxide-provoked prophage induction. Reverse transcription quantitative real time PCR (RT-qPCR) was used to assess abundance of particular transcripts. We have measured expression levels of genes from the *exo-xis* region (*ea8.5, ea22, orf73, orf61, orf60a*) and some key regulatory genes of λ and Φ24$_B$, i.e., *N, cro,* cII, *Q, R.* We found that expression of all tested genes was significantly impaired in Δ*orf63* mutants of both λ and Φ24$_B$ relative to wild-type phages at all tested times after prophage induction (**Figures 7A,B**, respectively). These results confirm that prophage induction is significantly impaired in the absence of *orf63*, and suggest a regulatory role for the *orf63* gene product in the control of expression of phage genes. In the case of phage λ, complementation of the Δ*orf63* mutation by overexpression of wild-type *orf63* from a plasmid was successful, at least at certain times after prophage induction (**Figure 7A**). However, we failed to obtain such a complementation in phage Φ24$_B$ (**Figure 7B**). This might suggest that specific ratio(s) of Orf63 is/are required for accurate regulation of phage development.

DISCUSSION

Considering that the complete 48,502 bp genome of λ was achieved in 1983, it is astounding that this well-investigated model virus still contains uncharacterized open reading frames, many of which lie between the *exo,* and *xis* genes. We began this investigation by demonstrating that *orf63* found within the *exo-xis* region, encodes a *bona fide* protein both in structural, and functional terms. Consistent with its name, this small 63 aa. protein is comprised of only two helices that cover most of the available sequence. Analytical gel filtration of purified Orf63 suggests that the oligomeric state is a trimer. If Orf63 deviates significantly from a globular shape, it is possible that the molecular weight may be overestimated by the gel filtration assay, the oligomeric state could be a dimer. However, given the significant line broadening observed during a series of initial NMR based surveys performed at 25°C that could only be alleviated by performing all of the studies subsequently at 37°C, the NMR data tend to corroborate the gel filtration findings.

Since Orf63 appears to be a functional protein, we tested effects of deletion of *orf63* on development of bacteriophages λ and Φ24$_B$. In the absence of functional Orf63, we observed a significant increase in the efficiency of lysogenization, and considerable lower efficiency of hydrogen peroxide-mediated prophage induction. These results may suggest that Orf63 is involved in the regulation of expression of specific phage genes. Studies with the use of RT-qPCR revealed that expression of vast majority of crucial regulatory genes, as well as genes from the *exo-xis* region, is significantly influenced by the absence of *orf63.* Moreover, perhaps specific ratio of Orf63 to other regulators is required, as it was impossible to obtain complementation with the wild-type *orf63* expressed from a plasmid in Φ24$_B$, though it was successful in λ. The hypothesis about the requirement of specific Orf63 ratio(s) for accurate regulation of phage development is supported by impaired expression of Φ24$_B$ genes during overexpression of *orf63*.

A previous yeast two-hybrid study identified the transcription factor YqhC as a possible protein partner of Orf63 (Blasche et al., 2013). YqhC is interesting because it is a transcription factor that promotes the synthesis of YqhD, the major enzyme responsible for detoxifying compounds produced from glucose under conditions of oxidative stress (Lee et al., 2010). Among the compounds that YqhD acts upon are 2-oxoaldehydes, toxic and highly reactive products of oxidative stress on the bacterium formed from glucose. It has been proposed that oxoaldehydes are one class of compound that is capable of inducing a wider stress response through SoxRS (Benov and Fridovich, 2002). Since neutrophils mount a vigorous oxidative attack during a STEC infection, Orf63 may be beneficial to the bacteriophage by manipulating the microenvironment of the bacterial host. Orf63 by binding the transcriptional activator YqhC, may prevent it from attenuating the stress response created by neutrophil mediated attack on EHEC strains in the gut, and promoting a transition to the lytic phase commensurate with the activation of associated phage genes in that response. A putative mechanism for Orf63-mediated modulation of prophage induction and phage genes' expression is presented in **Figure 8**. Notwithstanding its role in protein-protein interactions, it is still possible that Orf63 itself could function as a transcriptional regulator, although its small size, lack of a known DNA binding domain, and relatively few basic amino acids argue against this possibility. A high-resolution structure of Orf63 alone, or in complex with a possible interactor like YqhC, will resolve these outstanding questions.

In conclusion, Orf63 is a folded protein and has structural properties suggesting its regulatory role. Analyses of mutants of bacteriophages λ and Φ24$_B$ devoid of *orf63* indicated its function in the control of bacteriophage development at the stages of lysis vs. lysogenization decision and prophage induction. Further studies will include determination of biochemical properties of Orf63 and its possible interactions with phage- and/or host-encoded proteins, as well as with phage DNA, to understand molecular mechanisms of its function as a modulator of phage development.

AUTHOR CONTRIBUTIONS

AD and SB contributed equally to this work. AD, SB, BN, GT, TG, and AN took part in the physiological studies on bacteria and phages and data processing. AD, SB, and BN participated also in writing the manuscript and planning of the study. AR, SP, and LD performed the protein analyses. GW, LD, and AW were engaged in writing the manuscript and took part in planning of the study and discussions.

ACKNOWLEDGMENTS

This work was supported by the National Science Center (Poland), grants no. UMO-2013/09/B/NZ2/02366 and UMO-2015/17/B/NZ9/01724 to AW, and by the Faculty of Biology, University of Gdansk, grant no. 538-L140-B539-17 to AD. The Natural Sciences and Engineering Research Council (NSERC) is acknowledged for support to LD (Discovery Operating Grant 238924), and AR/SP (USRA training awards).

REFERENCES

Allison, H. E. (2003). Immunity profiles of wild-type and recombinant Shiga-like toxin-encoding bacteriophages and characterization of novel double lysogens. *Infect. Immun.* 71, 3409–3418. doi: 10.1128/IAI.71.6.3409-3418.2003

Appleyard, R. K. (1954). Segregation of new lysogenic types during growth of a doubly lysogenic strain derived from *Escherichia coli* K12. *Genetics* 39, 440–452.

Arber, W., Enquist, L., Hohn, B., Murray, N. E., and Murray, K. (1983). "Lambda II," in *Experimental Methods for Use with λ*, eds J. W. Roberts, F.W. Stahl, and R. A. Weisberg (New York, NY: Cold Spring Harbor Laboratory Press), 433–466.

Benov, L., and Fridovich, I. (2002). Induction of the soxRS regulon of *Escherichia coli* by glycolaldehyde. *Arch. Biochem. Biophys.* 407, 45–48. doi: 10.1016/S0003-9861(02)00498-8

Berjanskii, M. V., Neal, S., and Wishart, D. S. (2006). PREDITOR: a web server for predicting protein torsion angle restraints. *Nucleic Acids Res.* 34, W63–W69. doi: 10.1093/nar/gkl341

Blasche, S., Wuchty, S., Rajagopala, S. V., and Uetz, P. (2013). The protein interaction network of bacteriophage lambda with its host, *Escherichia coli. J. Virol.* 87, 12745–12755. doi: 10.1128/JVI.02495-13

Bloch, S. K., Felczykowska, A., and Nejman-Falenczyk, B. (2012). *Escherichia coli* O104:H4 out break- have we learnt a lesson from it? *Acta Biochim. Pol.* 59, 483–488.

Bloch, S., Nejman-Falenczyk, J., Łoś, J. M., Baranska, S., Lepek, K., Felczykowska, A., et al. (2013). Genes from the exo-xis region of λ and Shiga toxin-converting bacteriophages influence lysogenization and prophage induction. *Arch. Microbiol.* 195, 693–703. doi: 10.1007/s00203-013-0920-8

Bloch, S., Nejman-Falenczyk, B., Dydecka, A., Łoś, J. M., Felczykowska, A., Wegrzyn, A., et al. (2014). Different expression patterns of genes from the exo-xis region of bacteriophage λ and Shiga toxin-converting bacteriophage Φ24B following infection or prophage induction in *Escherichia coli. PLoS ONE* 9:e108233. doi: 10.1371/journal.pone.0108233

Bloch, S., Nejman-Falenczyk, B., Topka, G., Dydecka, A., Licznerska, K., Narajczyk, M., et al. (2015). UV-Sensitivity of Shiga Toxin-Converting Bacteriophage Virions Φ24B, 933W, P22, P27 and P32. *Toxins* 7, 3727–3739. doi: 10.3390/toxins7093727

Delaglio, F., Grzesiek, S., Vuister, G. W., Zhu, G., Pfeifer, J., and Bax, A. (1995). NMRPipe: a multidimensional spectral processing system based on UNIX pipes. *J. Biomol. NMR* 6, 277–293. doi: 10.1007/BF00197809

Hendrix, R. W., and Duda, R. L. (1992). Bacteriophage lambda PaPa: not a mother of all lambda phages. *Science* 258, 1145–1148. doi: 10.1126/science.1439823

Hyberts, S. G., Milbradt, A. G., Wagner, A. B., Arthanari, H., and Wagner, G. (2012). Application of iterative soft thresholding for fast reconstruction of NMR data non-uniformly sampled with multidimensional Poisson Gap scheduling. *J. Biomol. NMR* 52, 315–327. doi: 10.1007/s10858-012-9611-z

Jensen, K. F. (1993). The *Escherichia coli* K-12 wild types W3110 and MG1655 have an rph frameshift mutation that leads to pyrimidine starvation due to low pyrE expression levels. *J. Bact.* 175, 3401–3407. doi: 10.1128/jb.175.11.3401-3407.1993

Kwan, J. J., Smirnova, E., Khazai, S., Evanics, F., Maxwell, K. L., and Donaldson, L. W. (2013). The solution structures of two prophage homologues of the bacteriophage λ Ea8.5 protein reveal a newly discovered hybrid homeodomain/zinc-finger fold. *Biochemistry* 52, 3612–3614. doi: 10.1021/bi400543w

Lee, C., Kim, I., Lee, J., Lee, K. L., Min, B., and Park, C. (2010). Transcriptional activation of the aldehyde reductase YqhD by YqhC and its implication in glyoxal metabolism of *Escherichia coli* K-12. *J. Bacteriol.* 192, 4205–4214. doi: 10.1128/JB.01127-09

Licznerska, K., Dydecka, A., Bloch, S., Topka, G., Nejman-Falenczyk, B., Wegrzyn, A., et al. (2016a). The Role of the exo-xis region in oxidative stress-mediated induction of Shiga toxin-converting prophages. *Oxid. Med. Cell. Longev.* 2016:8453135. doi: 10.1155/2016/8453135

Licznerska, K., Nejman-Falenczyk, B., Bloch, S., Dydecka, A., Topka, G., Gasior, T., et al. (2016b). Oxidative stress in Shiga toxin production by enterohemorrhagic *Escherichia coli. Oxid. Med. Cell. Longev.* 2016:3578368. doi: 10.1155/2016/3578368

Loś, J. M., Golec, P., Wegrzyn, G., Wegrzyn, A., and Loś, M. (2008a). Simple method for plating *Escherichia coli* bacteriophages forming very small plaques or no plaques under standard conditions. *Appl. Environ. Microbiol.* 74, 5113–5120. doi: 10.1128/AEM.00306-08

Loś, J. M., Loś, M., Wegrzyn, A., and Wegrzyn, G. (2008b). Role of the bacteriophage λ exo-xis region in the virus development. *Folia Microbiol.* 53, 443–450. doi: 10.1007/s12223-008-0068-0

McGuffin, L. J., Bryson, K., and Jones, D. T. (2000). The PSIPRED protein structure prediction server. *Bioinformatics* 16, 404–405. doi: 10.1093/bioinformatics/16.4.404

Mizutani, S., Nakazono, N., and Sugino, Y. (1999). The so-called chromosomal verotoxin genes are actually carried by defective prophages. *DNA Res.* 6, 141–143. doi: 10.1093/dnares/6.2.141

Muniesa, M., Hammerl, J. A., Hertwig, S., Appel, B., and Brüssow, H. (2012). Shiga toxin-producing *Escherichia coli* O104:H4: a new challenge for microbiology. *Appl. Environ. Microbiol.* 78, 4065–4073. doi: 10.1128/AEM.00217-12

Nejman-Falenczyk, B., Bloch, S., Licznerska, K., Dydecka, A., Felczykowska, A., Topka, G., et al. (2015). A small, microRNA-size, ribonucleic acid regulating gene expression and development of Shiga toxin-converting bacteriophage $\Phi 24_B$. *Sci. Rep.* 5:10080. doi: 10.1038/srep1008

Nataro, J. P., and Kaper, J. B., (1998). Diarrheagenic *Escherichia coli. Clin. Microbiol. Rev.* 11, 142–201.

Riley, L. W., Remis, R. S., Helgerson, S. D., McGee, H. B., Wells, J. G., and Davis, B. R. (1983). Hemorrhagic colitis associated with a rare *Escherichia coli* serotype. *N. Engl. J. Med.* 308, 681–685. doi: 10.1056/NEJM198303243081203

Sambrook, J., and Russell, D. W. (2001). *Molecular Cloning: A Laboratory Manual, 3rd Edn.* New York, NY: Cold Spring Harbor Laboratory Press.

Schmidt, H. (2001). Shiga-toxin-converting bacteriophages. *Res. Microbiol.* 152, 687–695. doi: 10.1016/S0923-2508(01)01249-9

Skinner, S. P., Goult, B. T., Fogh, R. H., Boucher, W., Stevens, T. J., Laue, E. D., et al. (2015). Structure calculation, refinement and validation using CcpNmr Analysis. *Acta Crystallogr. D Biol. Crystallogr.* 71, 154–161. doi: 10.1107/S1399004714026662

Strauch, E., Hammerl, J. A., Konietzny, A., Schneiker-Bekel, S., Arnold, W., et al. (2008). Bacteriophage 2851 is a prototype phage for dissemination of the Shiga toxin variant gene 2c in *Escherichia coli* O157:H7. *Infect. Immun.* 76, 5466–5477. doi: 10.1128/IAI.00875-08

Wegrzyn, G., Glass, R. E., and Thomas, M. S. (1992). Involvement of the *Escherichia coli* RNA polymerase α subunit in transcriptional activation by the bacteriophage λ CI and CII proteins. *Gene* 122, 1–7. doi: 10.1016/0378-1119(92)90025-K

Wegrzyn, G., Wegrzyn, A., Konieczny, I., Bielawski, K., Konopa, G., Obuchowski, M., et al. (1995). Involvement of the host initiator function dnaA in the replication of coliphage lambda. *Genetics* 139, 1469–1481.

Wegrzyn, G., Licznerska, K., and Wegrzyn, A. (2012). Phage λ–new insights into regulatory circuits. *Adv. Virus. Res.* 82, 155–178. doi: 10.1016/B978-0-12-394621-8.00016-9

5

Fic Proteins of *Campylobacter fetus* subsp. *venerealis* Form a Network of Functional Toxin–Antitoxin Systems

Hanna Sprenger[1,2,3†], Sabine Kienesberger[1,2,4†], Brigitte Pertschy[1], Lisa Pöltl[1], Bettina Konrad[1], Priya Bhutada[1], Dina Vorkapic[1], Denise Atzmüller[1], Florian Feist[5], Christoph Högenauer[3], Gregor Gorkiewicz[2,4] and Ellen L. Zechner[1,4]*

[1] Institute of Molecular Biosciences, University of Graz, Graz, Austria, [2] Institute of Pathology, Medical University of Graz, Graz, Austria, [3] Division of Gastroenterology and Hepatology, Medical University of Graz, Graz, Austria, [4] BioTechMed-Graz, Graz, Austria, [5] Vehicle Safety Institute, Graz University of Technology, Graz, Austria

Edited by:
*Manuel Espinosa,
Centro de Investigaciones Biológicas (CSIC), Spain*

Reviewed by:
*Damian Lobato-Marquez,
Imperial College London,
United Kingdom
Ramon Diaz Orejas,
Consejo Superior de Investigaciones Científicas (CSIC), Spain*

***Correspondence:**
*Ellen L. Zechner
ellen.zechner@uni-graz.at*

† These authors have contributed equally to this work.

Enzymes containing the FIC (filamentation induced by cyclic AMP) domain catalyze post-translational modifications of target proteins. In bacteria the activity of some Fic proteins resembles classical toxin–antitoxin (TA) systems. An excess of toxin over neutralizing antitoxin can enable bacteria to survive some stress conditions by slowing metabolic processes and promoting dormancy. The cell can return to normal growth when sufficient antitoxin is present to block toxin activity. *Fic* genes of the human and animal pathogen *Campylobacter fetus* are significantly associated with just one subspecies, which is specifically adapted to the urogenital tract. Here, we demonstrate that the *fic* genes of virulent isolate *C. fetus* subsp. *venerealis* 84-112 form multiple TA systems. Expression of the toxins in *Escherichia coli* caused filamentation and growth inhibition phenotypes reversible by concomitant antitoxin expression. Key active site residues involved in adenylylation by Fic proteins are conserved in Fic1, Fic3 and Fic4, but degenerated in Fic2. We show that both Fic3 and the non-canonical Fic2 disrupt assembly and function of *E. coli* ribosomes when expressed independently of a trans-acting antitoxin. Toxicity of the Fic proteins is controlled by different mechanisms. The first involves intramolecular regulation by an inhibitory helix typical for Fic proteins. The second is an unusual neutralization by heterologous Fic–Fic protein interactions. Moreover, a small interacting antitoxin called Fic inhibitory protein 3, which appears unrelated to known Fic antitoxins, has the novel capacity to bind and neutralize Fic toxins encoded in *cis* and at distant sites. These findings reveal a remarkable system of functional crosstalk occurring between Fic proteins expressed from chromosomal and extrachromosomal modules. Conservation of *fic* genes in other bacteria that either inhabit or establish pathology in the urogenital tract of humans and animals underscores the significance of these factors for niche-specific adaptation and virulence.

Keywords: post-translational modification, adenylylation, toxin–antitoxin module, bacterial effector protein, bacterial evolution, niche adaptation, urogenital tract, virulence

INTRODUCTION

The genus *Campylobacter* comprises ecologically diverse species that colonize humans and animals. *Campylobacter jejuni* is known as the leading cause of human bacterial diarrhea worldwide. Other *Campylobacter* species, including *Campylobacter fetus,* are increasingly recognized as important human and animal pathogens (Lastovica and Allos, 2008; Man, 2011; Bullman et al., 2013). *C. fetus* is intriguing because although the two subspecies associated with mammals, *C. fetus* subsp. *fetus* and *C. fetus* subsp. *venerealis*, are highly related at the genome level, they exhibit quite different niche adaptations. *C. fetus* subsp. *fetus* has a broad host range (Skirrow and Benjamin, 1980; Harvey and Greenwood, 1985; Logue et al., 2003). In humans it causes gastrointestinal disease and belongs to the *Campylobacter* spp. most frequently associated with bacteremia (Lastovica and Allos, 2008; Man, 2011). By contrast, *C. fetus* subsp. *venerealis* is a host-restricted veterinary pathogen adapted to the urogenital tract of cattle (Blaser et al., 2008). Current understanding of the pathogenesis of emerging *Campylobacter* spp. is quite limited.

Recent comparative genomics of *C. fetus* subspecies revealed genetic determinants potentially contributing to this species' niche preferences and pathogenicity (Kienesberger et al., 2014; Graaf-van Bloois et al., 2016). Strikingly, *C. fetus* genomes encode multiple bacterial type IV secretion systems (T4SS), which generally contribute to pathogenicity by transferring specific protein and DNA substrates to recipient cells (Christie et al., 2014). One *C. fetus* T4SS has been evaluated experimentally and linked to virulence (Gorkiewicz et al., 2010). The conserved T4SS-encoding regions of *C. fetus* genomes fit into three phylogenetically different groups: one located exclusively on the chromosome, one observed exclusively on plasmids and a third located on both (Graaf-van Bloois et al., 2016). These authors further showed that both genes encoding T4SS components and genes encoding FIC domain proteins are significantly associated with the *C. fetus* subsp. *venerealis*. In the current study we focus on the function of the *fic* genes.

The Fido domain superfamily is composed of members of the FIC (filamentation induced by cAMP) and the Doc (death on curing) protein families and is common in all domains of life (Kinch et al., 2009). Proteins of the combined family contain a conserved motif [HPFx(D/E)GN(G/K)R]. Work in recent years has revealed that enzymes of the family catalyze post-translational modifications of proteins by addition of AMP, other nucleoside monophosphates, phosphocholine, or phosphate to a functionally critical amino acid (as reviewed in Cruz and Woychik, 2014; Garcia-Pino et al., 2014; Roy and Cherfils, 2015; Harms et al., 2016b). Since activities of the target proteins are typically altered as a result, Fido proteins are recognized as important regulators of metabolic functions.

Phylogenetic analysis of the superfamily places the paradigm Doc toxin of bacteriophage P1 in subfamily I (Garcia-Pino et al., 2014). Doc toxin is structurally similar to Fic proteins (Garcia-Pino et al., 2008); however, variation in the catalytic motif (K in place of the second G) confers kinase activity in place of NMP transfer activity (Castro-Roa et al., 2013; Cruz et al., 2014). Phosphorylation of translation elongation factor Tu by Doc leads to rapid translation arrest in *Escherichia coli* (Garcia-Pino et al., 2008; Liu et al., 2008; Castro-Roa et al., 2013).

Interest in the FIC protein subfamily has been fueled by the observation that bacterial pathogens secrete Fic enzymes to modify host proteins (Pan et al., 2008; Worby et al., 2009; Yarbrough et al., 2009; Mukherjee et al., 2011). Cell to cell transfer can be direct via type III or type IV secretion (Roy and Cherfils, 2015). In the host, Fic effector proteins contribute to bacterial pathogenicity by modifying proteins important to signaling (Roy and Mukherjee, 2009; Woolery et al., 2010). Fic effectors contain the canonical catalytic motif [HxFx(D/E)GNGRxxR] and initial studies showed that a typical reaction inactivates host GTPases by nucleotidyl transfer to a hydroxyl group of the protein side chain (Garcia-Pino et al., 2014; Roy and Cherfils, 2015). Several secreted FIC proteins transfer AMP in a reaction called adenylylation (Worby et al., 2009; Yarbrough et al., 2009; Zekarias et al., 2010; Palanivelu et al., 2011), but variation within the canonical core motif can alter enzyme activity (Mukherjee et al., 2011; Engel et al., 2012).

The targets of FIC enzyme modification are not restricted to proteins expressed by the host. However, their functions and regulation in producing bacteria are still poorly understood. *E. coli* has been used as a surrogate producer to gain insights into the activities of FIC proteins in bacteria. One function that has emerged from these studies is that Fic proteins act as toxin–antitoxin (TA) modules (Harms et al., 2016b). Bacterial TA systems play a major role in cellular adaptation to stress and persistence (Hayes and Van Melderen, 2011; Goeders and Van Melderen, 2014; Harms et al., 2016a). Activation of the toxin can cause slow cell growth or arrest the cell cycle allowing bacteria to enter a dormant state. Mechanistic understanding of TA activity has been developed with prototypic modules such as *phd-doc* of bacteriophage P1 (Lehnherr et al., 1993; Castro-Roa et al., 2013). Generally the toxin component is directed against the producing cell and interferes with bacterial physiology. Cellular processes inhibited by type II TA toxins include protein synthesis, cell wall synthesis, assembly of cytoskeletal structures and DNA topoisomerase action (Hayes and Van Melderen, 2011; Yamaguchi and Inouye, 2011; Goeders and Van Melderen, 2014; Harms et al., 2016b). The antitoxin component reversibly inactivates the toxin and/or regulates its expression. Unlike the toxin, the antitoxin is biochemically unstable so that, unless the antitoxin is continuously expressed, the free toxin forces the bacterial cell into a reversible dormant state or even kills the cell (Leplae et al., 2011; Goeders and Van Melderen, 2014). The TA system of bacteriophage P1 helps to maintain the lysogen through post-segregational killing of cells that are cured of the prophage (Lehnherr et al., 1993). Homologous *phd-doc* modules are also present on bacterial chromosomes and evidence thus far suggests a role for these systems in the formation of persister cells under stress conditions (Maisonneuve and Gerdes, 2014). Moreover, evidence is emerging that TA modules help bacteria overcome stress imposed by host colonization, early stages of infection and survival within host cells (Norton and Mulvey, 2012; Ren et al., 2012, 2014; De la Cruz et al., 2013; Lobato-Marquez et al., 2015). They can stabilize mobile genetic elements encoding virulence factors and contribute directly to virulence (see Lobato-Marquez et al., 2016 for a comprehensive review).

Campylobacter genomes generally lack homologs of prototypical TA systems (Shao et al., 2011). To date, only two TA systems (both located on a plasmid in *C. jejuni*) have been described (Shen et al., 2016). Given the general importance of TA systems in bacteria we asked whether the multiple *fic* genes in *C. fetus* fulfill this important role. Here, we show that the Fic proteins of *C. fetus* subsp. *venerealis* 84-112 indeed form TA systems with the capacity to disrupt the bacterial translational machinery. We further show that *fic* modules located on the chromosome and extrachromosomal DNA functionally interact. *Fic* homologs are genetically conserved in *C. fetus* subsp. *venerealis* isolates and in other human and animal urogenital pathogens, underscoring the significance of these factors for niche-specific adaptation.

MATERIALS AND METHODS

Bacteria

Strains used in this study are listed in Supplementary Table S1. *E. coli* and *Campylobacter* strains were grown as previously described (Kienesberger et al., 2007). Antibiotics were added to final concentrations of 100 $\mu g\ ml^{-1}$ ampicillin, 75 $\mu g\ ml^{-1}$ nalidixic acid, and either 12.5 or 25 $\mu g\ ml^{-1}$ chloramphenicol, 40 or 25 $\mu g\ ml^{-1}$ kanamycin for *E. coli* or *Campylobacter* cultivation, respectively.

Construction of Plasmids

Plasmids and oligonucleotides used in this study are listed in Supplementary Tables S1, S2. For expression in *E. coli*, genes of interest were amplified with PCR and the fragments were ligated to pBAD24 vector derivatives with distinct antibiotic resistance genes (see Supplementary Tables S1, S2).

Structure Predictions

For 3D structure prediction amino acid sequences (CDF65254.1, CDF65253.1, CDF65920.1, and CDF65967.1) were analyzed using the Phyre2 web portal (Rollins and Colwell, 1986). The output files were rendered with PyMOL (Schrodinger, 2010). Templates for Fic protein fold recognition via Phyre2 are listed in Supplementary Table S3.

E. coli Growth/Rescue Assays

Escherichia coli DH5α harboring pBAD plasmids with *fic* genes were grown with shaking overnight at 37°C in LB-broth supplemented with appropriate antibiotics and 0.2% glucose to repress expression, or 0.05% arabinose to induce expression via the P_{BAD} promoter. Bacterial growth was either monitored by survival plating of bacteria grown in 100 ml LB-broth or in 24-well plates with a culture volume of 1 ml per well and starting optical density measured at 600 nm (OD_{600}) of 0.05. Plates were incubated at 37°C under shaking at 180 rpm. OD_{600} was measured hourly in triplicate. Determination of colony forming units (CFUs) normally corresponded with OD_{600} measurements except at late time points where the CFU count of filamentous cells remained low.

Microscopy

Cultures of *E. coli* DH5alpha harboring pBAD derivatives with a starting OD_{600} of 0.05 were grown in LB-broth with 0.05% arabinose for 2 h. Cells were harvested, suspended in 1x phosphate buffered saline (PBS), pH 7.4 and incubated with Nile red for up to 60 min at room temperature in the dark. For immediate microscopy, 1 μl of the pellet was applied to an agar slide (1% agar solution poured on microscopy slide) to immobilize the cells. For later imaging, cells were fixed with 0.4% formaldehyde, before collected by centrifugation, resuspended in 1x PBS, pH 7.4 and stained as described before. Confocal microscopy was performed on a LEICA AOBS SP2 MP microscope (380 nm extinction, 510 nm emission).

Co-immunoprecipitation and Western Analysis

Pairs of FLAG-tagged and hemagglutinin (HA)-tagged proteins were co-expressed in *E. coli* C41(DE3) from respective plasmids (Supplementary Table S1). 100 ml LB broth supplemented with 0.2% glucose was inoculated with overnight cultures to OD_{600} of 0.1. When cultures reached $OD_{600} = 0.5–0.8$ protein expression was induced with 0.05% arabinose for 2 h. Thirty OD of cells were pelleted and washed with 50 ml buffer A (50 mM Tris-HCl pH 6.8, 100 mM NaCl). Cell lysis was performed as previously described (Gruber et al., 2016) except that the formaldehyde crosslinking step was omitted for this study. All further steps were as in Gruber et al. (2016). For protein detection, OD_{600} 0.015–0.05 equivalents of lysate and pull-down fractions were mixed with sample buffer containing DTT (0.09%) and SDS (0.1%), heated at 95°C for 10 min and resolved on SDS-PAGE (12.5%, Hoefer) or NUPAGE (12%, MES buffer, Invitrogen) gels. Proteins were transferred for 1.5 h onto PVDF membranes. Blocking was done overnight at 4°C in TST: (0.5 M Tris-HCl pH 7.5, 1.5 M NaCl, 1% Tween-20) supplemented with 3% milk powder. FLAG-tagged proteins were detected with HRP-conjugated α-FLAG antibody (A8592, Sigma) and HA-tagged proteins with HRP-conjugated α-HA antibody (12013819001, Roche). After washing (3 min × 10 min) with 1x TST blots were developed with ECL (Bio-Rad) according to the manufacturer's instructions (1:5 dilution in ddH_2O of substrate for FLAG detection).

Ribosome Profiles

Escherichia coli DH5α with plasmids were grown in 100 ml LB-broth to an OD_{600} of 0.4 to 0.6, then shifted to medium containing 0.05% arabinose for 1 h to induce expression. 30 s prior to cell harvest, chloramphenicol was added to a final concentration of 200 or 300 $\mu g\ ml^{-1}$. Cells were harvested by centrifugation for 10 min at 14,300 × g and 4°C. The cell pellet was resuspended in 500 μl cell lysis buffer (10 mM Tris-HCl pH 7.5, 10 mM $MgCl_2$, 30 mM NH_4Cl, 100 or 150 $\mu g\ ml^{-1}$ chloramphenicol) and either mixed with an equal volume of glass beads (300 μm in diameter) and vortexed for 5 min at 4°C or immediately frozen in liquid nitrogen (Bronowski et al., 2014). Suspensions with glass beads were centrifuged for 10 min at 6,400 × *g* at 4°C. The supernatant was collected, centrifuged for 3 min at 17,649 × *g* at 4°C and immediately applied to sucrose density centrifugation or stored at −70°C. Frozen suspensions were thawed in an ice bath, frozen again in liquid nitrogen and stored at −70°C.

The protocol for the sucrose density centrifugation was adapted from Jiang et al. (2007). An A_{260} of 8 of the cleared cell lysate was loaded onto a gradient of 5–45% sucrose in buffer (10 mM Tris-HCl pH 7.5, 10 mM $MgCl_2$, 100 mM NH_4Cl). Ultracentrifugation was carried out for 4 h at 4°C and 253,483 × *g* in a Beckman SW-41Ti rotor. Fractions of the gradient were collected using an UA-6 system (Teledyne ISCO) with continuous monitoring at A_{254}.

Numerical and Statistical Analysis of Ribosome Profiles

Ribosome profiles were scanned, traced in CorelDraw to increase contrast and xy-coordinates were extracted using DataThief III (Pascoe et al., 2015). To increase the reliability of calculations, in addition to peak values we calculated the Xgrad-values using a code (XSpan) written in VisualBasic. The code is available to interested readers upon request. The program 'XSpan' places the largest rectangular surface with a predefined width (= X) under individual peaks of a given curve and then calculates the Xgrad-values, which are height (H) and area (A) of the surfaces. XSpan can also extrapolate clipped curves (e.g., when a maximum value exceeds the measurement range) by fitting a cubic function such that it tangents the two flanks of a clipped peak. This option was utilized in this study to obtain the 70S heights (H). The Xgrad-value H is the highest amplitude (absorbance) measured for the surface of predefined width describing fractions of the analyzed sedimentation gradient. X, the width of the rectangular surfaces, was selected such that it covers approximately 1.2% of the gradient fractions analyzed. Statistical analysis was performed using the H values only. For each profile, the H values of the 30S and 50S subunit peaks were normalized to the H values of the first polysome peak and statistical significance was calculated using the paired Student's *t*-test. Statistical significance was assumed with *p*-values below 0.05.

PCR Screening, DNA Sequencing, and Sequence Alignment of *fic* Genes

Prevalence of TA genes was surveyed among *C. fetus* isolates via PCR using chromosomal DNA as template. We applied primer pairs 1/2 for *fic1*, 26/27 for *fic2*, 5/6 for *fic3*, 7/8 for *fic4*, and 9/10 for *fti3* (Supplementary Table S2). Sequencing of *fic2* amplicons from *C. fetus* subsp. *venerealis* strains V9, V20, V32, V60, V62, and V69 was performed with primers 28/29.

Phylogenetic Analysis of *Campylobacter* spp. Fic Proteins

The conserved Fido motif sequence HPFXXGNXR and full length Fic1-4 of *C. fetus* subsp. *venerealis* 84-112 were used in BlastP analysis to identify Fido proteins in whole genomes of *Campylobacter* species (if possible, finished whole genomes, if not available, genomes with low scaffold numbers were used). BlastP analysis was also performed with full-length Fics and the Fic2 specific motif sequence [HPFREGNTRTIA] under exclusion of epsilon-proteobacteria to screen for hits outside this class. Selected *Campylobacter* proteins, selected Fic reference proteins, as well as other bacteria from the urogenital tract identified by the BlastP were then used to generate the phylogeny tree. Retrieved proteins were aligned with MEGA6.06 using the BLOSUM matrix. The Neighbor joining tree was constructed with MEGA6.06 (Tamura et al., 2013). The tree was rooted to the translated ORF of housekeeping gene *glnA* of *C. fetus* subsp. *venerealis* 84-112. Protein accession numbers are listed in Supplementary Table S4.

RESULTS

FIC Domain Proteins of *C. fetus* subsp. *venerealis* 84-112

fic1 and *fic2* genes (**Figure 1A**) are chromosomally encoded and form part of a pathogenicity island (PAI) that harbors additionally a functional T4SS (Gorkiewicz et al., 2010). *C. fetus* subsp. *venerealis* 84-112 also carries extra-chromosomal DNA with features of an integrative conjugative element (ICE_84-112) (Kienesberger et al., 2014). Two additional *fic* gene homologs, *fic3* and *fic4*, were identified on the ICE (**Figure 1B**). Residues of the Fido superfamily core motif that enable FIC-containing enzymes to act as AMP transferases have been defined as HxFx(D/E)GNGRxxR (Kinch et al., 2009; Worby et al., 2009; Xiao et al., 2010; Engel et al., 2012). Fic1, Fic3, and Fic4 contain the complete signature of invariant residues [protein accession numbers CDF65254.1 (Fic1); CDF65920.1 (Fic3); CDF65967.1 (Fic4)]. In contrast, in Fic2 (CDF65253.1), the second conserved glycine at position 191 is replaced with threonine and the final arginine of the signature motif (R195A) is absent, suggesting that Fic2 does not have adenylylation activity. Fic1 and Fic4 also contain a conserved inhibitory motif (S/T)xxxE(G/N), which was shown to suppress adenylylation in well-studied systems (Harms et al., 2016b) (**Figure 1C**). Fic proteins containing this inhibitory helix (inh) are classified depending on whether the inh is part of the FIC fold as an N-terminal helix (class II) or a C-terminal helix (class III) (Engel et al., 2012). Fic1 thus belongs to class II, and Fic4 to class III. Class I Fic proteins do not contain the inhibitory motif themselves, but have an interaction partner that provides the inh *in trans*. Fic2 and Fic3 lack a motif with this overall consensus, thus they belong to class I. Fic1 may act as antitoxin for the degenerated toxin Fic2. Alternatively, mutations of the core motif in Fic2 may have altered enzyme activity and thus have bypassed the need for an inh motif. We also note that the 78 amino acid ORF (protein accession number CDF65919.1) upstream and partially overlapping *fic3* includes residues GHAIEN, which might provide the invariable glutamate (Engel et al., 2012; Goepfert et al., 2013) of a poorly conserved inhibitory motif (**Figure 1B**). Presence of a 48% identical homolog, *fti4* (protein accession number CDF65966.1), upstream and partially overlapping *fic4* strengthens the hypothesis that

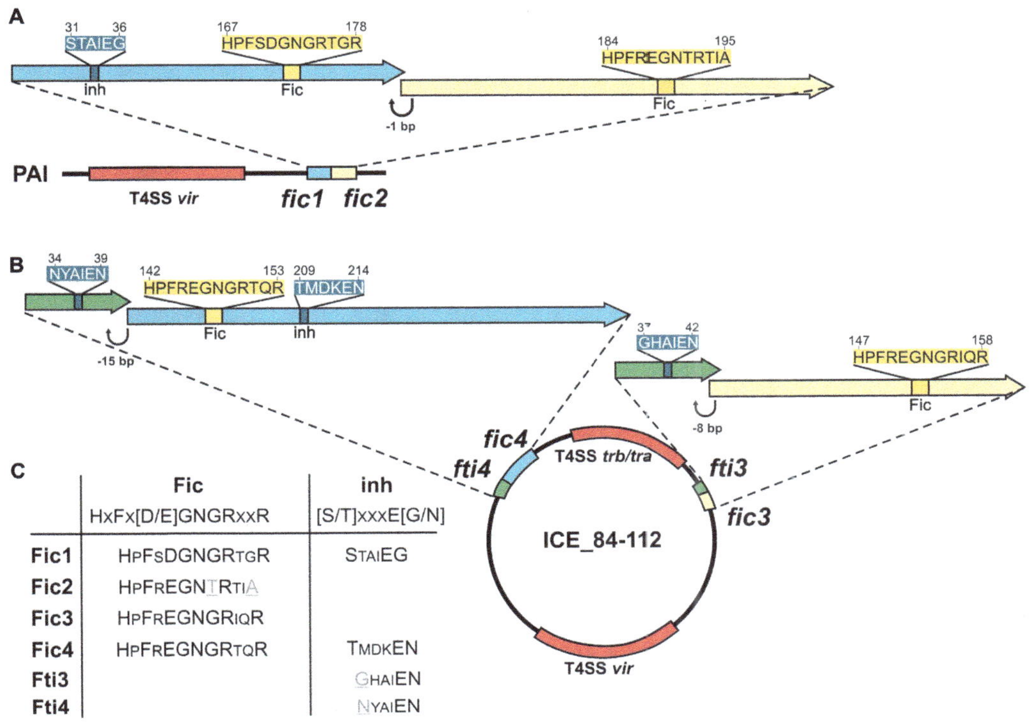

FIGURE 1 | Genetic organization of *Campylobacter fetus* subsp. *venerealis* 84-112 encoded *fic* genes. **(A)** Tandem *fic1* and *fic2* genes on the chromosomal pathogenicity island (PAI) overlap by 1 bp. **(B)** The two loci carried by the extrachromosomal element ICE 84-112 include *fti3* and *fic3,* which overlap by 8 bp, and *fti4* and *fic4* which overlap by 15 bp. *fic* gene clusters are located near loci encoding type IV secretion systems (T4SS), shown in red. Genes encoding solely a Fic motif are drawn as yellow arrows. Genes encoding an additional inhibitory motif (inh) are blue. Genes with only the inhibitory motifs are shown in green. **(C)** Conserved residues of each Fic protein are compared with signature Fic and inh motifs. Non-canonical amino acids in the Fic2 motif and Fic4 inh are indicated in gray and underlined.

each ORF encodes a small interacting protein to control the cognate FIC enzyme.

Protein Structure Prediction

Alignment of the predicted proteins shows strong conservation of the Fic core motif but low general similarity (not shown). A structure prediction was performed with the Phyre2 server using templates listed in Supplementary Table S4. All *C. fetus* subsp. *venerealis* homologs are predicted to share a similar FIC domain fold (**Figures 2A–D**). The set of common α-helices are colored from blue (N-terminal) to red (C-terminal), according to the core FIC domain secondary structure topology (Kinch et al., 2009). The predicted active site loops with the conserved core motif including the catalytic histidine are highlighted in black. Fic proteins typically carry a β hairpin close to the active site (gray). This structure, also called "the flap," constitutes the major target-protein docking site (Kinch et al., 2009; Xiao et al., 2010; Palanivelu et al., 2011; Garcia-Pino et al., 2014). The inhibitory motifs of Fic1 and Fic4, expected to prevent the adenylylation reaction by active site obstruction, are shown in pink. The remaining protein structure outside of each FIC core domain is shown in white. One additional shared feature we noted is the conserved KEKE motif (asterisks) at the C-termini of Fic1 and Fic2 that is reiterated in Fic4, once internally, and again at the C-terminus. Bacterial effector proteins secreted via a given T4SS typically display a short C-terminal stretch of conserved residues that mediates their specific recognition by the transfer machinery (Zechner et al., 2012; Christie et al., 2014). The conserved KEKE motif may represent such a dedicated translocation signal, but this has not been validated experimentally.

Fic1 and Fic2 Form a Functional TA Module

To gain insights to the function of the *C. fetus* proteins we expressed these in a heterologous bacterial host and asked whether the *fic1-fic2* module acts as a TA system. In that case, the inhibitory domain of Fic1 would be required to act both intra- and intermolecularly to regulate the enzymatic activity of Fic1 and Fic2. The *fic* genes of *C. fetus* were placed under transcriptional control of the P_{BAD} promoter and their effects on growth of *E. coli* were investigated. Shifting *E. coli* cells from LB broth with glucose to medium containing arabinose induced synthesis of the Fic proteins and culture density was monitored over time. Induction of *fic2* expression delayed growth of *E. coli* severely compared to the vector control, demonstrating that Fic2 is toxic despite its degenerate core motif (**Figure 3A**). Exchange of the catalytic histidine in variant Fic2_H184A eliminated

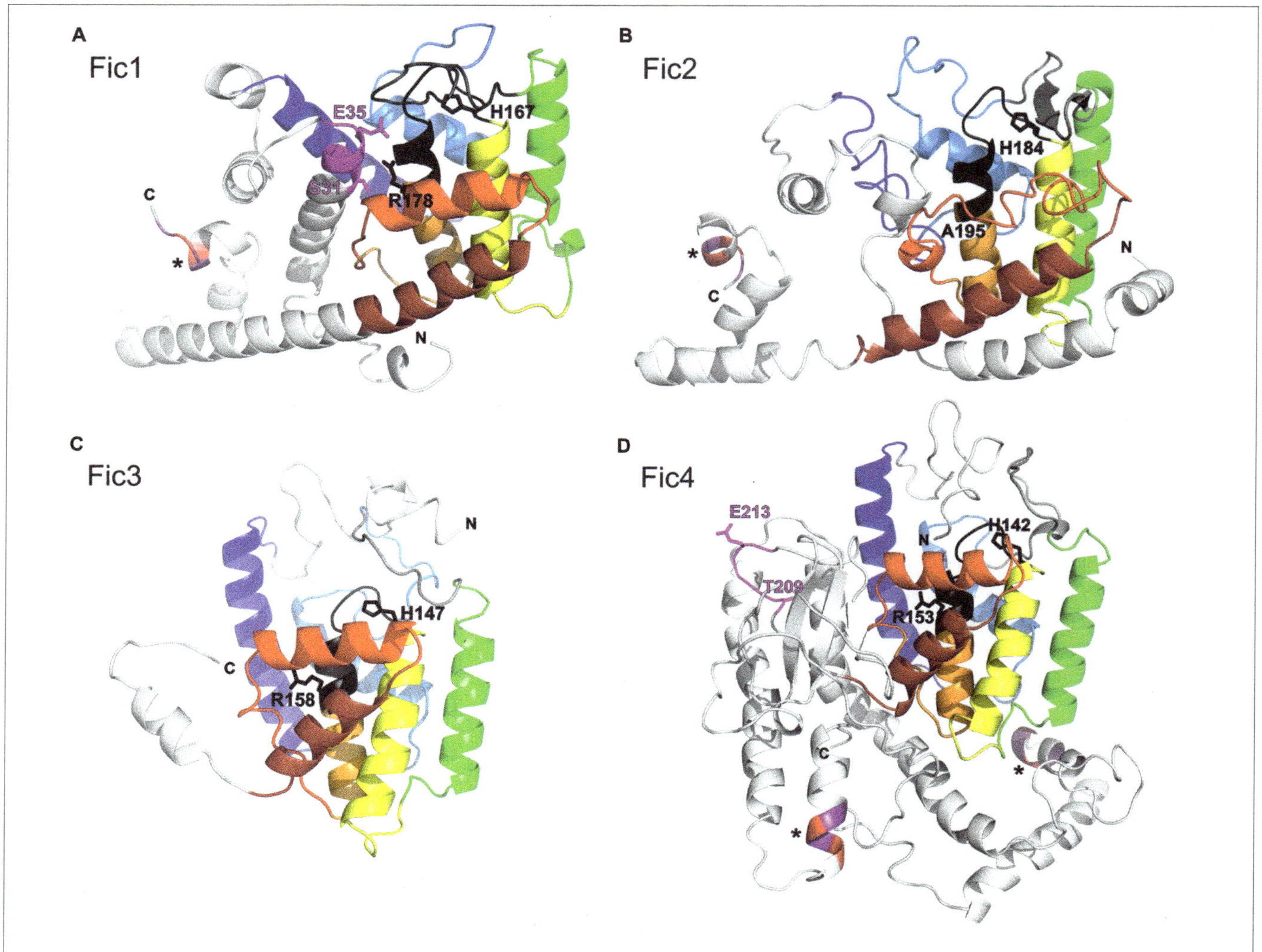

FIGURE 2 | Structural models of *C. fetus* subsp. *venerealis* Fic proteins. The structure of each Fic protein **(A–D)** was predicted with Phyre2. The rainbow colors from blue (N-termini; N) to red (C-termini; C) illustrate the core FIC domain secondary structure topology, as defined in Kinch et al. (2009). Black designates each FIC core motif. Pink specifies the inhibitory motifs (Fic1 and Fic4). Key residues within these motifs are numbered. Protein extensions outside the Fic fold are shown in white. Asterisk designates putative C-terminal translocation signals containing conserved KEKE residues.

toxicity and allowed the host to grow comparably to the vector control strain. Loss of phenotype could occur either because the histidine is indeed important to the activity of the enzyme, as predicted, or because the mutant variant is unstable. To exclude the latter possibility we purified the mutated protein and verified its stability during overexpression in *E. coli* and in isolated form (not shown). *E. coli* expressing *fic1* displayed logarithmic growth, but culture densities obtained after 8 h were lower than cells carrying the empty vector. To examine the role of the conserved inh of Fic1, key residues Ser31 and Glu35 were exchanged for alanine. The substitution apparently disrupted the protective function of this motif, since expression of Fic1_S31A/E35A was incompatible with cell growth. Cells were rescued from Fic2-induced growth arrest by co-expression of wild type Fic1, suggesting that Fic1 can act as an antitoxin for Fic2. The importance of the inh module in toxin neutralization was again shown when co-expression of Fic1_S31A/E35A and Fic2 arrested growth fully. The data imply that Fic1 catalyzes an activity detrimental to bacterial growth, but which is normally blocked intramolecularly by the protein's inh helix. Moreover, the bacterial cytotoxicity of Fic2 depends on the enzyme core motif and is neutralized by antitoxin Fic1.

To characterize the proposed toxin–antitoxin activities, we next compared the impact of Fic protein production on cellular morphology. Fic2 alone caused an extreme filamentous phenotype (**Figure 3B**) via a mechanism requiring the catalytic histidine since; by comparison, cells expressing Fic2H184A were similar to wild type. Cells expressing Fic1 appeared normal but formed filaments when the Fic1 inh motif was mutated. Coexpression of wild type Fic1 reversed the filamentous phenotype caused by Fic2 consistent with the neutralization observed during growth (**Figure 3A**).

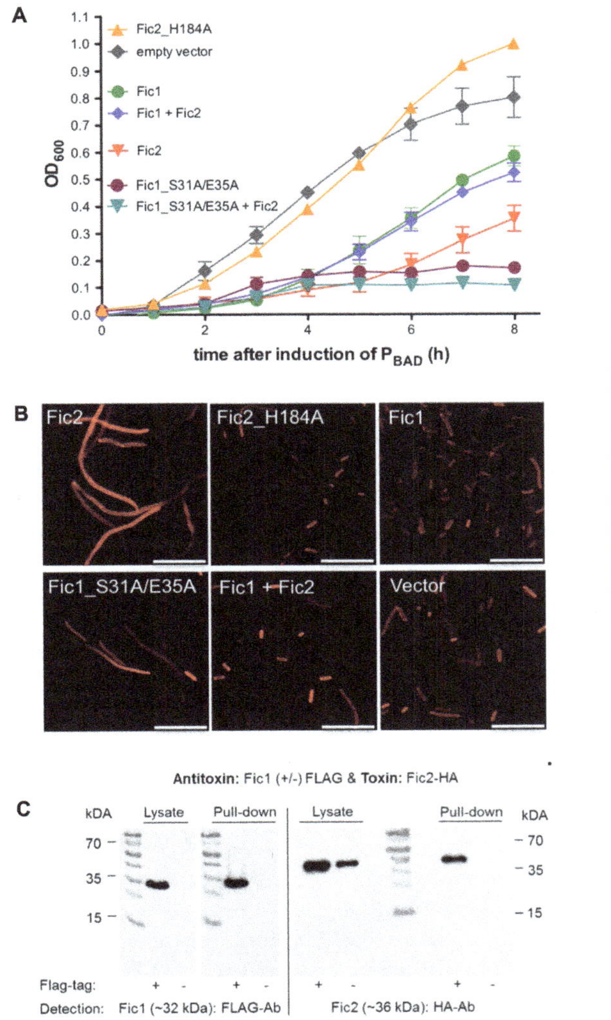

FIGURE 3 | Toxicity of Fic2 in *Escherichia coli* is neutralized by coexpression of Fic1. **(A)** Comparison of growth profiles of *E. coli* expressing *fic1* or *fic2* wild type alleles, mutant alleles, respective combinations, or the empty vector control. Results are mean values of OD_{600} ± standard deviation of three independent experiments. **(B)** Confocal microscopy of Nile red stained *E. coli* expressing indicated *fic* proteins or derivatives alone or in combination. Vector control (lower right panel), scale bars (25 μm). **(C)** Co-immunoprecipitation of Fic1 (32 kDa) and Fic2 (36 kDa). Either FLAG-tagged (+) or native (-) Fic1 were expressed in *E. coli* together with HA-tagged Fic2 and incubated with FLAG-affinity beads. Cell lysates and elution fractions (pull-down) were loaded on gels and proteins were detected by Western analysis using anti-FLAG- or anti-HA-antibodies (Ab).

Antitoxins similar to the PhD-Doc paradigm frequently inactivate the toxin by forming a stable complex. We asked whether inactivation of Fic2 toxicity by Fic1 might involve binding of the two proteins. Codons for a FLAG epitope were added to *fic1* and the hemagglutinin (HA) tag was added to *fic2*. Lysates of *E. coli* cells expressing both fusion proteins were incubated with FLAG-affinity beads. After elution of bound proteins, lysates and eluates were analyzed by western immunoblotting (**Figure 3C**). Anti-FLAG antibodies confirmed the presence of FLAG-tagged Fic1 in cell lysates and the absence of signal in control samples expressing native Fic1. Antibody to HA detected Fic2-HA fusion protein in the same cell lysates. HA signal in the pull down fraction indicated retention of Fic2 by Fic1. The specificity of this interaction was confirmed by the absence of signal when partner protein Fic1 lacked the FLAG epitope. These properties indicate that Fic1 and Fic2 of *C. fetus* subsp. *venerealis* 84-112 form a functional toxin–antitoxin system. Given that the enzymatic activity of antitoxin Fic1 is autoregulated via inh, this protein exhibits a mode of concomitant intra- and intermolecular- toxin neutralization novel for bacterial Fic proteins.

Fti3 Acts as an Antitoxin for Fic3

Fic3, like Fic2, carries the enzyme core motif but lacks an inh motif (**Figure 1**). Similar to the result of *fic2* expression, *E. coli* carrying *fic3* failed to grow under inducing conditions (**Figure 4A**). However, dual expression of *fic3* and its neighboring gene encoding the putative inhibitor protein restored *E. coli* growth completely. *E. coli* cells expressing the inhibitor alone grew indistinguishably from cells carrying the vector control. Since, this protein acts as an antitoxin for Fic3 we named the gene *fti3* (Fic toxin inhibitor 3). Microscopy of the toxin/antitoxin expressing *E. coli* revealed that inhibitor protein alone had no impact on cell morphology (**Figure 4B**). By contrast we observed extreme filamentation due to Fic3 that could be reversed by either mutation of the catalytic histidine in variant Fic3H147A, or co-expression of wild type Fic3 with Fti3. To assess the viability of cells expressing Fic3, samples of a culture before and after 4 or 8 h of arabinose-induced expression were plated on LB agar without arabinose. Viability of the culture dropped by several orders of magnitude after Fic3 induction (**Figure 4C**). By contrast cells producing the non-toxic variant Fic3H147A exhibited similar viability as the vector control strain. Direct interaction between Fti3 and toxin Fic3 was tested following coexpression of FLAG-tagged Fti3 and HA-tagged Fic3. FLAG-tagged Fti3 from the cell lysate bound the affinity matrix and specifically retained Fic3-HA in the pull down reaction (**Figure 4D**). No retention of Fic3 was detected when Fti3 lacked the FLAG epitope. We conclude that Fti3-Fic3 form another TA module on the extrachromosomal ICE in addition to the chromosomal system *fic1-fic2*.

Putative Antitoxin Fti4 Interacts with Fic4

Expression of *fic4* for up to 8 h had no effect on cell growth (**Figures 5A,C**) and affected cell morphology only mildly (**Figure 5B**). To test whether the protein's inhibitory motif was suppressing the predicted enzyme activity, residues Thr209 and Glu213 were exchanged for alanine. Expression of the mutant variant was compatible with normal growth comparable to cells expressing wild type Fic4 or the vector control (**Figure 5A**) but a filamentous phenotype was observed upon Fic4_T209A/E213A-HA expression (**Figure 5B**). Since cells expressing wild type Fic4 are phenotypically normal under these conditions we used the mutant variant to test for a possible antitoxin activity for the adjacent ORF, Fti4. Coexpression of Fti4

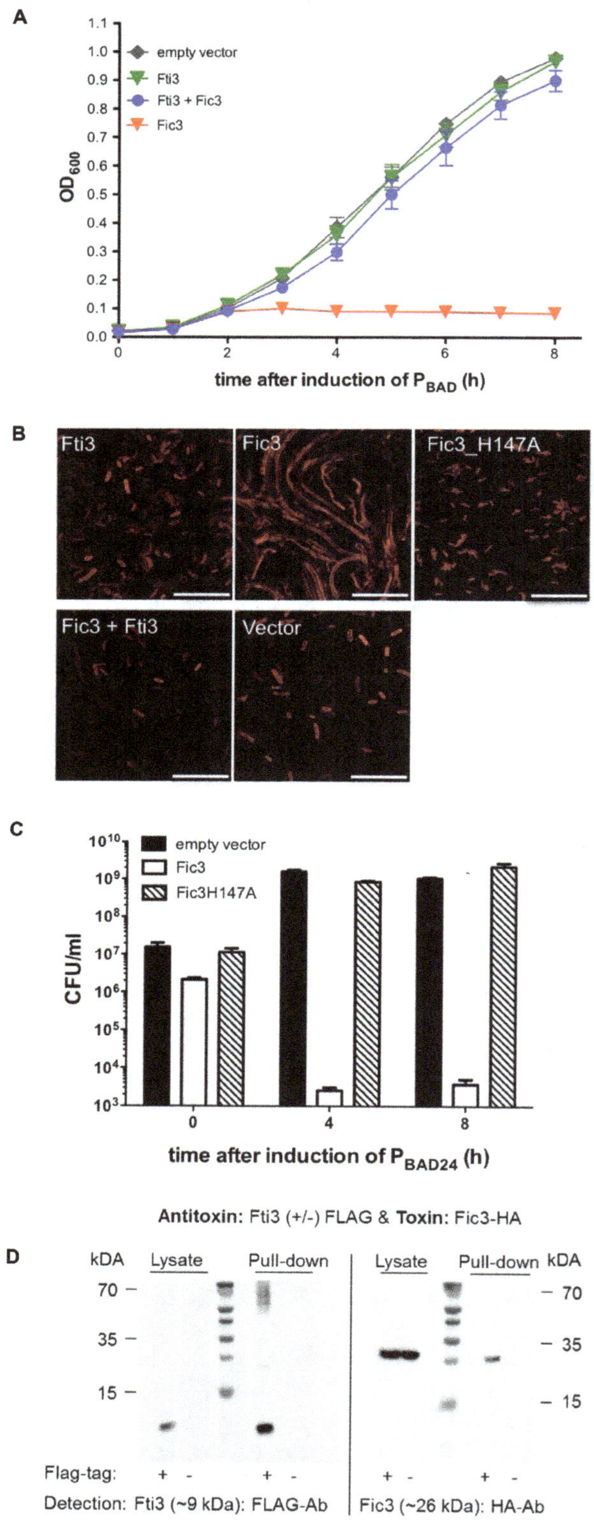

FIGURE 4 | Toxicity of Fic3 in *E. coli* is neutralized by coexpression of Fti3. **(A)** Growth profiles of *E. coli* expressing *fti3* or *fic3* or the combinations compared to the empty vector control. Results are mean values of

(Continued)

FIGURE 4 | Continued

OD_{600} ± standard deviation of three independent experiments. **(B)** Confocal microscopy of Nile red stained *E. coli* expressing indicated *fti* or *fic* genes or derivatives, respectively, alone or in combination. Vector control (lower right panel), scale bars (25 μm). **(C)** Colonies formed (CFU per ml) 0, 4, and 8 h post-induction for *E. coli* expressing *fic3*, *fic3*_H147A or the vector control. Results are mean values of three independent experiments.
(D) Co-immunoprecipitation of Fti3 (9 kDa) and Fic3 (26 kDa). Either FLAG-tagged (+) or native (–) Fti3 were expressed in *E. coli* together with HA-tagged Fic3 and incubated with FLAG-affinity beads. Cell lysates and elution fractions (pull-down) were loaded on gels and proteins were detected by Western analysis using indicated anti-FLAG- or anti-HA-antibodies (Ab).

and mutant Fic4 lessened filamentation substantially (**Figure 5B**) but had no impact on growth or survival (**Figures 5A,C**).

To test whether Fti4 and Fic4 physically interact, fusion proteins with epitope tags were created and simultaneously produced in *E. coli* as described above. FLAG-Fti4 (~10 kDa) was not directly detectable in the cell lysates but was visible after enrichment of the protein on the affinity matrix (**Figure 5D**). Fic4-HA was detected in lysates of both test and control strains. Fic4-HA was also retained on the FLAG affinity beads in a manner dependent on FLAG-Fti4. Since the functional tests described above showed phenotypes for Fti4 only when combined with the mutant derivative of Fic4, we also assayed for protein binding using the mutant allele. Similar to wild-type Fic4, co-retention of Fic4_T209A/E213A by FLAG-Fti4 was observed (**Figure 5D**, right panel).

In summary, some of the observed characteristics of Fic4 are consistent with the function of a toxin, yet the toxicity of the mutant form was quite mild compared to the inh-deficient Fic1 derivative and the wild type class I proteins Fic2 and Fic3. It is possible that evolution has introduced mutations outside of the Fic4 active site that impair enzyme activity. It is further possible that the surrogate host *E. coli* simply lacks the specific protein targeted by Fic4. Another hypothesis that we could test was to ask whether Fic4 might actually function as an antitoxin for a distinct locus (below).

Fic2 Toxicity Is Inactivated *in Trans*

To explore potential *in trans* interactions involving components of the distinct systems each toxin was expressed pairwise with every putative antitoxin. We found that the ICE_84-112 encoded antitoxin Fti3 reversed the growth defect caused by toxin Fic2 (**Figure 6A**). In contrast co-expression of Fic4 with Fic2 had no effect. Neutralization of Fic2 toxicity by Fti3 was confirmed by plating samples of the induced cultures (**Figure 6B**). Cells survived 4 and 8 h of Fic2 expression when co-expressing Fti3, but Fic4 was not able to counteract the toxicity of Fic2. Functional interaction between Fic2 and Fti3 was also supported by the normalized morphology of cells following co-expression compared to the filamentous phenotype caused by Fic2 alone (**Figure 6C**). A partial reversal of the Fic2-induced filamentation was observed with Fic4.

To test for direct binding interactions between the protein pairs, pull-down assays were performed with cells expressing Fic2-HA with either FLAG-tagged Fti3 or FLAG-tagged Fic4.

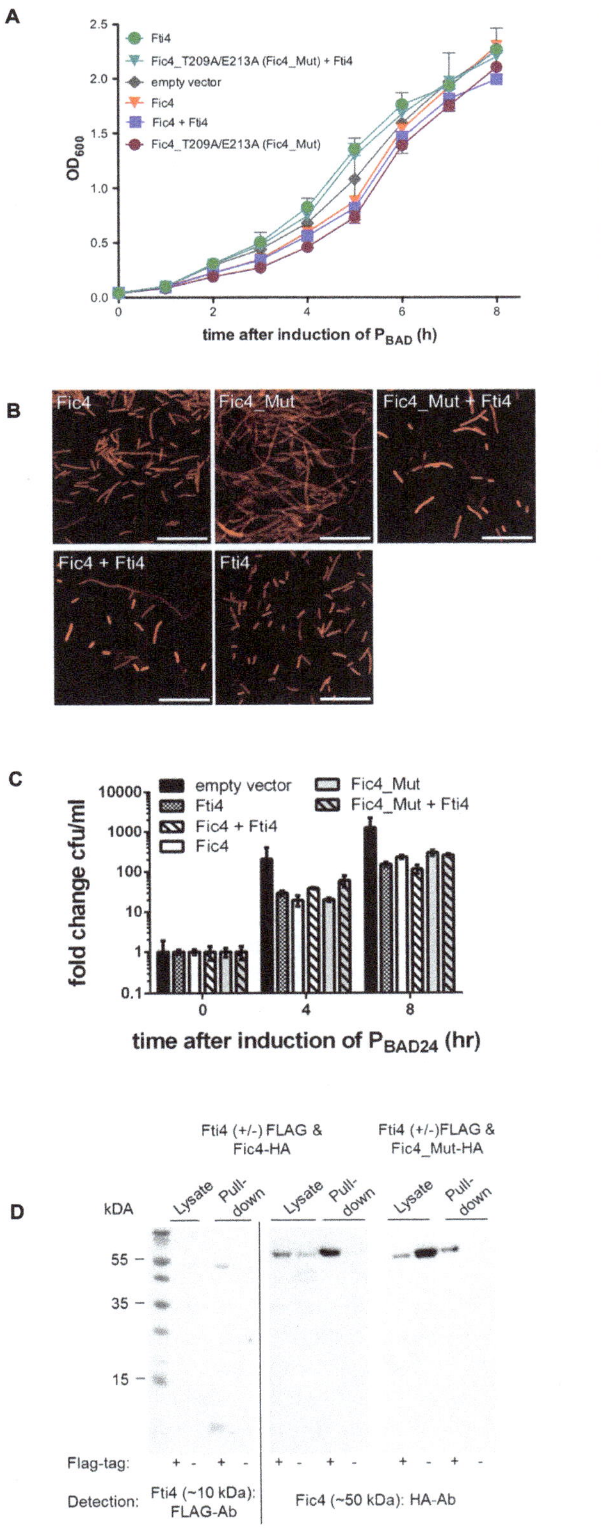

FIGURE 5 | Fti4 interacts with the non-toxic Fic4. **(A)** Growth profiles of *E. coli* expressing indicated *fti* and *fic* genes, or derivatives, alone or in combination. Results are mean values of OD_{600} ± standard deviation of three *(Continued)*

FIGURE 5 Continued

independent experiments. **(B)** Confocal microscopy of Nile red stained cells cultured in **(A)** as indicated. Fic4_Mut refers to Fic4_T209A/E213A. Scale bars (25 μm). **(C)** Colonies formed (CFU per ml) 0, 4, and 8 h post-induction for *E. coli* expressing *fic4* and *fic4*_209A/E213A with or without Fti4, or vector control. **(D)** Co-immunoprecipitation of Fti4 (10 kDa) and Fic4 (50 kDa). Either FLAG-tagged (+) or native (–) Fti4 were expressed in *E. coli* together with HA-tagged Fic4 and incubated with FLAG-affinity beads. Cell lysates and elution fractions (pull-down) were loaded on gels and proteins were detected by Western analysis using indicated anti-FLAG- or anti-HA-antibodies (Ab).

Consistent with the functional results shown above, Fti3 was able to retain Fic2 (**Figure 6D**). Remarkably, although Fic4 showed little antitoxin activity for Fic2, a complex of these proteins was detected nonetheless (**Figure 6E**).

To complete the analyses for toxin Fic2, the same tests were performed for the last putative antitoxin Fti4. No neutralizing activity by Fti4 was observed during cellular growth or by monitoring cell morphology. The pull down assay combining FLAG-tagged Fti4 with Fic2-HA also failed to detect interaction between these proteins (all data not shown). Given that Fti4 is 48% identical to Fti3, the lack of activity observed for Fti4 shows that the ability of antitoxin Fti3 to inactivate Fic2 is specific. As a final specificity check we also tested whether the toxic form of Fic1 (Fic1_S31A/E35A) was affected by co-production of either Fti3 or Fti4. No reversion of the poor growth, reduced survival or filamentous phenotypes caused by Fic1_S31A/E35A was observed (data not shown).

The sum of these data demonstrate that the bacterial growth phenotype caused by Fic2 is counteracted by the *cis* encoded antitoxin, Fic1, and independently by Fti3 *in trans*. Fic4 partially reversed the toxic effect of Fic2 in *E. coli*. Both the *cis* acting antitoxin Fic1 and the ICE-encoded proteins Fti3 and Fic4 were shown to bind toxin Fic2. These findings support a model of functional crosstalk occurring between chromosomally and ICE_84-112 encoded Fic proteins that act to control the toxin Fic2.

Fic4 Interacts with Toxin Fic3

To test for potential regulatory crosstalk occurring between Fic3 and antitoxins of the distinct systems, we again performed phenotypic tests following dual expression of each protein pair. Both Fic1 and Fic4 were unable to neutralize the extreme growth phenotype caused by Fic3 (**Figure 7A**). Dual expression of Fic3 with Fic1 did not revert the filamentation induced by Fic3, but partial recovery was apparent upon co-expression of Fic3 with Fic4, suggesting some neutralizing interactions (**Figure 7B**). Consistent with these results the pull-down assay was clearly negative for binding between Fic1 and Fic3 (**Figure 7C**), but a small yield of co-purified Fic3 was detected using FLAG-tagged Fic4 (**Figure 7D**). We performed the same analyses with cells coexpressing Fti4 with Fic3. Again despite its similarity to antitoxin Fti3, Fti4 had no affect on Fic3 toxicity and the proteins failed to bind under these conditions (not shown). In summary, we conclude that Fic3 is effectively neutralized by the *cis* encoded Fti3. Moreover, modest levels of complex formation

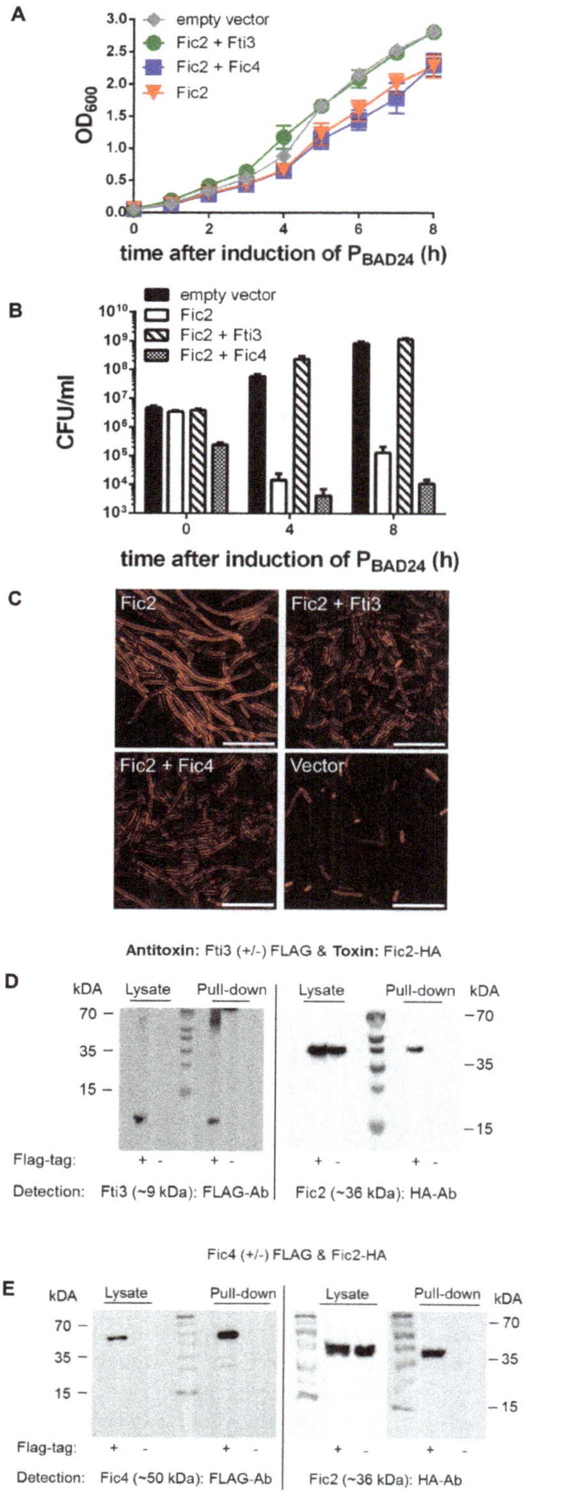

FIGURE 6 | Fic2 interacts with Fti3 and Fic4 and Fic2 toxicity is neutralized by Fti3. Growth profiles over time **(A)** and corresponding colonies formed (CFU per ml) **(B)** at 0, 4, and 8 h post-induction of *E. coli* expressing *fic2* alone or in combination with either *fti3* or *fic4*. Results are mean values of OD_{600} or *(Continued)*

FIGURE 6 | Continued
CFU/ml ± standard deviation of three independent experiments. **(C)** Confocal microscopy of Nile red stained *E. coli* expressing indicated *fic2* alone or in combination with *fti3* or *fic4*. Vector control (lower right panel), scale bars (25 μm). **(D)** Co-immunoprecipitation of Fti3 (9 kDa) and Fic2 (36 kDa) and **(E)** Fic4 (50 kDa) and Fic2 (36 kDa). Either FLAG-tagged (+) or native (–) Fti3 **(D)** or Fic4 ± FLAG **(E)** were expressed in *E. coli* together with HA-tagged Fic2 and incubated with FLAG-affinity beads. Cell lysates and elution fractions (pull-down) were loaded on gels and proteins were detected by Western analysis using indicated anti-FLAG- or anti-HA-antibodies (Ab).

with trans-acting factor Fic4 may contribute to regulation of this enzyme.

Fic2 or Fic3 Expression Inhibits Translation in *E. coli*

The identity of specific protein targets modified by Fic enzymes in bacteria is difficult to predict. It is known, however, that the activities of many TA toxins interfere with the translation process either directly, e.g., by cleavage of mRNA or tRNA, or as a downstream effect (Rajashekara et al., 2009; Park et al., 2013). To measure translation in *E. coli* cells expressing *C. fetus* Fic proteins, we performed sucrose gradient centrifugation of cell lysates and recorded polysome profiles (**Figures 8A,B**). In these analyses, the height of the polysome peaks is directly proportional to the translation levels, therefore translation defects can be faithfully detected by the reduction of polysome peaks. Moreover, because free ribosomal subunits, 70S monosomes and polysomes can be resolved, changes in the ratios between these different ribosomal (sub-) complexes can give additional information on the type of defect causing reduced translation. We recorded profiles from *fic*-expressing *E. coli* cells and compared them to those of the vector control strain. The signal corresponding to ribosomal subunits and translating ribosomes is indicated for each gradient. To make *fic*-dependent shifts in the relative abundance of these populations more apparent, profiles from different expressing strains were overlaid in the figure. Expression of Fic3 inhibited translation severely, as obvious from the massive reduction of polysome levels (**Figure 8A**, red trace; **Table 1**) compared to the vector control strain (black trace) or cells expressing just antitoxin Fti3 (green trace). Concomitantly, a strong increase of the 70S peak was detected, suggesting that ribosomal subunits are competent for joining into 70S ribosomes, but fail to enter into translation. We conclude that Fic3 blocks a step after subunit joining but before translation elongation. Dual expression of Fic3 and antitoxin Fti3 (blue trace) largely restored translation to normal levels. Moreover, the abnormally high 70S peak observed upon Fic3 overexpression was partially reduced upon Fti3 co-expression.

Fic2 expression also had a mild inhibitory effect on translation, as obvious from an accumulation of free 30S and 50S ribosomal subunits relative to the amount of 70S ribosomes and polysomes (**Figure 8B**). We compared the free subunit accumulation relative to the polysome abundance in multiple independent experiments ($n = 7$) and determined a quantitatively significant increase (**Table 1**). Notably, in contrast to Fic3 expression, Fic2 expression did not result in increased 70S levels, and even reduced 70S amounts compared to the vector control. Reduced 70S and increased free subunit levels are indicative of inefficient subunit joining. *E. coli* cells expressing antitoxin Fic1 (green trace) showed no significant variation in the ratio of free subunits versus polysomes compared to profiles from the vector control strain (**Figure 8B** and **Table 1**). The relative abundance of 70S species was mildly reduced in *fic1*-expressing vs. vector control cells however (**Table 1**), consistent with the observation that Fic1 expression slightly inhibits cell growth (**Figure 3**). In line with the neutralizing activity observed in our previous functional tests, simultaneous expression of antitoxin Fic1 with Fic2 (blue trace) resulted in a profile similar to the empty vector control.

Prevalence of *fic* Genes within *C. fetus* Strains

The sum of our findings suggests that the *C. fetus fic* genes act as TA systems. In that case the loci should be well-conserved within the species. We used PCR to survey the prevalence of the *fic* genes and *fti3* in 102 *C. fetus* isolates from geographically and ecologically diverse sources (summary in **Table 2**; detailed information in Supplementary Table S5). All of the *C. fetus* subsp. *venerealis* strains ($n = 62$) were positive for *fic1* and 59 out of 62 (95%) were positive for *fic2*. This finding is consistent with genetic linkage of the Fic2 toxin to the Fic1 antitoxin. Sequence analysis of full-length *fic2* amplicons randomly selected from our strain collection ($n = 6$) showed complete conservation for this subspecies (data not shown). In contrast, only 5 out of 40 (12.5%) *C. fetus* subsp. *fetus* strains harbor *fic1* and only two carry the *fic2* gene, whereby strain *C. fetus* subsp. *fetus* 98/v445 (F37) lacks the corresponding *fic1* antitoxin gene. Sequence analysis of this solitary *fic2* allele revealed 36 nucleotide changes, corresponding to 13 amino acid substitutions. Expression of the F37 *fic2* gene in *E. coli* confirmed that the mutated toxin is functionally impaired (data not shown). The *fic3* and *fic4* genes were detected exclusively in *C. fetus* subsp. *venerealis*, *fic3* in 11.3% (7/62) and *fic4* in 4.8% (3/62) of the isolates. Gene *fti3* shows higher abundance: 59.7% (37 of 62) *C. fetus* subsp. *venerealis* isolates and four *C. fetus* subsp. *fetus* isolates (4/40) carry the gene. Consistent with the predicted selective pressure for co-existence, all strains positive for *fic3* additionally encode the corresponding antitoxin Fti3. Moreover the high prevalence of *fic2* in *C. fetus* subsp. *venerealis* may select for stable maintenance of *fti3* even in the absence of the cognate toxin *fic3*.

In summary, we conclude that the presence of a *fic* toxin gene in *C. fetus* is typically linked to carriage of the paired antitoxin gene. The chromosomal TA system is highly conserved in *C. fetus* subsp. *venerealis*. The ICE-associated loci are also unique for *C. fetus* subsp. *venerealis* but are comparatively rare in the strains surveyed.

Phylogenetic Analysis

The significant association of these *fic* genes with *C. fetus* subsp. *venerealis* and their relative absence in *C. fetus* subsp. *fetus* led us to next ask whether they are present in other *Campylobacter*

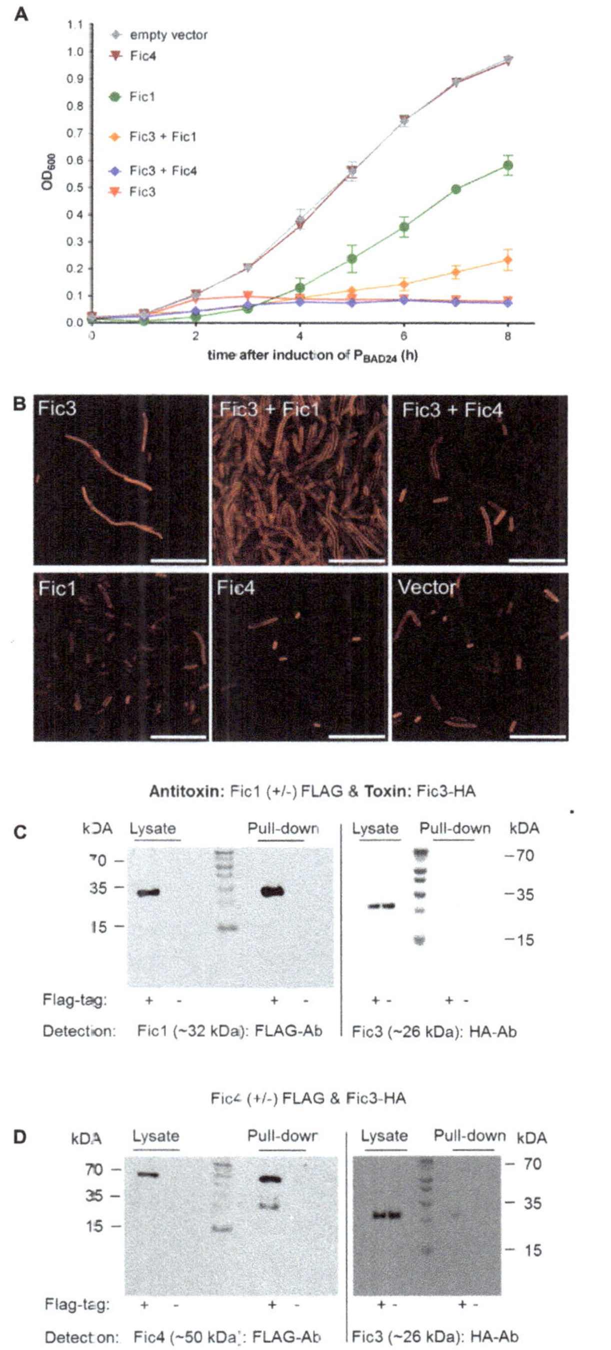

FIGURE 7 | Fic3 toxicity is not relieved by Fic1 or Fic4. **(A)** Growth profiles of *E. coli* expressing *fic3 or fic1* alone or *fic3* in combination with either *fic1* or *fic4*. Results are mean values of OD_{600} ± standard deviation of three independent experiments. **(B)** Confocal microscopy of Nile red stained *E. coli* expressing indicated *fic* proteins alone or in combination. Vector control (lower right panel), scale bars (25 μm). **(C)** Co-immunoprecipitation results for Fic1 (32 kDa) and Fic3 (26 kDa) or **(D)** Fic4 (50 kDA) and Fic3. Either FLAG-tagged (+) or native (–) Fic1 or Fic4 were expressed in *E. coli* together with HA-tagged Fic3 and incubated with FLAG-affinity beads. Cell lysates and elution fractions (pull-down) were loaded on gels and proteins were detected by Western analysis using indicated anti-FLAG- or anti-HA-antibodies (Ab).

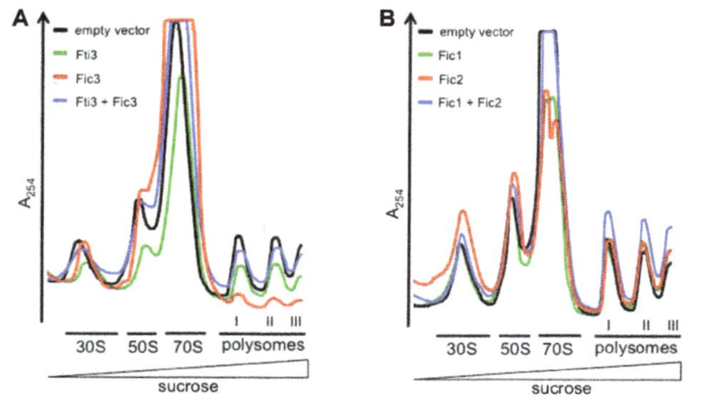

FIGURE 8 | Fic2 and Fic3 disrupt the translational machinery. Comparisons of ribosomal profiles of *E. coli* expressing **(A)** *fti3*, *fic3*, the combination (*fti3* + *fic3*) compared to the vector control, or **(B)** *fic1*, *fic2*, or the combination (*fic1* + *fic2*) compared the vector control. Absorption peaks corresponding to free 30S and 50S ribosomal subunits, 70S ribosomes, and polysomes are designated. The deficit in polysomes relative to free 30S subunits provoked by *fic2* and *fic3* expression is quantified in **Table 1**.

species and/or whether they are conserved in bacteria which inhabit the urogenital tract. Using the HPFXXGNXR motif in a BlastP analysis revealed that genomes of several *Campylobacter* species encode from one to four Fido proteins. BlastP analyses were then performed with the full-length *C. fetus* proteins Fic1-4 to identify related Fido proteins from epsilon-proteobacteria or distantly related bacteria. Interestingly, BlastP analysis using full-length Fic2 or the degenerated motif of Fic2 consequently retrieved proteins of bacterial species linked to human fertility complications (Moreno et al., 2016; Pelzer et al., 2017). We used these proteins, related *Campylobacter* proteins and selected reference Fic-proteins (Harms et al., 2016b) to generate the Neighbor joining tree shown in **Figure 9**. The tree architecture placed the proteins in two main branches. Cluster A includes Fic3 and Fic4 of *C. fetus* subsp. *venerealis* 84-112. Reference FIC-domain proteins of *Bartonella*, *Yersinia enterocolitica*, and *E. coli* (marked with asterisks) were also placed in this cluster.

TABLE 1 | Numerical and statistical analysis of ribosomal subunit peaks.

		Ratio subunits vs. polysome (I)		
		30S	**50S**	**70S***
ASet 1	Vector	1.06	1.42	6.76
	Fic1	1.08	1.52	3.15
	Fic2	1.47	1.70	2.64
	Fic1 + Fic2	0.86	1.26	3.07
BSet 2	Vector	1.06	1.40	3.21
	Fti3	1.16	1.34	3.38
	Fic3	1.94	nd	11.84
	Fti3 + Fic3	1.10	1.67	1.67
		Vector	Fic2	*p*-Values**
Statistic ($n = 7$)	30S	1.10 ± 0.02	1.50 ± 0.19	0.0005
	50S	1.49 ± 0.11	2.03 ± 0.38	0.0137

*A,B see **Figure 8**; Numbers represent ratios of 30S or 50S subunit versus first polysome peak (I) values. Numbers were calculated based on XSpan generated "Heights" (see section "Materials and Methods"); nd, 50S peaks are not separated enough from 70S peaks for numerical analysis with XSpan; *Height was generated by XSpan after extrapolation of clipped curves. **p-Values were calculated from seven profiles using the paired Student's t-test.*

TABLE 2 | Prevalence of *fic* genes in *C. fetus* subspecies.

		Chromosomal			ICE_84-112			
	#	***fic1***	***fic2***	***fic1+fic2***	***fic3***	***fic4***	***fti3***	***fti3+fic3***
Cff[a]	40	5	2	1	0	0	4	0
Cfv[a]	62	62	59	59	7	3	37	7

[a]*Cff C. fetus subsp. fetus, Cfv C. fetus subsp. Venerealis.*

Fic1 and Fic2 are both grouped in the B branch and resolve in separate subclusters. Interestingly, *Fusobacterium* spp. also harbor FIC proteins closely related to Fic1. *Fusobacteria* inhabit mucous membranes of humans and animals and both *Fusobacterium nucleatum* and *Fusobacterium necrophorum* cause abortion in cattle (Kirkbride et al., 1989; Otter, 1996). Moreover *F. nucleatum* causes intra-amniotic inflection and premature delivery in humans and mice (Han et al., 2004; Gauthier et al., 2011). We also note that *C. ureolyticus* ACS-301-V-Sch3b, isolated from the human vaginal tract, encodes a protein related to Fic1 and another related to Fic3 and Fic4 of cluster A.

Fic2 clusters with a hypothetical protein of *Campylobacter upsaliensis* JV21 and proteins of *C. helveticus* and the more recently described novel species *C. cuniculorum* and *C. corgagiensis*. *C. upsaliensis* is a human enteropathogen that typically causes diarrhea, bacteremia and sepsis, but human infection with *C. upsaliensis* has also been associated with spontaneous abortion (Gurgan and Diker, 1994). Moreover, proteins of bacterial species suspected or confirmed to play a role in human fertility and pregnancy outcome cluster in the same branch (indicated with a plus sign).

In conclusion, our findings show conserved FIC protein sequences in a variety of bacteria that either inhabit the urogenital tract of humans and animals or are able to establish pathology in this niche.

DISCUSSION

Campylobacter fetus subsp. *venerealis* 84-112 expresses a group of Fic proteins that are conserved in various veterinary and human pathogens of the urogenital tract. This study shows that the *fic* modules are prevalent and strongly conserved in *C. fetus* subsp. *venerealis* isolates but generally lacking in *C. fetus* subsp. *fetus*. The data also provide the first experimentally validated examples of TA activity for Fic proteins in *Campylobacter*. Bacterial genomes generally harbor multiple TA modules and it is becoming increasingly clear that the TA-associated toxins perform discrete, multipurpose functions (Ramage et al., 2009; Leplae et al., 2011; Lobato-Marquez et al., 2016; Diaz-Orejas et al., 2017). *C. fetus* subsp. *venerealis* 84-112 carries one functional module on the chromosome. Two additional systems are located on the extra chromosomal element

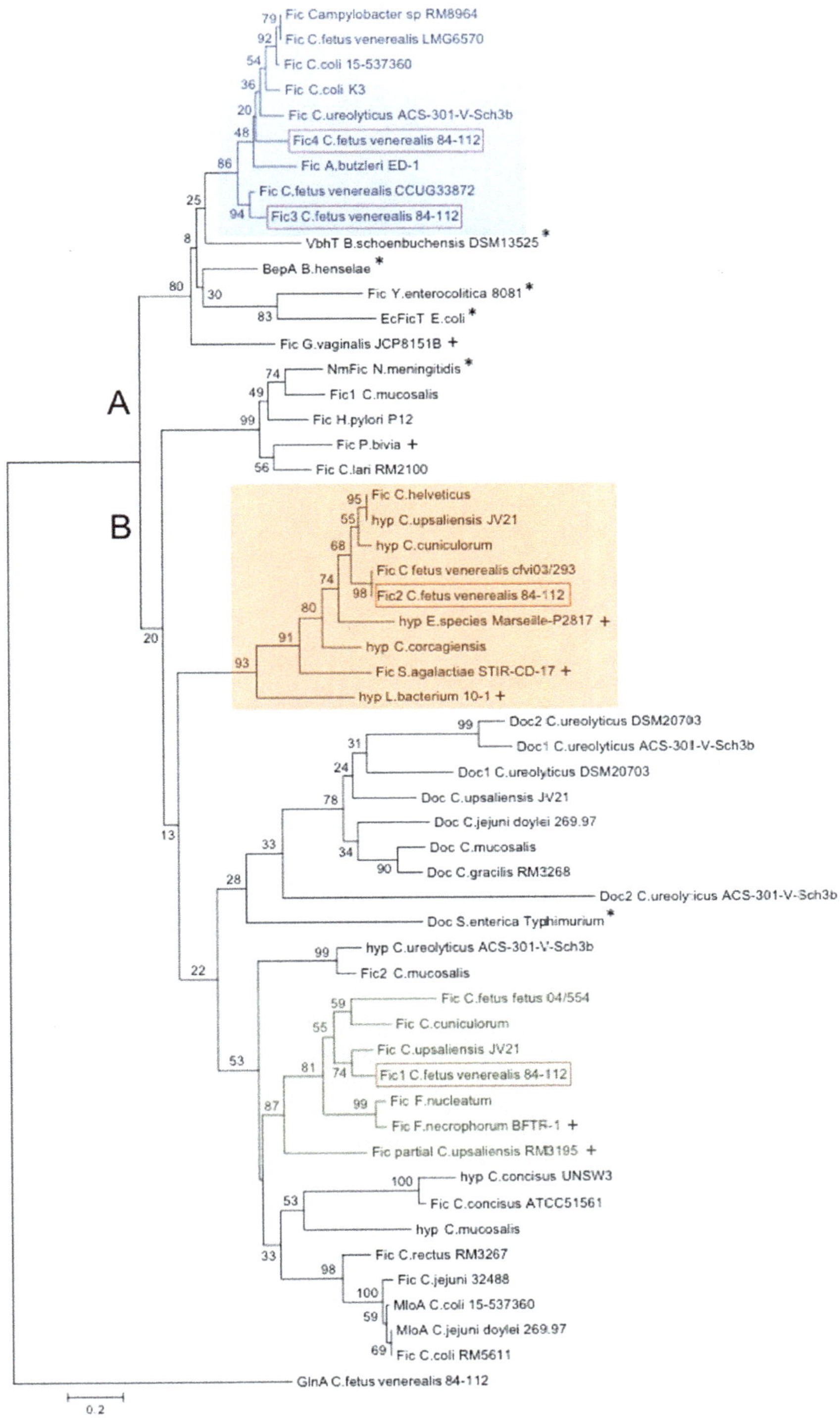

FIGURE 9 | Neighbor joining tree showing phylogenetic relationship of Fido proteins in *Campylobacter* spp. BlastP search of the conserved motif HPFXXGNXR, Fic2 motif HPFREGNTRTIA, and full-length Fic1-Fic4 proteins of *C. fetus* subsp. *venerealis* 84-112 revealed Fido proteins in *Campylobacter* spp., in closely related epsilon-proteobacteria, including *Arcobacter* (*A. butzleri*), *Helicobacter* (*H. pylori*), as well as in more distant species as indicated. The model template for protein structure prediction of Fic4, VbhT of *B. schoenbuchensis*, as well as other well-described reference Fic proteins were included (asterisk). The Neighbor joining tree contains 54 proteins and is rooted to the *C. fetus* subsp. *venerealis* 84-112 housekeeping protein GlnA. Protein and organism names are shown. Fic1-4 of *C. fetus* subsp. *venerealis* 84-112 are highlighted (red box). Proteins from distant species linked to infertility or abortion are indicated with pluses. Protein accession numbers are listed in Supplementary Table S5. The two obtained clusters (A and B) are indicated. Bootstrap values (1,000 replicates) are shown at the tree nodes. The scale bar represents 0.2 substitutions per amino acid position.

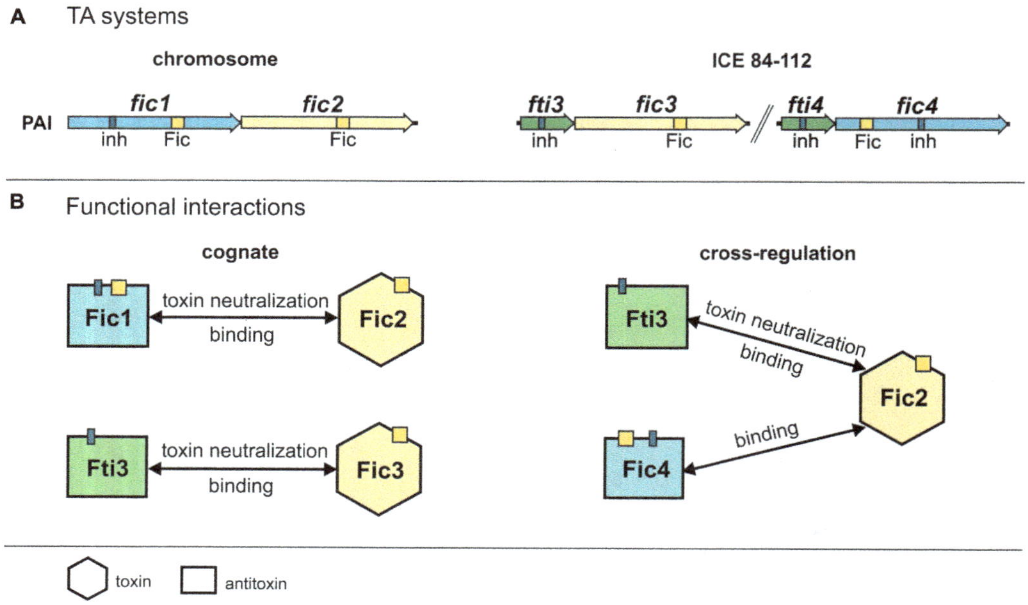

FIGURE 10 | Summary of cognate and non-cognate rescue and interaction data. **(A)** TA modules of *C. fetus* subsp *venerealis* strain 84-112 are depicted. The chromosomal locus encodes AT-Fic1/T-Fic2. The extrachromosomal ICE carries two additional independent modules AT-Fti3/T-Fic3 and the module encoding the putative TA system AT-Fti4/T-Fic4. Conserved active site motifs (Fic) and inhibitory helices (inh) are indicated. **(B)** The functional interaction and physical binding of cognate TA components detected in this study are shown (left). Fti4 binds Fic4 (not shown), but Fic4 has no toxic activity in *E. coli*. Cross-regulation of toxin Fic2 by non-cognate components is shown (right). Two unrelated ICE-encoded proteins, Fic4 and Fti3, bind T-Fic2. Heterologous AT-Fti3 counteracted Fic2 toxicity.

ICE_84-112 (Kienesberger et al., 2014). It has been speculated that TA components encoded by horizontally acquired DNA and chromosomal loci evolve toward functional cooperation (Saavedra De Bast et al., 2008; Makarova et al., 2009). Here, we demonstrate that Fti3 encoded by ICE_84-112 provides immunity for the chromosomal toxin Fic2 in addition to the cognate antitoxin Fic1 (see summary of results, **Figure 10**). We also show that the ICE-encoded Fic4 binds Fic2 and may therefore contribute to toxin regulation by complex formation or sequestration. Oligomerization was shown to play a role in regulating toxin activity for the class III Fic protein NmFic from *Neisseria meningitidis* (Stanger et al., 2016). In that case, activation of the NmFic toxin is blocked by tetramer formation. Control of *C. fetus* toxin Fic2 by Fic1 and possibly Fic4 implies that inhibitory strategies involving heteromeric complexes of different toxins are also possible. The structural similarities of the *C. fetus* Fic proteins may support physical interactions between non-cognate components as shown with the *ccd* and *parD* systems (Smith et al., 2012). Compared to well-studied Fic proteins or paradigm type II TA systems evidence for physical and functional interactions between non-cognate toxin–antitoxin systems is still relatively rare but clearly emerging as shown for *Mycobacterium tuberculosis* (Yang et al., 2010; Zhu et al., 2010). TA systems can also be interconnected through transcriptional regulation, for example positive feedback regulation can allow production of toxins to induce transcription of other TA systems (Kasari et al., 2013). As another example, toxin MqsR of the type II TA system MqsR/MqsA regulates GhoT toxin of a type V TA system via post-transcriptional differential mRNA cleavage. This activity results in a regulatory hierarchy where one TA system controls another (Wang et al., 2013). Wessner et al. (2015) have also established that regulatory crosstalk occurs between modules of type-I and -II TA families in the human pathogen *Enterococcus faecalis*. Taken together these data support the notion that bacteria harboring multiple TA systems may develop a complex hierarchy. Cooperative regulation of their activities would support concerted physiological responses in the cell. Moreover, interplay between TAs may enable bacteria to create heterogeneous populations and survive stress under a wider range of environmental conditions (Fasani and Savageau, 2013).

The ability to enter prolonged dormancy is an important factor in the epidemiology and spread of *Campylobacter* (Rollins and Colwell, 1986; Bronowski et al., 2014). Dormancy requires the bacterial cells to switch from normal activities to a static state (Harms et al., 2016a). Enzymes belonging to the Fido family are well-suited to control this process as they can modify a broad range of cellular proteins post-translationally (Garcia-Pino et al., 2014; Roy and Cherfils, 2015; Harms et al., 2016b). The general lack of similarity outside of the active loop implies that Fic1-4 bind to distinct protein targets. The core motif conserved in Fic1, Fic3, and Fic4 suggests these are competent for adenylylation. The modification reaction catalyzed by Fic2 remains unclear because the core motif deviates from the adenylylation consensus. Regardless of the biochemistry involved, we found that expression of both the canonical

Fic3 and the degenerate Fic2 toxins in *E. coli* interferes with translation. The inhibitory effect of Fic3 on translation was very pronounced. Fic3 expressing cells show a very high 70S peak and drastic reduction of polysome levels suggesting that ribosomal subunits are produced and can join, but are incapable of entering into translation. In contrast, Fic2 reduced 70S levels, while the amounts of free ribosomal subunits were increased. These properties are consistent with an early defect that impairs subunit joining. We conclude that Fic2 either directly affects subunit joining, or alternatively, that it inhibits a step of ribosome biogenesis, thereby causing the synthesis of aberrant, joining defective ribosomal subunits. Fic2 might therefore modify an rRNA processing factor or an assembly cofactor. We note with interest that ribosome profiles of cells overexpressing YfjG (RatA), a toxin of the *yfjG-yfjF* operon on the *E. coli* chromosome are similar to those expressing Fic2. YfjG inhibits 70S ribosome association and blocks the translation initiation step (Zhang and Inouye, 2011). Work in other laboratories has shown that mutation or depletion of ribosome assembly GTPases, but also inhibition of translation, is associated with filamentous cell morphology (Karbstein, 2007). These attributes resemble the phenotypes we observed. To better understand the mechanism of bacterial cytotoxicity we are characterizing each of the *C. fetus* subsp. *venerealis* toxins biochemically and structurally.

Although we can generally conclude that the TA systems described here exist in *C. fetus* subsp. *venerealis* 84-112 to control the switch between normal and static metabolic states, details about the biological context of that activity remain unknown. Emerging data from animal models of uropathogenic *E. coli* infection establish that TA systems are important for niche-specific colonization and survival, and a contribution to virulence was described in *Salmonella* Typhimurium (Norton and Mulvey, 2012; De la Cruz et al., 2013; Lobato-Marquez et al., 2015). Thus, presence of multiple Fic proteins in *C. fetus* subsp. *venerealis* may enhance long term survival under hostile conditions within the host or in response to stress during its environment–animal host infectious cycle (Man, 2011). Fasani and Savageau (2013) have proposed a general model of TA systems in which redundancy of the systems is important for increasing the frequency of persister cells. Evidence is further emerging that TA modules can contribute directly to the virulence repertoire of bacteria (Lobato-Marquez et al., 2016). Conservation of related Fic proteins in isolates of *Arcobacter, Bartonella, Fusobacterium, Streptococcus, Lachnospiraceae, Prevotella, Gardnerella,* and *Enterococcus* that colonize or cause disease in urogenital or feto-placental tissue in humans and livestock underscores the probable importance of this group of Fic proteins for niche adaptation and pathogenicity. Characterizing the protein-interaction networks of the Fic proteins of *C. fetus* and analogs from other urogenital pathogens will be the next step in understanding these complex multipurpose toxins.

AUTHOR CONTRIBUTIONS

HS, SK, GG, and EZ designed the research. CH contributed study resources. HS, SK, EP, LP, BK, PB, DV, and DA performed experiments. HS, SK, BP, PF, and EZ analyzed the data. HS, SK, BP, GG, and EZ wrote the paper. All authors read and approved the final manuscript.

ACKNOWLEDGMENTS

We thank S. Raffl for technical assistance, K. Gruber for generating the protein structure models and H. Wolinski and K. Hellauer for expert assistance in microscopy.

REFERENCES

Blaser, M. J., Newell, D. G., Thompson, S. A., and Zechner, E. L. (2008). "Pathogenesis of *Campylobacter fetus* infections," in *Campylobacter*, 3rd Edn, eds I. Nachamkin, C. M. Szymanski, and M. J. Blaser (Washington, DC: American Society for Microbiology), 401–428.

Bronowski, C., James, C. E., and Winstanley, C. (2014). Role of environmental survival in transmission of *Campylobacter jejuni*. *FEMS Microbiol. Lett.* 356, 8–19. doi: 10.1111/1574-6968.12488

Bullman, S., Lucid, A., Corcoran, D., Sleator, R. D., and Lucey, B. (2013). Genomic investigation into strain heterogeneity and pathogenic potential of the emerging gastrointestinal pathogen *Campylobacter ureolyticus*. *PLOS ONE* 8:e71515. doi: 10.1371/journal.pone.0071515

Castro-Roa, D., Garcia-Pino, A., De Gieter, S., van Nuland, N. A., Loris, R., and Zenkin, N. (2013). The Fic protein Doc uses an inverted substrate to phosphorylate and inactivate EF-Tu. *Nat. Chem. Biol.* 9, 811–817. doi: 10.1038/nchembio.1364

Christie, P. J., Whitaker, N., and Gonzalez-Rivera, C. (2014). Mechanism and structure of the bacterial type IV secretion systems. *Biochim. Biophys. Acta* 1843, 1578–1591. doi: 10.1016/j.bbamcr.2013.12.019

Cruz, J. W., Rothenbacher, F. P., Maehigashi, T., Lane, W. S., Dunham, C. M., and Woychik, N. A. (2014). Doc toxin is a kinase that inactivates elongation factor Tu. *J. Biol. Chem.* 289, 7788–7798. doi: 10.1074/jbc.M113.544429

Cruz, J. W., and Woychik, N. A. (2014). Teaching Fido new ModiFICation tricks. *PLOS Pathog.* 10:e1004349. doi: 10.1371/journal.ppat.1004349

De la Cruz, M. A., Zhao, W., Farenc, C., Gimenez, G., Raoult, D., Cambillau, C., et al. (2013). A toxin-antitoxin module of *Salmonella* promotes virulence in mice. *PLOS Pathog.* 9:e1003827. doi: 10.1371/journal.ppat.1003827

Diaz-Orejas, R., Espinosa, M., and Yeo, C. C. (2017). The importance of the expendable: toxin-antitoxin genes in plasmids and chromosomes. *Front. Microbiol.* 8:1479. doi: 10.3389/fmicb.2017.01479

Engel, P., Goepfert, A., Stanger, F. V., Harms, A., Schmidt, A., Schirmer, T., et al. (2012). Adenylylation control by intra- or intermolecular active-site obstruction in Fic proteins. *Nature* 482, 107–110. doi: 10.1038/nature10729

Fasani, R. A., and Savageau, M. A. (2013). Molecular mechanisms of multiple toxin-antitoxin systems are coordinated to govern the persister phenotype. *Proc. Natl. Acad. Sci. U.S.A.* 110, E2528–E2537. doi: 10 1073/pnas.1301023110

Garcia-Pino, A., Christensen-Dalsgaard, M., Wyns, L., Yarmolinsky, M., Magnuson, R. D., Gerdes, K., et al. (2008). Doc of prophage P1 is inhibited by its antitoxin partner Phd through fold complementation. *J. Biol. Chem.* 283, 30821–30827. doi: 10.1074/jbc.M805654200

Garcia-Pino, A., Zenkin, N., and Loris, R. (2014). The many faces of Fic: structural and functional aspects of Fic enzymes. *Trends Biochem. Sci.* 39, 121–129. doi: 10.1016/j.tibs.2014.01.001

Gauthier, S., Tetu, A., Himaya, E., Morand, M., Chandad, F., Rallu, F., et al. (2011). The origin of *Fusobacterium nucleatum* involved in intra-amniotic infection and preterm birth. *J. Matern. Fetal Neonatal Med.* 24, 1329–1332. doi: 10.3109/14767058.2010.550977

Goeders, N., and Van Melderen, L. (2014). Toxin-antitoxin systems as multilevel interaction systems. *Toxins* 6, 304–324. doi: 10.3390/toxins6010304

Goepfert, A., Stanger, F. V., Dehio, C., and Schirmer, T. (2013). Conserved inhibitory mechanism and competent ATP binding mode for adenylyltransferases with Fic fold. *PLOS ONE* 8:e64901. doi: 10.1371/journal.pone.0064901

Gorkiewicz, G., Kienesberger, S., Schober, C., Scheicher, S. R., Gully, C., Zechner, R., et al. (2010). A genomic island defines subspecies-specific virulence features of the host-adapted pathogen *Campylobacter fetus* subsp. *venerealis*. *J. Bacteriol.* 192, 502–517. doi: 10.1128/JB.00803-09

Graaf-van Bloois, L., Miller, W. G., Yee, E., Gorkiewicz, G., Forbes, K. J., Zomer, A. L., et al. (2016). *Campylobacter fetus* subspecies contain conserved type IV secretion systems on multiple genomic islands and plasmids. *PLOS ONE* 11:e0152832. doi: 10.1371/journal.pone.0152832

Gruber, C. J., Lang, S., Rajendra, V. K., Nuk, M., Raffl, S., Schildbach, J. F., et al. (2016). Conjugative DNA transfer is enhanced by plasmid R1 partitioning proteins. *Front. Mol. Biosci.* 3:32. doi: 10.3389/fmolb.2016.00032

Gurgan, T., and Diker, K. S. (1994). Abortion associated with *Campylobacter upsaliensis*. *J. Clin. Microbiol.* 32, 3093–3094.

Han, Y. W., Redline, R. W., Li, M., Yin, L., Hill, G. B., and McCormick, T. S. (2004). *Fusobacterium nucleatum* induces premature and term stillbirths in pregnant mice: implication of oral bacteria in preterm birth. *Infect. Immun.* 72, 2272–2279. doi: 10.1128/IAI.72.4.2272-2279.2004

Harms, A., Maisonneuve, E., and Gerdes, K. (2016a). Mechanisms of bacterial persistence during stress and antibiotic exposure. *Science* 354, aaf4268.

Harms, A., Stanger, F. V., and Dehio, C. (2016b). Biological diversity and molecular plasticity of FIC domain proteins. *Annu. Rev. Microbiol.* 70, 341–360. doi: 10.1146/annurev-micro-102215-095245

Harvey, S., and Greenwood, J. R. (1985). Isolation of *Campylobacter fetus* from a pet turtle. *J. Clin. Microbiol.* 21, 260–261.

Hayes, F., and Van Melderen, L. (2011). Toxins-antitoxins: diversity, evolution and function. *Crit. Rev. Biochem. Mol. Biol.* 46, 386–408. doi: 10.3109/10409238.2011.600437

Jiang, M., Sullivan, S. M., Walker, A. K., Strahler, J. R., Andrews, P. C., and Maddock, J. R. (2007). Identification of novel *Escherichia coli* ribosome-associated proteins using isobaric tags and multidimensional protein identification techniques. *J. Bacteriol.* 189, 3434–3444.

Karbstein, K. (2007). Role of GTPases in ribosome assembly. *Biopolymers* 87, 1–11. doi: 10.1002/bip.20762

Kasari, V., Mets, T., Tenson, T., and Kaldalu, N. (2013). Transcriptional cross-activation between toxin-antitoxin systems of *Escherichia coli*. *BMC Microbiol.* 13:45. doi: 10.1186/1471-2180-13-45

Kienesberger, S., Gorkiewicz, G., Joainig, M. M., Scheicher, S. R., Leitner, E., and Zechner, E. L. (2007). Development of experimental genetic tools for *Campylobacter fetus*. *Appl. Environ. Microbiol.* 73, 4619–4630. doi: 10.1128/AEM.02407-06

Kienesberger, S., Sprenger, H., Wolfgruber, S., Halwachs, B., Thallinger, G. G., Perez-Perez, G. I., et al. (2014). Comparative genome analysis of *Campylobacter fetus* subspecies revealed horizontally acquired genetic elements important for virulence and niche specificity. *PLOS ONE* 9:e85491. doi: 10.1371/journal.pone.0085491

Kinch, L. N., Yarbrough, M. L., Orth, K., and Grishin, N. V. (2009). Fido, a novel AMPylation domain common to fic, doc, and AvrB. *PLOS ONE* 4:e5818. doi: 10.1371/journal.pone.0005818

Kirkbride, C. A., Gates, C. E., and Libal, M. C. (1989). Ovine and bovine abortion associated with *Fusobacterium nucleatum*. *J. Vet. Diagn. Invest.* 1, 272–273. doi: 10.1177/104063878900100318

Lastovica, A. J., and Allos, B. M. (2008). "Clinical significance of *Campylobacter* and related species other than *Campylobacter jejuni* and *Campylobacter coli*," in *Campylobacter*, 3 Edn, eds I. Nachamkin, C. M. Szymanski, and M. J. Blaser (Washington, DC: ASM), 123–149.

Lehnherr, H., Maguin, E., Jafri, S., and Yarmolinsky, M. B. (1993). Plasmid addiction genes of bacteriophage P1: doc, which causes cell death on curing of prophage, and phd, which prevents host death when prophage is retained. *J. Mol. Biol.* 233, 414–428. doi: 10.1006/jmbi.1993.1521

Leplae, R., Geeraerts, D., Hallez, R., Guglielmini, J., Dreze, P., and Van Melderen, L. (2011). Diversity of bacterial type II toxin-antitoxin systems: a comprehensive search and functional analysis of novel families. *Nucleic Acids Res.* 39, 5513–5525. doi: 10.1093/nar/gkr131

Liu, M., Zhang, Y., Inouye, M., and Woychik, N. A. (2008). Bacterial addiction module toxin Doc inhibits translation elongation through its association with the 30S ribosomal subunit. *Proc. Natl. Acad. Sci. U.S.A.* 105, 5885–5890. doi: 10.1073/pnas.0711949105

Lobato-Marquez, D., Diaz-Orejas, R., and Garcia-Del Portillo, F. (2016). Toxin-antitoxins and bacterial virulence. *FEMS Microbiol. Rev.* 40, 592–609. doi: 10.1093/femsre/fuw022

Lobato-Marquez, D., Moreno-Cordoba, I., Figueroa, V., Diaz-Orejas, R., and Garcia-del Portillo, F. (2015). Distinct type I and type II toxin-antitoxin modules control *Salmonella* lifestyle inside eukaryotic cells. *Sci. Rep.* 5:9374. doi: 10.1038/srep09374

Logue, C. M., Sherwood, J. S., Elijah, L. M., Olah, P. A., and Dockter, M. R. (2003). The incidence of *Campylobacter* spp. on processed turkey from processing plants in the Midwestern United States. *J. Appl. Microbiol.* 95, 234–241. doi: 10.1046/j.1365-2672.2003.01969.x

Maisonneuve, E., and Gerdes, K. (2014). Molecular mechanisms underlying bacterial persisters. *Cell* 157, 539–548. doi: 10.1016/j.cell.2014.02.050

Makarova, K. S., Wolf, Y. I., and Koonin, E. V. (2009). Comprehensive comparative-genomic analysis of type 2 toxin-antitoxin systems and related mobile stress response systems in prokaryotes. *Biol. Direct* 4:19. doi: 10.1186/1745-6150-4-19

Man, S. M. (2011). The clinical importance of emerging *Campylobacter* species. *Nat. Rev. Gastroenterol. Hepatol.* 8, 669–685. doi: 10.1038/nrgastro.2011.191

Moreno, I., Codoner, F. M., Vilella, F., Valbuena, D., Martinez-Blanch, J. F., Jimenez-Almazan, J., et al. (2016). Evidence that the endometrial microbiota has an effect on implantation success or failure. *Am. J. Obstet. Gynecol.* 215, 684–703. doi: 10.1016/j.ajog.2016.09.075

Mukherjee, S., Liu, X., Arasaki, K., McDonough, J., Galan, J. E., and Roy, C. R. (2011). Modulation of Rab GTPase function by a protein phosphocholine transferase. *Nature* 477, 103–106. doi: 10.1038/nature10335

Norton, J. P., and Mulvey, M. A. (2012). Toxin-antitoxin systems are important for niche-specific colonization and stress resistance of uropathogenic *Escherichia coli*. *PLOS Pathog.* 8:e1002954. doi: 10.1371/journal.ppat.1002954

Otter, A. (1996). *Fusobacterium necrophorum* abortion in a cow. *Vet. Rec.* 139, 318–319.

Palanivelu, D. V., Goepfert, A., Meury, M., Guye, P., Dehio, C., and Schirmer, T. (2011). Fic domain-catalyzed adenylylation: insight provided by the structural analysis of the type IV secretion system effector BepA. *Protein Sci.* 20, 492–499. doi: 10.1002/pro.581

Pan, X., Luhrmann, A., Satoh, A., Laskowski-Arce, M. A., and Roy, C. R. (2008). Ankyrin repeat proteins comprise a diverse family of bacterial type IV effectors. *Science* 320, 1651–1654. doi: 10.1126/science.1158160

Park, S. J., Son, W. S., and Lee, B. J. (2013). Structural overview of toxin-antitoxin systems in infectious bacteria: a target for developing antimicrobial agents. *Biochim. Biophys. Acta* 1834, 1155–1167. doi: 10.1016/j.bbapap.2013.02.027

Pascoe, B., Meric, G., Murray, S., Yahara, K., Mageiros, L., Bowen, R., et al. (2015). Enhanced biofilm formation and multi-host transmission evolve from divergent genetic backgrounds in *Campylobacter jejuni*. *Environ. Microbiol.* 17, 4779–4789. doi: 10.1111/1462-2920.13051

Pelzer, E., Gomez-Arango, L. F., Barrett, H. L., and Nitert, M. D. (2017). Review: maternal health and the placental microbiome. *Placenta* 54, 30–37. doi: 10.1016/j.placenta.2016.12.003

Rajashekara, G., Drozd, M., Gangaiah, D., Jeon, B., Liu, Z., and Zhang, Q. (2009). Functional characterization of the twin-arginine translocation system in *Campylobacter jejuni*. *Foodborne Pathog. Dis.* 6, 935–945. doi: 10.1089/fpd.2009.0298

Ramage, H. R., Connolly, L. E., and Cox, J. S. (2009). Comprehensive functional analysis of *Mycobacterium tuberculosis* toxin-antitoxin systems: implications for pathogenesis, stress responses, and evolution. *PLOS Genet.* 5:e1000767. doi: 10.1371/journal.pgen.1000767

Ren, D., Kordis, A. A., Sonenshine, D. E., and Daines, D. A. (2014). The ToxAvapA toxin-antitoxin locus contributes to the survival of nontypeable *Haemophilus influenzae* during infection. *PLOS ONE* 9:e91523. doi: 10.1371/journal.pone.0091523

Ren, D., Walker, A. N., and Daines, D. A. (2012). Toxin-antitoxin loci vapBC-1 and vapXD contribute to survival and virulence in nontypeable *Haemophilus influenzae*. *BMC Microbiol.* 12:263. doi: 10.1186/1471-2180-12-263

Rollins, D. M., and Colwell, R. R. (1986). Viable but nonculturable stage of *Campylobacter jejuni* and its role in survival in the natural aquatic environment. *Appl. Environ. Microbiol.* 52, 531–538.

Roy, C. R., and Cherfils, J. (2015). Structure and function of Fic proteins. *Nat. Rev. Microbiol.* 13, 631–640. doi: 10.1038/nrmicro3520

Roy, C. R., and Mukherjee, S. (2009). Bacterial FIC proteins AMP up infection. *Sci. Signal.* 2:pe14. doi: 10.1126/scisignal.262pe14

Saavedra De Bast, M., Mine, N., and Van Melderen, L. (2008). Chromosomal toxin-antitoxin systems may act as antiaddiction modules. *J. Bacteriol.* 190, 4603–4609. doi: 10.1128/JB.00357-08

Schrodinger, L. L. C. (2010). *The PyMOL Molecular Graphics System, Version 1.3r1.*

Shao, Y., Harrison, E. M., Bi, D., Tai, C., He, X., Ou, H. Y., et al. (2011). TADB: a web-based resource for Type 2 toxin-antitoxin loci in bacteria and archaea. *Nucleic Acids Res.* 39, D606–D611. doi: 10.1093/nar/gkq908

Shen, Z., Patil, R. D., Sahin, O., Wu, Z., Pu, X. Y., Dai, L., et al. (2016). Identification and functional analysis of two toxin-antitoxin systems in *Campylobacter jejuni*. *Mol. Microbiol.* 101, 909–923. doi: 10.1111/mmi.13431

Skirrow, M. B., and Benjamin, J. (1980). '1001' *Campylobacters*: cultural characteristics of intestinal campylobacters from man and animals. *J. Hyg.* 85, 427–442.

Smith, A. B., Lopez-Villarejo, J., Diago-Navarro, E., Mitchenall, L. A., Barendregt, A., Heck, A. J., et al. (2012). A common origin for the bacterial toxin-antitoxin systems *parD* and *ccd*, suggested by analyses of toxin/target and toxin/antitoxin interactions. *PLOS ONE* 7:e46499. doi: 10.1371/journal.pone.0046499

Stanger, F. V., Burmann, B. M., Harms, A., Aragao, H., Mazur, A., Sharpe, T., et al. (2016). Intrinsic regulation of FIC-domain AMP-transferases by oligomerization and automodification. *Proc. Natl. Acad. Sci. U.S.A.* 113, E529–E537. doi: 10.1073/pnas.1516930113

Tamura, K., Stecher, G., Peterson, D., Filipski, A., and Kumar, S. (2013). MEGA6: molecular evolutionary genetics analysis version 6.0. *Mol. Biol. Evol.* 30, 2725–2729. doi: 10.1093/molbev/mst197

Wang, X., Lord, D. M., Hong, S. H., Peti, W., Benedik, M. J., Page, R., et al. (2013). Type II toxin/antitoxin MqsR/MqsA controls type V toxin/antitoxin GhoT/GhoS. *Environ. Microbiol.* 15, 1734–1744. doi: 10.1111/1462-2920.12063

Wessner, F., Lacoux, C., Goeders, N., Fouquier d'Herouel, A., Matos, R., Serror, P., et al. (2015). Regulatory crosstalk between type I and type II toxin-antitoxin systems in the human pathogen *Enterococcus faecalis*. *RNA Biol.* 12, 1099–1108. doi: 10.1080/15476286.2015.1084465

Woolery, A. R., Luong, P., Broberg, C. A., and Orth, K. (2010). AMPylation: something old is new again. *Front. Microbiol.* 1:113. doi: 10.3389/fmicb.2010.00113

Worby, C. A., Mattoo, S., Kruger, R. P., Corbeil, L. B., Koller, A., Mendez, J. C., et al. (2009). The fic domain: regulation of cell signaling by adenylylation. *Mol. Cell.* 34, 93–103. doi: 10.1016/j.molcel.2009.03.008

Xiao, J., Worby, C. A., Mattoo, S., Sankaran, B., and Dixon, J. E. (2010). Structural basis of Fic-mediated adenylylation. *Nat. Struct. Mol. Biol.* 17, 1004–1010. doi: 10.1038/nsmb.1867

Yamaguchi, Y., and Inouye, M. (2011). Regulation of growth and death in *Escherichia coli* by toxin-antitoxin systems. *Nat. Rev. Microbiol.* 9, 779–790. doi: 10.1038/nrmicro2651

Yang, M., Gao, C., Wang, Y., Zhang, H., and He, Z. G. (2010). Characterization of the interaction and cross-regulation of three *Mycobacterium tuberculosis* RelBE modules. *PLOS ONE* 5:e10672. doi: 10.1371/journal.pone.0010672

Yarbrough, M. L., Li, Y., Kinch, L. N., Grishin, N. V., Ball, H. L., and Orth, K. (2009). AMPylation of Rho GTPases by *Vibrio* VopS disrupts effector binding and downstream signaling. *Science* 323, 269–272. doi: 10.1126/science.1166382

Zechner, E. L., Lang, S., and Schildbach, J. F. (2012). Assembly and mechanisms of bacterial type IV secretion machines. *Philos. Trans. R. Soc. Lond. B Biol. Sci.* 367, 1073–1087. doi: 10.1098/rstb.2011.0207

Zekarias, B., Mattoo, S., Worby, C., Lehmann, J., Rosenbusch, R. F., and Corbeil, L. B. (2010). *Histophilus somni* IbpA DR2/Fic in virulence and immunoprotection at the natural host alveolar epithelial barrier. *Infect. Immun.* 78, 1850–1858. doi: 10.1128/IAI.01277-09

Zhang, Y., and Inouye, M. (2011). RatA (YfjG), an *Escherichia coli* toxin, inhibits 70S ribosome association to block translation initiation. *Mol. Microbiol.* 79, 1418–1429. doi: 10.1111/j.1365-2958.2010.07506.x

Zhu, L., Sharp, J. D., Kobayashi, H., Woychik, N. A., and Inouye, M. (2010). Noncognate *Mycobacterium tuberculosis* toxin-antitoxins can physically and functionally interact. *J. Biol. Chem.* 285, 39732–39738. doi: 10.1074/jbc.M110.163105

6

Toxin ζ Triggers a Survival Response to Cope with Stress and Persistence

*María Moreno-del Álamo, Mariangela Tabone †, Virginia S. Lioy † and Juan C. Alonso **

Department of Microbial Biotechnology, Centro Nacional de Biotecnología (CSIC), Madrid, Spain

Edited by:
Manuel Espinosa,
Centro de Investigaciones Biológicas (CSIC), Spain

Reviewed by:
Ramon Diaz Orejas,
Consejo Superior de Investigaciones Científicas (CSIC), Spain
Nadia Berkova,
Institut National de la Recherche Agronomique (INRA), France

***Correspondence:**
Juan C. Alonso
jcalonso@cnb.csic.es

Bacteria have evolved complex regulatory controls in response to various environmental stresses. Protein toxins of the ζ superfamily, found in prominent human pathogens, are broadly distributed in nature. We show that ζ is a uridine diphosphate-N-acetylglucosamine (UNAG)-dependent ATPase whose activity is inhibited *in vitro* by stoichiometric concentrations of ε_2 antitoxin. *In vivo*, transient ζ expression promotes a reversible multi-level response by altering the pool of signaling purine nucleotides, which leads to growth arrest (dormancy), although a small cell subpopulation persists rather than tolerating toxin action. High c-di-AMP levels (absence of phosphodiesterase GdpP) decrease, and low c-di-AMP levels (absence of diadenylate cyclase DisA) increase the rate of ζ persistence. The absence of CodY, a transition regulator from exponential to stationary phase, sensitizes cells to toxin action, and suppresses persisters formed in the Δ*disA* context. These changes, which do not affect the levels of stochastic ampicillin (Amp) persistence, sensitize cells to toxin and Amp action. Our findings provide an explanation for the connection between ζ-mediated growth arrest (with alterations in the GTP and c-di-AMP pools) and persistence formation.

Keywords: toxin-antitoxin system, cell wall inhibition, c-di-AMP, CodY, (p)ppGpp, DisA

INTRODUCTION

The toxin-antitoxin (TA) systems are widely distributed in free-living bacteria, in their extrachromosomal elements, and in archaea (Gerdes, 2013; Unterholzner et al., 2013). The toxins of all known TA systems are proteins while the antitoxins are either proteins or non-coding RNAs. The TA systems are classified into five different TA types (Yamaguchi et al., 2011), being the most broadly distributed the type II TA system, where both the toxin and the antitoxin are proteins (Leplae et al., 2011; Gerdes, 2013). The type II toxins use different strategies to regulate growth control and cellular processes related to the general stress response. Toxins of the ζ/PezT superfamily, which are among the most broadly distributed in nature, are found in major human pathogens and in environmentally important bacteria of the phylum Firmicutes (Mutschler and Meinhart, 2013). The plasmid-borne ζ gene product from *Streptococcus pyogenes*, *Streptococcus agalactiae* or *Enterococcus faecalis* and the chromosome-encoded ζ toxin from *Clostridium perfringens* or *Staphylococcus aureus* (∼285 amino acids) share ∼43% sequence identity with chromosome-encoded *Streptococcus pneumoniae* or *Streptococcus suis* PezT toxin (∼255 amino acids) (reviewed in Mutschler and Meinhart, 2013). When free in solution, these toxins interact with uridine diphosphate-N-acetylglucosamine (UNAG), ATP-Mg^{2+} or GTP-Mg^{2+} (denoted ATP and GTP), and with their cognate dimeric ε/PezA antitoxin (ε/$PezA_2$) (Meinhart et al., 2001, 2003; Khoo et al., 2007; Mutschler et al., 2011). A non-toxic heterotetrameric complex (ζε$_2$ζ/PezT-PezA$_2$-PezT) interacts with UNAG, but not with ATP/GTP (Meinhart et al., 2001, 2003; Khoo et al., 2007; Mutschler et al., 2011).

Enzymes of the ζ/PezT toxin superfamily have a common fold core with phosphotransferases (Meinhart et al., 2001, 2003; Khoo et al., 2007; Mutschler et al., 2011). Toxin ζ/PezT transfers the ATP/GTP γ-phosphate to the 3′-hydroxyl group of the UNAG amino sugar, rendering UNAG-3P unreactive and thus reducing cell wall biosynthesis (Mutschler et al., 2011). Although, a quantitative analysis of this reaction showed that in the presence of limiting UNAG and ATP, toxin ζ mainly hydrolyzed ATP and only traces of the γ-phosphate are transferred to UNAG (Tabone et al., 2014a).

The fine mechanisms of bacterial responses to toxin action are not generally conserved among different bacterial phyla (Gerdes, 2013). The evolutionary distance between *Escherichia coli* and *Bacillus subtilis*, which is larger than the time divergence between yeasts and humans, reflects the notable differences made by the purine nucleotides in the stringent response (Potrykus and Cashel, 2008; Liu et al., 2015). In *E. coli* (a representative of the γ-proteobacteria class), toxin-mediated persister formation is linked to high levels of guanosine (penta)tetraphosphate ([p]ppGpp), which inhibits the PPX phosphatase; dropping of PPX increases polyphosphate levels that activate Lon protease degradation of the antitoxins, with subsequent release of active toxins (Maisonneuve et al., 2013). These free mRNase toxins contribute to persistence to some, but not all antibiotics (Harms et al., 2016). The role of toxin action in bacteria of the phylum Firmicutes, and whether these toxins induce persistence or tolerance, is poorly understood. We therefore examined the role of *S. pyogenes* pSM19035-encoded ζ toxin in growth arrest (dormancy), alone or with antibiotic in *B. subtilis* cells (representative of the Firmicutes), by controlling expression of the toxin at or near physiological concentrations. In our analysis, we did not study the role of (p)ppGpp in antitoxin degradation and free toxin release. We found that transient expression of a short-lived toxin ζ variant (ζY83C) induced different temporal sets of cell responses and growth arrest, but a small cell subpopulation (5×10^{-5} to 1×10^{-4}) exits the dormant state, leading to persistent or tolerant *B. subtilis* cells (Lioy et al., 2012).

Analysis of the metabolic changes induced by the free toxin showed that within the first 5 min, ζY83C expression decreased the intracellular GTP pool and dysregulated transcription of 78 genes, of which 28 with reduced expression are essential for cell proliferation (Lioy et al., 2012). Induction of genes involved in the SOS response was not observed, but the expression was documented of genes that could modulate toxin action, such as increased *comGA* and *relA* expression or decreased *glmS* gene expression (Lioy et al., 2012). It is likely that by altering ATP:GTP ratios, toxin ζY83C modifies availability of the initiating nucleotides; this in turn changes promoter preferences by RNA polymerase, and the intracellular signaling (Krasny and Gourse, 2004; Pedley and Benkovic, 2017).

Within the first 15 min of ζY83C expression, the intracellular ATP concentration decreases and that of (p)ppGpp increases (Lioy et al., 2012). The contribution of increased *comGA* and *relA* expression lead to higher (p)ppGpp levels (Potrykus and Cashel, 2008; Hahn et al., 2015; Liu et al., 2015), which directly inhibit both salvage and *de novo* GTP synthesis (Lopez et al., 1981; Kriel et al., 2012; Pedley and Benkovic, 2017). In *B. subtilis*, low GTP levels lead to derepression of CodY, a global transcriptional regulator from exponential to stationary phase (Handke et al., 2008; Kriel et al., 2012; Bittner et al., 2014; Brinsmade et al., 2014).

Downregulation of GlmS contributes indirectly to reducing the pool of UNAG synthesis, and a small UNAG pool increases levels of the essential cyclic 3,5-diadenosine monophosphate (c-di-AMP) second messenger (Witte et al., 2008; Zhu et al., 2016). Changes in the intracellular level of c-di-AMP, which play an essential role in K^+ transport and cell wall homeostasis (Gundlach et al., 2017), indirectly increase the intracellular (p)ppGpp pool (Rao et al., 2010; Corrigan et al., 2015). The relationship between the effective levels of c-di-AMP and bacterial persisters is nonetheless poorly characterized.

At later stages of toxin ζY83C expression, synthesis of macromolecules (DNA, RNA, proteins) is inhibited and membrane potential is impaired (30–90 min; Lioy et al., 2012). Direct interaction of (p)ppGpp with DNA primase inhibits DNA replication (Wang et al., 2007; Srivatsan and Wang, 2008), (p)ppGpp-mediated low levels of GTP decrease mRNA transcription (Krasny and Gourse, 2004), and the essential GTPases decrease the amount of mature 70S ribosomes and reduce translation (Corrigan et al., 2016). Within 60–120 min, cell wall biosynthesis is reduced by ζ-mediated phosphorylation of a UNAG fraction, leading to accumulation of unreactive UNAG-3P (Mutschler et al., 2011; Lioy et al., 2012), and by (p)ppGpp inhibition of peptidoglycan metabolism (Eymann et al., 2002). All these metabolic changes are reversible, however, because when the stress condition is relieved (or after artificial induction of antitoxin expression), the antitoxin ε$_2$ reverses the ζ-induced dormant state and the cell population "awakens" (Tabone et al., 2014a,b).

When bacterial growth is challenged by addition of antibiotic, susceptible cells stop growing, but a small subpopulation shows persistence (a biphasic time-inactivation curve) or tolerance to the drug (a linear time-inactivation curve; see **Figure 1**; Lewis, 2010; Amato et al., 2013; Brauner et al., 2016). These complex phenotypes have been attributed to diverse stochastically induced stresses, with the toxin reducing the activity of the antibiotic or enhancing efflux activities to form persisters or tolerant cells (Lewis, 2010; Balaban et al., 2013; Brauner et al., 2016; Harms et al., 2016) or to produce cells susceptible to antibiotic action, as in *B. subtilis* (Wu et al., 2011; Tabone et al., 2014b). Toxin ζ increases (p)ppGpp and decreases GTP pools, thus decreasing antibiotic persistence/tolerance formation; in contrast, low, dysregulated (p)ppGpp levels (in the Δ*relA* context) increase toxin and antibiotic persistence/tolerance (Tabone et al., 2014b).

To analyze how toxin ζ helps to induce a growth arrest state (dormancy), how antitoxin ε$_2$ promotes exit from this state, and to learn about the interconnection between toxin action and the persister/tolerant state, we have studied the metabolic activities of purine nucleotides on persister/tolerant bacterial. Transient controlled expression experiments with toxin and antitoxin showed that toxin ζ induced a reversible growth-arrested state

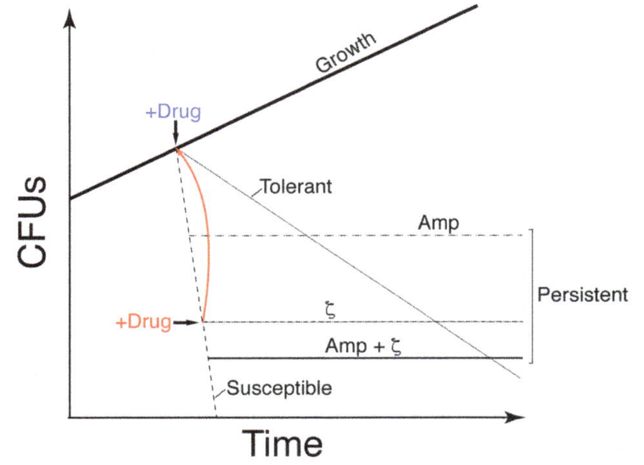

FIGURE 1 | Graphic illustration showing the difference in growth of the different stress survival strategies. Proliferation of susceptible clonal cells is halted by transient toxin ζ expression (IPTG addition) or Amp addition (2x MIC) (+Drug, blue). A large fraction of cells is susceptible to the drug (dashed line); a subpopulation persists and forms colonies, leading to a biphasic time-inactivation curve (ζ [dotdashed] or Amp [twodotted dashed] persisters) rather than a linear time-inactivation curve (tolerants; dotted line). Transient expression of antitoxin ε_2 (+Drug, red) awakens the susceptible cells to toxin ζ action (solid red line). Transient toxin ζ expression and Amp addition yield distinct persister subpopulations.

in a large fraction of proliferating, susceptible *B. subtilis* cells, but that a small subpopulation persists rather than tolerating toxin action (see **Figure 1**). Controlled upregulation of antitoxin ε_2 reversed growth arrest *in vivo* and inhibited the UNAG-dependent ATPase activity of toxin ζ *in vitro*. GdpP- or DisA-dependent alteration of the c-di-AMP pool and CodY-dependent responses revealed that ampicillin (Amp) persisters and ζ-mediated persisters are distinct subpopulations, perhaps with different exit control, and that Amp enhanced killing of ζ-mediated persisters.

MATERIALS AND METHODS

Bacterial Strains and Plasmids

The bacterial strains and plasmids used in this study are listed in **Table 1**. All *B. subtilis* strains are isogenic with BG214. *Escherichia coli* BL21(DE3) cells harboring pBT290-borne ε gene under the transcriptional control of the T7 RNA polymerase-dependent promoter (P_{T7}), or pCB920-borne wild type (*wt*) ζ gene under the control of P_{T7} and ε gene under its native RNA polymerase σ^A-dependent promoter (P_ω) were used for protein purification as described (Camacho et al., 2002; Tabone et al., 2014b).

Growth Conditions

The BG214 derivatives were grown to mid-exponential phase ($\sim 5 \times 10^7$ cells ml^{-1}) at 37°C in minimal medium S7 (MMS7) supplemented with the necessary amino acid (Lioy et al., 2006). Except for Δ*relA*, strains were grown in MMS7 with methionine and tryptophan at 50 μg ml^{-1} each (Lioy et al., 2006). The Δ*relA*

TABLE 1 | Bacterial strains.

Strains	Relevant genotype	References
BG1125[a,b]	+ *lacI*, P_{hsp} ζ, *aadA* [pCB799-*xylR*, P_{xylA} ε, *cat*]	Lioy et al., 2006
BG689[a]	+ *xylR*, P_{xylA} ζY83C, *cat*	Lioy et al., 2006
BG1145[a]	+ Δ*relA, xylR*, P_{xylA} ζY83C, *cat*	Lioy et al., 2012
BG1325[a]	+ Δ*gdpP*, *xylR*, P_{xylA} ζY83C, *cat*	This study
BG1323[a]	+ Δ*disA*, *xylR*, P_{xylA} ζY83C, *cat*	This study
BG1525[a]	+ *codY*::(*erm*::*spc*), *xylR*, P_{xylA}ζY83C, *cat*	This study
BG1527[a]	+ *codY*::(*erm*::*spc*), Δ*disA*, *xylR*, P_{xylA} ζY83C, *cat*	This study
BL21(DE3)[c]	+ [pCB920, P_{T7} ζ gene, P_ω ω and ε genes, *bla*]	Tabone et al., 2014a
BL21(DE3)[c]	+ [pBT290, P_{T7} ε gene, *bla*]	Camacho et al., 2002

[a]*All Bacillus subtilis strains are isogenic with BG214 (trpCE metA5 amyE1 ytsJ1 rsbV37 xre1 xkdA1 att^{SPß} att^{ICEBs1}).*
[b]*BG1125 cells bearing pCB799-borne ε gene were grown in MMS7 medium containing 0.05% xylose to titrate basal expression of the wt ζ toxin.*
[c]*Escherichia coli BL21(DE3) genotype (ompT gal [λ DE3, int::lacI::Plac$_{UV5}$::T7 gene 1] fhuA2 [dcm] ΔhsdS).*

strain shows an "auxotrophy phenotype" for valine, leucine, isoleucine and threonine, and was also supplemented with these amino acids (25 μg ml^{-1} each) (Roche, Germany; Lioy et al., 2006).

BG1125 bearing *lacI*-P_{hsp} *wt* ζ and pCB799-borne *xylR*-P_{xylA} *wt* ε (**Table 1**), in which ζ gene expression (transcribed by P_{hsp}) is regulated by IPTG (Calbiochem, Spain) addition and the ε gene (transcribed by P_{xylA}) is regulated by xylose (Xyl, Sigma, USA) addition (Lioy et al., 2012), was grown in MMS7 supplemented with Xyl (0.05%). In the absence of IPTG [Sigma, USA] there are ~40 ζ toxin monomers/colony-forming units (CFU), which lead to genetic rearrangement. To titrate basal ζ toxin levels, traces of Xyl (0.05%) were added to allow synthesis of low but marked ε_2 antitoxin levels by the pCB799-borne ε gene. After IPTG addition, toxin ζ concentration increased in a very short time (10 min) up to ~1,500 ζ monomers/CFU, and its steady-state level remained for at least 240 min; these toxin levels are considered the "physiological concentration" (Lioy et al., 2012). At indicated times, 0.5% Xyl was added to induce antitoxin ε_2 expression, and the culture was incubated 15 min before being plated without inductor or with 0.5% Xyl (Lioy et al., 2012).

In BG689 or BG1145 bearing the *xylR*-P_{xylA} ζY83C cassette (**Table 1**), expression of the toxin ζY83C variant was induced by addition of 0.5% Xyl. BG689 or BG1145 cells were grown in MMS7 to $\sim 5 \times 10^7$ cells ml^{-1} at 37°C. Xylose addition increased ζY83C levels to a plateau within the first 10 min, and the steady-state level of the toxin remained for at least 240 min (Tabone et al., 2014b).

Where indicated, toxin and/or antitoxin expression was induced by adding IPTG and/or Xyl. Before plating, cells were centrifuged and resuspended in fresh LB medium to remove the inductor or the antibiotic, and dilutions were plated on LB agar plates containing glucose (which switches off *xylR*-P_{xylA} cassette expression) or Xyl to express the ε_2 antitoxin. The survival rate was derived from the number of CFU in a given condition relative to CFU of the non-induced/non-antibiotic-treated control. Except Δ*relA*, cells grew in MMS7 with a doubling time of 50–60 min. The doubling time of Δ*relA* cells increased 1.4-fold compared to the BG689 strain. Normal-sized and small colonies were observed in the Δ*relA* and Δ*disA codY* contexts. All plates were incubated for 20 h at 37°C.

The minimum inhibitory concentration (MIC) of Amp [Sigma, USA] was estimated by exposing 1–3 × 10^6 cells ml^{-1} (16 h, 37°C) in MMS7 with shaking (240 rpm). The Amp concentration used (3 μg ml^{-1}) was twice the MIC (2x MIC). In the absence of inducer, the presence of the ζY83C (BG689 strain) or the ζ gene (BG1125 bearing pCB799) does not affect the MIC (Tabone et al., 2014b).

Protein Purification and Biochemical Assays

The *S. pyogenes* pSM19035-encoded ζ gene was overexpressed in *E. coli* BL21(DE3) cells from a rifampicin-resistant T7 RNAP-dependent promoter as reported (Tabone et al., 2014a). In short, IPTG was added to induce the expression of T7 RNAP that transcribed *wt* ζ toxin, and 30 min later rifampicin (Fluka, USA), was added to selectively block the expression of the ω and ε genes. After 120 min of incubation and full decay of the $ε_2$ antitoxin, the cells were harvested. The over-expressed long-living ζ toxin was purified in two steps as described (Tabone et al., 2014a). The fractions containing the ζ protein were dialyzed against buffer A (50 mM Tris-HCl pH 7.5, 80 mM NaCl) containing 50% glycerol and stored at −20°C. The ε gene was overexpressed in *E. coli* BL21(DE3) cells harboring pBT290 under the control of rifampicin-resistant P_{T7} (Ceglowski et al., 1993), and antitoxin $ε_2$ was overexpressed, and purified as described (Camacho et al., 2002). The purified protein was stored in buffer A containing 50% glycerol at −20°C (Camacho et al., 2002).

The ATPase, dATPase or GTPase activities of ζ toxin were measured using a (d)NTP/NADH-linked assay (De La Cruz et al., 2000; Yadav et al., 2012). Reactions (50 μl) contained the indicated concentration of ζ toxin and the NADH enzyme mix (310 μM NADH [Roche, Germany], 100 U ml^{-1} lactic dehydrogenase [Sigma, USA], 500 U ml^{-1} pyruvate kinase [Roche, Germany], and 2.5 mM phosphoenolpyruvate [Roche, Germany]) in buffer B (50 mM Tris-HCl pH 7.5, 50 mM NaCl, 10 mM MgOAc, 1 mM DTT, 50 μg/ml BSA) with the indicated concentration of ATP, GTP or dATP, and 10 mM UNAG or uridine diphosphate-N-acetylgalactosamine (UNAGal) [Sigma, USA]. We determined the specific (d)NTPase activity (in μM) by measuring the (d)NDP production rate using a Shimadzu CPS-20A dual-beam spectrophotometer as described (Yadav et al., 2012). A standard curve with known amounts of NADH was obtained and used to convert the rate of ADP/GDP/dADP production from absorbance/time to concentration/rate (De La Cruz et al., 2000; Yadav et al., 2012).

RESULTS

Toxin ζ Preferentially Hydrolyzes ATP

Toxin ζ hydrolyses ATP, even in the presence of a 10- to 15-fold excess of cold GTP (Tabone et al., 2014a), suggesting that toxin ζ prefers ATP to GTP (Tabone et al., 2014a). To examine these reactions, we purified toxin ζ in the absence of its cognate antitoxin $ε_2$.

In the absence of UNAG, toxin ζ does not undergo autophosphorylation or hydrolyze NTP; with UNAG (2 mM) and 500 nM toxin ζ, only traces of the γ-phosphate of ATP (0.5 mM) were transferred to UNAG (Tabone et al., 2014a). We tested directly for nucleotide used preferentially by toxin ζ. Limiting ζ concentrations (60 nM) were used to analyze ζ-mediated ATP, GTP or dATP hydrolysis in physiological concentrations of UNAG and of nucleotides. The *B. subtilis* intracellular UNAG, ATP, GTP, and dATP pools approached ~10, ~10, ~5 and ~0.02 mM, respectively (Lopez et al., 1979; Lioy et al., 2012; Bittner et al., 2014).

Toxin ζ did not hydrolyze purine nucleotide when UNAG was omitted (**Figure 2A**). At physiological UNAG and ATP concentrations (10 mM each), toxin ζ (60 nM) hydrolyzed ATP in a reaction that rapidly reached saturation, which suggested that ζ is a UNAG-dependent NTPase. The final rate of ζ ATP hydrolysis approached the maximum rate (K_{cat}) of 1520 ± 120 min^{-1} (**Figure 2A**).

The UNAG-dependent ζ ATPase activity was then compared with a *bona fide* ATPase enzyme. When the single-stranded DNA-dependent RecA ATPase was measured in parallel, *B. subtilis* RecA hydrolyzed ATP at near the previously observed K_{cat} of 9 ± 0.3 min^{-1} (Yadav et al., 2014; Carrasco et al., 2015), which suggested that ζ is a very robust ATPase. UNAG-dependent ζ-mediated ATP hydrolysis was nonetheless sensitive to variations in ATP concentration, because when ATP was reduced to half (5 mM), the K_{cat} was reduced ~3-fold (510 ± 44 min^{-1}).

When ATP was replaced by physiological GTP concentrations (5 mM), ζ was able to hydrolyze GTP in a UNAG-dependent manner and the reaction reached saturation in ~7 min. The final steady state rate of GTP hydrolysis was reduced by ~5-fold (K_{cat} 280 ± 47 min^{-1}) compared with physiological ATP concentrations (**Figure 2A**). Increasing the GTP concentration to 10 mM did not improve the reaction.

We analyzed the potential role of dATP as a substrate (**Figure 2A**). In the presence of physiological UNAG and dATP concentrations (10 and 0.02 mM, respectively), we observed no ζ-mediated UNAG-dependent dATP hydrolysis (**Figure 2A**). To test whether ζ catalyzes dATP hydrolysis, we increased its concentration artificially. At a 10-fold excess of dATP (0.2 mM), ~3 min lag time was needed to reach the steady state rate of ζ-mediated dATP hydrolysis; however, saturation was not reached in 30 min reaction (**Figure 2A**). With a 10-fold excess of dATP, its hydrolysis was reduced by ~20-fold compared with ATP. ATP is probably the preferred ζ nucleotide cofactor.

UNAGal Is a Poor Inducer of the Toxin ζ ATPase

Toxin ζ interacts specifically with UNAG rather than UDP-glucose (Mutschler et al., 2011); in addition, *B. subtilis* GalE is able to interconvert UNAG and UDP-N-acetylgalactosamine (UNAGal), and the cell wall contains N-acetylglucosamine and N-acetylgalactosamine (Soldo et al., 2003). To determine whether UNAGal, a C-4 epimer of UNAG, can activate toxin ζ ATPase

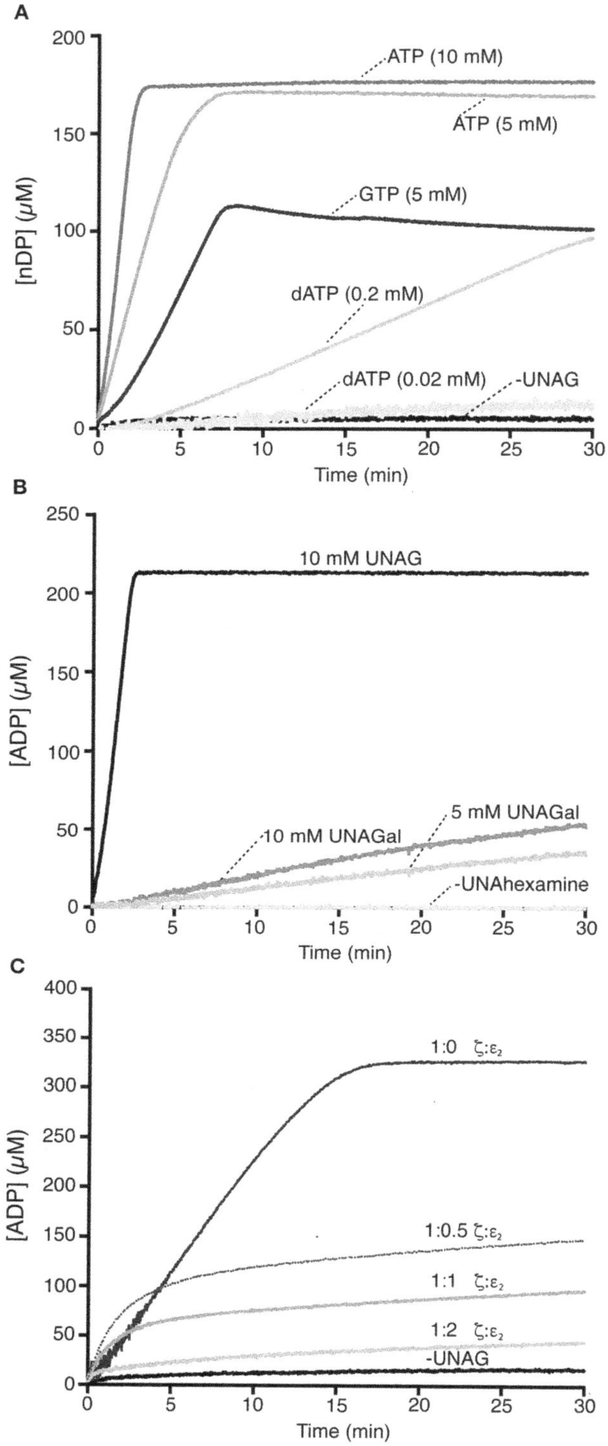

FIGURE 2 | Toxin ζ preferentially hydrolyzes ATP. **(A)** A fixed toxin ζ concentration (60 nM) was incubated (30 min, 37°C) in buffer A containing 10 mM UNAG, and ATP (10 or 5 mM) or GTP (5 mM), or dATP (0.02 or 0.2 mM). The control reaction lacked UNAG. **(B)** Toxin ζ is a UNAG-dependent ATPase. Toxin ζ (60 nM) was incubated as above in buffer A with 10 mM ATP and 10 mM UNAG, or UNAGal (5 or 10 mM). **(C)** Stoichiometric ε_2 antitoxin *(Continued)*

FIGURE 2 | Continued
inhibits the UNAG-dependent ζ ATPase. A fixed ζ toxin concentration (30 nM) and increasing antitoxin ε_2 concentrations (15-60 nM) were incubated (30 min, 37°C) in buffer A containing limiting concentrations of ATP (2 mM) and UNAG (4 mM). The amount of ATP hydrolyzed was calculated (see Section Materials and Methods). The control reaction lacks UNAG. All reactions were repeated three or more times with similar results.

activity, we carried out ATPase assays with increasing UNAGal concentrations.

In the absence of UNAG or UNAGal (minus UNAhexamines), ATP hydrolysis by toxin ζ was at background level (**Figure 2B**). Quantitative analysis of these reactions showed that at physiological UNAGal concentrations, the final ζ-mediated ATP hydrolysis rate was ~85-fold lower (K_{cat} 20 min^{-1}) than ζ in the presence of UNAG. In the presence of a UNAGal excess (10 mM), the K_{cat} was slightly increased (28 min^{-1}), but was still ~60-fold lower than that at physiological UNAG concentrations (**Figure 2B**); this result indicates that ζ ATPase activity is specifically stimulated by UNAG rather than by UNAGal. PetZ similarly accumulates UNAG-3P after 60 min, and UNAGal-3P after 720 min incubation (Mutschler et al., 2011).

Antitoxin ε_2 Inhibits UNAG-Dependent ζ-Mediated ATP Hydrolysis

In vitro, the $\zeta\varepsilon_2\zeta$ complex is reported to hydrolyze ATP and phosphorylate UNAG to form inactive UNAG-3P (Mutschler et al., 2011). In contrast, *in vivo* experiments showed that the ε_2 antitoxin inhibits the effect of toxin ζ, perhaps by forming the inactive $\zeta\varepsilon_2\zeta$ complex (Lioy et al., 2006, 2010). To test whether toxin ζ hydrolyzes ATP in the presence of the antitoxin ε_2, both proteins were purified separately (Camacho et al., 2002; Tabone et al., 2014a) and UNAG-dependent ATPase activity measured.

The antitoxin ε_2, alone or with UNAG, did not hydrolyze ATP (**Figure 2C**). In the presence of UNAG and ATP, the rate of UNAG-dependent ζ-mediated ATP hydrolysis was reduced by increasing antitoxin ε_2 concentrations (**Figure 2C**). At $\zeta{:}\varepsilon_2$ ratios of 1:0.5 or 1:1, the kinetics of ζ-mediated ATP hydrolysis was initially unaltered, but ATP hydrolysis was inhibited after 5 min. At a slight ε_2 excess (1:2 ratio), the antitoxin inhibited ζ ATPase activity (**Figure 2C**). Results were similar when both proteins were preincubated (5 min) at a 1:1 $\zeta{:}\varepsilon_2$ ratio ($\zeta\varepsilon_2\zeta$ complex; not shown), which suggests that when it interacts with ζ, the antitoxin occupies the ATP binding pocket (Meinhart et al., 2003) and inhibits toxin ATPase activity. This is consistent with the crystal structure of the biologically inactive $\zeta\varepsilon_2\zeta$ complex and with the interpretation that antitoxin ε_2 is necessary and sufficient to inactivate toxin ζ. It is likely that the long reaction incubation time (24 h) and/or low ε_2 stability could explain discrepancies with the previous report (Mutschler et al., 2011).

Toxin ζ Induces Reversible Growth Arrest But a Small Subpopulation Evades Its Action

The release of toxins from their cognate antitoxins [or induction of toxin expression (+Drug in blue in **Figure 1**)], should lead to a bimodal time-inactivation curve if persisters appeared (dotdashed line). This deviates from the simple decay, anticipated for a population of only susceptible cells (dashed line) or for a uniformly tolerant bacterial population (dotted line, in **Figure 1**; Brauner et al., 2016; Harms et al., 2016). Inactivation of the toxin by expression of the antitoxin (+Drug in red) should lead to recovery of the plating efficiency (red solid line) if the toxin is bacteriostatic (**Figure 1**). To test whether expression of physiological levels of free toxin ζ induce persistence (dotdashed line) or tolerance (dotted line) and to study the mechanism used for such a phenotype (bacteriostasis or bacteriolysis) we performed long term survival assays. Toxin ζ was induced for a long period, and then antitoxin ε_2 expression was induced (**Figure 3A**).

Bacillus subtilis BG1125 bearing the ζ gene under the control of IPTG induction is prone to rearrangement in the absence of IPTG (Lioy et al., 2012). To overcome this effect, the pCB799-borne ε gene under the control of Xyl was transferred into the background (**Table 1**; see Section Materials and Methods). BG1125 cells bearing pCB799 were grown in MMS7 supplemented with 0.05% Xyl, to $\sim 5 \times 10^7$ cells ml^{-1} ($OD_{560} = 0.2$), and expression of the ζ gene was induced by IPTG addition (time zero). Cells, which formed colonies after plating on LB agar without IPTG, showed a bimodal time-inactivation curve suggesting the presence of persisters (**Figure 3A**), rather than showing a uniform simple decay, expected for tolerant cells (**Figure 1**, dotted line).

To test whether the persisters are due to noise that causes instability in a bacterial population (a reduced cell fraction transiently insensitive to toxin action) or noisy gene expression (a reduced fraction with no toxin expression), we maintained IPTG induction up to 900 min, after which cells were plated in the absence of the inducer. In the former case, only a fraction of the non-replicating dormant cells would exit the arrest state and resume growth after plating without IPTG, whereas in the latter case, cell proliferation of persisters is predicted to increase 8- to 16-fold. After IPTG addition, the small persister subpopulation increased slightly (~3-fold) during the first 240 min, to later remain apparently constant (**Figure 3A**); this suggested negligible biological noise during the first 240 min, and persisters were transiently insensitive to toxin action.

Massive expression of toxin PezT or ζ triggers an irreversible bactericidal effect in *E. coli* grown in rich medium or *B. subtilis* grown in minimal medium, respectively (Zielenkiewicz and Ceglowski, 2005; Mutschler et al., 2011), but physiological concentrations of free toxin ζ induce a reversible bacteriostatic state (Lioy et al., 2012). To identify the source of these discrepancies, we tested whether IPTG-induced growth arrest in *B. subtilis* cells is fully reversible after antitoxin ε_2 expression triggered by 0.5% Xyl (15 min), followed by plating in LB agar with 0.5% Xyl but lacking IPTG. Antitoxin ε_2 expression was sufficient to reverse growth arrest, and most cells recovered proliferation, even after 900 min of toxin ζ action (**Figure 3A**). Although, a reduced fraction (10 to 15% of total cells) were stained with propidium iodide, suggesting a membrane compromise in these cells. It is likely that toxin ζ induces a reversible inhibition of cell growth, and that antitoxin ε_2 expression is necessary and sufficient to switch off toxin-induced responses, with cells awakening and forming colonies even after 15 h of toxin incubation (see **Figure 1**, solid red line), but 10 to 15% of total cells might loss cell viability.

Dysregulated (p)ppGpp Levels Increase the Rate of ζY83C Persisters

Bacillus subtilis encodes one bifunctional RelA synthase-hydrolase and two mono-functional SasA (also termed

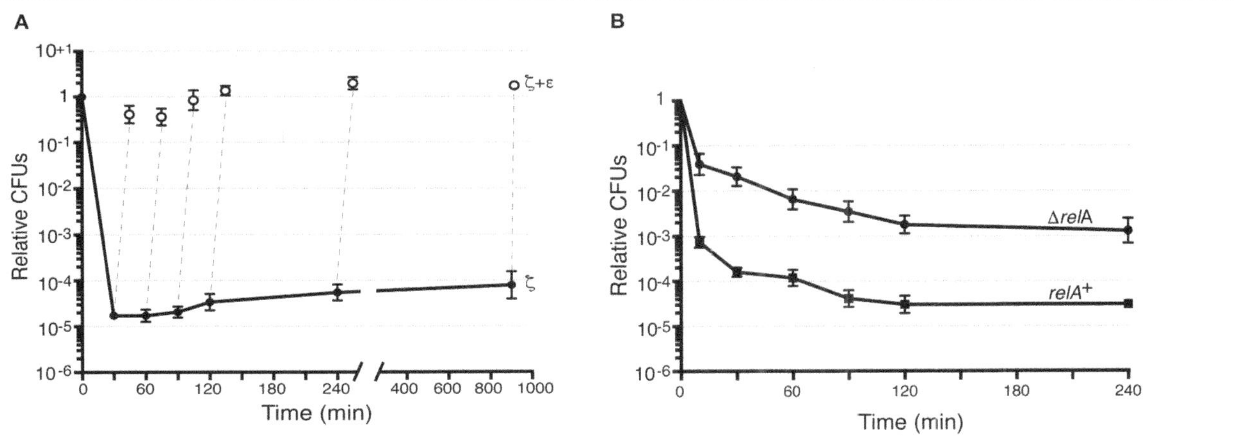

FIGURE 3 | Toxin ζ induces reversible dormancy and selects for pre-existing persisters. **(A)** BG1125 cells (*lacI-P*$_{hsp}$ ζ *spc* cassette) bearing pCB799-borne *xylR-P*$_{XylA}$ε cassette were grown in MMS7 medium containing traces of xylose (Xyl; 0.05%) to $\sim 5 \times 10^7$ cells ml^{-1} (37°C). IPTG (2 mM) was added to half the culture to induce ζ expression (time 0) and the culture was further incubated. At various times, samples were withdrawn and plated in LB agar plates (●, ζ-expressing) or to allow antitoxin expression, 0.5% Xyl was added, the culture incubated (15 min) and plated onto LB agar plates (○, antitoxin ε_2 induction). **(B)** The effect of ζY83C expression on CFU was measured. BG689 (■) or BG1145 (●) cells were cultured in MMS7 to $\sim 5 \times 10^7$ cells ml^{-1} (37°C). Xyl (0.5%) was added to half of the culture to induce ζY83C expression (time 0). At various times, samples were withdrawn and plated onto LB agar plates. Data are shown as mean ± standard error of the mean (SEM), from >4 independent experiments.

YwaC/RelP/Sas1) and SasB (YjbM/RelQ/Sas2) synthases [see Nanamiya et al., 2008; Srivatsan and Wang, 2008]. In the absence of RelA, an excess of GTP over GDP as well as baseline levels of (p)ppGpp "dysregulated" by the SasA and SasB synthases, increase toxin persistence or tolerance by >150-fold (Tabone et al., 2014b). This effect correlates with dysregulated (p)ppGpp levels. Lowering the GTP levels without affecting (p)ppGpp, by treating cells with decoyinine (a GMP synthetase inhibitor), the persistent rate was indistinguishable between treated or untreated Δ*relA* cells (Lioy et al., 2012).

To test whether the CFU increase correlates with a simple decay curve predicted from a uniform tolerant bacterial population or with a biphasic time-inactivation curve due to persistence (**Figure 1**), we analyzed toxin expression in the *relA*$^+$ (BG689) or Δ*relA* (BG1145) cells bearing the *xylR*-P_{xylA} ζY83C cassette (**Table 1**). The *relA*$^+$ and Δ*relA* cells were grown in MMS7 to ~5 × 10^7 ml^{-1}, expression of the ζY83C gene was induced with 0.5% Xyl, and the time-inactivation curve was analyzed. In the absence of RelA, a typical biphasic curve was observed upon expression of physiological concentrations of the toxin, with an ~160-fold (~5 × 10^{-3}) increase in the rate of persisters after plating on LB agar without Xyl (**Figure 3B**).

Varying the c-di-AMP Pool Alters the Rate of Toxin But Not of Amp Persistence

The second messenger c-di-AMP, which is at the heart of cell wall homeostasis, is produced mainly by Gram-positive bacteria of the phyla Firmicutes and Actinobacteria, and by some species of the δ-Proteobacteria class (Corrigan and Grundling, 2013). In Firmicutes, high or low c-di-AMP levels indirectly increase (p)ppGpp (Rao et al., 2010; Corrigan et al., 2015), whereas in *Staphylococcus aureus*, they, respectively, increase or decrease β-lactam tolerance/resistance (Corrigan et al., 2011, 2015). It is not known whether c-di-AMP has a role in toxin ζ responses to stress.

In Firmicutes, intracellular c-di-AMP levels are precisely regulated by two sets of enzymes with opposite effects and by two purine nucleotides. The diadenylate cyclases (DAC) synthesize c-di-AMP from two ATP molecules, the phosphodiesterases (PDE) degrade c-di-AMP into pApA; (p)ppGpp and pApA inhibit PDE enzyme activity (Rao et al., 2010; Corrigan and Grundling, 2013; Huynh and Woodward, 2016), which predicts that c-di-AMP levels increase during starvation. Exponentially growing *B. subtilis* cells express two DAC (DisA and CdaA) and two PDE enzymes (GdpP and PgpH; Rao et al., 2010; Corrigan et al., 2011; Corrigan and Grundling, 2013; Commichau et al., 2015; Huynh and Woodward, 2016). The absence of both DAC or of both PDE causes aberrant physiology and synthetic lethality when the medium contained high K^+ (5 mM KCl), but one representative of each family can be deleted with no apparent effect (Corrigan et al., 2011; Corrigan and Grundling, 2013; Commichau et al., 2015; Huynh and Woodward, 2016; Gundlach et al., 2017). C-di-AMP levels vary marginally (2- to 3-fold) in cells lacking DisA, CdaA or GdpP compared to the *wt* strain (Oppenheimer-Shaanan et al., 2011; Gándara and Alonso, 2015).

To determine how *B. subtilis* cells respond to toxin-mediated stress, we induced transient toxin ζY83C expression and studied the effect of disturbing the c-di-AMP metabolic balance by deleting one DAC (DisA) or one PDE (GdpP) enzyme on an isogenic background (**Table 1**). In parallel, Amp was used as a second stressor at twice the MIC, in anticipation that toxin expression and Amp would respond to different physiological cues. The MIC of Amp was similar in all strains tested. After Amp exposure, a subpopulation of clonal cells yielded a biphasic time-kill curve (twodotted dashed line), which indicated that they persisted rather than becoming Amp-tolerant (dotted line; **Figure 1**). Similar biphasic curves are reported for other bacterial species after Amp exposure (Lewis, 2010; Amato et al., 2013; Brauner et al., 2016; Harms et al., 2016).

Bacillus subtilis cells that lack GdpP show intracellular c-di-AMP levels ~2-fold higher than those of the *wt* strain (Gándara and Alonso, 2015). Absence of GdpP indirectly increases (p)ppGpp pools (Gundlach et al., 2015; Zhu et al., 2016), which suggests that a small number of specific signaling nucleotides integrate and coordinate key metabolic intersections in response to variation of the intracellular c-di-AMP pool. We constructed and analyzed a strain bearing the *xylR*-P_{xylA} ζY83C cassette in the context of Δ*gdpP* (**Table 1**). Regulated ζY83C expression in the *wt* or Δ*gdpP* contexts produced a typical biphasic survival curve, with an initial rapid decrease in CFU and a persistent subpopulation with a stable number of CFU (between 10 and 300 min; not shown); for direct comparison of the various strains, data at 120 min are shown (**Figure 4**). In the *wt* strain, ζY83C expression used (p)ppGpp to mediate rapid inhibition of cell proliferation, and a cell subpopulation entered a toxin- (~7.2 × 10^{-5}) or Amp-persistent state (~2.1 × 10^{-3}; **Figure 4**), as reported (Lioy et al., 2012). In the Δ*gdpP* strain, after transient toxin expression, the persistence rate decreased by ~10-fold (7 × 10^{-6}), but did not significantly affect the persistence rate after

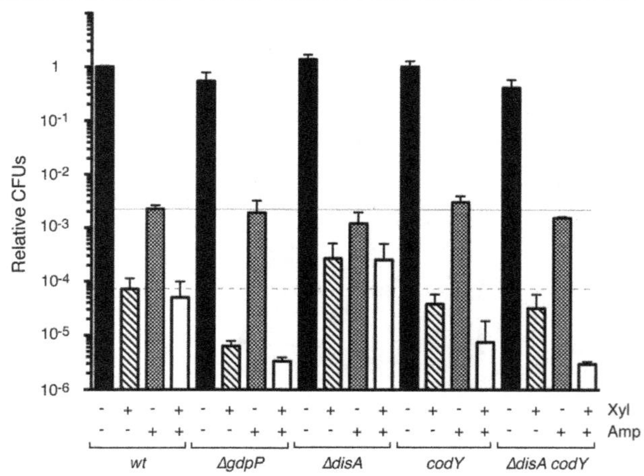

FIGURE 4 | Variation in the c-di-AMP or GTP pools alters the level of ζY83C toxin persisters, but not stochastic Amp persisters during exponential growth. The set of isogenic strains was grown in MMS7 to ~5 × 10^7 cells ml^{-1} (37°C), followed by addition of 0.5% Xyl and/or Amp (3 μg ml^{-1}). Cultures were incubated (120 min), then plated onto LB agar plates. The number of CFU is shown relative to the non-induced/non-Amp-treated control. + and − denote the presence or absence of the indicated compound. Data shown as mean ± SEM, from >4 independent experiments.

Amp addition (~1.7 × 10^{-3}). In the absence of GdpP, exposure to Amp and Xyl decreased the rate of persisters by ~16-fold (~3 × 10^{-6} survivals) compared to the *wt* strain (**Figure 4**), which suggests that a subset of toxin or Amp persisters randomly switched to the susceptible state and were targeted by Amp or the toxin.

Cells lacking DisA have ~2-fold lower levels of intracellular c-di-AMP than *wt* cells (Oppenheimer-Shaanan et al., 2011; Gándara and Alonso, 2015). We constructed the *xylR*-P_{xylA} ζY83C Δ*disA* strain (**Table 1**), and found that the persister cell rate was slightly affected by Amp addition (<2-fold, ~1.5 × 10^{-3} survivals). Transient toxin ζY83C expression increased persisters ~4-fold (~3 × 10^{-4}) in the Δ*disA* compared to the *wt* strain (**Figure 4**). Transient toxin ζY83C expression and Amp addition did not notably affect colony formation compared to addition of Xyl alone (**Figure 4**).

Absence of CodY Alters the Rate of Toxin ζY83C But Not of Amp Persistence

Transient toxin ζ expression decreased the GTP pool in exponentially growing *B. subtilis* cells (Lioy et al., 2012). Intracellular GTP levels have a central role in modulating the stringent response and in reprogramming gene regulation to allow appropriate adaptation to stress. CodY, a GTP-binding protein, is a pleiotropic regulator that senses intracellular branched chain amino acids and GTP levels (Sonenshein, 2007). Low GTP levels, as found during acute stress, release CodY from DNA, leading to deregulation of genes involved in adaptation to nutrient limitation (Ratnayake-Lecamwasam et al., 2001; Belitsky and Sonenshein, 2013; Bittner et al., 2014). The role of CodY in toxin ζ stress responses is unknown. To test whether CodY modulates toxin and/or antibiotic persistence, we constructed the *xylR*-P_{xylA} ζY83C *codY* strain (**Table 1**). Lack of CodY did not markedly alter the Amp persister rate (~2.5 × 10^{-3} survivals; **Figure 4**). After Xyl exposure, however, we observed a slight decrease in the toxin persister rate (~2-fold) compared to *wt* cells (**Figure 4**). Transient Xyl and Amp addition decreased CFU (~8 × 10^{-6} survivals) and the level of persisters decreased by ~5-fold compared to the *wt* strain (**Figure 4**), which suggest that cells lacking CodY adapt poorly to toxin and Amp stress.

Lack of CodY Suppresses Toxin Persistence Triggered by Low c-di-AMP Levels

The absence of DisA increased, and of CodY or GdpP decreased the rate of toxin persistence (**Figure 4**). Since lack of both CodY and GdpP strongly affected cell recovery, but the combined absence of CodY and DisA showed a less stringent phenotype, we constructed the *xylR*-P_{xylA} ζY83C Δ*disA codY* strain (BG1527; **Table 1**). The BG1527 strain yielded colonies with diffuse borders, a 3:1 normal:small size ratio, and viability reduced by ~1.4-fold compared to parental BG689 strains, but lack of CodY and DisA did not notably alter the rate of Amp persisters (~2.5 × 10^{-3} survivals). Following toxin ζY83C expression, we observed a moderate decrease (~2-fold) in the toxin persister rate (~3.5 × 10^{-5} survivals) compared to *wt* cells, similar to the *codY* strain (**Figure 4**). Addition of Xyl and Amp greatly decreased the persistence rate (~3 × 10^{-6} survivals; **Figure 4**). Different clonal subpopulations of persisting cells thus probably evolved differently in response to the toxin and Amp.

DISCUSSION

Toxin ζ represents a class of UNAG-dependent ATPases (**Figure 2A**). As another mechanism to halt cell proliferation, toxin ζ also catalyzes the transfer of part of the ATP γ-phosphate generated upon ATP hydrolysis to a fraction of UNAG, to yield unreactive UNAG-3P (Mutschler et al., 2011; Tabone et al., 2014a). Stoichiometric concentrations of purified antitoxin ε_2 are necessary and sufficient to inactivate toxin ζ action, which suggests that no other factor contributes to ζ inactivation *in vitro*.

Using a set of isogenic *B. subtilis* strains, we tested how purine nucleotide signaling integrates and coordinates the toxin mode of action *in vivo*. Toxin ζ expression induced a biphasic time-inactivation curve with initial rapid, reversible growth arrest of the bulk of susceptible cells; a minor cell subpopulation showed non-inheritable toxin persistence rather than tolerance. Subsequent expression of the ε_2 antitoxin reversed ζ-induced dormancy, and the cells formed colonies even after 900 min of growth arrest. After accumulation of the $\zeta\varepsilon_2\zeta$ complex, the heterogeneous dormancy state is nearly fully reversible; but a reduced subpopulation (up to 15%) of total cells is still stained with propidium iodide. It is likely that the ζ phosphotransferase might compromise the awakening of these cells, which may have a poor fitness or be maladapted. Persisters are formed through redundant mechanisms, and both the biological basis of persistence and the mechanisms that lead to persister formation are poorly understood in Firmicutes. Direct comparison with the well-characterized *E. coli* system could introduce some noise. For example, in both *E. coli* and *B. subtilis* cells, physiological (p)ppGpp levels are necessary for toxin-induced persistence (Korch et al., 2003; Nguyen et al., 2011; Lioy et al., 2012; Amato et al., 2013; Maisonneuve et al., 2013). In the absence of hydrolase-synthase SpoT, *E. coli* cells are not viable (Xiao et al., 1991), but in the *spoT1* context (attenuated hydrolase activity), high levels of dysregulated (p)ppGpp give rise to hypertolerance (Amato et al., 2013; Maisonneuve et al., 2013). In *B. subtilis* cells, lack of the hydrolase-synthase RelA leads to undetectable levels of dysregulated (p)ppGpp, which in turn do not inhibit GTP synthesis and contribute indirectly to hyperpersistence (~160-fold increase). In the absence of (p)ppGpp, there is no persistence signal in *B. subtilis* or *S. aureus* cells, but reduction of GTP and ATP (Tabone et al., 2014b) or ATP levels (Conlon et al., 2016), respectively, leads to cell susceptibility to distinct antibiotics. Indeed, the artificial reduction of the GTP level sensitizes the cells to different antibiotics in the absence of (p)ppGpp (Tabone et al., 2014b). In *E. coli* cells that lack the 10 host-encoded mRNA interferases, levels of persisters to certain antibiotics decrease (Maisonneuve et al., 2011), whereas in *B. subtilis* cells, absence of the single mRNAase NdoA (MazF) increases antibiotic lethality rather than inducing persister cell formation (Wu et al., 2011).

A very small fraction of *E. coli* cells (~0.01%) is reported to have a high (p)ppGpp concentration, which triggers entry into the persistent state (Maisonneuve et al., 2013). Our study addressed the mechanism of persister formation in *B. subtilis* cells in conditions in which toxin/antitoxin expression were controlled by external inducers; antitoxin degradation thus had no role, which rendered unnecessary the analysis of (p)ppGpp in toxin release. If the stochastic switch to produce (p)ppGpp is the sole factor that triggers persister formation, the proportion of toxin and Amp persisters should be similar, and transient toxin expression and Amp addition would not further decrease cell viability. This was not observed. To explain our results, we must assume that a certain cell fraction switches stochastically to the persistent state prior to environmental challenges (Amp persisters), but sensing the metabolic state is of key importance for responsive induction of persistence. Alterations in the GTP (*codY*) or c-di-AMP pools (*gdpP* or *disA*) indicated a constant "awakening" rate after Amp addition, but a variable proportion of toxin persisters (**Figure 4**). Toxin ζ temporarily and reversibly increases the (p)ppGpp pool, reduces the ATP and GTP pools, and modulates c-di-AMP and UNAG levels to allow cells to readjust their metabolism from logarithmic growth to "growth arrest," enabling cells to cope with environmental stress. The pattern of toxin persistence was varied by altering the c-di-AMP pool, which acts as two opposite mechanisms that are negatively and positively controlled by toxin expression. The subpopulation of bet-hedging persister cells that arises before changes in the environment and those triggered by toxin-induced metabolic changes both lead to toxin persisters.

Responsive strategies based on environmental sensing alter phenotypic switching between growth-arrested and persister cells. By varying the intracellular pool of signaling nucleotide, the stochastic subpopulation of toxin ζ persisters varied up to 40-fold (Δ*disA* vs. Δ*gdpP*). When both stress sources (Amp and free toxin) were present, however, a fraction of Amp (or toxin) persisters might awaken and become sensitive to the second stressor, decreasing the rate of persisters by up to 200-fold (Δ*disA* vs. Δ*gdpP* background).

Based on these results and our previous work (Lioy et al., 2006, 2012; Tabone et al., 2014a,b), we propose that the *modus operandi* of toxin ζ-induced growth arrest is to reduce the ATP (by direct hydrolysis) and GTP (by conversion to [p]ppGpp) pools. As a consequence of this, (p)ppGpp levels are increased, and a fraction of UNAG becomes phosphorylated. High (p)ppGpp directly inhibits regeneration and *de novo* GTP synthesis; it positively and negatively regulates the c-di-AMP pool, and decreases the proton-motive force (lowering the ATP pool) as well as UNAG (Kriel et al., 2012). These imbalances induce diverse transient, reversible states to ensure population survival in adverse conditions. Except (p)ppGpp dysregulation on the Δ*relA* background, there is no direct information that a discrete metabolite increases persister formation. ATP depletion is thought to be a general mechanism of persister formation in bacteria (Conlon et al., 2016; Shan et al., 2017), although we found that a reduction in ATP levels leads to ζ-induced growth arrest rather than to persister formation. We propose that the interrelationship between ATP, GTP, (p)ppGpp, c-di-AMP, and UNAG contribute, via a poorly characterized mechanism, to ζ-induced cell growth arrest and persister formation.

AUTHOR CONTRIBUTIONS

MM, VL, MT, and JA conceived and designed the experiments for this study. MM, VL, and MT performed the experiments. JA wrote the manuscript. All authors discussed the data and made comments on the manuscript.

ACKNOWLEDGMENTS

We thank Boris R. Belitsky (Tufts University School of Medicine, Boston, MA USA) for providing the BB1043 (*codY*::[*erm*::*spc*]) mutant strain, Silvia Ayora (CNB-CSIC) for critical comments, and Catherine Mark (CNB-CSIC) for editorial assistance. MT is a PhD fellow of the La Caixa Foundation International Fellowship Programme (La Caixa/CNB). This study was supported by the Spanish Ministerio de Economía y Competividad and the European Union (MINECO-FEDER) BFU2015-67065-P to JA.

REFERENCES

Amato, S. M., Orman, M. A., and Brynildsen, M. P. (2013). Metabolic control of persister formation in *Escherichia coli*. *Mol. Cell* 50, 475–487. doi: 10.1016/j.molcel.2013.04.002

Balaban, N. Q., Gerdes, K., Lewis, K., and McKinney, J. D. (2013). A problem of persistence: still more questions than answers? *Nat. Rev. Microbiol.* 11, 587–591. doi: 10.1038/nrmicro3076

Belitsky, B. R., and Sonenshein, A. L. (2013). Genome-wide identification of *Bacillus subtilis* CodY-binding sites at single-nucleotide resolution. *Proc. Natl. Acad. Sci. U.S.A.* 110, 7026–7031. doi: 10.1073/pnas.1300428110

Bittner, A. N., Kriel, A., and Wang, J. D. (2014). Lowering GTP level increases survival of amino acid starvation but slows growth rate for *Bacillus subtilis* cells lacking (p)ppGpp. *J. Bacteriol.* 196, 2067–2076. doi: 10.1128/JB.01471-14

Brauner, A., Fridman, O., Gefen, O., and Balaban, N. Q. (2016). Distinguishing between resistance, tolerance and persistence to antibiotic treatment. *Nat. Rev. Microbiol.* 14, 320–330. doi: 10.1038/nrmicro.2016.34

Brinsmade, S. R., Alexander, E. L., Livny, J., Stettner, A. I., Segre, D., Rhee, K. Y., et al. (2014). Hierarchical expression of genes controlled by the *Bacillus subtilis* global regulatory protein CodY. *Proc. Natl. Acad. Sci. U.S.A.* 111, 8227–8232. doi: 10.1073/pnas.1321308111

Camacho, A. G., Misselwitz, R., Behlke, J., Ayora, S., Welfle, K., Meinhart, A., et al. (2002). *In vitro* and *in vivo* stability of the $\varepsilon_2\zeta_2$ protein complex of the broad host-range *Streptococcus pyogenes* pSM19035 addiction system. *Biol. Chem.* 383, 1701–1713. doi: 10.1515/bc.2002.191

Carrasco, B., Yadav, T., Serrano, E., and Alonso, J. C. (2015). *Bacillus subtilis* RecO and SsbA are crucial for RecA-mediated recombinational DNA repair. *Nucleic Acids Res.* 43, 5984–5997. doi: 10.1093/nar/gkv545

Ceglowski, P., Boitsov, A., Chai, S., and Alonso, J. C. (1993). Analysis of the stabilization system of pSM19035-derived plasmid pBT233 in *Bacillus subtilis*. *Gene* 136, 1–12. doi: 10.1016/0378-1119(93)90441-5

Commichau, F. M., Dickmanns, A., Gundlach, J., Ficner, R., and Stulke, J. (2015). A jack of all trades: the multiple roles of the unique essential second messenger cyclic di-AMP. *Mol. Microbiol.* 97, 189–204. doi: 10.1111/mmi.13026

Conlon, B. P., Rowe, S. E., Gandt, A. B., Nuxoll, A. S., Donegan, N. P., Zalis, E. A., et al. (2016). Persister formation in *Staphylococcus aureus* is associated with ATP depletion. *Nat. Microbiol* 1, 16051. doi: 10.1038/nmicrobiol.2016.51

Corrigan, R. M., Abbott, J. C., Burhenne, H., Kaever, V., and Grundling, A. (2011). c-di-AMP is a new second messenger in *Staphylococcus aureus* with a role in controlling cell size and envelope stress. *PLoS Pathog.* 7:e1002217. doi: 10.1371/journal.ppat.1002217

Corrigan, R. M., Bellows, L. E., Wood, A., and Grundling, A. (2016). ppGpp negatively impacts ribosome assembly affecting growth and antimicrobial tolerance in Gram-positive bacteria. *Proc. Natl. Acad. Sci. U.S.A.* 113, E1710–E1719. doi: 10.1073/pnas.1522179113

Corrigan, R. M., Bowman, L., Willis, A. R., Kaever, V., and Grundling, A. (2015). Cross-talk between two nucleotide-signaling pathways in *Staphylococcus aureus*. *J. Biol. Chem.* 290, 5826–5839. doi: 10.1074/jbc.M114.598300

Corrigan, R. M., and Grundling, A. (2013). Cyclic di-AMP: another second messenger enters the fray. *Nat. Rev. Microbiol.* 11, 513–524. doi: 10.1038/nrmicro3069

De La Cruz, E. M., Sweeney, H. L., and Ostap, E. M. (2000). ADP inhibition of myosin V ATPase activity. *Biophys. J.* 79, 1524–1529. doi: 10.1016/S0006-3495(00)76403-4

Eymann, C., Homuth, G., Scharf, C., and Hecker, M. (2002). *Bacillus subtilis* functional genomics: global characterization of the stringent response by proteome and transcriptome analysis. *J. Bacteriol.* 184, 2500–2520. doi: 10.1128/JB.184.9.2500-2520.2002

Gándara, C., and Alonso, J. C. (2015). DisA and c-di-AMP act at the intersection between DNA-damage response and stress homeostasis in exponentially growing *Bacillus subtilis* cells. *DNA Repair* 27, 1–8. doi: 10.1016/j.dnarep.2014.12.007

Gerdes, K. (2013). *Prokaryotic Toxin-antitoxins*. Berlin: Springer. doi: 10.1007/978-3-642-33253-1

Gundlach, J., Herzberg, C., Kaever, V., Gunka, K., Hoffmann, T., Weiss, M., et al. (2017). Control of potassium homeostasis is an essential function of the second messenger cyclic di-AMP in *Bacillus subtilis*. *Sci. Signal.* 10:eaal3011. doi: 10.1126/scisignal.aal3011

Gundlach, J., Mehne, F. M., Herzberg, C., Kampf, J., Valerius, O., Kaever, V., et al. (2015). An essential poison: synthesis and degradation of cyclic Di-AMP in *Bacillus subtilis*. *J. Bacteriol.* 197, 3265–3274. doi: 10.1128/JB.00564-15

Hahn, J., Tanner, A. W., Carabetta, V. J., Cristea, I. M., and Dubnau, D. (2015). ComGA-RelA interaction and persistence in the *Bacillus subtilis* K-State. *Mol. Microbiol.* 97, 454–471. doi: 10.1111/mmi.13040

Handke, L. D., Shivers, R. P., and Sonenshein, A. L. (2008). Interaction of *Bacillus subtilis* CodY with GTP. *J. Bacteriol.* 190, 798–806. doi: 10.1128/JB.01115-07

Harms, A., Maisonneuve, E., and Gerdes, K. (2016). Mechanisms of bacterial persistence during stress and antibiotic exposure. *Science* 354:aaf4268. doi: 10.1126/science.aaf4268

Huynh, T. N., and Woodward, J. J. (2016). Too much of a good thing: regulated depletion of c-di-AMP in the bacterial cytoplasm. *Curr. Op. Microbiol.* 30, 22–29. doi: 10.1016/j.mib.2015.12.007

Khoo, S. K., Loll, B., Chan, W. T., Shoeman, R. L., Ngoo, L., Yeo, C. C., et al. (2007). Molecular and structural characterization of the PezAT chromosomal toxin-antitoxin system of the human pathogen *Streptococcus pneumoniae*. *J. Biol. Chem.* 282, 19606–19618. doi: 10.1074/jbc.M701703200

Korch, S. B., Henderson, T. A., and Hill, T. M. (2003). Characterization of the hipA7 allele of *Escherichia coli* and evidence that high persistence is governed by (p)ppGpp synthesis. *Mol. Microbiol.* 50, 1199–1213. doi: 10.1046/j.1365-2958.2003.03779.x

Krasny, L., and Gourse, R. L. (2004). An alternative strategy for bacterial ribosome synthesis: *Bacillus subtilis* rRNA transcription regulation. *EMBO J.* 23, 4473–4483. doi: 10.1038/sj.emboj.7600423

Kriel, A., Bittner, A. N., Kim, S. H., Liu, K., Tehranchi, A. K., Zou, W. Y., et al. (2012). Direct regulation of GTP homeostasis by (p)ppGpp: a critical component of viability and stress resistance. *Mol. Cell* 48, 231–241. doi: 10.1016/j.molcel.2012.08.009

Leplae, R., Geeraerts, D., Hallez, R., Guglielmini, J., Dreze, P., and Van Melderen, L. (2011). Diversity of bacterial type II toxin-antitoxin systems: a comprehensive search and functional analysis of novel families. *Nucleic Acids Res.* 39, 5513–5525. doi: 10.1093/nar/gkr131

Lewis, K. (2010). Persister cells. *Annu. Rev. Microbiol.* 64, 357–372. doi: 10.1146/annurev.micro.112408.134306

Lioy, V. S., Machon, C., Tabone, M., Gonzalez-Pastor, J. E., Daugelavicius, R., Ayora, S., et al. (2012). The ζ toxin induces a set of protective responses and dormancy. *PLoS ONE* 7:e30282. doi: 10.1371/journal.pone.0030282

Lioy, V. S., Martin, M. T., Camacho, A. G., Lurz, R., Antelmann, H., Hecker, M., et al. (2006). pSM19035-encoded ζ toxin induces stasis followed by death in a subpopulation of cells. *Microbiology* 152, 2365–2379. doi: 10.1099/mic.0.28950-0

Lioy, V. S., Rey, O., Balsa, D., Pellicer, T., and Alonso, J. C. (2010). A toxin-antitoxin module as a target for antimicrobial development. *Plasmid* 63, 31–39. doi: 10.1016/j.plasmid.2009.09.005

Liu, K., Bittner, A. N., and Wang, J. D. (2015). Diversity in (p)ppGpp metabolism and effectors. *Curr. Op. Microbiol.* 24, 72–79. doi: 10.1016/j.mib.2015.01.012

Lopez, J. M., Dromerick, A., and Freese, E. (1981). Response of guanosine 5'-triphosphate concentration to nutritional changes and its significance for *Bacillus subtilis* sporulation. *J. Bacteriol.* 146, 605–613.

Lopez, J. M., Marks, C. L., and Freese, E. (1979). The decrease of guanine nucleotides initiates sporulation of *Bacillus subtilis*. *Biochim. Biophys. Acta* 587, 238–252. doi: 10.1016/0304-4165(79)90357-X

Maisonneuve, E., Castro-Camargo, M., and Gerdes, K. (2013). (p)ppGpp controls bacterial persistence by stochastic induction of toxin-antitoxin activity. *Cell* 154, 1140–1150. doi: 10.1016/j.cell.2013.07.048

Maisonneuve, E., Shakespeare, L. J., Jorgensen, M. G., and Gerdes, K. (2011). Bacterial persistence by RNA endonucleases. *Proc. Natl. Acad. Sci. U.S.A.* 108, 13206–13211. doi: 10.1073/pnas.1100186108

Meinhart, A., Alings, C., Strater, N., Camacho, A. G., Alonso, J. C., and Saenger, W. (2001) Crystallization and preliminary X-ray diffraction studies of the εζ addiction system encoded by *Streptococcus pyogenes* plasmid pSM19035. *Acta Crystallogr. D Biol. Crystallogr.* 57, 745–747. doi: 10.1107/S0907444901004176

Meinhart, A., Alonso, J. C., Strater, N., and Saenger, W. (2003). Crystal structure of the plasmid maintenance system ε/ζ: functional mechanism of toxin ζ and inactivation by $\varepsilon_2\zeta_2$ complex formation. *Proc. Natl. Acad. Sci. U.S.A.* 100, 1661–1666. doi: 10.1073/pnas.0434325100

Mutschler, H., Gebhardt, M. Shoeman, R. L., and Meinhart, A. (2011). A novel mechanism of programmed cell death in bacteria by toxin-antitoxin systems corrupts peptidoglycan synthesis. *PLoS Biol.* 9:e1001033. doi: 10.1371/journal.pbio.1001033

Mutschler, H., and Meinhart, A. (2013). Type II Toxin-Antitoxin Loci: The ε/ζ Family," in *Prokaryotic Toxin-Antitoxins*, ed K. Gerdes (Berlin; Heidelberg: Springer-Verlag), 205–223.

Nanamiya, H., Kasai, K., Nozawa, A., Yun, C. S., Narisawa, T., Murakami, K., et al. (2008). Identification and functional analysis of novel (p)ppGpp synthetase genes in *Bacillus subtilis*. *Mol. Microbiol.* 67, 291–304. doi: 10.1111/j.1365-2958.2007.06018.x

Nguyen, D., Joshi-Datar, A., Lepine, F., Bauerle, E., Olakanmi, O., Beer, K., et al. (2011). Active starvation responses mediate antibiotic tolerance in biofilms and nutrient-limited bacteria. *Science* 334, 982–986. doi: 10.1126/science.1211037

Oppenheimer-Shaanan, Y., Wexselblatt, E., Katzhendler, J., Yavin, E., and Ben-Yehuda, S. (2011). c-di-AMP reports DNA integrity during sporulation in *Bacillus subtilis*. *EMBO Rep.* 12, 594–601. doi: 10.1038/embor.2011.77

Pedley, A. M., and Benkovic S. J. (2017). A new view into the regulation of purine metabolism: the purinosome. *Trends Biochem. Sci.* 42, 141–154. doi: 10.1016/j.tibs.2016.09.009

Potrykus, K., and Cashel, M. (2008). (p)ppGpp: still magical? *Annu. Rev. Microbiol.* 62, 35–51. doi: 10.1146/annurev.micro.62.081307.162903

Rao, F., See, R. Y., Zhang, D., Toh, D. C., Ji, Q., and Liang, Z. X. (2010). YybT is a signaling protein that contains a cyclic dinucleotide phosphodiesterase domain and a GGDEF domain with ATPase activity. *J. Biol. Chem.* 285, 473–482. doi: 10.1074/jbc.M109.040238

Ratnayake-Lecamwasam, M., Serror, P., Wong, K. W., and Sonenshein, A. L. (2001). *Bacillus subtilis* CodY represses early-stationary-phase genes by sensing GTP levels. *Genes Dev.* 15, 1093–1103. doi: 10.1101/gad.874201

Shan, Y., Brown Gandt, A., Rowe, S. E., Deisinger, J. P., Conlon, B. P., and Lewis, K. (2017). ATP-dependent persister formation in *Escherichia coli*. *MBio* 8, e02267–e02216. doi: 10.1128/mbio.02267-16

Soldo, B., Scotti, C., Karamata, D., and Lazarevic, V. (2003). The *Bacillus subtilis* Gne (GneA, GalE) protein can catalyse UDP-glucose as well as UDP-N-acetylglucosamine 4-epimerisation. *Gene* 319, 65–69. doi: 10.1016/S0378-1119(03)00793-5

Sonenshein, A. L. (2007). Control of key metabolic intersections in *Bacillus subtilis*. *Nat. Rev. Microbiol.* 5, 917–927. doi: 10.1038/nrmicro1772

Srivatsan, A., and Wang, J. D. (2008). Control of bacterial transcription, translation and replication by (p)ppGpp. *Curr. Op. Microbiol.* 11, 100–105. doi: 10.1016/j.mib.2008.02.001

Tabone, M., Ayora, S., and Alonso, J. C. (2014a). Toxin ζ reversible induces dormancy and reduces the UDP-N-acetylglucosamine pool as one of the protective responses to cope with stress. *Toxins (Basel)* 6, 2787–2803. doi: 10.3390/toxins6092787

Tabone, M., Lioy, V. S., Ayora, S., Machon, C., and Alonso, J. C. (2014b). Role of toxin ζ and starvation responses in the sensitivity to antimicrobials. *PLoS ONE* 9:e86615. doi: 10.1371/journal.pone.0086615

Unterholzner, S. J., Poppenberger, B., and Rozhon, W. (2013). Toxin-antitoxin systems: biology, identification, and application. *Mob. Genet. Elements* 3:e26219. doi: 10.4161/mge.26219

Wang, J. D., Sanders, G. M., and Grossman, A. D. (2007). Nutritional control of elongation of DNA replication by (p)ppGpp. *Cell* 128, 865–875. doi: 10.1016/j.cell.2006.12.043

Witte, G., Hartung, S., Buttner, K., and Hopfner, K. P. (2008). Structural biochemistry of a bacterial checkpoint protein reveals diadenylate cyclase activity regulated by DNA recombination intermediates. *Mol. Cell* 30, 167–178. doi: 10.1016/j.molcel.2008.02.020

Wu, X., Wang, X., Drlica, K., and Zhao, X. (2011). A toxin-antitoxin module in *Bacillus subtilis* can both mitigate and amplify effects of lethal stress. *PLoS ONE* 6:e23909. doi: 10.1371/journal.pone.0023909

Xiao, H., Kalman, M., Ikehara, K., Zemel, S., Glaser, G., and Cashel, M. (1991). Residual guanosine 3',5'-bispyrophosphate synthetic activity of *relA* null mutants can be eliminated by *spoT* null mutations. *J. Biol. Chem.* 266, 5980–5990.

Yadav, T., Carrasco, B., Myers, A. R., George, N. P., Keck, J. L., and Alonso, J. C. (2012). Genetic recombination in *Bacillus subtilis*: a division of labor between two single-strand DNA-binding proteins. *Nucleic Acids Res.* 40, 5546–5559. doi: 10.1093/nar/gks173

Yadav, T., Carrasco, B., Serrano, E., and Alonso, J. C. (2014). Roles of *Bacillus subtilis* DprA and SsbA in RecA-mediated genetic recombination. *J. Biol. Chem.* 289, 27640–27652. doi: 10.1074/jbc.M114.577924

Yamaguchi, Y., Park, J. H., and Inouye, M. (2011). Toxin-antitoxin systems in bacteria and archaea. *Annu. Rev. Genet.* 45, 61–79. doi: 10.1146/annurev-genet-110410-132412

Zhu, Y., Pham, T. H., Nhiep, T. H., Vu, N. M., Marcellin, E., Chakrabortti, A., et al. (2016). Cyclic-di-AMP synthesis by the diadenylate cyclase CdaA is modulated by the peptidoglycan biosynthesis enzyme GlmM in *Lactococcus lactis*. *Mol. Microbiol.* 99, 1015–1027. doi: 10.1111/mmi.13281

Zielenkiewicz, U., and Ceglowski, P. (2005). The toxin-antitoxin system of the streptococcal plasmid pSM19035. *J. Bacteriol.* 187, 6094–6105. doi: 10.1128/JB.187.17.6094-6105.2005

7

Features of CRISPR-Cas Regulation Key to Highly Efficient and Temporally-Specific crRNA Production

Andjela Rodic [1,2], Bojana Blagojevic [3], Magdalena Djordjevic [3], Konstantin Severinov [4,5] and Marko Djordjevic [1]*

[1] Faculty of Biology, Institute of Physiology and Biochemistry, University of Belgrade, Belgrade, Serbia, [2] Multidisciplinary PhD Program in Biophysics, University of Belgrade, Belgrade, Serbia, [3] Institute of Physics Belgrade, University of Belgrade Belgrade, Serbia, [4] Waksman Institute of Microbiology, Rutgers University, Piscataway, NJ, United States, [5] Skolkovo Institute of Science and Technology, Skolkovo, Russia

Edited by:
Tatiana Venkova,
University of Texas Medical Branch, United States

Reviewed by:
Jintao Liu,
University of California, San Diego, United States
Robert Martin Blumenthal,
University of Toledo, United States
Andrea Ciliberto,
IFOM - The FIRC Institute of Molecular Oncology, Italy

***Correspondence:**
Marko Djordjevic
dmarko@bio.bg.ac.rs

Bacterial immune systems, such as CRISPR-Cas or restriction-modification (R-M) systems, affect bacterial pathogenicity and antibiotic resistance by modulating horizontal gene flow. A model system for CRISPR-Cas regulation, the Type I-E system from *Escherichia coli*, is silent under standard laboratory conditions and experimentally observing the dynamics of CRISPR-Cas activation is challenging. Two characteristic features of CRISPR-Cas regulation in *E. coli* are cooperative transcription repression of *cas* gene and CRISPR array promoters, and fast non-specific degradation of full length CRISPR transcripts (pre-crRNA). In this work, we use computational modeling to understand how these features affect the system expression dynamics. Signaling which leads to CRISPR-Cas activation is currently unknown, so to bypass this step, we here propose a conceptual setup for *cas* expression activation, where *cas* genes are put under transcription control typical for a restriction-modification (R-M) system and then introduced into a cell. Known transcription regulation of an R-M system is used as a proxy for currently unknown CRISPR-Cas transcription control, as both systems are characterized by high cooperativity, which is likely related to similar dynamical constraints of their function. We find that the two characteristic CRISPR-Cas control features are responsible for its temporally-specific dynamical response, so that the system makes a steep (switch-like) transition from OFF to ON state with a time-delay controlled by pre-crRNA degradation rate. We furthermore find that cooperative transcription regulation qualitatively leads to a cross-over to a regime where, at higher pre-crRNA processing rates, crRNA generation approaches the limit of an infinitely abrupt system induction. We propose that these dynamical properties are associated with rapid expression of CRISPR-Cas components and efficient protection of bacterial cells against foreign DNA. In terms of synthetic applications, the setup proposed here should allow highly efficient expression of small RNAs in a narrow time interval, with a specified time-delay with respect to the signal onset.

Keywords: CRISPR-Cas activation, pre-crRNA processing, CRISPR regulation, crRNA generation, biophysical modeling

INTRODUCTION

CRISPR-Cas are adaptive immune systems, which defend prokaryotic cells against foreign DNA, including viruses and plasmids. A CRISPR-Cas system consists of a CRISPR (Clustered Regularly Interspaced Short Palindromic Repeats) array and associated *cas* genes (Makarova et al., 2006; Barrangou et al., 2007; Brouns et al., 2008; Hille and Charpentier, 2016). CRISPR arrays consist of identical direct repeats (R) of about 30 bp in length, interspaced with spacers (S) of similar length and variable sequence. Spacer sequences are often complementary to fragments of viral or plasmid DNA. A match between a CRISPR spacer and invading phage (bacterial virus) sequence provides immunity to infection (Barrangou et al., 2007; Hille and Charpentier, 2016). The entire CRISPR locus is initially transcribed as a long transcript (called pre-crRNA) (Pougach et al., 2010; Pul et al., 2010), which is further processed by Cas proteins to small protective CRISPR RNAs (called crRNAs) (Brouns et al., 2008; Pougach et al., 2010; Djordjevic et al., 2012). crRNAs are responsible for recognition and, together with Cas proteins, inactivation of invading foreign genetic elements (Brouns et al., 2008; Al-Attar et al., 2011). Cas proteins also take part in CRISPR adaptation, which is a process in which new spacers from viral genomes are inserted in CRISPR array. **Figure 1** shows a schematic gene diagram for Type I-E CRISPR-Cas from *E. coli*, (Mojica and Diez-Villasenor, 2010; Patterson et al., 2017), which we consider in this paper. The *cas* genes and the CRISPR array are transcribed from separate promoters, which are located inside of the intergenic regions here denoted by IGLB and L (the leader sequence), respectively (see **Figure 1**; Pougach et al., 2010; Pul et al., 2010).

Promoters for *cas* operon and the CRISPR array are repressed in Type I-E CRISPR-Cas in *E. coli* (Pougach et al., 2010; Pul et al., 2010; Westra et al., 2010), which makes this system silent under standard conditions. Consequently, to generate crRNAs that can protect the bacterial cell, CRISPR-Cas has to be activated. Thus, to understand the system function it is crucial to understand the main features that control dynamics of CRISPR-Cas activation (Mojica and Diez-Villasenor, 2010; Richter et al., 2012; Patterson et al., 2017). However, approaching this problem experimentally is complicated due to the following:

i. It requires direct experimental observation of *in vivo* dynamics of molecular species (proteins or RNA) in a cell (see e.g., Morozova et al., 2015).
ii. The signaling which leads to system induction is currently unclear (Ratner et al., 2015; Patterson et al., 2017), e.g., even a viral infection, an obvious trigger, is not sufficient to activate the system.
iii. To understand the roles of the key system features in its response/dynamics these features would have to be perturbed, which may require extensive reengineering of the system.

A complementary approach is to use mathematical/biophysical modeling to assess how different features of CRISPR-Cas expression affect system dynamics. Moreover, *in silico* analysis allows one to study alternative system architectures, and/or to perturb the natural system (see e.g., Rodic et al., 2017), which in turn allows understanding the role of its key regulatory features.

Experimental research has led to a consistent picture of the main CRISPR-Cas regulatory features in closely related *E. coli* and *Salmonella enterica* (Pul et al., 2010; Westra et al., 2010; Medina-Aparicio et al., 2011). Under standard conditions, promoters for both CRISPR array and *cas* genes are repressed by global regulators (H-NS and LRP). Repression by these regulators is highly cooperative, as their binding is nucleated at certain position, and then extends along the DNA through cooperative interactions between repressor molecules (Bouffartigues et al., 2007). Additional regulators, such as CRP, may also be involved in the repression of *cas* operon (Yang et al., 2014). While the exact signaling mechanism remains unclear, this repression must be relieved upon appropriate external signal (e.g., envelope stress that may signal bacteriophage invasion), through the action of transcription activators (LexA, LeuO, and BaeR-S are likely involved) (Richter et al., 2012; Patterson et al., 2017). In particular, for Type I-E CRISPR-Cas in *E. coli*, it was shown that cooperative repression by H-NS can be relieved by elevated amount of LeuO (Pul et al., 2010; Westra et al., 2010). Thus, highly cooperative repression, which is abolished by transcription activators, emerges as a major feature of CRISPR-Cas transcription control in *E. coli* and its relatives.

Another crucial mechanism in CRISPR-Cas expression is pre-crRNA transcript processing (Brouns et al., 2008; Pougach et al., 2010). Experiments in *E. coli*, reported that overexpression of

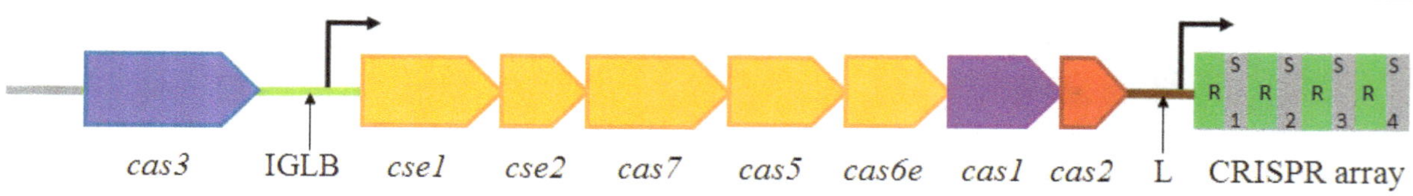

FIGURE 1 | A scheme of a Type I-E CRISPR-Cas system from *E coli* (Al-Attar et al., 2011; Makarova et al., 2011, 2015). The *cas* genes and the CRISPR array are indicated. R and S within the CRISPR array correspond, respectively, to repeats and spacers; note that the spacer sequences differ from each other, and are labeled by consecutive numbers (1, 2, 3...). IGLB and L correspond to intergenic regions where promoters for the Cascade complex genes (*cse1,2*, *cas7,5,6e*) and the Cas1,2 adaptation proteins (IGLB) and the CRISPR array (L) are located. The two promoters within IGLB and L are indicated by arrows. One of the Cas proteins (Cas6e) is responsible for processing pre-crRNA to crRNA. The effector Cascade complex is composed of proteins encoded by genes marked with yellow color. It binds crRNA, which recognizes invading DNA. Once recognized, foreign DNA is destroyed by the product of *cas3* (Brouns et al., 2008).

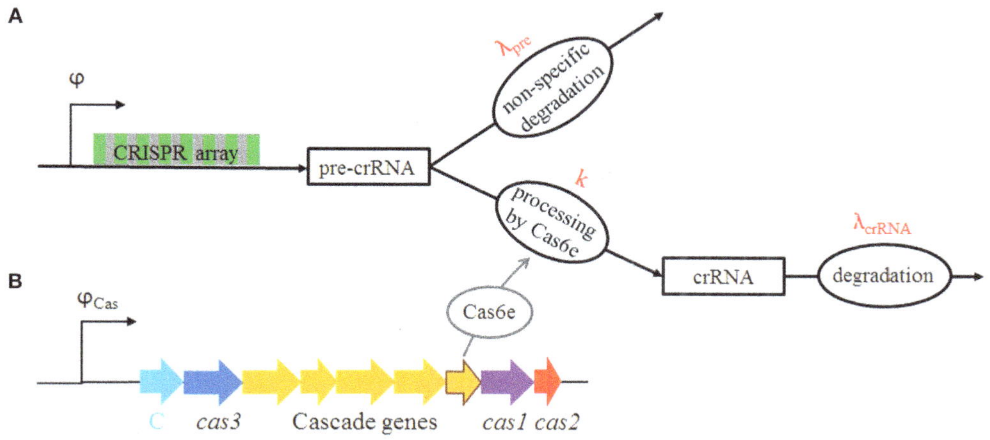

FIGURE 2 | (A) A scheme of CRISPR transcript processing: CRISPR array is transcribed (i.e., pre-crRNA is generated) with rate φ, and the transcript is either (non-specifically) degraded with rate λ_{pre}, or processed to crRNAs by Cas6e with rate k; individual crRNAs are then degraded with rate λ_{crRNA} (Djordjevic et al., 2012). **(B)** The proposed model system for CRISPR-Cas activation: *cas* genes (including *cas6e*, whose product processes pre-crRNA to crRNA), and the transcription factor (C), are transcribed from φ_{Cas} promoter. To reproduce the same qualitative features of transcription regulation as in a native CRISPR-Cas system (cooperative regulation), φ_{Cas} is put under control of C protein, in the same manner as in a well-studied AhdI R-M system (Bogdanova et al., 2008). The system is induced when the plasmid expressing *cas* genes and C protein enters a bacterial cell, as indicated in **Figure 3**. Gradual expression of *cas* genes, leading to Cas6e protein synthesis (gray oval), then increases k (this is indicated by the full arrow in the figure), which in turn results in crRNA generation.

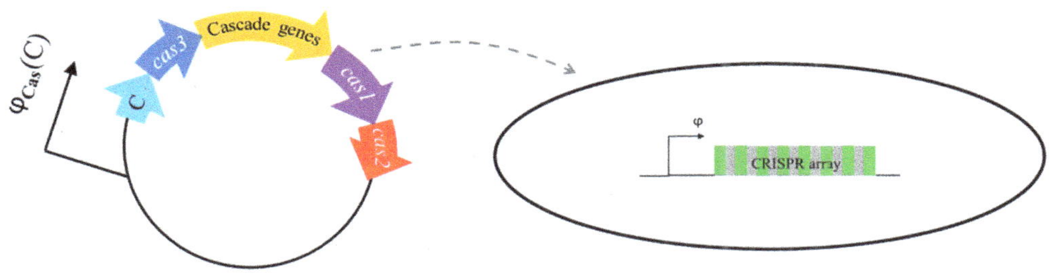

FIGURE 3 | A scheme of the basic setup: pre-crRNA is transcribed in a cell from CRISPR array with rate φ. Transcription of *cas* genes occurs from a plasmid, which enters the cell, inducing crRNA generation. A transcription regulator (C) is transcribed together with *cas* genes, regulating transcription of the φ_{Cas} promoter in a same qualitative manner as exhibited in native CRISPR-Cas system. To achieve this, C protein controls transcription of φ_{Cas} promoter in the same way as in a well-studied R-M system. This leads to a gradual synthesis of Cas6e and C protein, leading to system activation, as schematically shown in **Figure 2B**.

Cas6e (which is responsible for pre-crRNA processing) generates highly abundant crRNAs from pre-crRNA which is present at low abundance (Pougach et al., 2010). We previously showed that a simple quantitative model—whose relevant kinetic scheme is shown in **Figure 2A**—explains this observation (Djordjevic et al., 2012), so that a small decrease in pre-crRNA abundance leads to a much larger (around two orders of magnitude) increase in crRNA abundance. Interestingly, the main mechanism responsible for this strong amplification is fast non-specific degradation of pre-crRNA (see **Figure 2**) by unidentified nuclease(s). In particular, when cas genes expression increases, processing of pre-crRNA by Cas6e is favored and diverts the entire pre-crRNA molecule away from the path of non-specific degradation. Therefore, the fast non-specific degradation of pre-crRNA should be considered as a second major regulatory feature of CRISPR-Cas expression.

The modeling described in Djordjevic et al. (2012) took into account only the transcript processing step, i.e., it was assumed that there is an infinitely abrupt (stepwise) increase of pre-crRNA to crRNA processing rate, and pre-crRNA generation rate. This is, however, a clear idealization of the induction mechanism, as transcription regulation of *cas* genes and CRISPR array promoters is neglected. That is, in reality, pre-crRNA processing rate can be increased only gradually, as it takes time to synthesize the needed Cas proteins. The rate of Cas proteins synthesis is in turn directly related to the transcription control of the *cas* gene promoter in the IGLB region (see **Figure 1**). Similarly, the rate by which pre-crRNA is synthesized is determined by the transcription control of the CRISPR array promoter (L region).

Consequently, a more realistic model of CRISPR-Cas expression dynamics has to take into account both the regulation of CRISPR array and Cas protein synthesis, and CRISPR transcript processing. However, a major obstacle in achieving such model is that signaling which leads to the system induction, and detailed mechanism of CRISPR-Cas transcription regulation, is still unclear. We here propose a model system for CRISPR-Cas induction by assuming that activation of crRNA production is put under transcriptional control exhibited in a restriction-modification (R-M) immune system (Pingoud et al., 2014). As argued below, such model system would have qualitative features of transcription regulation expected for a CRISPR-Cas, and will keep the same transcript processing mechanism as that described for native system. On the other hand, this model system allows bypassing the currently unknown signaling that leads to CRISPR-Cas activation, and can be readily analyzed *in silico*, since transcription regulation of a well-studied R-M system (AhdI, see Bogdanova et al., 2008)—for which we previously showed that it can be reliably modeled (see below)—is used as a proxy for transcription regulation of CRISPR-Cas system.

Through this approach, we expect to:

i. Obtain quantitatively more realistic model of CRISPR-Cas induction dynamics, in which the transcription regulation, i.e., the gradual synthesis of relevant enzymes and transcription regulators is explicitly taken into account.
ii. Qualitatively understand the main features of CRISPR-Cas induction, in particular the roles of cooperative transcription regulation, and of fast non-specific degradation of pre-crRNA.
iii. Propose an experimental setup for CRISPR-Cas induction that mimics the main qualitative features of the native system.

The setup of the model will be explicitly considered in the next subsection.

RESULTS

In silico Experiment Setup

The Model System

We start from a CRISPR transcript processing scheme, which is shown in **Figure 2**. According to this scheme, pre-crRNA is generated with rate φ, and subsequently either non-specifically degraded (due to activity of an unspecified nuclease) with rate λ_{pre}, or is processed by Cas6e to crRNAs with rate k. crRNAs are subsequently degraded with rate λ_{crRNA}. All the parameters in the scheme are experimentally determined in (Djordjevic et al., 2012) (for Type I-E CRISPR-Cas in *E. coli*) and explicitly stated in Methods. In particular, the main feature of the transcript processing is a large (non-specific) pre-crRNA degradation rate (with $\lambda_{pre} \sim 1$ 1/min), which is much larger than crRNA degradation rate (with $\lambda_{crRNA} \sim 1/100$ 1/min). In the experiments, crRNA production is artificially activated, by overexpressing Cas6e from a plasmid, which increases pre-crRNA processing rate (k) for between one and two orders of magnitude (between $10\lambda_{pre}$ and $100\lambda_{pre}$). While the repression of the *cas* promoter in IGLB region (see **Figure 1**) is very strong, with very small amount of Cas6e synthesized when the system is uninduced, the repression of the CRISPR array promoter is significantly weaker, with rather strong basal rate of pre-crRNA generation ($\varphi \sim 10$ 1/min) (Pougach et al., 2010; Pul et al., 2010; Westra et al., 2010; Djordjevic et al., 2012).

As indicated in the Introduction, we previously modeled the transcript processing mechanism (Djordjevic et al., 2012), where we took that k is increased abruptly, i.e., as a step function at $t = 0$. This neglects the transcription regulation of *cas* and CRISPR array promoters. Such abrupt increase of k will provide a baseline for our predictions, which will now take into account that Cas6e (the enzyme which processes pre-crRNA to crRNA) is synthesized gradually. While in the experiments crRNA generation is activated by overexpressing Cas6e from a plasmid (see e.g., Pougach et al., 2010), it is likely that in the native system the expression of CRISPR array is activated as well (Pul et al., 2010). Consequently, we will also take into account a gradual synthesis of the regulator [in our case, a C-protein (Tao et al., 1991; Bogdanova et al., 2008)], which can activate CRISPR array transcription by increasing the basal rate φ to a higher value.

To include transcription regulation of the *cas* promoter, i.e., the gradual synthesis of Cas6e and C transcriptional regulator, we here propose the model system whose setup is schematically shown in **Figures 2**, **3**. This setup includes a CRISPR array which is expressed from a promoter with basal transcription activity φ (**Figure 3**). The second component is a vector (plasmid, virus) which expresses *cas* genes and the control protein C that are jointly transcribed from a promoter with transcription activity φ_{Cas}. While Cas3 is not directly relevant for the problem considered here (dynamics of crRNA generation), as it does not take part in crRNA biogenesis, it is necessary for CRISPR interference (Hille and Charpentier, 2016). We therefore include it in the setup to allow expression of all *cas* genes, i.e. to have a fully functional CRISPR-Cas system.

As detailed below, φ_{Cas} is regulated by C. To mimic the qualitative features of transcription regulation in native CRISPR-Cas system, we employ the transcription regulation found in some R-M systems, as explained in the next subsection. The system is activated when the vector enters a bacterial cell lacking its own *cas* genes, which leads to a gradual synthesis of Cas proteins (including Cas6e), therefore increasing the processing rate k, which in turn leads to crRNA generation (see **Figure 2B**—the full arrow) by pre-crRNA processing. Gradual increase of pre-crRNA generation rate can be also considered through this model, through activation of CRISPR array promoter by gradually synthesized C.

Note that the setup above, where *cas* genes are introduced in a cell on a vector, allows bypassing the unknown signaling step in CRISPR-Cas induction. That is, the vector entering the cell marks the start of the system activation (setting zero time in the dynamics simulations), and mimics the signaling which starts synthesis of the transcription activator. Therefore, the key regulatory features which characterize the downstream steps (CRISPR array transcription and transcript processing) can be studied both *in silico* (which will be done here), and also potentially experimentally. In terms of experimental implementation, introducing *cas* genes in a cell on a virus also allows synchronizing the cell population, which is an approach previously implemented to visualize R-M protein kinetics (Mruk and Blumenthal, 2008).

Putting CRISPR-Cas under Transcription Control of an R-M System

As discussed above, *cas* promoter will be put under transcription control exhibited by R-M systems. Below, the main elements necessary for modeling the system transcription regulation are introduced.

R-M systems are often mobile, and can spread from one bacterial host to the other (Mruk and Kobayashi, 2013). When a plasmid carrying R-M system genes enters a naive bacterial host, the host genome is initially unmethylated, and can consequently be cut by the restriction enzyme. It is, therefore, evident that expression of the restriction enzyme and methyltransferase must be tightly regulated in order to ensure that bacterial genome is protected by the methyltransferase ("antidote"), before it is cut by the restriction enzyme. This tight regulation is often achieved through a dedicated control (C) proteins (Tao et al., 1991; Vijesurier et al., 2000).

We here concentrate on the AhdI R-M system, whose transcription control by C protein has been well-studied (Bogdanova et al., 2008). The activation of AhdI by C protein is reminiscent of CRISPR-Cas activation, as strong cooperative interactions are involved in both cases. In particular, C proteins bound at promoter-proximal and promoter-distal operators interact with high binding cooperativity, so that configuration in which only one operator is occupied cannot be observed in the absence of RNA polymerase (RNAP). At lower C protein concentrations, RNAP can outcompete C protein bound at promoter-proximal operator, leading to transcriptionally active configuration (Bogdanova et al., 2009). Moreover, another feature exhibited in AhdI transcription control, i.e., autoregulation by C protein, is also likely found in CRISPR-Cas transcription regulation. That is, LeuO that activates CRISPR-Cas expression (Westra et al., 2010) also regulates its own transcription. In particular, similarly to transcription regulation of *cas* genes, *leuO* is repressed by H-NS, while this repression is abolished by LeuO (Chen et al., 2001). At high concentrations, C protein is bound at both promoter-proximal and promoter-distal position, leading to the promoter repression—see **Figure 5** in (Bogdanova et al., 2009) and the scheme of the transcription configurations shown in **Figure 5** (framed in the figure). Negative autoregulation is also exhibited by LeuO, as it inhibits transcription activation of its gene by BglJ-RcsB (Stratmann et al., 2012). Therefore, putting *cas* genes under transcription control found in AhdI mimics the main qualitative features of CRISPR-Cas transcription regulation, namely, gradual synthesis of Cas proteins, cooperativity in transcription regulation, and putative autoregulation.

Another advantage of this setup is that we previously showed that biophysical modeling can be used to:(i) explain *in vitro* measurements of the *wild type* and mutant R-M system transcription control (Bogdanova et al., 2008), (ii) explain *in vivo* measurements of the system dynamics (Morozova et al., 2015), (iii) effectively perturb the main R-M system features and relate these perturbations with the system dynamics (Rodic et al., 2017). Consequently, transcription control of a well-studied AhdI R-M system, whose transcription regulation can be reliably modeled (Bogdanova et al., 2008), will serve as a proxy for the transcription control of a much less understood CRISPR-Cas system.

In silico Analysis of the Main System Features

The baseline for our predictions will be provided by a model in which the increase of pre-crRNA to crRNA processing rate k is infinitely abrupt—we will call this the baseline model. Comparing the baseline model with predictions that take into account the system transcription regulation (as schematically shown in **Figures 2, 3**), allows analyzing how gradual synthesis of Cas6e affects kinetics of crRNA generation.

While in the native CRISPR-Cas both *cas* genes and CRISPR array promoters are repressed by global regulators, the repression of *cas* genes was found to be much stronger (Pul et al., 2010; Westra et al., 2010)—consequently, when the system is (experimentally) artificially induced, this is commonly done by expressing only *cas* genes (Pougach et al., 2010; Semenova et al., 2016; Musharova et al., 2017). However, in the native system, it is likely that expression of both CRISPR array and *cas* genes is activated when the appropriate induction signal(s) is received (Pul et al., 2010). We will therefore investigate the system dynamics when only *cas* genes are activated (i.e., only pre-crRNA processing rate is gradually increased), and when *cas* genes and CRISPR array promoter transcription are jointly (and gradually) increased. Consequently, in both of the models introduced below (constitutive and cooperative), we will consider two options. First, when only transcription of *cas* genes is activated, while transcription activity of CRISPR array remains constant. Second, we will consider the case when the transcription activity of CRISPR array is increased as well.

We further introduce two models of *cas* gene and CRISPR array transcription regulation:

i *The constitutive model* (**Figure 4**). In this model *cas* genes are expressed from a constitutive promoter, so that they are transcribed with the constant rate once the plasmid enters a cell. In the case when we consider that the system is activated by only increasing pre-crRNA processing rate, the transcription activity φ is kept constant. When CRISPR array transcription rate is increased as well, increasing φ is exhibited in the simplest manner, by binding of a single C protein activator. Note that, in accordance with its name, no cooperativity is exhibited for transcription regulation described by this model.
ii *The cooperative model* (**Figure 5**). In this model, C protein regulates the transcription of *cas* genes, and its own transcription, in the same manner as in AhdI R-M system. As noted above, such transcription regulation is characterized by strong cooperative interactions. CRISPR array transcription rate is either kept constant, or in the case when it is increased, we take that it is exhibited in the same way as for *cas* promoter transcription (the dashed arrow in **Figure 5**).

Studying of the two models allows one to assess how the cooperative transcription regulation (which also characterizes the native CRISPR-Cas system) compares to the activation in which no cooperativity is exhibited, and therefore allows us to assess the role of this key system feature. Also, considering the two models when φ is first kept constant, and then increased together with k, allows assessing significance of CRISPR array transcription

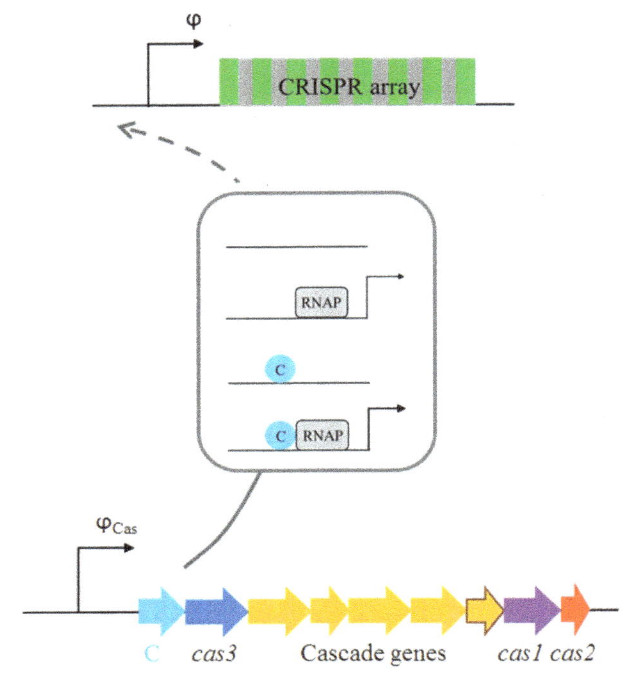

FIGURE 4 | Transcription regulation of *cas* and CRISPR array promoters in the *constitutive model*. C and *cas* genes are transcribed from a constitutive promoter of constant strength φ_{cas}. The CRISPR array promoter is either considered constitutive, with constant transcription activity (φ), or is regulated by C protein (indicated by the dashed arrow), where a scheme corresponding to this regulation is framed. The scheme shows possible configurations of CRISPR array promoter, where activation of CRISPR array transcription is achieved in the simplest manner, through the binding of a single C protein which acts as a transcription activator to the CRISPR array promoter. Transcriptionally active configurations are denoted by arrows, with thicker arrow indicating larger transcription activity.

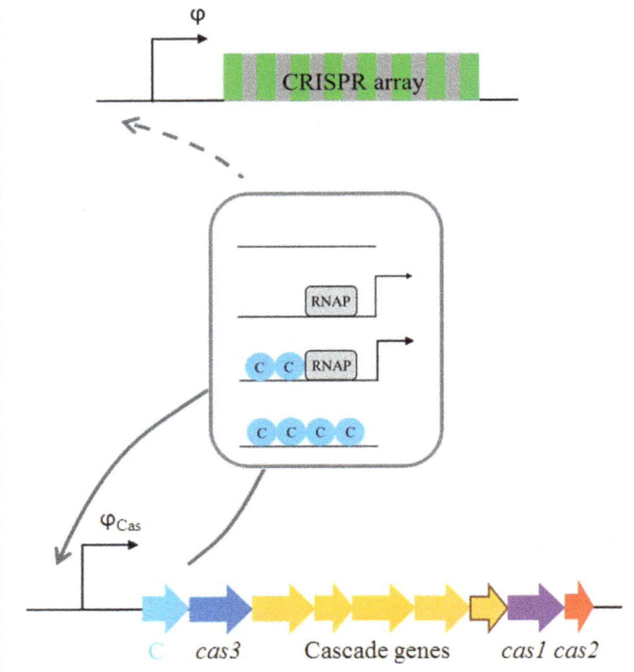

FIGURE 5 | Transcription regulation of *cas* and CRISPR array promoters in the *cooperative model*. The framed scheme shows promoter configurations, where transcription regulation is exhibited in the same manner as for *AhdI* system, through cooperative interactions. Arrows in the scheme denote transcriptionally active configurations, with thicker arrow indicating larger promoter transcription activity. The full gray arrow indicates that *cas* promoter is regulated as described by the scheme, with the same parameters as in *AhdI* R-M system (Pougach et al., 2010). The dashed arrow indicates that the same transcription regulation is also exhibited for CRISPR array promoter, in the case when its transcription activity φ is not assumed constant.

control. To allow a direct comparison of models dynamics, the overall strength of φ_{Cas} is adjusted so that the same value of maximal pre-crRNA processing rate is achieved. Similarly, when the transcription rate of CRISPR array is increased, the interaction parameters are adjusted so that the same equilibrium increase of φ is achieved in both models (see Methods).

Modeling Results

Kinetics of Pre-crRNA and crRNA Production

We first consider the situation in which crRNA generation is activated by expressing Cas proteins, such that the processing rate k is gradually increased, while the CRISPR array transcription activity remains constant. In this case, we compare the system dynamics for: (i) *baseline model*, in which the processing rate k is increased as a step function, which corresponds to the limit of infinitely fast system induction, (ii) constitutive model (see **Figure 4**), and (iii) cooperative model (see **Figure 5**).

In constitutive and cooperative models, the gradual synthesis of Cas6e leads to gradual change of transcript processing rate k (k^* is a processing constant):

$$k(t) = [Cas6e](t) \cdot k^* \tag{1}$$

Figure 6 illustrates how the processing rate (k) changes with time, when the baseline, constitutive, and cooperative models of *cas* gene expression are assumed. For the constitutive model (the dash-dotted curve), the processing rate uniformly increases and reaches an equilibrium value, for all values of k_{eq} considered in three panels of **Figure 6**. On the other hand, for cooperative model (the dashed curve) and at higher values of k_{eq} (**Figures 6B,C**), we see a rapid increase of k at initial times, followed by a fast return to the equilibrium value due to repression at higher C protein concentrations.

In **Figure 7**, we address how different k dynamics (shown in **Figure 6**), affects pre-crRNA and crRNA generation. Specifically, φ is held constant at its initial value (10 1/min), while k changes according to the baseline, constitutive, or cooperative models until reaching the same equilibrium value of $10\lambda_{pre}$, $100\lambda_{pre}$, and $1{,}000\lambda_{pre}$ (left, central, and right columns of **Figure 7**, respectively). The model of abrupt Cas6e expression serves as a baseline for assessing the dynamics in the other two models (constitutive and cooperative), in which Cas6e is realistically (gradually) expressed.

In **Figures 7A–D**, we see that cooperative model leads to the steepest transition from ON to OFF state (in the case

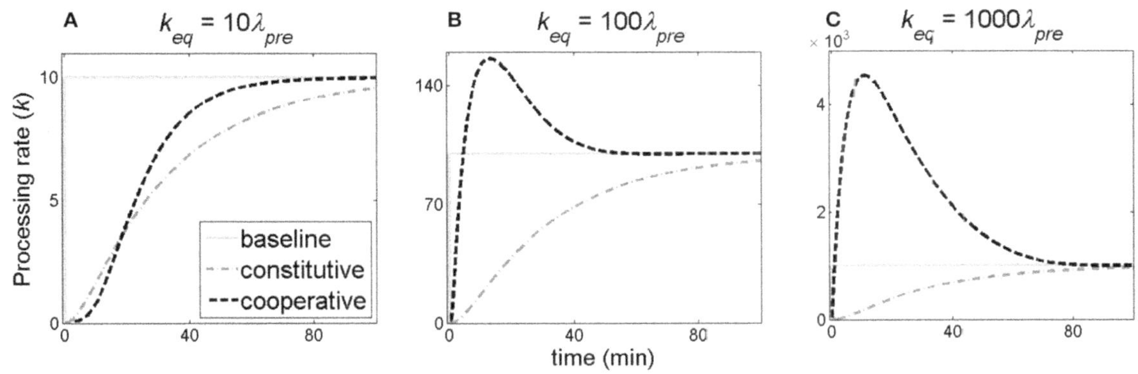

FIGURE 6 | Comparing the dynamics of the pre-crRNA processing rate change. The change of the processing rate k with time is shown for: the baseline model (solid gray curve), the constitutive model (dash-dotted gray curve) and the cooperative model (dashed black curve). **(A–C)** correspond to different k_{eq} values ($k_{eq} = 10\lambda_{pre}$, $100\lambda_{pre}$, $1{,}000\lambda_{pre}$, respectively). CRISPR transcription activity is constant (10 1/min).

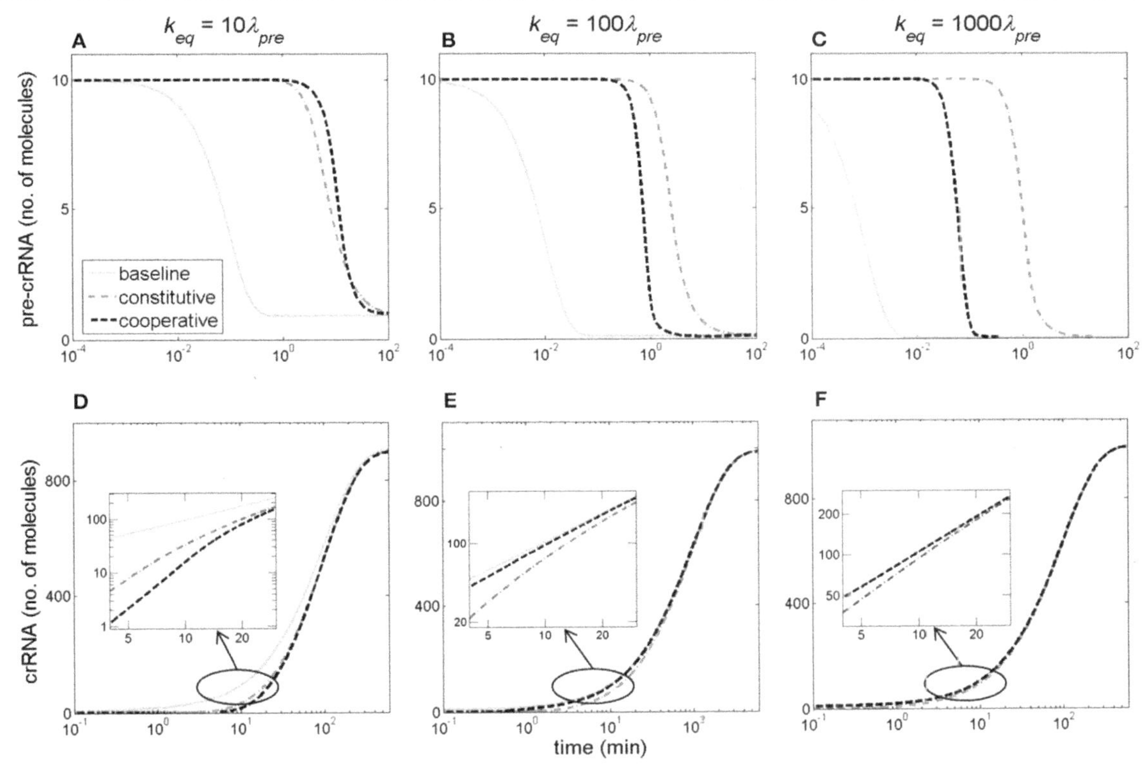

FIGURE 7 | Kinetics of pre-crRNA and crRNA generation. The columns correspond to k_{eq} values of $10\lambda_{pre}$ **(A,D)**, $100\lambda_{pre}$ **(B,E)**, and $1{,}000\lambda_{pre}$ **(C,F)**, which are reached through the baseline model (the gray solid curve), the constitutive model (the gray dash-dotted curve) or cooperative model (the black dashed curve). The upper **(A–C)** and the lower **(D–F)** rows correspond, respectively, to pre-crRNA and crRNA kinetics. CRISPR array promoter transcription activity is kept constant at 10 1/min.

of pre-crRNA), and from OFF to ON state (in the case of crRNA). Furthermore, we can distinguish between two different regimes in **Figure 7**. At lower k_{eq} (left column in **Figure 7**), there is a noticeably slower accumulation of crRNA at early times in both cooperative and constitutive models compared to the baseline model of infinitely abrupt processing rate (k) increase (**Figure 7D**). On the other hand, at higher k_{eq} ($k_{eq} \geq 100$ 1/min, the central and right columns in **Figure 7**), the dynamics of crRNA accumulation for cooperative model becomes faster compared to constitutive model dynamics at early times, and approaches the limit of infinitely abrupt k increase (see the inserts in **Figures 7E,F**). The faster kinetics of crRNA increase in cooperative model is due to the fast increase of k at early times in this model (**Figures 6B,C**).

Effects of *cas* Genes Regulation

From **Figure 7**, we observe that transcripts reach their steady-state levels quite late, i.e., >100 min post-induction. Such late time is, however, not relevant for cell response to phage infection, since infected *E. coli* lyse ~20 min post-infection, while shut-off of essential cell functions happens earlier (Kruger and Schroeder, 1981). Therefore, in **Figure 8** we estimate pre-crRNA and crRNA levels for all three models at 20 min post-induction, as the maximal value of pre-crRNA processing rate k_{eq} is changed from very low to high values ($>100\lambda_{pre}$, characteristic for artificial Cas6e induction), while keeping the level of CRISPR array transcription constant ($\varphi = 10$ 1/min).

The following features emerge from **Figure 8**:

i. A switch-like system behavior for both pre-crRNA and crRNA curves in the cooperative model, while the constitutive and baseline models yield much more gradual responses to changes in k_{eq}. For crRNA, the cooperative model leads to a rapid transition from the OFF state (with essentially no crRNA generated at 20 min), to the ON state (with high abundance of crRNA), and reciprocal situation for pre-crRNA. Consequently, for small amounts of synthesized Cas6e (i.e., small k_{eq} values), which can be caused by leaks in *cas* promoter activity, the system remains in OFF state. On the other hand, once the system is activated when the processing rate (directly related to the amount of Cas6e available) reaches a certain threshold ($k_{eq} \gtrsim 50$), a large amount of crRNA is generated at early times, which should allow protection from foreign DNA invasion. The significance of this behavior is considered in Discussion.

ii. An interesting cross-over behavior in the cooperative model, where at low k_{eq} values crRNA amounts are low, while at high k_{eq} values the synthesized crRNA amounts become larger than in the constitutive model, and approach the baseline model curve. Therefore, at high k-values (~100 1/min), which are encountered in experiments, (Pougach et al., 2010; Djordjevic et al., 2012) the model of cooperative *cas* gene expression leads to accumulation of protective crRNA amounts close to those achievable in the limit of infinitely abrupt k increase. Consequently, the high cooperativity in transcription regulation, characteristic for native CRISPR-Cas system regulation, leads to a highly efficient crRNA generation at high transcript processing rates.

iii. Sufficient crRNA levels are generated to protect host cell against bacteriophage infection, at early times post-induction, even at relatively low values of pre-crRNA processing rate. That is, k_{eq} somewhat larger than 11/min leads to ~10 crRNAs which already corresponds to the amount that negatively affects phage development (Pougach et al., 2010); moreover, a small additional k_{eq} increase leads to a large increase in generated crRNAs in the cooperative model, due to the rapid transition from OFF to ON state.

iv. A saturation in generated crRNA amounts at early times post-induction. That is, for k_{eq}~100 1/min the amount of generated crRNAs at 20 min stops significantly increasing with further increase in k_{eq}. This saturation can be relieved (leading to increase in the amount of generated crRNA), if CRISPR array transcription activity is increased, which is further analyzed below.

Perturbing Pre-crRNA Degradation Rate

We next perturb the second key feature of CRISPR-Cas regulation—fast non-specific degradation of pre-crRNA. The consequence of pre-crRNA degradation rate λ_{pre} decrease at constant φ was next investigated for all three models. The

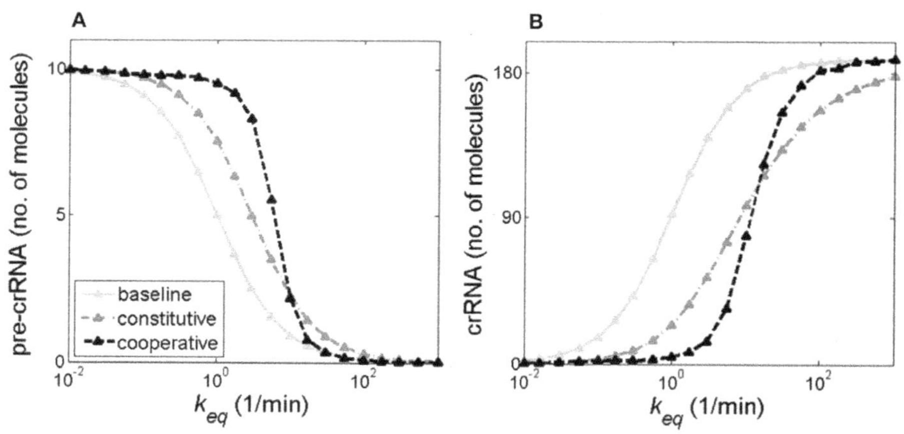

FIGURE 8 | Pre-crRNA and crRNA amounts early post-induction for different models of *cas* gene transcription regulation. The figure shows **(A)** pre-crRNA and **(B)** crRNA amounts 20 min post-induction (i.e., 20 min after introduction of the vector expressing *cas* genes), as a function of the maximal (equilibrium) value of the transcript processing rate *k*. CRISPR promoter transcription activity is kept constant ($\varphi = 10$ 1/min). The gray solid, the gray dash-dotted, and the black dashed curves correspond, respectively, to baseline, constitutive, and cooperative models of *cas* regulation.

decrease was followed at different k_{eq} values (i.e., at different levels of Cas6e activity), where φ is held constant.

The effects of λ_{pre} decrease are similar for all three models, so in **Figure 9** we show the results only for the cooperative model. For all k_{eq} values we see that abolishing the fast decay of pre-crRNA (decreasing λ_{pre}), significantly decreases the time delay of the onset of crRNA generation. This effect is most pronounced at high k_{eq} values (**Figure 9C**). Also, perturbing the degradation rate deforms crRNA dynamics curve with respect to the standard Hill (sigmoidal) shape that is exhibited at high λ_{pre} such as $\lambda_{pre} = 1/50$. Furthermore, analogously to **Figure 8**, in Figure S1 (Supplementary Material), we show how crRNA amount at 20 min after induction depends on pre-crRNA degradation rate λ_{pre}. One can clearly observe that as λ_{pre} decreases, the amount of generated crRNA early post-induction significantly increases, consistently with the decrease of the time delay of onset of crRNA generation observed in **Figure 9**.

Relieving crRNA Production Saturation by Increasing Pre-crRNA Generation

In addition to *cas* genes, CRISPR array promoter is also repressed (though more weakly) by global transcription regulators (Pul et al., 2010; Westra et al., 2010). Consequently, crRNA generation can be also augmented by increasing CRISPR array transcription activity. Therefore, we next assess how joint increase of k (achieved by activating *cas* gene transcription) and φ (achieved by increasing CRISPR array transcription) affects generated crRNA amount 20 min post-induction for all three regulatory models.

As can be seen from **Figure 10**, increasing φ robustly relieves crRNA saturation (see also discussion of **Figure 8**). Moreover, one can see that a relatively modest, factor of two increase of φ (from 10 1/min to 20 1/min) can abolish the need of a significant, order of magnitude, k increase to produce the same amount or crRNA. As above, we observe a switch-like behavior for the cooperative model (compare **Figure 10C** with **Figures 10A,B**), with cooperative model curves exhibiting the steepest transition from OFF to ON state for all φ values.

Regulation of CRISPR Array Transcription Activity

We next consider how different models of regulation of CRISPR array transcription affect crRNA dynamics. For all three models, the transcription activity φ is increased by an order of magnitude (from $\varphi = 10$ 1/min to $\varphi = 100$ 1/min), for different k_{eq} values ($k_{eq} = \lambda_{pre}$, $10\lambda_{pre}$, and $100\lambda_{pre}$), see Figure S2 (Supplementary Material). We obtain that the cooperative model leads to a more controlled (attenuated) pre-crRNA dynamics, which is due to the presence of repressing mechanism at high C protein amounts (see Figure S3). For crRNA dynamics, we observe that the cooperative model exhibits the steepest transition from OFF to ON state. Moreover, this model leads to the largest delay in crRNA generation. Consequently, in addition to pre-crRNA degradation rate, the cooperative transcription regulation also contributes to the delay between the activating signal and the onset of crRNA generation.

We previously (**Figure 9**) perturbed pre-crRNA degradation rate while keeping the transcription rate φ constant. Finally, we now also decrease λ_{pre} under the conditions when both *cas* genes and CRISPR array transcription is activated according to all three models (see Figure S4). The results are qualitatively similar to **Figure 9** (where φ is constant), i.e., decreasing λ_{pre} diminishes the switch-like system response and/or decreases the time-delay in the onset of pre-crRNA generation.

DISCUSSION AND SUMMARY

One of the most prominent problems in understanding CRISPR-Cas function is assessing dynamics of the system activation, i.e., understanding the roles of the key features of CRISPR-Cas regulation. Addressing this problem is complicated by the fact that exact conditions for system activation remain unclear. In fact, for Type I-E CRISPR-Cas system in *E. coli*, even bacteriophage infection itself is not sufficient

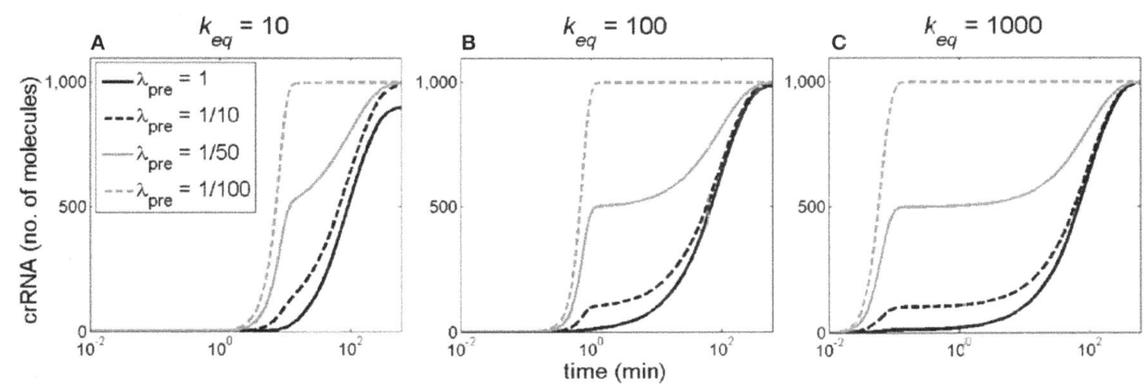

FIGURE 9 | The effect of perturbing pre-crRNA degradation rate on the dynamics of crRNA generation. The pre-crRNA processing rate increases to its equilibrium value through the cooperative model, while φ is held constant (at 10 1/min). Different curves correspond to different λ_{pre} values: 1 (solid black), 1/10 (dashed black), 1/50 (solid gray), and 1/100 1/min (dashed gray). **(A–C)**, correspond to different k_{eq} values indicated at the top of each panel.

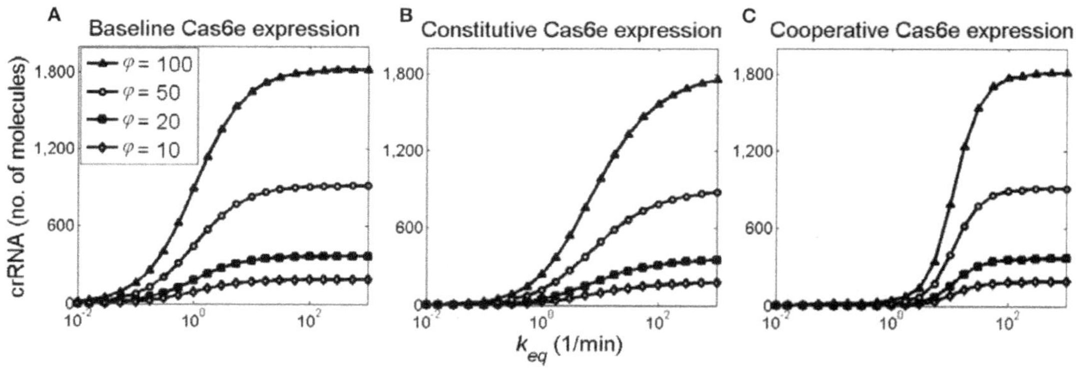

FIGURE 10 | Relieving saturation in generated crRNA amounts through joint k and φ increase. crRNA amount as a function of k_{eq} at 20 min post-induction, obtained for **(A)** baseline, **(B)** constitutive, and **(C)** cooperative models. Curves marked with diamonds, filled squares, circles, and filled triangles, correspond, respectively, to φ of 10, 20, 50, and 100 1/min.

to induce the system. We here proposed a synthetic setup which allows inducing CRISPR-Cas with qualitative features that correspond to native system regulation, while bypassing currently unclear conditions under which the system is activated. This setup involves putting *cas* genes and/or CRISPR array under transcription control found in a well-studied R-M system, which exhibits cooperative transcription regulation that is also characteristic of CRISPR-Cas regulation (Bouffartigues et al., 2007; Westra et al., 2010). A major advantage of the setup is that it can be readily experimentally implemented, e.g., by introducing *cas* genes and the regulator (C protein) in a cell on a virus. This would allow synchronizing the cell population, and experimentally observing the system dynamics, where such measurements could be directly compared with the predictions provided here. Another advantage is that major parameters in the setup have been inferred from experimental data, as both CRISPR transcript processing, and AhdI transcription regulation, have been experimentally well-studied (Bogdanova et al., 2008; Pougach et al., 2010; Djordjevic et al., 2012).

Consequently, this setup allows us to directly (*in silico*) address how the system regulation contributes to its dynamical response. In particular, previous experimental and computational work point to cooperative regulation of *cas* gene and CRISPR array transcription, and fast non-specific degradation of pre-crRNA, as two main system regulatory features (Pougach et al., 2010; Pul et al., 2010; Westra et al., 2010; Djordjevic et al., 2012). We therefore investigated two alternative regulatory architectures, one with constitutive, and the other with cooperative *cas* gene regulation. The dynamics corresponding to these two architectures was then compared with the baseline model, in which pre-crRNA processing rate is increased infinitely abruptly. We assessed the dynamics in the case when only *cas* genes are activated (i.e., only pre-crRNA processing rate is gradually increased), and when *cas* genes and CRISPR array promoter transcription is jointly increased. We focused on early system dynamics (within the first 20 min post-induction), as this period is most relevant for defending the cell against invading viruses. Finally, we also perturbed the high pre-crRNA non-specific degradation rate, under different system conditions described above, and assessed what effect such perturbation has on system dynamics.

The main result of the analysis is that the system regulation leads to a clear switch-like behavior, characterized by an initial delay of crRNA synthesis, followed by a steep transition from OFF to ON state. Unexpectedly, it is not only the cooperative transcription regulation, but also fast non-specific pre-crRNA degradation, which leads to such dynamics. That is, decreasing the high pre-crRNA degradation rate effectively abolishes the delay in crRNA generation, and deforms the crRNA kinetics from the standard sigmoidal (Hill) shape (Hill, 2013) typical for switch-like system response (**Figure 9**). Interestingly, we also found that, when pre-crRNA processing rate and CRISPR array transcription rate are jointly (and gradually) increased, as likely exhibited in the native system, the system is more robust to perturbations in the degradation rate (Figure S4).

The cooperative transcription regulation leads to an interesting cross-over behavior in the early system dynamics. At low pre-crRNA processing rates, cooperative regulation leads to much smaller crRNA amounts at early times compared to constitutive expression. On the other hand, at higher processing rates, there is a large increase in synthesized crRNA amounts, which approach the limit of infinitely abrupt system induction. Interestingly, when the system is artificially activated by overexpressing *cas* genes, pre-crRNA processing rates correspond to the regime of the highly enhanced crRNA production (Djordjevic et al., 2012). While the parameters of the native system induction are unclear, it is tempting to hypothesize that they may also reach this cross-over, allowing the system to generate crRNAs with the rate close to the limit of infinitely fast induction at times when they are needed.

The rapid transition of the system from OFF to ON state is straightforward to interpret in terms of its function in immune response. When a potential signal indicating infection is received by the cell, CRISPR-Cas has a very short time to generate sufficient crRNA amounts to protect the cell, as bacteriophages are typically highly efficient in shutting-down essential cell functions. Thus, there is a question whether enough crRNA can be generated in a model which accounts for gradual synthesis of proteins that process pre-crRNA and/or are responsible for gradual CRISPR array activation. We robustly obtained that enough crRNA can be generated at early times, even when the system is activated by only increasing the pre-crRNA processing rate. Moreover, a much smaller increase of the processing rate is needed to achieve certain crRNA amount, if CRISPR array transcription is activated as well. Therefore, these results may explain the relatively inefficient repression of CRISPR array promoter, since even a small increase of CRISPR array transcription rate efficiently increases generated crRNA amounts. In fact, the need to rapidly produce large amounts of crRNAs may be a major constraint on system dynamics.

In distinction to the rapid transition of the system from "OFF" to "ON" state, interpretation of the delay in crRNA generation, which comes as a model prediction, is less straightforward. One possibility is that such delay is related with primed adaptation in CRISPR-Cas, which relies on a pre-existing (priming) spacer that enables a biased uptake of new spacers—therefore serving to minimize infection by phage escape mutants that would otherwise evade the interference (Sternberg et al., 2016). In particular, it has been found that priming is facilitated by slow or delayed CRISPR interference, leading to a steady-state flux of substrates from which new spacers can be acquired (Kunne et al., 2016; Severinov et al., 2016; Musharova et al., 2017). Such delay in CRISPR interference can clearly be achieved by a delay in crRNA generation that is predicted in our work.

It has been proposed that Type I-E CRISPR-Cas in *E. coli* may have functions other than immunity. For example, it was found by bioinformatics analysis that the system is changing very slowly, in distinction to rapid diversification of CRISPR arrays in other species, indicating that the system is not taking an active role in defense against immediate viral threats (Touchon et al., 2011). In this respect, it may be useful to view the dynamical properties inferred here in a more general terms, namely of a capability of expressing a large number of molecules in a narrow time interval, with a specific time-delay with respect to reception of an external signal. It is clear that such highly efficient, and temporally specific response, may be highly desirable for multiple cellular functions. It would be very interesting to find out how functions of *E. coli* Type I-E CRISPR-Cas, yet to be discovered in the future, would fit within the dynamical properties inferred here.

METHODS

We start from a previously introduced model of CRISPR transcript processing by Cas proteins (Djordjevic et al., 2012). In this model (see **Figure 2A**), a short-living transcript [pre-crRNA] is synthesized with a promoter transcription activity φ, and further, either quickly degraded with a degradation rate λ_{pre}, or processed (cut) into shorter, long-living RNAs [crRNA] with a processing rate k. Processed transcripts are degraded with a rate λ_{crRNA}. In the equations below, we assume that the processing rate depends linearly on the substrate (pre-crRNA) amount, since the amount of pre-crRNA is small [<10 molecules per cell (Pougach et al., 2010)], so that the corresponding kinetic equations are:

$$\frac{d[\text{pre}-\text{crRNA}]}{dt} = \varphi - (\lambda_{\text{pre}} + k) \cdot [\text{pre}-\text{crRNA}] \qquad (2)$$

$$\frac{d[\text{crRNA}]}{dt} = k \cdot [\text{pre}-\text{crRNA}] - \lambda_{\text{crRNA}} \cdot [\text{crRNA}] \qquad (3)$$

The equations above are further solved deterministically, as both CRISPR array and *cas* genes are expressed from promoters with strong basal transcription. Furthermore, the small pre-crRNA amount is due to fast non-specific degradation, i.e., due to the transcript processing step. With respect to this, note that there is an access of enzyme (Cas6e) over substrate (pre-crRNA) (Djordjevic et al., 2012), so the equations describing the transcript processing are linear. Therefore, their deterministic solution accurately describes the mean of the stochastic simulations.

In the previous study (Djordjevic et al., 2012), we considered a model in which transcription regulation is neglected, so that k and φ increase in an idealized manner, i.e., infinitely abruptly. We now introduce models where the relevant enzymes and transcription regulators are synthesized in a realistic (i.e., gradual) manner. Specifically, k in Equation now explicitly depends on time, and is proportional to the enzyme (the processing protein, Cas6e) concentration, i.e., $k = [Cas6e] \cdot k^*$, where k^* is processing constant. We here consider that this processing rate k can change with time in the following ways:

1) Infinitely abruptly, from 0 to its equilibrium value, k_{eq} at $t = 0$, which we refer to as the baseline model.
2) Gradually, with [Cas6e](t), where Cas6e is expressed from a constitutive promoter (promoter with constant transcription activity), see **Figure 4**.
3) Also gradually with [Cas6e](t), where Cas6e is now expressed from an AhdI-like regulated promoter (see **Figure 5**).

As noted above, we either keep the CRISPR array transcription rate φ constant (which allows us investigating the dynamics in response to changing only pre-crRNA processing rate), or allow φ to change:

1) Infinitely abruptly (the baseline model), so that at $t = 0$ it increases from its starting value (10 1/min) to the equilibrium value.
2) Gradually, through the simplest activation mechanism, where a single C protein activates transcription from the CRISPR array promoter (the dashed arrow in **Figure 4**).
3) Also gradually with C(t), where the same transcription regulation as in AhdI RM system is exhibited (the dashed arrow in **Figure 5**).

In constructing Cas6e and CRISPR expression models, we refer to our existing model of AhdI restriction-modification (RM) system control (Bogdanova et al., 2008), which describes expression of the control protein (C) and the restriction endonuclease (R)—C and R are co-transcribed in AhdI RM system. We here use a thermodynamical model of CR operon transcription regulation, and a dynamical model of transcript and protein expression.

For $t = 0$ we take the moment when plasmid carrying C and *cas* genes enters the naïve host. Thus, all initial conditions are set to zero, except for $[\text{pre-crRNA}](t = 0) = \varphi/\lambda_{pre} = 10\,(1/\min)$(Djordjevic et al., 2012), as extracted from the Equation in equilibrium. Note that while C and *cas* genes enter the cell on a plasmid, CRISPR array is expressed within the cell, with the transcription rate φ.

Constitutive Model of *cas* Gene and CRISPR Array Expression

We assume that C and *cas* genes are co-transcribed from a constitutive (unregulated) *cas* promoter (see above and **Figure 4**). C and *cas* transcript and protein concentrations change with time:

$$\frac{d[c-cas](t)}{dt} = \varphi_{Cas} - \lambda_{Cas}\cdot[\text{c}-\text{cas}](t) \quad (4)$$

$$\frac{dC(t)}{dt} = k_C\cdot[\text{c}-\text{cas}](t) - \lambda_C\cdot C(t) \quad (5)$$

$$\frac{d[Cas6e](t)}{dt} = k_{Cas6e}\cdot[\text{c}-\text{cas}](t) - \lambda_{Cas6e}\cdot[\text{Cas6e}](t). \quad (6)$$

Note that all the notation (including in the equation above), is introduced in **Table 1**. The first terms on the right-hand side represent transcript/protein synthesis by transcription/translation, while the second terms represent transcript/protein decay by degradation. The parameter values are as in AhdI RM system model (with Cas6e now replacing R in AhdI system), and are also provided in the table at the end of the methods. Since C and Cas6e protein degradation rates are taken to be the same, it follows:

$$[\text{Cas6e}](t) = \frac{k_{Cas6e}}{k_C}C(t), \quad (7)$$

So that the differential equation for Cas6e dynamics can be omitted. We set the value of φ_{Cas} to one (see the next subsection) so that the equilibrium processing rate is the same for the constitutive and the cooperative models (see e.g., **Figure 6**), which allows a direct comparison of the dynamics in these two models. Consequently, we set k^* so that $k_{eq} = [Cas6e]_{eq}\cdot k^* = 10\,(1/\min)$. Regarding CRISPR array transcription φ, we keep it constant, in the case when we consider the system activation by overexpression of *cas* genes. In the case when we also consider activation of CRISPR transcription, we introduce a simple model of CRISPR expression regulation (the dashed arrow in **Figure 4**), where CRISPR promoter, apart from being unoccupied, can be found in the following three configurations, which are represented by the reactions shown below: (i) RNAP alone bound to the promoter (8), (ii) a C monomer alone bound to its binding site (9), and (iii) RNAP recruited by a C monomer bound to its binding site, acting as a transcription activator —note that these configurations correspond to the second, third and fourth line in the framed part of **Figure 4**, respectively.

$$DNA + RNAP \underset{K_{1A}}{\rightleftharpoons} RNAP - DNA \quad (8)$$

$$DNA + C \underset{K_{2A}}{\rightleftharpoons} C - DNA \quad (9)$$

$$C - DNA + RNAP \underset{K_{3A}}{\rightleftharpoons} C - DNA - RNAP \quad (10)$$

TABLE 1 | Notations used in model equations.

Variables	Description	
φ_{Cas}	Transcription activity of *cas* promoter	
φ	Transcription activity of CRISPR promoter	
[c-cas]	Concentration of *cas* operon transcript	
[pre-crRNA]	Concentration of unprocessed CRISPR array transcript	
[crRNA]	Concentration of processed CRISPR array transcript	
C	Concentration of control protein	
[Cas6e]	Concentration of processing protein	
KINETIC MODEL CONSTANTS		
k^*	CRISPR transcript processing constant	0.02
k_C	Translation constant for control protein	0.60
k_{Cas6e}	Translation constant for processing protein	3.00
λ_{Cas}	Rate of *cas* transcript decay	0.20
λ_{pre}	Rate of unprocessed CRISPR transcript decay	1.00
λ_{crRNA}	Rate of processed CRISPR transcript decay	0.01
λ_C	Rate of control protein decay	0.033
λ_{Cas6e}	Rate of Cas6e processing protein decay	0.033
TRANSCRIPTION REGULATION MODELS CONSTANTS		
α	Proportionality constants	1.663
γ		110
α'		110
a	Constants which absorb the relevant equilibrium dissociation constants and RNA polymerase concentration	1.60×10^{-1}
p		9.25×10^{-1}
q		1.41×10^{-5}
d		1.00×10^{-1}
e		Adjusted
f		2.00×10^{2}
a'		1.00×10^{-1}
p'		Adjusted
q'		2.50×10^{-5}
K_D		6.50×10^{2}

The equilibrium dissociation constants of the above reactions are given by:

$$K_{1A} = [DNA]\,[RNAP]\,/\,[RNAP - DNA] \quad (11)$$

$$K_{2A} = [DNA]\,[C]\,/\,[C - DNA] \quad (12)$$

$$K_{3A} = [C - DNA]\,[RNAP]\,/\,[C - DNA - RNAP]. \quad (13)$$

Using the Shea-Ackers based approach, i.e. assuming that the transcription activity is proportional to the equilibrium promoter occupancy by RNAP, we derive the expression for CRISPR promoter transcriptional activity:

$$\varphi = \gamma\frac{Z_{RNAP} + Z_{C-RNAP}}{1 + Z_{RNAP} + Z_C + Z_{C-RNAP}} \quad (14)$$

where γ is a proportionality constant, while configuration statistical weights correspond to: $Z_{RNAP} = [RNAP - DNA]\,/\,[DNA]$ – RNAP alone bound to the promoter, $Z_C = [C - DNA]\,/\,[DNA]$–C monomer alone bound to its binding site, $Z_{C-RNAP} = [C - DNA - RNAP]\,/\,[DNA]$ – RNAP recruited to the promoter by a bound C monomer. We can obtain φ dependence on C concentration:

$$\varphi(C) = \gamma \frac{d + def\,[C]}{1 + d + e\,[C] + def\,[C]} \tag{15}$$

If we introduce parameters expressed in terms of the equilibrium binding constants and RNAP concentration:

$$d = [RNAP]/K_{1A} \tag{16}$$
$$e = 1/K_{2A} \tag{17}$$
$$f = K_{1A}/K_{3A}. \tag{18}$$

To estimate the parameters, we use a condition:

$$\varphi(0) = 10\frac{1}{\text{min}} \tag{19}$$

which corresponds to the value in Djordjevic et al. (2012), and:

$$\varphi(C_{eq}) = 100\frac{1}{\text{min}} \tag{20}$$

Another (evident) condition is that the fraction, which appears on the right-hand side of the Equation (15), has to be smaller than 1. By adjusting the parameters to satisfy the conditions (19) and (20), we obtain $d < 1/9$, which allows setting the values of d and γ. Further, we notice that $e = 99/\left([C]_{eq} \cdot (f - 100)\right)$ and, having fixed the value of f, we can adjust e with respect to $[C]_{eq}$.

The unprocessed [pre-crRNA] and processed [crRNA] transcript amounts change with time according to the Equations (2) and (3), where φ is given by .

Cooperative Model of *cas* and CRISPR Expression

As opposed to the constitutive *cas* operon expression, we here assume that the *cas* promoter is regulated by C as in the wild type AhdI RM system (Bogdanova et al., 2008), through cooperative interactions (see **Figure 5**). The following set of reactions describes the transcriptional regulation of the *cas* promoter by the C protein (note the promoter configurations shown in **Figure 5**):

$$C + C \underset{K_1}{\rightleftarrows} D \tag{21}$$
$$DNA + RNAP \underset{K_2}{\rightleftarrows} RNAP - DNA \tag{22}$$
$$D + DNA \underset{K_3}{\rightleftarrows} D - DNA \tag{23}$$
$$D - DNA + D \underset{K_4}{\rightleftarrows} T - DNA \tag{24}$$
$$D - DNA + RNAP \underset{K_5}{\rightleftarrows} D - DNA - RNAP \tag{25}$$

where C and D stand for C protein monomers and dimers, respectively.

The reactions (21)–(25) represent:

- (21) C monomers dimerization;
- (22) *RNAP* binding to the *cas* promoter forming *RNAP-DNA* complex;
- (23) *D* binding to the distal binding site forming *D-DNA* complex;
- (24) *D* recruitment to the proximal binding site forming *T-DNA* complex;
- (25) *RNAP* recruitment to the *cas* promoter forming *D-DNA-RNAP* complex.

In equilibrium the above reactions lead to the following equations of the equilibrium dissociation constants:

$$K_1 = \frac{[C]^2}{[D]} \tag{26}$$
$$K_2 = \frac{[DNA][RNAP]}{[RNAP - DNA]} \tag{27}$$
$$K_3 = \frac{[D][DNA]}{[D - DNA]} \tag{28}$$
$$K_4 = \frac{[D][D - DNA]}{[T - DNA]} \tag{29}$$
$$K_5 = \frac{[RNAP][D - DNA]}{[D - DNA - RNAP]} \tag{30}$$

Taking into account the aforementioned Shea-Ackers assumption we obtain:

$$\varphi_{Cas} = \alpha \frac{Z_{RNAP} + Z_{D-RNAP}}{1 + Z_{RNAP} + Z_{D-RNAP} + Z_T}, \tag{31}$$

α is a proportionality constant, $Z_{RNAP} = [RNAP-DNA]/[DNA]$, $Z_{D-RNAP} = [D - DNA - RNAP]/[DNA]$ and $Z_T = [T - DNA]/[DNA]$ denote the statistical weights of only RNAP bound to the promoter, RNAP recruited to the promoter by a C dimer bound to the distal binding site, and a C tetramer repressing transcription, respectively.

By using Equations (26)–(30), the Equation (31) can be rewritten in terms of C monomer concentration (following the notation in Bogdanova et al., 2008; Rodic et al., 2017):

$$\varphi_{Cas}(C) = \alpha \frac{a - b[C]^2}{1 + a + b[C]^2 + c[C]^4} \tag{32}$$

which can be expressed, by using the redefined parameters, in the following form:

$$\varphi_{Cas}(C) = \alpha \frac{a + ap[C]^2}{1 + a + ap[C]^2 + p^2q[C]^4}. \tag{33}$$

We set α so that the equilibrium value of *cas* transcription activity corresponds to one (adapted from Bogdanova et al., 2008). Parameters a, p, and q depend on the equilibrium dissociation constants and RNAP concentration and are given by:

$$a = [RNAP]/K_2 \tag{34}$$
$$p = \frac{K_2}{K_1K_3K_5} \tag{35}$$
$$q = \frac{1}{K_1^2K_3K_4p^2} = \frac{K_3K_5^2}{K_2^2K_4} \tag{36}$$

While their values are deduced from the already determined a, b, and c, that correspond to the best fit to the AhdI experimentally measured transcription activity vs. C (Bogdanova et al., 2008).

Regarding the dynamics, note that C and Cas6e transcript and protein amounts change with time according to the Equations (4)–(6), where φ_{Cas} is given by .

Similarly as for the constitutive model, we keep φ constant, in the case when we consider inducing the system through increasing pre-crRNA processing rate. When we also consider regulation of CRISPR array transcription, we assume that CRISPR promoter is regulated by C in the same way as *cas* promoter. Thus, following the same procedure we obtain for the CRISPR promoter transcription activity:

$$\varphi = \alpha' \frac{a' + a'p'[C]^2}{1 + a' + a'p'[C]^2 + p'^2 q'[C]^4} \quad (37)$$

where constants α', a', p', and q' are determined by imposing the same constraints on φ as above (-). Specifically, these constraints lead to the condition $a' < \frac{1}{9}$, which allows setting parameters a' and α'. Further, from Equation (20) we express p' in terms of q' and get $q' < \frac{1}{400*99}$ (deduced from the real roots criterion of quadratic equation), based on which we set q', and subsequently obtain the relation for adjusting p' with respect to k_{eq} (i.e., C_{eq}). Again, the unprocessed [pre-crRNA] and processed [crRNA] transcript amounts change with time according to the Equations (2) and (3), where φ is replaced with (37).

Changing Pre-crRNA Processing Rate

From Equation (1) we have that

$$k_{eq} = [Cas6e]_{eq} \cdot k^*, \quad (38)$$

where we adjust the equilibrium value of k in the constitutive and the cooperative case by varying the concentration of Cas6e in equilibrium. The equilibrium Cas6e concentration can be derived from the steady-state conditions for Equations and :

$$[Cas6e]_{eq} = \frac{k_{Cas6e}}{\lambda_{Cas}\lambda_{Cas6e}} \varphi_{Cas}(C_{eq}). \quad (39)$$

In the model of constitutive C and Cas6e expression, the equilibrium concentration of Cas6e is adjusted through the change of φ_{Cas} (being constant with time). In the case of cooperative C and Cas6e expression, $[Cas6e]_{eq}$ is adjusted through the change of α in Equation (33), i.e., through the change of overall *cas* promoter strength, taking into account that $[C]_{eq}$ is proportional to $[Cas6e]_{eq}$ according to (7).

Joint Change of k and φ

We here investigate how the joint change of k and φ, which corresponds to the joint increase of *cas6e* and CRISPR array gene expression, affects the dynamics of [pre-crRNA] and [crRNA] transcripts. We start from the baseline model of infinitely abrupt increase of k and φ. We then compare the baseline model to the more realistic case of constitutive and the cooperative models. We take φ change from the initial value of 10 1/min to 100 1/min in equilibrium, while k_{eq} takes on values λ_{pre}, $10\lambda_{pre}$, and $100\lambda_{pre}$. Note that the change in k_{eq}, implies joint change of φ_{Cas} in Equation (4) and e in Equation (15) in the constitutive case; in the cooperative case it implies joint change of α and p in Equation (33) and p' in Equation (37), which ensures the same functional dependency $\varphi(t)$, for different values of k_{eq}.

Perturbing Pre-crRNA Degradation Rate λ_{pre}

The pre-crRNA degradation rate λ_{pre} is perturbed (decreased) in the following two cases:

i. With the transcription rate φ (10 1/min) held constant. The equilibrium value of k is then adjusted by varying φ_{Cas} in the constitutive, and α in the cooperative model.
ii. When both φ and the processing rate k reach the equilibrium value (100 1/min) gradually, with the effect of the change assessed in all three models (baseline, constitutive and cooperative). k_{eq} reaches the value 100 1/min through the change of φ_{Cas} in the constitutive, and α and p in the cooperative model, while φ increases from $\varphi(0) = 10$ 1/min to $\varphi\left(C_{eq}\right) = 1,001$/min through adjusting the parameters e in the constitutive, and p' in the cooperative model.

Note that changing λ_{pre} affects the initial amount of pre-crRNA (which is an initial condition for the differential equations) according to the relation $[pre - crRNA]_{eq}(t = 0) = \varphi(t = 0)/\lambda_{pre}$ (see Equation 2), which follows from the steady-state condition for pre-crRNA when the system is not activated.

AUTHOR CONTRIBUTIONS

All authors have given approval to the final version of the manuscript. MarD conceived and coordinated the work, with the help of KS and MagD. AR and BB performed calculations and the analysis. All the authors interpreted the results and contributed to writing the manuscript.

ACKNOWLEDGMENTS

We thank Ekaterina Semenova for carefully reading the paper and useful suggestions.

REFERENCES

Al-Attar, S., Westra, E. R., van der Oost, J., and Brouns, S. J. (2011). Clustered regularly interspaced short palindromic repeats (CRISPRs): the hallmark of an ingenious antiviral defense mechanism in prokaryotes. *Biol. Chem.* 392, 277–289. doi: 10.1515/bc.2011.042

Barrangou, R., Fremaux, C., Deveau, H., Richards, M., Boyaval, P., Moineau, S., et al. (2007). CRISPR provides acquired resistance against viruses in prokaryotes. *Science* 315, 1709–1712. doi: 10.1126/science.1138140

Bogdanova, E., Djordjevic, M., Papapanagiotou, I., Heyduk, T., Kneale, G., and Severinov, K. (2008). Transcription regulation of the type II restriction-modification system AhdI. *Nucleic Acids Res.* 36, 1429–1442. doi: 10.1093/nar/gkm1116

Bogdanova, E., Zakharova, M., Streeter, S., Taylor, J., Heyduk, T., Kneale, G., et al. (2009). Transcription regulation of restriction-modification system Esp1396I. *Nucleic Acids Res.* 37, 3354–3366. doi: 10.1093/nar/gkp210

Bouffartigues, E., Buckle, M., Badaut, C., Travers, A., and Rimsky, S. (2007). H-NS cooperative binding to high-affinity sites in a regulatory element results in transcriptional silencing. *Nat. Struct. Mol. Biol.* 14, 441–448. doi: 10.1038/nsmb1233

Brouns, S. J., Jore, M. M., Lundgren, M., Westra, E. R., Slijkhuis, R. J., Snijders, A. P., et al. (2008). Small CRISPR RNAs guide antiviral defense in prokaryotes. *Science* 321, 960–964. doi: 10.1126/science.1159689

Chen, C. C., Fang, M., Majumder, A., and Wu, H. Y. (2001). A 72-base pair AT-rich DNA sequence element functions as a bacterial gene silencer. *J. Biol. Chem.* 276, 9478–9485. doi: 10.1074/jbc.M010501200

Djordjevic, M., Djordjevic, M., and Severinov, K. (2012). CRISPR transcript processing: a mechanism for generating a large number of small interfering RNAs. *Biol. Direct* 7:24. doi: 10.1186/1745-6150-7-24

Hill, T. L. (2013). *Cooperativity Theory in Biochemistry: Steady-State and Equilibrium Systems.* New York, NY: Springer Science & Business Media.

Hille, F., and Charpentier, E. (2016). CRISPR-Cas: biology, mechanisms and relevance. *Proc. Natl. Acad. Sci. U.S.A.* 371:20150496. doi: 10.1098/rstb.2015.0496

Kruger, D. H., and Schroeder, C. (1981). Bacteriophage T3 and bacteriophage T7 virus-host cell interactions. *Microbiol. Rev.* 45, 9–51.

Kunne, T., Kieper, S. N., Bannenberg, J. W., Vogel, A. I., Miellet, W. R., Klein, M., et al. (2016). Cas3-derived target DNA degradation fragments fuel primed CRISPR adaptation. *Mol. Cell* 63, 852–864. doi: 10.1016/j.molcel.2016.07.011

Makarova, K. S., Grishin, N. V., Shabalina, S. A., Wolf, Y. I., and Koonin, E. V. (2006). A putative RNA-interference-based immune system in prokaryotes: computational analysis of the predicted enzymatic machinery, functional analogies with eukaryotic RNAi, and hypothetical mechanisms of action. *Biol. Direct* 1:7. doi: 10.1186/1745-6150-1-7

Makarova, K. S., Haft, D. H., Barrangou, R., Brouns, S. J., Charpentier, E., Horvath, P., et al. (2011). Evolution and classification of the CRISPR-Cas systems. *Nat. Rev. Microbiol.* 9, 467–477. doi: 10.1038/nrmicro2577

Makarova, K. S., Wolf, Y. I., Alkhnbashi, O. S., Costa, F., Shah, S. A., Saunders, S. J., et al. (2015). An updated evolutionary classification of CRISPR-Cas systems. *Nat. Rev. Microbiol.* 13, 722–736. doi: 10.1038/nrmicro3569

Medina-Aparicio, L., Rebollar-Flores, J., Gallego-Hernandez, A., Vazquez, A., Olvera, L., Gutierrez-Rios, R., et al. (2011). The CRISPR/Cas immune system is an operon regulated by LeuO, H-NS, and leucine-responsive regulatory protein in *Salmonella enterica* serovar Typhi. *J. Bacteriol.* 193, 2396–2407. doi: 10.1128/JB.01480-10

Mojica, F. J., and Diez-Villasenor, C. (2010). The on-off switch of CRISPR immunity against phages in *Escherichia coli. Mol. Microbiol.* 77, 1341–1345. doi: 10.1111/j.1365-2958.2010.07326.x

Morozova, N., Sabantsev, A., Bogdanova, E., Fedorova, Y., Maikova, A., Vedyaykin, A., et al. (2015). Temporal dynamics of methyltransferase and restriction endonuclease accumulation in individual cells after introducing a restriction-modification system. *Nucleic Acids Res.* 44, 790–800. doi: 10.1093/nar/gkv1490

Mruk, I., and Blumenthal, R. M. (2008). Real-time kinetics of restriction-modification gene expression after entry into a new host cell. *Nucleic Acids Res.* 36, 2581–2593. doi: 10.1093/nar/gkn097

Mruk, I., and Kobayashi, I. (2013). To be or not to be: regulation of restriction-modification systems and other toxin-antitoxin systems. *Nucleic Acids Res.* 42, 70–86. doi: 10.1093/nar/gkt711

Musharova, O., Klimuk, E., Datsenko, K. A., Metlitskaya, A., Logacheva, M., Semenova, E., et al. (2017). Spacer-length DNA intermediates are associated with Cas1 in cells undergoing primed CRISPR adaptation. *Nucleic Acids Res.* 45, 3297–3307. doi: 10.1093/nar/gkx097

Patterson, A. G., Yevstigneyeva, M. S., and Fineran, P. C. (2017). Regulation of CRISPR-Cas adaptive immune systems. *Curr. Opin. Microbiol.* 37, 1–7. doi: 10.1016/j.mib.2017.02.004

Pingoud, A., Wilson, G. G., and Wende, W. (2014). Type II restriction endonucleases—a historical perspective and more. *Nucleic Acids Res.* 42, 7489–7527. doi: 10.1093/nar/gku447

Pougach, K., Semenova, E., Bogdanova, E., Datsenko, K. A., Djordjevic, M., Wanner, B. L., et al. (2010). Transcription, processing and function of CRISPR cassettes in *Escherichia coli. Mol. Microbiol.* 77, 1367–1379. doi: 10.1111/j.1365-2958.2010.07265.x

Pul, U., Wurm, R., Arslan, Z., Geissen, R., Hofmann, N., and Wagner, R. (2010). Identification and characterization of *E. coli* CRISPR-cas promoters and their silencing by H-NS. *Mol. Microbiol.* 75, 1495–1512. doi: 10.1111/j.1365-2958.2010.07073.x

Ratner, H. K., Sampson, T. R., and Weiss, D. S. (2015). I can see CRISPR now, even when phage are gone: a view on alternative CRISPR-Cas functions from the prokaryotic envelope. *Curr. Opin. Infect. Dis.* 28:267. doi: 10.1097/QCO.0000000000000154

Richter, C., Chang, J. T., and Fineran, P. C. (2012). Function and regulation of clustered regularly interspaced short palindromic repeats (CRISPR) / CRISPR associated (Cas) systems. *Viruses* 4, 2291–2311. doi: 10.3390/v4102291

Rodic, A., Blagojevic, B., Zdobnov, E., Djordjevic, M., and Djordjevic, M. (2017). Understanding key features of bacterial restriction-modification systems through quantitative modeling. *BMC Syst. Biol.* 11, 1–15. doi: 10.1186/s12918-016-0377-x

Semenova, E., Savitskaya, E., Musharova, O., Strotskaya, A., Vorontsova, D., Datsenko, K. A., et al. (2016). Highly efficient primed spacer acquisition from targets destroyed by the *Escherichia coli* type IE CRISPR-Cas interfering complex. *Proc. Natl Acad. Sci. U.S.A.* 113, 7626–7631. doi: 10.1073/pnas.1602639113

Severinov, K., Ispolatov, I., and Semenova, E. (2016). The influence of copy-number of targeted extrachromosomal genetic elements on the outcome of CRISPR-Cas defense. *Proc. Natl. Acad. Sci. U.S.A.* 3:45. doi: 10.3389/fmolb.2016.00045

Sternberg, S. H., Richter, H., Charpentier, E., and Qimron, U. (2016). Adaptation in CRISPR-Cas Systems. *Mol. Cell* 61, 797–808. doi: 10.1016/j.molcel.2016.01.030

Stratmann, T., Pul, U., Wurm, R., Wagner, R., and Schnetz, K. (2012). RcsB-BglJ activates the *Escherichia coli* leuO gene, encoding an H-NS antagonist and pleiotropic regulator of virulence determinants. *Mol. Microbiol.* 83, 1109–1123. doi: 10.1111/j.1365-2958.2012.07993.x

Tao, T., Bourne, J. C., and Blumenthal, R. M. (1991). A family of regulatory genes associated with type II restriction-modification systems. *J. Bacteriol.* 173, 1367–1375. doi: 10.1128/jb.173.4.1367-1375.1991

Touchon, M., Charpentier, S., Clermont, O., Rocha, E. P., Denamur, E., and Branger, C. (2011). CRISPR distribution within the *Escherichia coli* species is not suggestive of immunity-associated diversifying selection. *J. Bacteriol.* 193, 2460–2467. doi: 10.1128/JB.01307-10

Vijesurier, R. M., Carlock, L., Blumenthal, R. M., and Dunbar, J. C. (2000). Role and mechanism of action of C · *PvuII*, a regulatory protein conserved among restriction-modification systems. *J. Bacteriol.* 182, 477–487. doi: 10.1128/JB.182.2.477-487.2000

Westra, E. R., Pul, U., Heidrich, N., Jore, M. M., Lundgren, M., Stratmann, T., et al. (2010). H-NS-mediated repression of CRISPR-based immunity in *Escherichia coli* K12 can be relieved by the transcription activator LeuO. *Mol. Microbiol.* 77, 1380–1393. doi: 10.1111/j.1365-2958.2010.07315.x

Yang, C. D., Chen, Y. H., Huang, H. Y., Huang, H. D., and Tseng, C. P. (2014). CRP represses the CRISPR/Cas system in *Escherichia coli*: evidence that endogenous CRISPR spacers impede phage P1 replication. *Mol. Microbiol.* 92, 1072–1091. doi: 10.1111/mmi.12614

8

Scoring Targets of Transcription in Bacteria Rather than Focusing on Individual Binding Sites

Marko Djordjevic[1], Magdalena Djordjevic[2] and Evgeny Zdobnov[3]*

[1] Institute of Physiology and Biochemistry, Faculty of Biology, University of Belgrade, Belgrade, Serbia, [2] Institute of Physics Belgrade, University of Belgrade, Belgrade, Serbia, [3] Swiss Institute of Bioinformatics and Department of Genetic Medicine and Development, University of Geneva, Geneva, Switzerland

Edited by:
Tatiana Venkova,
Fox Chase Cancer Center,
United States

Reviewed by:
Alexandre V. Morozov,
Rutgers University, The State University of New Jersey,
United States
Yuriy L. Orlov,
Institute of Cytology and Genetics (RAS), Russia
Anastasia Anashkina,
Engelhardt Institute of Molecular Biology (RAS), Russia

***Correspondence:**
Marko Djordjevic
dmarko@bio.bg.ac.rs

Reliable identification of targets of bacterial regulators is necessary to understand bacterial gene expression regulation. These targets are commonly predicted by searching for high-scoring binding sites in the upstream genomic regions, which typically leads to a large number of false positives. In contrast to the common approach, here we propose a novel concept, where overrepresentation of the scoring distribution that corresponds to the entire searched region is assessed, as opposed to predicting individual binding sites. We explore two implementations of this concept, based on Kolmogorov–Smirnov (KS) and Anderson–Darling (AD) tests, which both provide straightforward *P*-value estimates for predicted targets. This approach is implemented for pleiotropic bacterial regulators, including σ^{70} (bacterial housekeeping σ factor) target predictions, which is a classical bioinformatics problem characterized by low specificity. We show that KS based approach is both faster and more accurate, departing from the current paradigm of AD being slower, but more accurate. Moreover, KS approach leads to a significant increase in the search accuracy compared to the standard approach, while at the same time straightforwardly assigning well established *P*-values to each potential target. Consequently, the new KS based method proposed here, which assigns *P*-values to fixed length upstream regions, provides a fast and accurate approach for predicting bacterial transcription targets.

Keywords: direct target gene predictions, transcription factor binding site predictions, transcription regulation, position specific weight matrices, transcription targets, transcription start starts, sigma70, bacterial gene expression regulation

INTRODUCTION

Identifying targets of transcription regulators (transcription targets), such as genes that are directly regulated by a given transcription factor, or transcribed by a certain σ factor, is a crucial step toward understanding bacterial gene expression regulation. Such knowledge is in turn crucial for both biotechnology applications and fundamental understanding of how bacteria respond to changing environment (e.g., during host pathogen interactions).

The task of identifying transcription targets is typically exhibited by starting from either: (i) large scale *in vivo* binding experiments such as ChIP-Seq (Wade et al., 2007; Park, 2009), (ii) large scale *in vitro* binding data, such as high-throughput SELEX (Roulet et al., 2002; Jagannathan et al., 2006) and protein binding microarrays (PBM) (Bulyk, 2006; Newburger and Bulyk, 2009), and (iii) smaller scale experiments, such as SELEX, primer extension (for σ factors) and DNA footprinting (Green et al., 1989; Tuerk and Gold, 1990), which are typically assembled in databases such as TRANSFAC (Wingender, 2008), JASPAR (Mathelier et al., 2016), or RegulonDB (Gama-Castro et al., 2008). From these binding experiments, specificity of a given transcription factor (TF) is then extracted through some of the numerous methods that have been developed for this purpose. Those methods can be based on either information theory considerations (Stormo, 2000; Bulyk, 2004; Favorov et al., 2005; Ozoline and Deev, 2006; Levitsky et al., 2014; Korostelev et al., 2016), or on biophysical models (Stormo and Fields, 1998; Djordjevic et al., 2003; Djordjevic and Sengupta, 2006; Stormo and Zhao, 2010; Vilar, 2010; Djordjevic, 2013; Vilar and Saiz, 2013; Locke and Morozov, 2015), but in either case the inferred DNA binding specificity is represented in a form of a matrix, often called position specific weight matrix (PSWM). Note that, in the case of biophysics based approaches, these PSWMs in fact correspond to the so-called energy matrix (Djordjevic et al., 2003; Stormo and Zhao, 2010). These methods, up to now, have been shown to be able to extract the binding specificity with a reasonable accuracy, particularly when the data are coming from (controlled) high-throughput *in vitro* experiments (Bulyk, 2004, 2006; Djordjevic and Sengupta, 2006).

Once PSWMs are inferred, in prokaryotes they are used to scan genomic regions upstream of potential targets (e.g., the upstream intergenic regions), to find putative direct regulatory targets (Kim and Ren, 2006). These putative targets are next typically compared with the results of high-throughput experiments, such as DNA microarray data, or crosschecked with results of *in vivo* binding experiments (e.g., with the locations of binding peaks from ChIP-Seq experiments). This crosschecking may provide comprehensive information on the underlying regulatory mechanism, e.g., to what extent binding of the regulator under the given experimental conditions matches with the putative list of the genomic regions to which it is expected to bind. Such information is particularly useful when the binding specificity is inferred from *in vitro* binding studies, and is then crosschecked with independent experiments coming from *in vivo* binding measurements (Kim and Ren, 2006; Stormo and Zhao, 2010).

Despite the importance of accurately predicting direct targets for a given regulator, the bulk of the research efforts concentrate on more accurately inferring PSWM. On the other hand, a typical procedure for identifying putative direct targets in bacteria is rather simple, and involves scanning the upstream genomic regions by the inferred PSWM (Kim and Ren, 2006; Wade et al., 2007; Stormo and Zhao, 2010; de Jong et al., 2012). The sites with maximal PSWM scores are then identified, and those above certain thresholds are classified as putative targets. This procedure, however, often results in low search accuracy, in particular, in a very large number of false positives (Robison et al., 1998; Stormo, 2000). In eukaryotes, methods that predict clusters of transcription factor binding sites (TFBS) are also used, in addition to predicting individual TFBS. However, to successfully apply these methods, one often has to know which TFs functionally interact (Hannenhalli, 2008). Also, a recent evaluation shows that the clustering methods lead to lower accuracy compared to individual TFBS predictions (Jayaram et al., 2016). The major reason behind the apparent low accuracy in the search of direct target genes is that individual high-scoring binding sites can easily appear by random chance in a sufficiently long genomic sequence, leading to so called non-sites (Kim and Ren, 2006). While this problem may be, to some extent, alleviated by negative selection acting on these non-sites, this negative selection is likely small. Furthermore, another problem, accurately assigning statistical significance to the targets predicted in such approach is also not well explored. That is, the maximal scoring sites are located in the tale of the weight matrix score distribution, and accurately calculating this tale requires doing an inverse Laplace transform of the corresponding partition function, which, in itself, is an ill-resolved numerical problem (Hertz and Stormo, 1999). Consequently, putative targets above certain threshold are typically reported without assigning statistical significance to the corresponding hits.

To address the problem of accurate transcription target predictions, we here develop a new concept which is based on the following hypothesis. We propose that, rather than identifying individual sites with high weight matrix scores, a better measure is assessing enrichment of the high scoring sites over a certain background in the entire region that is searched. This proposal then does not depend on individual high-scoring sites (which can easily emerge by random), but instead on comparing the weight matrix score distribution for the entire searched region with a certain background distribution. Note that this automatically accounts for the random occurrence of high-scoring binding sites, since such random occurrences (non-sites) would also appear in the background distribution. Moreover, this hypothesis directly couples with elegant statistical methods that allow determining statistical significance of a difference between the two distributions, such as Kolmogorov–Smirnov (KS) or Anderson–Darling (AD) tests. Therefore, these statistical tests also allow straightforwardly assigning a well-established statistical significance to the predicted direct targets, which also addresses the other major deficiency of the usual approach discussed above. Consequently, in contrast to the previous approaches, we will here develop a method which is based on assigning P values to fixed length upstream regions (e.g., the upstream intergenic regions in bacteria), rather than picking up only the best scoring PSWM matches (or their clusters).

However, significant questions emerge with regard to our proposed novel concept:

(i) Can search based on this hypothesis indeed identify direct targets with high accuracy? Does the classification threshold, based on statistical significance assigned through this approach, lead to high prediction accuracy?
(ii) What is the appropriate background (null distribution)?
(iii) What statistical method is more optimal to implement in the problem, Kolmogorov–Smirnov, Anderson–Darling, or perhaps a combination (hybrid) of these two approaches?

In this proof-of-the-concept paper, we will explore this new method by predicting direct targets for bacterial pleiotropic regulators (σ^{70}, CRP, FNR), which present a classical (currently unresolved) bioinformatics problem characterized by low prediction specificity. On the other hand, accurately predicting transcription targets of bacterial regulators is crucial for understanding bacterial gene expression regulation. Considering bacterial regulators also allows a more straightforward interpretation of the obtained results, as complicating issues such as chromatin state/accessibility (Forties et al., 2011; Chen and Bundschuh, 2014; Chereji and Morozov, 2014) that are present in eukaryotes are largely absent here.

RESULTS AND DISCUSSION

Overrepresentation of PSWM Scoring Distributions

We start by exploring the basic concept behind our hypothesis that the distribution of PSWM scores is overrepresented in the regions where binding of transcription regulators is expected, and that the overrepresentation is absent in the regions where they do not bind. This concept is illustrated by the upper panel of **Figure 1**, where binding of a pleiotropic *Escherichia coli* transcription factor CRP (also known as CAP) to the convergent intergenic regions, and to the rest of the intergenic regions (here called the "other intergenic regions"), is assessed. Note that the convergent intergenic regions are located downstream of both of the adjacent genes, while the other intergenic regions are located upstream of at least one of the adjacent genes. Therefore, there should be no CRP binding sites in the convergent intergenic regions, while CRP binding sites should be located in a subset of the other intergenic regions, which are upstream of its regulatory targets. Accordingly, in the upper left panel of **Figure 1**, we observe a significant overrepresentation of CRP PSWM scores in the other intergenic regions, while such overrepresentation is absent in the convergent intergenic regions. Note that, in **Figure 1**, the background distribution corresponds to randomized intergenic regions, with the sequences randomized so as to preserve trinucleotide frequencies. We obtain similar results (the middle panels) for another *E. coli* pleiotropic transcription factor (FNR), i.e., we also observe an overrepresentation in the other intergenic regions (though now smaller compared to CRP), and an

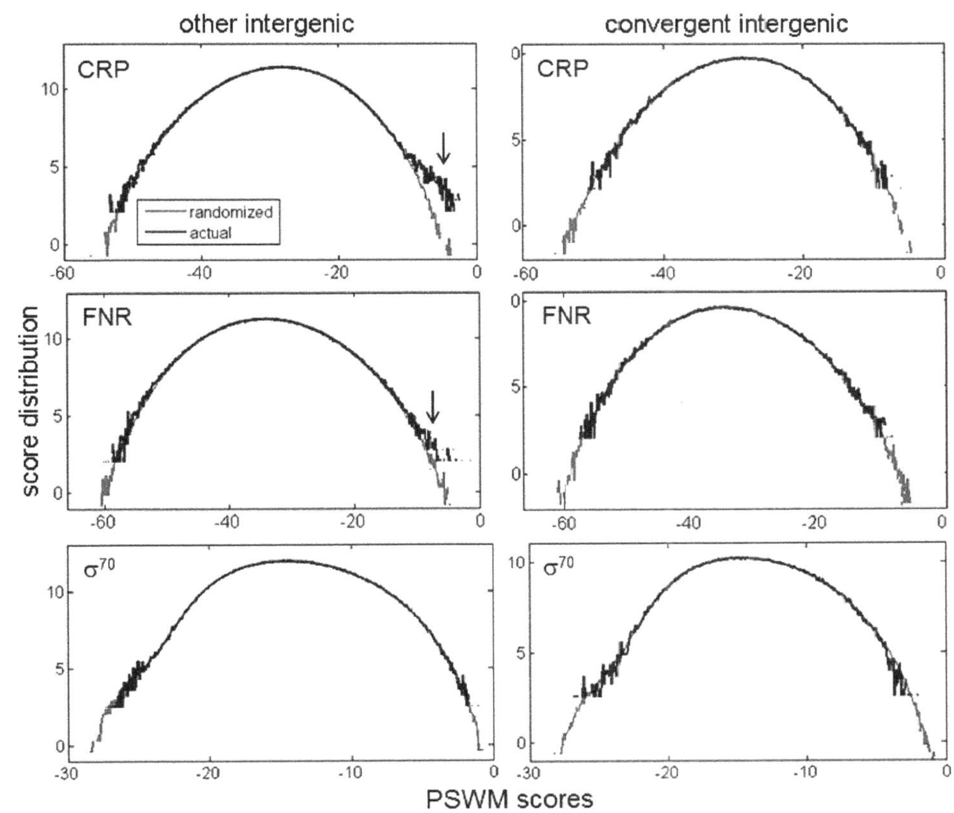

FIGURE 1 | The score distributions for CRP and σ^{70} transcription regulators. The **(upper, middle, lower)** correspond to CRP, FNR, and σ^{70}, respectively. The left panels correspond to the other intergenic regions (where functional binding is expected to appear) while the right panels correspond to the convergent intergenic regions (where functional binding is not expected to appear). Other intergenic regions are located upstream of at least one of the adjacent genes, while convergent intergenic regions are located downstream of both of the adjacent genes (by intergenic region we consider the entire sequence between the two adjacent genes). In each figure, the actual and the randomized PSWM distributions are shown in black and gray, respectively. Note that the higher binding scores (closer to zero), correspond to stronger predicted binders. The overrepresentation in the other intergenic regions for CRP and FNR is indicated by arrows.

absence of overrepresentation in the convergent intergenic regions.

On the other hand, a more complex case is presented in the lower panels of **Figure 1**. Here, binding of the *E. coli* σ^{70} factors to the other intergenic (the left panel) and the convergent intergenic (the right panel) regions is assessed. Note that bacterial σ factors ensure transcription initiation (i.e., provide signal for transcription start sites), and different σ factors are related with transcription exhibited under different conditions in a bacterium (Paget and Helmann, 2003; Feklístov et al., 2014). In particular, σ^{70} is the housekeeping σ factor in *E. coli*, which is associated with transcribing a large number of bacterial genes under normal conditions (therefore having a large regulon). We observe an absence of overrepresentation in both the other and the convergent intergenic regions, in fact a small underrepresentation in the high scoring tail for the convergent intergenic regions can be observed. The absence of the overrepresentation is likely a consequence of significant negative selection on σ^{70} non-sites, as a subset of the other intergenic regions (from which transcription of the downstream genes is directed) has to be enriched with σ^{70} binding sites.

σ^{70} binding, in which no global overrepresentation is observed, evidently corresponds to a more complex case of the regulatory target recognition. Consequently, in the results below, we will first concentrate on σ^{70}, to demonstrate utility of the method even in a more complicated scenario. In addition, prediction of σ factor binding sites, and their corresponding direct targets (i.e., genes that they transcribe) is a classical (unresolved) bioinformatics problem that is considered notoriously hard (Stormo, 2000; Towsey et al., 2008; Purtov et al., 2014), but one that is crucial for understanding bacterial transcription. Predictions of σ^{70} targets are moreover important since RNA-seq experiments (which can map transcription start-sites) are still rare in bacteria, and a number of transcription start sites are active under non-standard conditions, which likely differ from those used in the experiments (Feklístov et al., 2014). Therefore, an additional motivation is to investigate whether our approach can lead to reasonable predictions for such a difficult problem. We will then come back to analyzing two other *E. coli* pleiotropic regulators (CRP and FNR), which display the more standard/expected binding score distributions.

Kolmogorov–Smirnov Based Approach

The main idea behind the new approach is to observe an overrepresentation of PSWM score distribution for the entire upstream genomic region of interest, with respect to a chosen background (null) distribution. We then need to provide a measure of the difference between the two scoring distributions (corresponding to the upstream genomic regions, and the background distribution), as well as a measure of statistical significance for this difference. Assessing this difference can be directly implemented through Kolmogorov–Smirnov (KS) test, which is illustrated in **Figure 2**.

In the left panel, an example of an upstream intergenic region, which is clearly enriched by σ^{70} binding sites, is shown. The solid curve corresponds to the cumulative distribution function (CDF), corresponding to PSWM scores of this intergenic region.

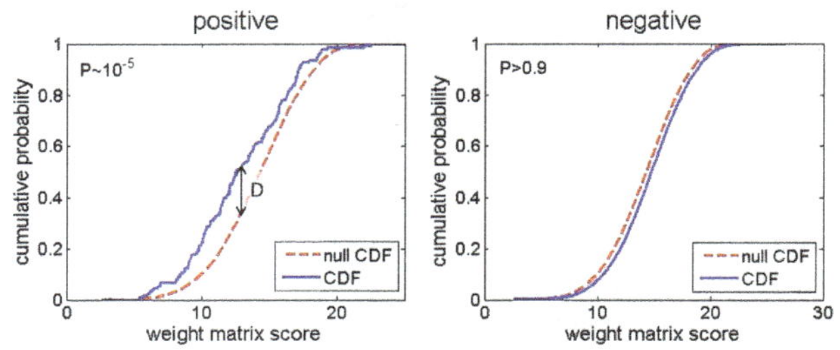

FIGURE 2 | Cumulative distribution functions corresponding to PSWM scores for the upstream genomic regions (the solid blue curve), and the background distribution (the dashed red curve) are shown in the figure. In the **(left)**, an intergenic region that is strongly enriched by σ^{70} binding sites, and that would be reported as a direct target in the search, is shown. In the **(right)**, an extreme case of the upstream intergenic region, which is depleted of σ^{70} binding sites, is shown. *P*-values from KS are indicated in both panels.

Note that the usual KS measure of the difference between the two distributions (which we here denote as *D* score) is indicated in the figure. With respect to the *D* score, note that we here use the one-sided KS test, i.e., we impose the condition that CDF of the upstream genomic regions has to be above the background distribution CDF, which is the condition that corresponds to overrepresentation – i.e., the case of significant underrepresentation being reported as a hit is excluded. KS test also directly provides the *P*-value corresponding to this *D* score, which in turn allows assessing statistical significance of the potential target. On the other hand, the right panel presents an example where the upstream intergenic region is depleted of σ^{70} binding sites. In this case, CDF of PSWM scores corresponding to this depleted intergenic region is actually below the background distribution, so that the gene downstream of this intergenic region is clearly not reported as a direct target of σ^{70} (*D* score is very close to zero in this case). Note that CDF of the upstream intergenic region does not have to be below the background CDF (as happens in the extreme case shown in the right panel), to be excluded as a hit. That is, all hits with small *D* values, which are statistically non-significant, are not reported as putative targets.

Enrichment of *D* Scores

To implement the KS based method, the choice of the background (null) distribution becomes important. This is actually already indicated in **Figure 1**, where we have seen that, due to the negative selection, the distribution corresponding to the randomized regions may not overlap with the distribution in the regions where no binding happens. We here test two choices of the background distributions: (A) the distribution corresponding to the randomized regions, where the intergenic regions are randomized, and their corresponding PSWM scoring distribution is used as the background, (B) genomic regions where functional binding is not expected, for which we use the convergent intergenic regions, as explained in **Figure 1**. Note that these two choices correspond, respectively, to the left (the randomized regions) and the right (the convergent intergenic regions) panel shown in **Figure 3**.

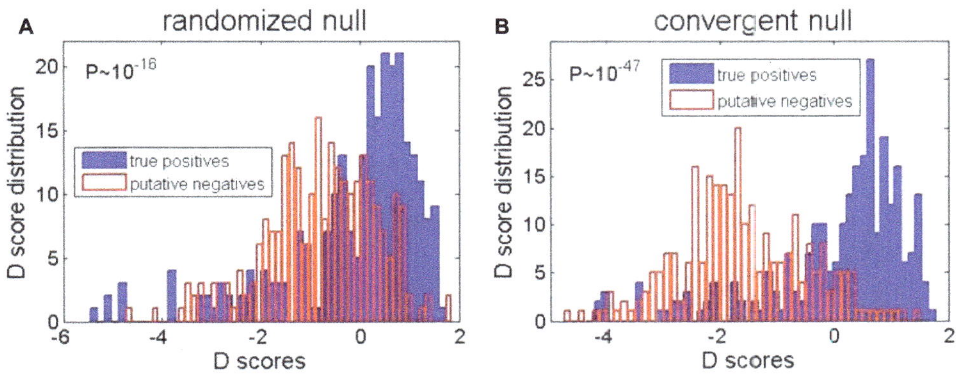

FIGURE 3 | Enrichment of *D* scores for two background distributions. The **(A,B)** represent, respectively, the background distributions that correspond to the randomized and the convergent intergenic regions. The open red and the solid blue histograms, respectively, correspond to the *D* score distributions for the positives (the upstream intergenic regions with experimentally detected binding sites) and the putative negatives (the genomic regions deep inside *E. coli* ORF, where functional σ^{70} binding does not appear). The difference between the blue and the red distributions is assessed by the *P*-value, indicated in each panel.

For each of these two choices of the background distributions, the *D* score distribution is calculated in the following two cases: (i) the red histogram, which corresponds to positives (i.e., the upstream intergenic regions, which are experimentally known to contain σ^{70} binding sites); (ii) the blue histogram: which corresponds to putative negatives, i.e., the genomic regions where σ^{70} binding should *not* appear. Specifically, we here use genomic sequences deep inside ORF (coding sequences), where we expect no initiation of transcription (i.e., no functional σ^{70} binding). We here mark such regions as putative negatives.

We see a significant enrichment of *D* scores in the true positive vs. putative negative regions, for both choices of the background distributions (i.e., for both **Figures 3A,B**). However, we see that the enrichment is clearly much higher when the background distribution corresponds to the convergent intergenic regions, as clearly indicated by the *P*-values in the **Figures 3A,B**. The most likely reason is that the randomized regions do not capture (possibly significant) negative selection that acts on σ^{70} binding sites. That is, the functional binding, which one needs to detect, comes on the 'top' of possibly a large number of non-sites that are 'deleted' by the negative selection. Consequently, in the further analysis, we will use the background distribution which corresponds to the convergent intergenic regions.

ROC Curves and Comparison with Anderson–Darling Test

Our next goal is to compare the accuracy of KS-based approach, with the standard method for identifying putative targets in bacteria. This method (which we further call "Max") involves scanning the upstream genomic regions by PSWM, and classifying as putative targets those regions that contain individual binding sites with PSWM scores above certain threshold. To this end, we use the same positives and putative negatives as introduced in the previous subsection, and the null distribution that corresponds to the convergent intergenic regions. In addition, as an alternative to KS test, the AD test can also be used to detect overrepresentation of the binding scores with respect to the null distribution. Consequently, we also address how accurately the two tests (AD and KS) can predict direct targets of σ^{70}. The corresponding prediction accuracies are assessed by ROC curves shown in **Figure 4**.

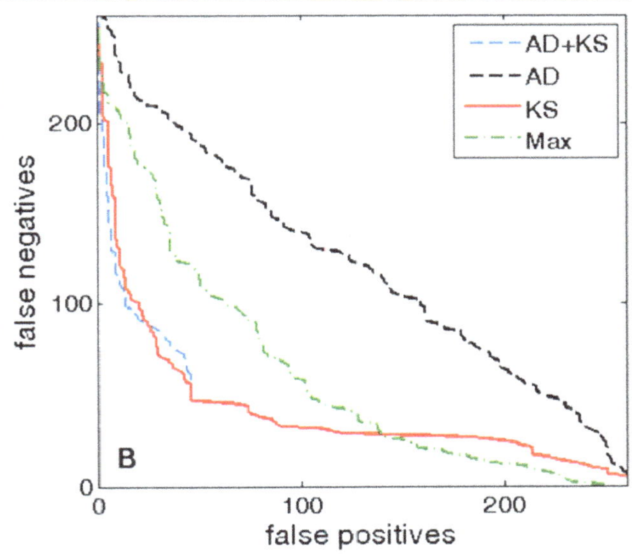

FIGURE 4 | ROC curves, σ^{70} binding. The ROC curves for σ^{70} direct targets, which correspond to different tests, are marked in the legend. The positive and the putative negative sets are the same as in **Figure 3**, and the false negatives and the false positives are estimated with respect to these two sets, where the convergent background distribution is used. In the legend, "Max" corresponds to the standard method for direct target identification (see the first paragraph of this subsection).

Importantly, we see that KS based approach (the solid red curve) shows a substantially better performance compared to the standard method (the dot-dashed green curve). In particular, note that for the fixed number of false positives there are up to three times fewer false negatives. Such a reduction in the

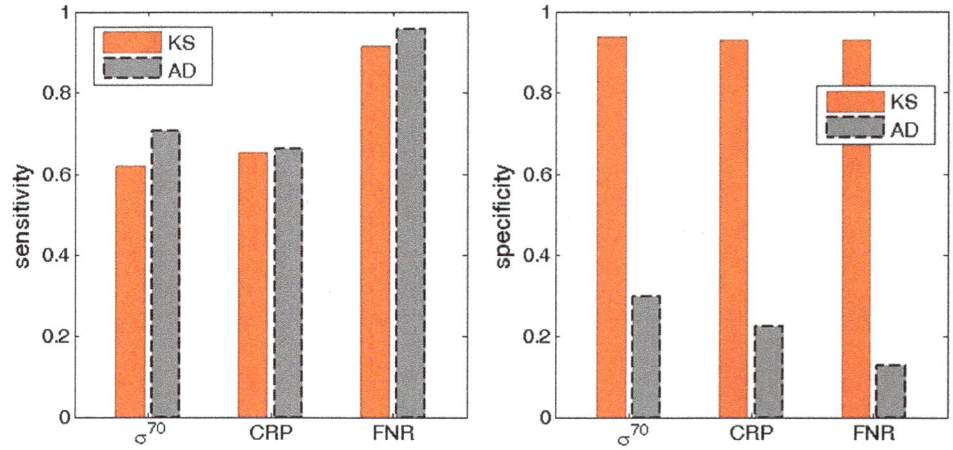

FIGURE 5 | Comparison of sensitivity and specificity for KS and AD methods. The **(left, right)** correspond, respectively, to the sensitivity and specificity estimates, obtained for the usual $P = 0.05$ confidence level. The red and the gray bars correspond to KS and AD methods respectively. The sensitivity and the specificity estimates are shown for σ^{70} (the left bars), CRP (the central bars), and FNR (the right bars). The sensitivity and specificity are calculated as, respectively, TP/P, and TN/N, where TP are true positives, TN true negatives, while P and N are the number of positives and negatives, respectively.

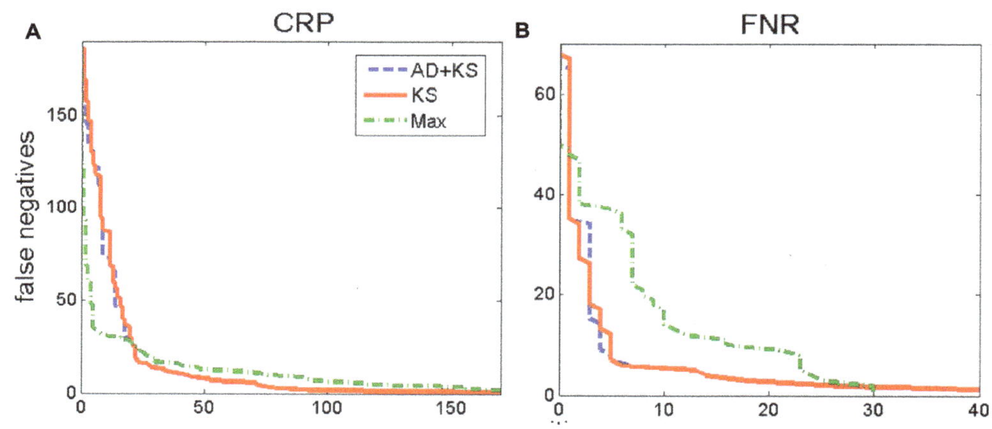

FIGURE 6 | ROC curves, CRP and FNR. The **(A)** and the **(B)** correspond, respectively, to the ROC curves for CRP and FNR. The true positive and the putative negatives are as in **Figure 3**, and the convergent intergenic regions are used for the background distribution. Different tests which correspond to the shown ROC curves are marked in the figure legend.

number of false positives is expected, i.e., in accordance with the hypothesis we presented above, since individual sites with high PSWM scores can easily appear by random. On the other hand, their appearance is automatically taken into account through the background distribution, i.e., a potential target will be classified as a hit only if the binding scores in the entire searched region are enriched (overrepresented) with respect to the background distribution.

Furthermore, we see that KS (the solid red curve) leads to a higher detection accuracy compared to AD (the dashed black curve). Moreover, KS test is also much (~400 times) faster in predicting the direct targets. Consequently, in this application, KS test is both faster and more accurate than AD. Note that this runs opposite to the common paradigm, according to which AD is slower, but more accurate compared to KS (Stephens, 1974).

To investigate the reason behind the (unexpected) significantly higher accuracy obtained with KS method, in **Figure 5** we compare the sensitivity (the left panel) and the specificity (the right panel) for KS and AD methods. The comparison corresponds to the standard classification threshold ($P < 0.05$) for both methods, and is provided for σ^{70} (analyzed in **Figure 4**) and for CRP and FNR transcription factors (analyzed in **Figures 6**, **7** below). We see that the sensitivity is high, and about the same, for both methods (with AD displaying even slightly larger sensitivity). On the other hand, in the right panel of **Figure 5**, it can be seen that the specificity is much smaller for AD method, which then leads to its lower accuracy compared to KS method. To interpret this result, one should note that we necessarily work with an approximation of the true null distribution, e.g., the negative selection on non-sites in the convergent intergenic regions is likely not the same as in the other intergenic regions, in which the target classification is exhibited. Consequently, the main general advantage of AD method, which is its large sensitivity, becomes a weakness in this application, as small differences with respect to the null distribution (that may also arise from its approximate nature), are (conveniently) not captured by KS, but are classified as statistically highly significant by AD test, leading to low AD specificity (a large number of false positives).

Next, there comes a question if a combination of AD and KS tests can provide an improved accuracy compared to either of the two tests alone. With that respect, we made an algorithm corresponding to a hybrid where the KS test is implemented first to filter-out those upstream regions with clearly insignificant *P*-values, i.e., regions where the actual distribution is clearly too close to the null distribution. Afterward, AD is applied to those regions with distributions that are more different to the null distribution, for which AD performs better. From **Figure 4**, we see that such AD-KS hybrid (the dashed blue curve) indeed shows a substantially better accuracy compared to AD test alone, and has a similar accuracy to KS test alone. This result is consistent with the discussion above, i.e., when AD test alone is used, a number of the upstream regions that are eliminated by KS test, are falsely classified as targets by AD (since, due to small specificity, AD proclaims even small difference between the distributions as being significant). On the other hand, when in AD-KS hybrid AD is applied only to those distributions that are more different with respect to the null distribution (therefore bypassing its main problem of low specificity), the accuracy becomes similar to KS test. As an outlook, note that AD-KS hybrid might be further improved by optimizing the threshold for KS selection. We will further concentrate only on KS and AD-KS tests, as they have a much better accuracy to AD test alone.

We next come back to assessing KS approach for two more standard binding score distributions (see **Figure 1**), exhibited by CRP and FNR transcription regulators. We here construct the positive and the putative negative sets in the same way as for σ^{70}, i.e., the positives correspond to the intergenic regions where the transcription regulator binding is experimentally shown, while the putative negatives correspond to the sequences deep inside the coding regions, where functional binding is not expected. The corresponding ROC curves are shown in **Figure 6**.

For FNR (**Figure 6B**), we obtain similar results as for σ^{70}, i.e., KS (and KS-AD hybrid) lead to a significantly better ROC curve performance compared to the standard method (e.g., for a fixed false negative number, there is a several times smaller number of false positives for KS). On the other hand, we see that for CRP the two curves (KS and the standard method) have apparently similar performances, i.e., while the standard method shows better performance at low false positive numbers, it is outperformed by KS at higher false positives. The similar performance of KS in the case of CRP is not surprising, i.e., is likely a consequence of the fact that functional binding dominates over non-sites in this case, as implied by the large PSWM score overrepresentation exhibited in such case (see the upper left panel in **Figure 1**). Consequently, the results in **Figure 5** are in line with

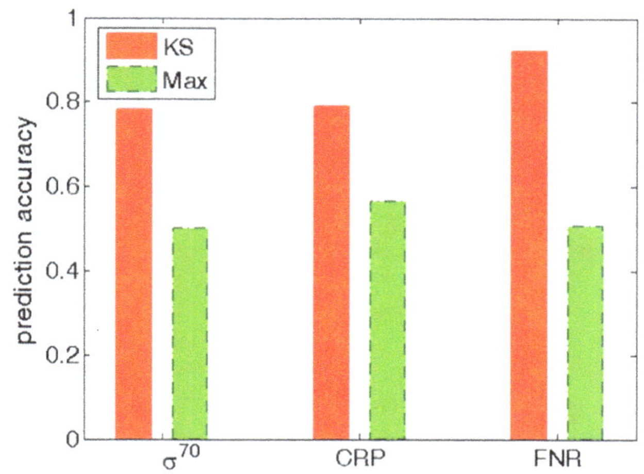

FIGURE 7 | Prediction accuracy for the standard classification thresholds. The threshold in KS search is based on the estimated statistical significance (the usual $P = 0.05$ threshold is taken). The threshold in the PSWM search corresponds to the standard choice where most (98%) of the experimentally determined binding examples would be recovered in the search. The prediction accuracy for these two thresholds is shown for σ^{70} (the left bars), CRP (the central bars), and FNR (the right bars). The prediction accuracy is calculated as (TP + TN)/(TP + FP + FN + TN), where TP (true positives), TN (true negatives), FP (false positives) and FN (false negatives) are calculated for the two methods at the corresponding threshold choices.

our main hypothesis that the main utility of KS approach is in accurate classification of non-sites.

Statistical Significance and the Classification Threshold

Independently from the ROC performance, KS has a significant advantage of straightforwardly assigning statistical significance to each predicted target, which is normally not available for standard PSWM search (see Introduction). We here explore the utility of such robust statistical significance estimate with the example of assigning a classification threshold. With the KS approach a natural threshold choice is provided by the *P*-value, typically set to $P = 0.05$. As such a natural choice is normally not available for standard PSWM search, the threshold is usually set so that almost all (~98%) of the experimentally determined binding sites from which PSWM is constructed are recovered in the search. In **Figure 7**, we explore the search accuracy associated with the two choices of the binding threshold, i.e., $P = 0.05$ for KS method and the standard threshold (see above) for PSWM search.

We see that the threshold based on KS significance estimate leads to much higher prediction accuracy for σ^{70} and FNR, which is expected based on the significantly better ROC performance of KS in these two cases (**Figures 4, 6B**). Moreover, in **Figure 7** we also see notably higher prediction accuracy in the case of CRP, where a similar ROC performance was observed for KS and standard PSWM search (**Figure 6A**). Consequently, the notably higher search accuracy for KS in the case of CRP observed in **Figure 7** is based on the more optimal choice of the classification threshold. This underlines the advantage of the threshold choice based on the robust statistical significance measure.

CONCLUSION AND OUTLOOK

We here proposed a new computational approach to direct regulatory target prediction. The approach is based on assessing the significance of the difference between PSWM scoring distributions, which correspond to the upstream genomic regions, and the background distribution. As a consequence, P-value is assigned to the entire upstream region, instead of searching for individual high-scoring binding sites. We implemented this approach through classical Kolmogorov–Smirnov and Anderson–Darling tests, as well as through a hybrid of these two approaches. Surprisingly, and contrary to the current paradigm, we have seen that the approach based on Kolmogorov–Smirnov test leads to a higher search accuracy compared to Anderson–Darling based approach, while also being (as expected) computationally less demanding. While the hybrid approach has a substantially higher accuracy compared to Anderson–Darling test, it does not outperform the simpler Kolmogorov–Smirnov test. We interpreted this result by Anderson–Darling test classifying small differences with respect to the background distribution as true binding targets, leading to low specificity of the approach.

We furthermore showed that the Kolmogorov–Smirnov based approach leads to a substantially higher accuracy compared to the standard approach, reducing the number of false positives for several times. Moreover, a clear advantage of Kolmogorov–Smirnov approach is that it straightforwardly assigns statistical significance to any tested upstream intergenic region. We demonstrated this advantage on the example of the classification threshold, where we have seen that the robust significance estimate provided by Kolmogorov–Smirnov leads to a much more optimal threshold choice. We find that genomic regions, where functional binding is not expected, provide better background compared to randomized genomic regions. We here, i.e., for analysis of prokaryotic transcription regulation, used convergent intergenic regions for background distribution. In eukaryotes the choice of background distribution would be more complicated and remains to be investigated, where one possibility would be to take genomic sequences far from coding regions (where there may not be much TFBS).

To prove this new concept in the direct regulatory target prediction, we tested it in the case of pleiotropic bacterial regulators. This allowed a more straightforward interpretation of the obtained results, while testing the method on some of the classical problems otherwise characterized by low prediction specificity. As an outlook, the method proposed here is of a general significance, and it will be in the future also implemented in the more complicated case of direct target prediction for eukaryotic transcriptional regulators. Moreover, while the model was here applied in the context of PSWM, more complex models which take into account interdependences of nucleotides in TFBS were also developed (Eggeling et al., 2015; Kulakovskiy et al., 2016; Nettling et al., 2017). While these methods lead to a better performance in some cases, more often (simpler) PSWMs perform better, which is likely due to overfitting, i.e., due to a limited number of TFBS from which the model is trained (Benos et al., 2002; Nguyen and Androulakis, 2009). Therefore, despite the limitations of PSWMs, they are still the leading approach in TFBS search (Nguyen and Androulakis, 2009; Fazius et al., 2011). In any case, the new approach proposed here does not depend on the scoring method (i.e., if a classical PSWM, or a higher order model, is used), since the approach is based on comparing the distributions of the scores (i.e., is not limited by how the actual scores are calculated). Consequently, the KS approach proposed here might present a general method of choice for efficiently and accurately predicting target loci of transcription regulators.

MATERIALS AND METHODS

Defining the Upstream Genomic Regions

The *E. coli* intergenic sequences are divided in two groups, where binding of transcription regulators is expected (other intergenic regions) and not expected (convergent intergenic regions). The other intergenic regions, and the convergent intergenic regions, include, respectively, those that are located upstream of at least one adjacent gene, and downstream of both of the adjacent genes.

For the positive set in σ^{70} case, in KS, AD and KS-AD hybrid searches, we take those *E. coli* intergenic sequences that contain σ^{70} binding sites with experimental evidence from RegulonDB database (Gama-Castro et al., 2011), which results in the total of 263 upstream genomic regions. Similarly, for the positive set in CRP and FNR case, we take these intergenic sequences with experimental evidence of the regulator binding from RegulonDB database. For the putative negative set, we use the same number of sequences, with the same length, as those in the positive set, but now sampled from ORF (coding sequences), where we exclude 50 bps at both 5′ and 3′ ends; this is done to exclude the flanking sequences, in which σ^{70} binding sites are sometimes located.

For obtaining the randomized distribution, an ensemble of randomized sequences was constructed, by sampling all trinucleotide probabilities in the intergenic regions. The randomized sequences were searched, and the corresponding randomized scoring distributions are obtained, in the same manner as for the upstream genomic regions, which is further described below.

PSWM Scoring Distributions

CRP and FNR PSWM were constructed from the binding sites assembled in DPInteract database (Robison et al., 1998), through the standard information-theory based procedure (Stormo, 2000). σ^{70} PSWM were constructed starting from recent *de novo* alignment (Djordjevic, 2011), where the promoter elements were systematically aligned starting directly from the experimentally determined TSS. Briefly, the alignment includes: −10 and −35 elements, spacer weights corresponding to variable spacer length (between 15 and 19 bps), conserved sequences upstream of −10 element. PSWMs for CRP, FNR and σ^{70} search are provided in Supplementary Table 1. Scores were assigned to each DNA segment in the upstream genomic and the randomized sequences using these PSWM, from which the corresponding scoring distributions were generated.

KS, AD, and KS-AD Hybrid Based Searches

For KS based search, one-sided Kolmogorov-Smirnov test was used. For each tested upstream intergenic region PSWM score distribution was generated as described above, and compared with an appropriate background distribution whose CDF was constructed. This comparison results in *P*-value and *D* score for each tested upstream genomic segment. The threshold on *D* scores was then moved in order to change the number of false positives and false negatives, and construct the ROC curves.

For AD based search, the MATLAB based routine 'adtest' was used, where PSWM score distributions corresponding to upstream genomic region, and the background distribution were compared. For each tested upstream genomic region, *P*-value and AD test statistics ('adstat') was sampled. The ROC curves were constructed based on 'adstat' scores.

For KS-AD hybrid search, KS and AD tests were implemented as described above, with KS test used first to exclude the upstream intergenic regions with low difference between the two distributions. A liberal *P*-value threshold of 0.5 was used in this exclusion, so that only the upstream genomic regions with very low significance are eliminated by KS test. The rest of the upstream regions are then subjected to AD test, which is used to calculate *P*-value and AD test statistics. The codes for KS, AD and KS-AD hybrid approaches are available upon request.

AUTHOR CONTRIBUTIONS

All authors have given approval to the final version of the manuscript. MarD conceived the work, with the help of MagD and EZ. MarD and MagD implemented the method and performed the analysis. All the authors interpreted the results. MarD wrote the paper, with the help of MagD and EZ.

ACKNOWLEDGMENTS

We thank Djordje Markovic and Marija Basic for help with **Figure 1**, and to Christopher Rands for critically reading the manuscript.

REFERENCES

Benos, P. V., Bulyk, M. L., and Stormo, G. D. (2002). Additivity in protein–DNA interactions: how good an approximation is it? *Nucleic Acids Res.* 30, 4442–4451. doi: 10.1093/nar/gkf578

Bulyk, M. L. (2004). Computational prediction of transcription-factor binding site locations. *Genome Biol.* 5:201. doi: 10.1186/gb-2003-5-1-201

Bulyk, M. L. (2006). DNA microarray technologies for measuring protein–DNA interactions. *Curr. Opin. Biotechnol.* 17, 422–430. doi: 10.1016/j.copbio.2006.06.015

Chen, C., and Bundschuh, R. (2014). Quantitative models for accelerated protein dissociation from nucleosomal DNA. *Nucleic Acids Res.* 42, 9753–9760. doi: 10.1093/nar/gku719

Chereji, R. V., and Morozov, A. V. (2014). Functional roles of nucleosome stability and dynamics. *Brief. Funct. Genomics* 14, 50–60. doi: 10.1093/bfgp/elu038

de Jong, A., Pietersma, H., Cordes, M., Kuipers, O. P., and Kok, J. (2012). PePPER: a webserver for prediction of prokaryote promoter elements and regulons. *BMC Genomics* 13:299. doi: 10.1186/1471-2164-13-299

Djordjevic, M. (2011). Redefining *Escherichia coli* σ^{70} promoter elements: -15 motif as a complement of the -10 motif. *J. Bacteriol.* 193, 6305–6314. doi: 10.1128/JB.05947-11

Djordjevic, M. (2013). Efficient transcription initiation in bacteria: an interplay of protein–DNA interaction parameters. *Integr. Biol.* 5, 796–806. doi: 10.1039/c3ib20221f

Djordjevic, M., and Sengupta, A. M. (2006). Quantitative modeling and data analysis of SELEX experiments. *Phys. Biol.* 3, 13–28. doi: 10.1088/1478-3975/3/1/002

Djordjevic, M., Sengupta, A. M., and Shraiman, B. I. (2003). A biophysical approach to transcription factor binding site discovery. *Genome Res.* 13, 2381–2390. doi: 10.1101/gr.1271603

Eggeling, R., Roos, T., Myllymäki, P., and Grosse, I. (2015). Inferring intra-motif dependencies of DNA binding sites from ChIP-seq data. *BMC Bioinformatics* 16:375. doi: 10.1186/s12859-015-0797-4

Favorov, A. V., Gelfand, M. S., Gerasimova, A. V., Ravcheev, D. A., Mironov, A. A., and Makeev, V. J. (2005). A Gibbs sampler for identification of symmetrically structured, spaced DNA motifs with improved estimation of the signal length. *Bioinformatics* 21, 2240–2245. doi: 10.1093/bioinformatics/bti336

Fazius, E., Shelest, V., and Shelest, E. (2011). SiTaR: a novel tool for transcription factor binding site prediction. *Bioinformatics* 27, 2806–2811. doi: 10.1093/bioinformatics/btr492

Feklístov, A., Sharon, B. D., Darst, S. A., and Gross, C. A. (2014). Bacterial sigma factors: a historical, structural, and genomic perspective. *Annu. Rev. Microbiol.* 68, 357–376. doi: 10.1146/annurev-micro-092412-155737

Forties, R. A., North, J. A., Javaid, S., Tabbaa, O. P., Fishel, R., Poirier, M. G., et al. (2011). A quantitative model of nucleosome dynamics. *Nucleic Acids Res.* 39, 8306–8313. doi: 10.1093/nar/gkr422

Gama-Castro, S., Jiménez-Jacinto, V., Peralta-Gil, M., Santos-Zavaleta, A., Peñaloza-Spinola, M. I., Contreras-Moreira, B., et al. (2008). RegulonDB (version 6.0): gene regulation model of *Escherichia coli* K-12 beyond transcription, active (experimental) annotated promoters and textpresso navigation. *Nucleic Acids Res.* 36(Suppl. 1), D120–D124. doi: 10.1093/nar/gkm994

Gama-Castro, S., Salgado, H., Peralta-Gil, M., Santos-Zavaleta, A., Muñiz-Rascado, L., Solano-Lira, H., et al. (2011). RegulonDB version 7.0: transcriptional regulation of *Escherichia coli* K-12 integrated within genetic sensory response units (Gensor Units). *Nucleic Acids Res.* 39(Suppl. 1), D98–D105. doi: 10.1093/nar/gkq1110

Green, P. J., Kay, S. A., Lam, E., and Chua, N.-H. (1989). "*In vitro* DNA footprinting," in *Plant Molecular Biology Manual*, eds S. B. Gelvin, R. A. Schilperoort, and B. P. S. Verma (Dordrecht: Springer), 415–436. doi: 10.1007/978-94-009-0951-9_21

Hannenhalli, S. (2008). Eukaryotic transcription factor binding sites—modeling and integrative search methods. *Bioinformatics* 24, 1325–1331. doi: 10.1093/bioinformatics/btn198

Hertz, G. Z., and Stormo, G. D. (1999). Identifying DNA and protein patterns with statistically significant alignments of multiple sequences. *Bioinformatics* 15, 563–577. doi: 10.1093/bioinformatics/15.7.563

Jagannathan, V., Roulet, E., Delorenzi, M., and Bucher, P. (2006). HTPSELEX–a database of high-throughput SELEX libraries for transcription factor binding sites. *Nucleic Acids Res.* 34, D90–D94. doi: 10.1093/nar/gkj049

Jayaram, N., Usvyat, D., and Martin, A. C. (2016). Evaluating tools for transcription factor binding site prediction. *BMC Bioinformatics* doi: 10.1186/s12859-016-1298-9 [Epub ahead of print].

Kim, T. H., and Ren, B. (2006). Genome-wide analysis of protein-DNA interactions. *Annu. Rev. Genomics Hum. Genet.* 7, 81–102. doi: 10.1146/annurev.genom.7.080505.115634

Korostelev, Y. D., Zharov, I. A., Mironov, A. A., Rakhmaininova, A. B., and Gelfand, M. S. (2016). Identification of position-specific correlations between DNA-binding domains and their binding sites. Application MerR family transcription factors. *PLOS ONE* 11:e0162681. doi: 10.1371/journal.pone.0162681

Kulakovskiy, I. V., Vorontsov, I. E., Yevshin, I. S., Soboleva, A. V., Kasianov, A. S., Ashoor, H., et al. (2016). HOCOMOCO: expansion and enhancement of the collection of transcription factor binding sites models. *Nucleic Acids Res.* 44, D116–D125. doi: 10.1093/nar/gkv1249

Levitsky, V. G., Kulakovskiy, I. V., Ershov, N. I., Oshchepkov, D. Y., Makeev, V. J., Hodgman, T., et al. (2014). Application of experimentally verified transcription factor binding sites models for computational analysis of ChIP-Seq data. *BMC Genomics* 15:80. doi: 10.1186/1471-2164-15-80

Locke, G., and Morozov, A. V. (2015). A biophysical approach to predicting protein–DNA binding energetics. *Genetics* 200, 1349–1361. doi: 10.1534/genetics.115.178384

Mathelier, A., Fornes, O., Arenillas, D. J., Chen, C.-Y., Denay, G., Lee, J., et al. (2016). JASPAR 2016: a major expansion and update of the open-access database of transcription factor binding profiles. *Nucleic Acids Res.* 44, D110–D115. doi: 10.1093/nar/gkv1176

Nettling, M., Treutler, H., Cerquides, J., and Grosse, I. (2017). Combining phylogenetic footprinting with motif models incorporating intra-motif dependencies. *BMC Bioinformatics* 18:141 doi: 10.1186/s12859-017-1495-1

Newburger, D. E., and Bulyk, M. L. (2009). UniPROBE: an online database of protein binding microarray data on protein–DNA interactions. *Nucleic Acids Res.* 37(Suppl. 1), D77–D82. doi: 10.1093/nar/gkn660

Nguyen, T. T., and Androulakis, I. P. (2009). Recent advances in the computational discovery of transcription factor binding sites. *Algorithms* 2, 582–605. doi: 10.3390/a2010582

Ozoline, O. N., and Deev, A. A. (2006). Predicting antisense RNAs in the genomes of *Escherichia coli* and *Salmonella typhimurium* using promoter-search algorithm PlatProm. *J. Bioinform. Comput. Biol.* 4, 443–454. doi: 10.1142/S0219720006001916

Paget, M. S. B., and Helmann, J. D. (2003). The σ^{70} family of sigma factors. *Gen. Biol.* 4, 203–208. doi: 10.1186/gb-2003-4-1-203

Park, P. J. (2009). ChIP–seq: advantages and challenges of a maturing technology. *Nat. Rev. Genet.* 10, 669–680. doi: 10.1038/nrg2641

Purtov, Y. A., Glazunova, O. A., Antipov, S. S., Pokusaeva, V. O., Fesenko, E. E., Preobrazhenskaya, E. V., et al. (2014). Promoter islands as a platform for interaction with nucleoid proteins and transcription factors. *J. Bioinform. Comput. Biol.* 12:1441006. doi: 10.1142/S0219720014410066

Robison, K., McGuire, A., and Church, G. (1998). A comprehensive library of DNA-binding site matrices for 55 proteins applied to the complete *Escherichia coli* K-12 genome. *J. Mol. Biol.* 284, 241–254. doi: 10.1006/jmbi.1998.2160

Roulet, E., Busso, S., Camargo, A., Simpson, A., Mermod, N., and Bucher, P. (2002). High-throughput SELEX–SAGE method for quantitative modeling of transcription-factor binding sites. *Nat. Biotechnol.* 20, 831–835. doi: 10.1038/nbt718

Stephens, M. A. (1974). EDF statistics for goodness of fit and some comparisons. *J. Am. Stat. Assoc.* 69, 730–737. doi: 10.1080/01621459.1974.10480196

Stormo, G. D. (2000). DNA binding sites: representation and discovery. *Bioinformatics* 16, 16–23. doi: 10.1093/bioinformatics/16.1.16

Stormo, G. D., and Fields, D. S. (1998). Specificity, free energy and information content in protein-DNA interactions. *Trends Biochem. Sci.* 23, 109–113. doi: 10.1016/S0968-0004(98)01187-6

Stormo, G. D., and Zhao, Y. (2010). Determining the specificity of protein–DNA interactions. *Nat. Rev. Genet.* 11, 751–760. doi: 10.1038/nrg2845

Towsey, M., Hogan, J., Mathews, S., and Timms, P. (2008). The in silico prediction of promoters in bacterial genomes. *Genome Inform.* 19, 178–189.

Tuerk, C., and Gold, L. (1990). Systematic evolution of ligands by exponential enrichment: RNA ligands to bacteriophage T4 DNA polymerase. *Science* 249, 505–510. doi: 10.1126/science.2200121

Vilar, J. M. (2010). Accurate prediction of gene expression by integration of DNA sequence statistics with detailed modeling of transcription regulation. *Biophys. J.* 99, 2408–2413. doi: 10.1016/j.bpj.2010.08.006

Vilar, J. M., and Saiz, L. (2013). Systems biophysics of gene expression. *Biophys. J.* 104, 2574–2585. doi: 10.1016/j.bpj.2013.04.032

Wade, J. T., Struhl, K., Busby, S. J., and Grainger, D. C. (2007). Genomic analysis of protein–DNA interactions in bacteria: insights into transcription and chromosome organization. *Mol. Microbiol.* 65, 21–26. doi: 10.1111/j.1365-2958.2007.05781.x

Wingender, E. (2008). The TRANSFAC project as an example of framework technology that supports the analysis of genomic regulation. *Brief. Bioinform.* 9, 326–332. doi: 10.1093/bib/bbn016

9

Ribonucleotide Reductases from Bifidobacteria Contain Multiple Conserved Indels Distinguishing them from all other Organisms: *In Silico* Analysis of the Possible Role of a 43 aa Bifidobacteria-Specific Insert in the Class III RNR Homolog

Seema Alnajar, Bijendra Khadka and Radhey S. Gupta*

Department of Biochemistry and Biomedical Sciences, McMaster University, Hamilton, ON, Canada

Edited by:
Manuel Espinosa, Centro de Investigaciones Biológicas (CSIC), Spain

Reviewed by:
Paloma López, Centro de Investigaciones Biológicas (CSIC), Spain
Paul Meyers, University of Cape Town, South Africa

***Correspondence:**
Radhey S. Gupta
gupta@mcmaster.ca

Bifidobacteria comprises an important group/order of bacteria whose members have widespread usage in the food and health industry due to their health-promoting activity in the human gastrointestinal tract. However, little is known about the underlying molecular properties that are responsible for the probiotic effects of these bacteria. The enzyme ribonucleotide reductase (RNR) plays a key role in all organisms by reducing nucleoside di- or tri- phosphates into corresponding deoxyribose derivatives required for DNA synthesis, and RNR homologs belonging to classes I and III are present in either most or all *Bifidobacteriales.* Comparative analyses of these RNR homologs have identified several novel sequence features in the forms of conserved signature indels (CSIs) that are exclusively found in bifidobacterial RNRs. Specifically, in the large subunit of the aerobic class Ib RNR, three CSIs have been identified that are uniquely found in the *Bifidobacteriales* homologs. Similarly, the large subunit of the anaerobic class III RNR contains five CSIs that are also distinctive characteristics of bifidobacteria. Phylogenetic analyses indicate that these CSIs were introduced in a common ancestor of the *Bifidobacteriales* and retained by all descendants, likely due to their conferring advantageous functional roles. The identified CSIs in the bifidobacterial RNR homologs provide useful tools for further exploration of the novel functional aspects of these important enzymes that are exclusive to these bacteria. We also report here the results of homology modeling studies, which indicate that most of the bifidobacteria-specific CSIs are located within the surface loops of the RNRs, and of these, a large 43 amino acid insert in the class III RNR homolog forms an extension of the allosteric regulatory site known to be essential for protein function. Preliminary docking studies suggest that this large CSI may be playing a role in enhancing the stability of the RNR dimer complex. The possible significances of the identified CSIs, as well as the distribution of RNR homologs in the *Bifidobacteriales,* are discussed.

Keywords: novel features of ribonucleotide reductases, probiotic bacteria, *Bifidobacteriales,* conserved signature inserts and deletions, homology modeling and protein docking studies, extended allosteric site, phylogenetic analysis

INTRODUCTION

The *Bifidobacteriales* constitute an important order of bacteria within the phylum Actinobacteria (Ventura et al., 2007; Zhi et al., 2009; Gao and Gupta, 2012). While some species belonging to this order are pathogenic (Smith et al., 1992; Bradshaw et al., 2006; Alves et al., 2014; Kenyon and Osbak, 2014) many *Bifidobacteriales* species belonging to the genus *Bifidobacterium* are known for their beneficial health-promoting effects in humans and other mammals (Gibson et al., 1995; Leahy et al., 2005; Masco et al., 2005; Ventura et al., 2009; Cronin et al., 2011). These probiotic bifidobacteria form a significant constituent in the microbiota of the human colon, and exert their effects as commensal microorganisms (Biavati et al., 2000; Turroni et al., 2008, 2009; Mills et al., 2011; Milani et al., 2014; Ventura et al., 2014). As a result, these bacteria are frequently exploited by the food industry to create consumable products that increase their relative proportion in the gut (Gibson et al., 1995; Sanders, 1998; Masco et al., 2005; Oberg et al., 2011; Ventura et al., 2014). Bifidobacteria are Gram-positive, anaerobic, saccharolytic organisms with a unique metabolic pathway known as the "bifid shunt" (Palframan et al., 2003; Biavati and Mattarelli, 2006; Milani et al., 2015). While many characteristics are known about this important group of bacteria, the biochemical and molecular properties contributing toward their probiotic effects, and adaptability in their respective environments, remain elusive (Ventura et al., 2009; Turroni et al., 2014).

The present study focuses on the enzyme ribonucleotide reductase (RNR), the sole enzyme capable of reducing nucleoside di- or tri- phosphates (NDPs or NTPs) into deoxyribonucleotides (dNDPs or dNTPs) (Eklund et al., 2001; Nordlund and Reichard, 2006; Torrents, 2014). There are currently three recognized classes of RNRs, named classes I, II, and III, sharing no more than 10% sequence identity across their lengths, which are distributed in different organisms (Logan et al., 1999; Sintchak et al., 2002; Torrents et al., 2002). Class I RNR is further divided into three subclasses viz. Ia, Ib, Ic (Jordan et al., 1996; Jiang et al., 2007; Bollinger et al., 2008). The distributions of these different classes of RNRs within the bacterial domain does not follow any specific pattern that can be correlated with the phylogenies of the bacterial phyla (Torrents et al., 2002; Lundin et al., 2009). However, since the different classes of RNR employ different mechanisms of action and require differing environmental prerequisites to function, we explore their distribution in bifidobacteria in an attempt to identify any unique characteristics that may distinguish them.

Each RNR is capable of reducing all four ribonucleotides into their corresponding deoxyribonucleotides by exhibiting a tightly regulated allosteric substrate specificity site, and employing a convoluted mechanism involving radical chemistry that ultimately results in the removal of a hydrogen from the 3′ carbon of the substrate (Brown and Reichard, 1969; Reichard, 1993, 2010; Eriksson et al., 1997; Eliasson et al., 1999). Some RNRs have an additional overall activity site, made possible by the existence of an ATP cone domain at the N-terminus (Thelander and Reichard, 1979). Class I RNRs use NDPs as their substrate, and are aerobic tetramers consisting of one large (R1) and one small (R2) homodimers (Nordlund and Reichard, 2006). The R2 dimer harbors a dinuclear metallocofactor where the radical is formed and subsequently transferred to the active site located at the R1 subunit. Classes Ia, Ib, and Ic differ in the type of metallocofactor in R2 (manganese and/or iron), as well as the different cofactors required for enzymatic function (Petersson et al., 1980; Nordlund and Reichard, 2006; Jiang et al., 2007; Bollinger et al., 2008; Torrents, 2014). Most bifidobacteria harbor a class Ib RNR (Lundin et al., 2009), whose large and small subunit are encoded by the *nrdE* and *nrdF* genes, respectively, and require NrdH as a reductant and NrdI as a cofactor; in contrast, classes Ia and Ic utilize thioredoxin and/or glutaredoxin as reductants, do not require additional cofactors, and their large and small subunits are encoded by *nrdA* and *nrdB*, respectively (Jordan et al., 1997; Cotruvo and Stubbe, 2008; Roca et al., 2008; Crona et al., 2011). Class II RNRs are not oxygen sensitive, use either NDPs or NTPs as their substrate, and are the only monomeric class of RNR (Tamao and Blakley, 1973; Larsson et al., 2010), however, their structural topology mimics a dimer (Sintchak et al., 2002). No known bifidobacteria harbor a class II homolog, but all bifidobacteria possess a class III RNR. Class III RNRs are encoded by *nrdD* and *nrdG* genes and function under strictly anaerobic conditions, with NTPs as their sole substrates (Garriga et al., 1996; Torrents et al., 2001). They consist of a large R1 subunit (NrdD) that is a dimer in its native state, and works concomitantly with a small activase (NrdG) which generates the radical utilizing a [4Fe-4S] cluster (Eliasson et al., 1992; Sun et al., 1995; Logan et al., 1999). This is unlike class I RNRs, where radical formation by the small subunit is required to induce dimer formation of the large subunit (Ollagnier et al., 1996; Torrents, 2014). Despite the described differences in the properties of the different classes of RNRs, the remarkable structural similarities seen across the three main RNRs strongly suggest a common evolutionary origin of them (Poole et al., 2002; Sintchak et al., 2002; Torrents et al., 2002). In all three RNRs, allosteric regulation involves binding of the dNDP/dNTP products at a 4-helix bundle, involving two helices from each monomeric subunit, at the dimer interface of the enzyme (Uhlin and Eklund, 1994; Larsson et al., 2001). The allosteric regulation causes conformational changes at a highly conserved 10 stranded α/β barrel where the active site "finger loop" structure resides in its center, or is brought to its center upon activation (Aurelius et al., 2015).

Although previous studies have significantly contributed to the current understanding of the structure and function of the different RNRs, in the present work we focus on the specific biochemical/molecular properties of the RNRs from bifidobacteria that may shed light on their unique physiological effects. Our earlier work describes a number of conserved signature indels (CSIs) in the homologs of many important proteins from that are uniquely found in all *Bifidobacteriales* (Zhang et al., 2016). These CSIs represent vertically transferred genetic changes that are indicated to have occurred in a common ancestor of the group in which they are found, thus asserting their value as highly specific molecular markers. In the present work, we have performed similar comparative genomic studies that have led to the identification of several novel CSIs in class Ib and III RNR homologs that are shared by all genome-sequenced *Bifidobacteriales* species that contain the respective protein homolog(s), but are absent in all other bacteria. We also describe the results of protein modeling which illustrate the structural location of these CSIs, as well as the results of preliminary *in silico* docking studies which suggest that one of the large CSIs [a 43 amino acid (aa) insertion in class III RNR] may be playing a role in NrdD complex stability.

MATERIALS AND METHODS

Identification of Conserved Signature Indels

The approach used to identify CSIs in RNR was as described in earlier work (Gupta, 2014; Zhang et al., 2016). Multiple sequence alignments (MSAs) were initially created using the Clustal_X 2.1 (Larkin et al., 2007; Goujon et al., 2010) program for the protein sequences of NrdE, NrdF, NrdH, NrdD, and NrdG homologs from about 10–15 *Bifidobacteriales* species, as well as 8–10 species from other groups/phyla of bacteria. These sequence alignments were examined for the presence of conserved indels that are limited to the *Bifidobacteriales* homologs and are flanked on both sides by at least five conserved residues in the neighboring 30–40 aa. A detailed Blastp search (Altschul et al., 1997) was then conducted on the sequence region containing the potential conserved indels to investigate the species-specificities of the identified indels. The indels that were not flanked by conserved regions were not further investigated in our work. The signature files shown here were created using SIG_CREATE and SIG_STYLE from the GLEANS.net program as described in earlier work (Gupta, 2014; Zhang et al., 2016). Unless otherwise indicated, all of the reported CSIs are specific for the *Bifidobacteriales* homologs and similar CSIs were not observed in homologs from any other bacterial species within the top 500–1000 blast hits examined.

Phylogenetic Tree Construction

In this study we have constructed three separate phylogenetic trees: (i) based on NrdE (large subunit of class Ib RNR) sequences, (ii) based on NrdD (large subunit of class III RNR) sequences, and (iii) based on the large subunit sequences from class I (NrdA, NrdE), II (NrdJ), and III (NrdD) RNRs. For these studies, NrdE and NrdD homologs from all genome sequenced bifidobacterial species were obtained from the NCBI GenBank sequence database (Benson et al., 2017). The species represented in the tree based on NrdD sequences included 49 of 58 validly published *Bifidobacterium* species, the two known *Scardovia* species, all three *Alloscardovia* species, and the single species known from the *Parascardovia* and *Gardnerella* genera. The tree based on NrdE sequences similarly included the subset of these *Bifidobacteriales* species where the protein homolog was detected. Sequences from members of the *Bifidobacteriales* genera *Aeriscardovia* and *Pseudoscardovia* were not available at the present time and were not included in our study. For each tree, a MSA of RNR homologs was created using the Clustal_X 2.1 (Larkin et al., 2007; Goujon et al., 2010) program. For each of these trees, we have additionally included a number of outgroup species (20 species for the NrdD tree, 23 species for the NrdE tree) from other orders in the Actinobacteria phylum, as well as Firmicutes species. For the tree concerning the sequences from large subunits of all RNR classes, we have included <10 NrdE and <10 NrdD sequences from representative *Bifidobacteriales*, in addition to several species across various bacterial phyla in order to depict the evolutionary history of RNR classes. The MEGA 6 program (Tamura et al., 2013) was used to construct a maximum likelihood (ML) tree based on 1000 bootstrap replicates for each alignment employing the Whelan and Goldman model substitution method (Whelan and Goldman, 2001). Gaps and regions with missing data from the sequence alignments were completely removed. In each case, a discrete Gamma distribution was used to model evolutionary rate differences among sites (five categories) and the Jones–Taylor–Thornton substitution method was used to compute the initial trees for the heuristic search using the Neighbor-joining method with a matrix of pairwise distances (Jones et al., 1992).

Homology Modeling of RNR Homologs and Structural Analysis of CSIs

The approach used to model the CSIs involves homology modeling based on previously crystallized class Ib and III RNR proteins. A Position-Specific-Iterated Blastp search (Altschul et al., 1997) was performed on *Bifidobacterium longum* NrdE (Accession no. EPE39971) and NrdD (Accession no. KXS29127) sequences against the PDB database which revealed that the class Ib RNR from *Salmonella typhimurium* (PDB ID: 1PEQ) (Uppsten et al., 2003) and class III RNR from Enterobacteria phage T4 (PDB ID: 1H7B) (Larsson et al., 2001) exhibited the highest sequence similarity to the *Bifidobacteriales* homologs and provided suitable templates for homology modeling of the RNR isoforms of *B. longum*. A conserved domain search (CD-Search) (Marchler-Bauer and Bryant, 2004) was conducted on the *B. longum* sequences. Homology modeling was performed using MODELLER v9.11 (Eswar et al., 2007) and the top 500 models were initially created and ranked on the basis of their Discrete Optimized Protein Energy (DOPE) scores (Shen and Sali, 2006). The selected models of RNR homologs with the highest DOPE score were then submitted to the GalaxyRefine server (Heo et al., 2013; Lee et al., 2016) to obtain atomic-level energy minimization and to improve the stereochemical quality of the model. The secondary structure elements in the regions containing CSIs were examined and compared with results of the PSIPRED and CONCORD analyses to ensure their reliability (Jones, 1999; Wei et al., 2011; Buchan et al., 2013). The stereochemical properties of the final models were assessed using four independent servers: RAMPAGE, ERRAT, PROSA and Verify3D (Bowie et al., 1991; Luthy et al., 1992; Colovos and Yeates, 1993; Sippl et al., 1999; Lovell et al., 2003; Wiederstein and Sippl, 2007). These validation tools utilize a dataset of highly refined solved structures to evaluate the statistical significance of models based on the conformation, location, and the environment of individual amino acids in the protein sequence, as well as the model's overall structural stability. The structural alignments of the models with the respective templates were carried out using PyMOL

TABLE 1 | Results of protein–protein docking studies for the *Bifidobacterium longum* RNR (NrdD subunit) either containing or lacking the large 43 aa CSI.

Docking server	Protein models of the NrdD		
	CSI-containing (Extended helix)	CSI-containing (based on PSIPRED/CONCORD)	CSI lacking
ClusPro	−1297.200	−1190.900	−953.000
ClusPro + ROSIE	−1092.851	−1110.039	−1042.091
PatchDock	*	16192.000	11234.000
PatchDock + ROSIE	*	−1109.473	−1033.714
"ROSIE"	−1105.550	−1114.689	−1043.235

The results of docking studies are shown for two different models of the CSI-containing protein (Extended helix and based on PSIPRED/CONCORD) and compared with those for the model for CSI-lacking protein. The monomers based on the indicated models were submitted to ClusPro and PatchDock servers. Results from the ClusPro and PatchDock servers were also submitted to the ROSIE (RosettaDock) server for local refinement. The results shown under the heading "ROSIE" were obtained by submitting model monomers that were superimposed onto the crystallized 1H7B (Larsson et al., 2001) template dimer. For the ClusPro and ROSIE servers, the lower binding energy (more negative values) is indicative of the increase in stability of the docking complexes, while for the PatchDock server higher (positive) scores indicate stronger binding affinity (see Materials and Methods section for further details). Asterisks (*) indicate that no biologically relevant complexes were obtained from the indicated docking servers.

Version 1.8 (Schrödinger, 2016) in order to analyze the location and the structural features of the CSIs in the protein structure. This procedure was followed to create the homology models of both of the RNR homologs found in bifidobacteria. In addition, a structural model of class III RNR was also generated using I-TASSER, an online server that uses threading to predict three dimensional protein structure (Zhang, 2008; Roy et al., 2010; Yang et al., 2015).

Protein–Protein Docking Analysis of the Class III RNR Homologs

Protein–protein docking studies were performed in order to assess the possible role of a large CSI in the formation or stabilization of the dimeric structure of class III RNR in bifidobacteria. A structural model of the class III RNR from *B. longum* was created by removing the CSI residues from its primary sequence, using the methods described above for other RNR homologs. An additional structural model of RNR was generated with the CSI region constructed as an extended helix. In this structural model, the CSI has a slightly different secondary structure than those of the models that followed PSIPRED/CONCORD, or I-TASSER predictions. This was done in an attempt to be inclusive of multiple possible structural conformations of the CSI. Four structures of the anaerobic RNR monomer viz. PSIPRED/CONCORD based model, I-TASSER generated model, model with extended helix, and the CSI-lacking model, were submitted to two independent web-based protein–protein docking programs using default parameters, viz. PatchDock Version B 1.3 (Schneidman-Duhovny et al., 2005) and ClusPro Version 2.0 (Comeau et al., 2004). PatchDock is an efficient molecular docking algorithm that employs a geometry-based shape complementarity approach which aims to yield refined atomic contacts of protein–protein complexes. Its scoring function takes into consideration both geometric fit and atomic desolvation energy (Duhovny et al., 2002). On the other hand, ClusPro utilizes PIPER, a rigid body docking program, which is based on a novel Fast-Fourier Transform (FTT) docking approach with pairwise potential. Its scoring function is thus based on pairwise interaction potentials (Comeau et al., 2004; Kozakov et al., 2006). The resulting top scoring dimer complex models of RNR from each server (if any) were then refined using the RosettaDock (ROSIE) server (Lyskov and Gray, 2008; Chaudhury et al., 2011; Lyskov et al., 2013). For the docking scores of ClusPro and RosettaDock, the lower (negative) binding energy value indicates improved stability of the docking complexes. In the case of PatchDock, the geometry shape complementarity score was utilized to determine rank, and higher (positive) scores indicate stronger binding affinity (see also notes in **Table 1**). In addition, the monomeric forms of each of the four models were structurally aligned with the established biological assembly of the RNR dimer, and the resulting dimer orientations were utilized as additional inputs for submission to the ROSIE server. The resulting refined structure from ROSIE with the lowest total score, maximum cluster size and the smallest RMSD with respect to the solved structure of RNR complex, was chosen as a representative structure for detail interface interaction analysis. To analyze the dimer interface, this class III RNR dimeric output structure was submitted to the PDBePISA Version 1.48 server, using default parameters (Krissinel and Henrick, 2007).

RESULTS

Identification of Conserved Signature Indels in Class I and Class III RNR Homologs and Their Phylogenetic Implications

Comparative analysis of the *Bifidobacteriales* genomes indicated that all of the sequenced species from this order contain an anaerobic class III RNR homolog. In addition, an aerobic class I RNR belonging to the class Ib group was also found in most species from this order except *B. adolescentis, B. angulatum, B. dentium, B. gallicum, B. cuniculi, B. lemurum, B. merycicum, B. moukalabense, B. ruminantium*, and members of the genera *Parascardovia* and *Scardovia*. The sequences of class Ib and III RNRs were examined for the presence of any CSIs that are specific for bifidobacteria. The results of these studies have identified three CSIs in the large subunit of the class Ib RNR (NrdE) homologs, which are specifically found in the bifidobacterial enzyme. Sequence information for two of these CSIs, which are comprised of 4 and 2 aa inserts in a conserved region of the NrdE protein, is shown in **Figure 1**. As seen in the figure, both of these CSIs are flanked on either side by conserved regions and while they are commonly shared by all of the bifidobacteria harboring the NrdE homolog, they are not present in any other bacterial species in the top 500 blast hits. Sequence information for one additional CSI in the NrdE protein, consisting of a 1 aa

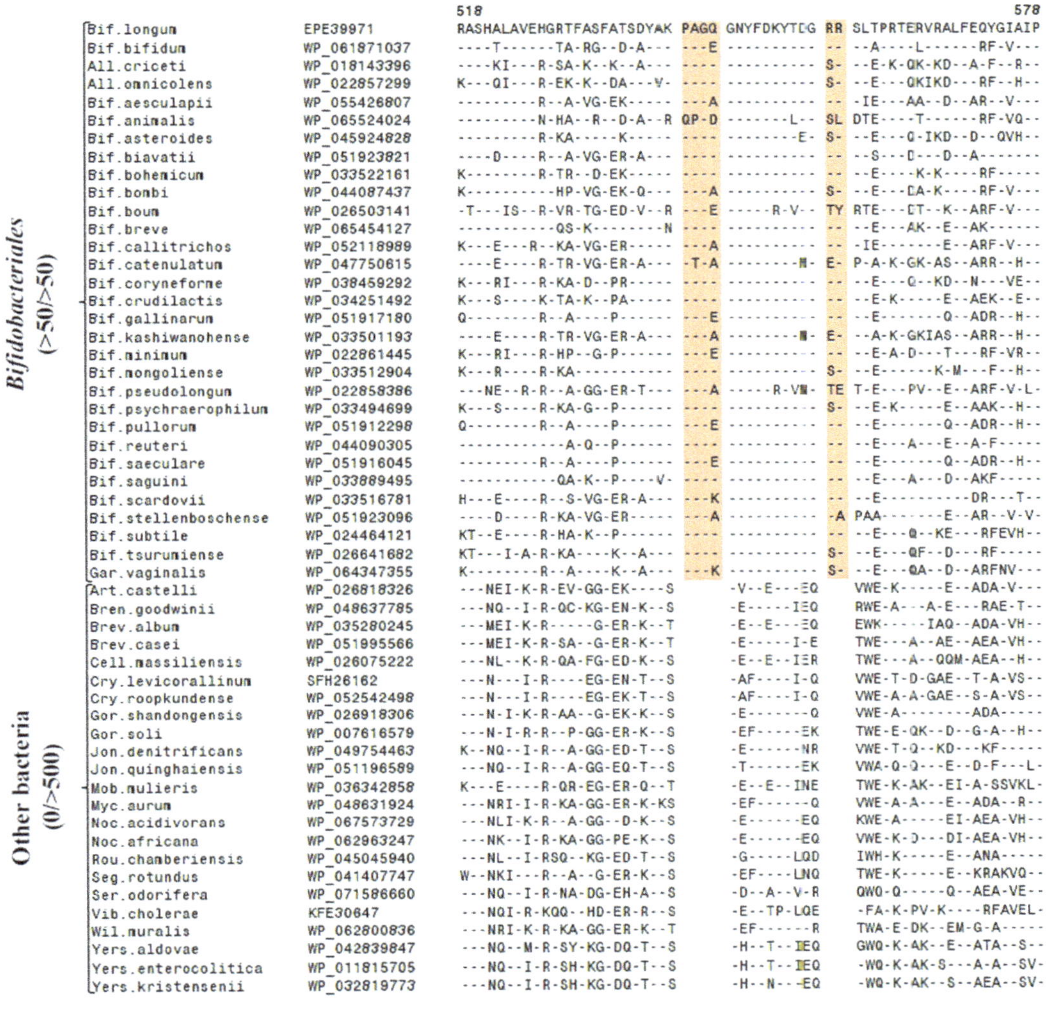

FIGURE 1 | Partial sequence alignment of the large subunit of the class Ib ribonucleotide reductase (NrdE) protein showing two conserved inserts (highlighted) that are exclusively found in all *Bifidobacteriales* members that carry the homolog, but absent in other bacteria. The dashes in the sequence alignment denote identity with the amino acid found on the top line. The Genbank accession numbers of the sequences are shown in the second column. The results are shown for only a limited number of species, however, other species not shown showed similar pattern as described here. Abbreviations used for the genus names are: *All., Alloscardovia; Art., Arthrobacter; Bif., Bifidobacterium; Bren., Brenneria; Brev., Brevibacterium; Cell., Cellulomonas; Cry., Cryobacterium; Gar, Gardnerella; Gor., Gordonia; Jon., Jonesia; Mob., Mobiluncus; Myc., Mycobacterium; Noc., Nocardia; Rou., Rouxiella; Seg., Segniliparus; Ser., Serratia; Vib., Vibrio; Wil., Williamsia; Yer., Yersinia.*

deletion also specific for all *Bifidobacteriales* strains that harbor the protein, is presented in Supplementary Figure 1, and is once again absent in other bacteria.

Similarly, analysis of the sequences from the NrdD homolog has also led to the identification of five CSIs that are specific for the *Bifidobacteriales* homologs. Sequence information for one large CSI, a 43 aa insertion, that is specifically found in the NrdD homolog from bifidobacteria, is presented in **Figure 2**. Sequence information for four other CSIs, consisting of a 1 aa deletion, two 1 aa insertions, and a 4 aa insertion, which are also either exclusively or mainly found in the *Bifidobacteriales* NrdD homologs, are presented in Supplementary Figures 2–5. Of these other CSIs, the 1 aa deletion (Supplementary Figure 3) is also present in *Coriobacteriales* species, which are also anaerobic and saccharolytic bacteria. Additionally, one of the CSIs consisting of a 1 aa insert is also shared by *Lactobacillus* species (Supplementary Figure 2), and the other single aa insert is shared by few other species from the Actinobacteria phylum (Supplementary Figure 5). Asides from these cases, the CSIs were not found in the additional 500–1000 bacterial outgroups examined. The locations of the different identified CSIs over the lengths of the NrdE and NrdD proteins and their respective domains are presented in Supplementary Figures 6, 7. All of the CSIs in the NrdE and NrdD subunits are located in the RNR domain of the respective proteins. In contrast to the NrdE and NrdD proteins, no specific CSIs were found in the NrdF, NrdH, or NrdG proteins. Due to the exclusive presence of most of these CSIs in the RNR homologs from bifidobacteria, they

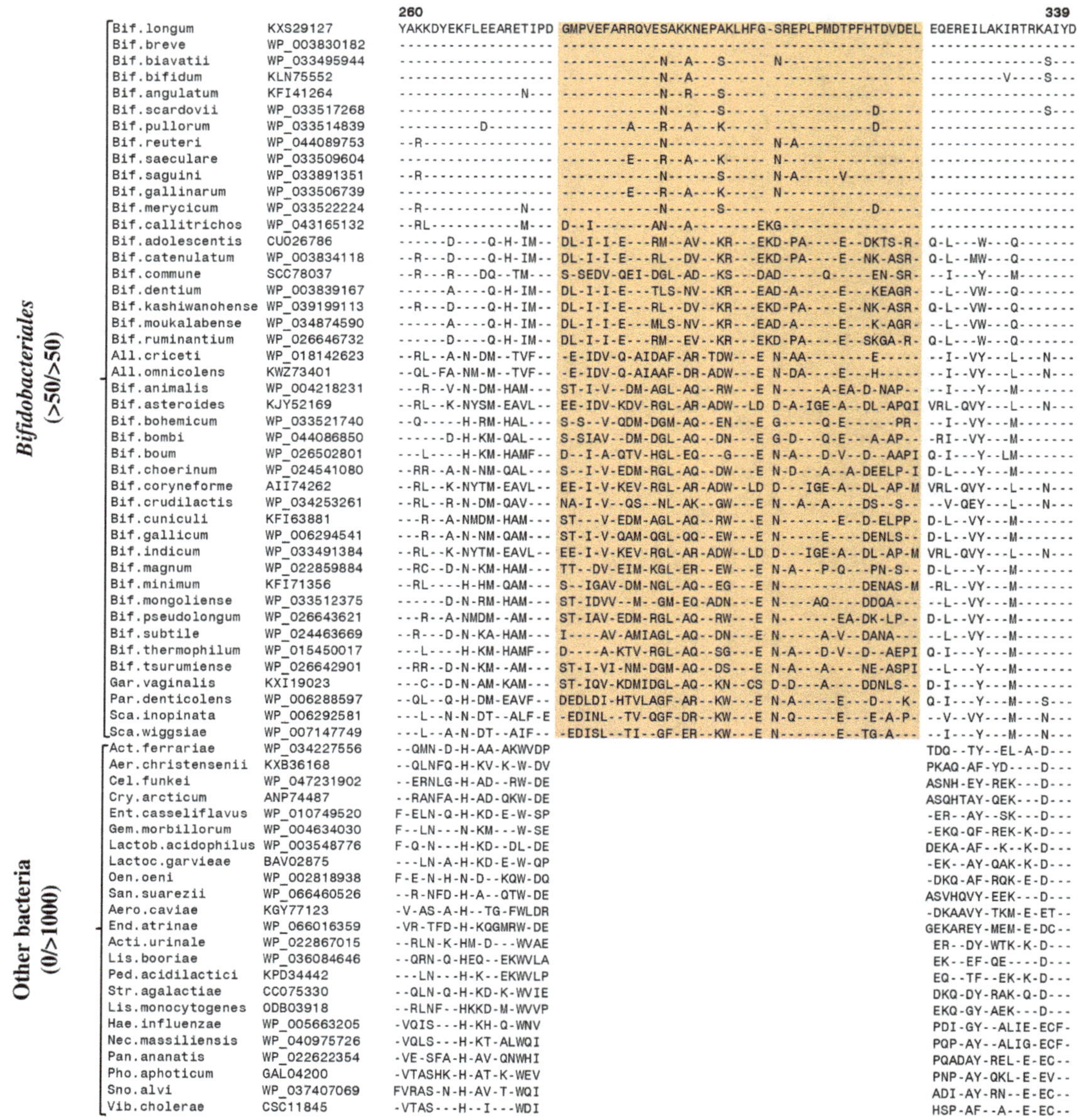

FIGURE 2 | Partial sequence alignment of the large subunit of the class III ribonucleotide reductase (NrdD) protein showing a 43 or 44 amino acid insertion (highlighted) that is exclusively found in all *Bifidobacteriales* members, and absent in other bacteria. Other details are as in **Figure 1**. Abbreviations used for the genus names are: *Act., Actinotalea; Acti., Actinotignum; Aer., Aerococcus; Aero., Aeromonas; All., Alloscardovia; Bif., Bifidobacterium; Cel., Cellulosimicrobium; Cry., Cryobacterium; End., Endozoicomonas; Ent., Enterococcus; Gar., Gardnerella; Gem., Gemella; Hae., Haemophilus; Lactob., Lactobacillus; Lactoc., Lactococcus; Lis., Listeria; Nec., Necropsobacter; Oen., Oenococcus; Pan., Pantoea; Par., Parascardovia; Ped., Pediococcus; Pho., Photobacterium; San., Sanguibacter; Sca., Scardovia; Sno., Snodgrassella; Str., Streptococcus; Vib., Vibrio.*

provide molecular markers for distinguishing members of the order *Bifidobacteriales* from other bacteria, and they may inform important differences in the molecular/biochemical properties of the RNR homologs from bifidobacteria. It should be mentioned that in addition to the described CSIs, the NrdE and NrdD homologs from bifidobacteria also harbor other genetic changes such as amino acid substitutions that appear specific for them. Although the evolutionary significance of these changes is not clear and was not studied in the present work, it is likely that some of them also play important role in conjunction with the CSIs in the novel functional aspect(s) of the RNRs from bifidobacteria.

Maximum-likelihood phylogenetic trees were constructed for the class Ib and class III RNRs protein sequences based on the NrdE and NrdD proteins, and these trees are shown in **Figures 3A,B**, respectively. In addition to the sequences from a large number of *Bifidobacteriales* species covering the order, the tree also contains information for several other Actinobacteria as well as a limited number of Firmicutes species; the sequences

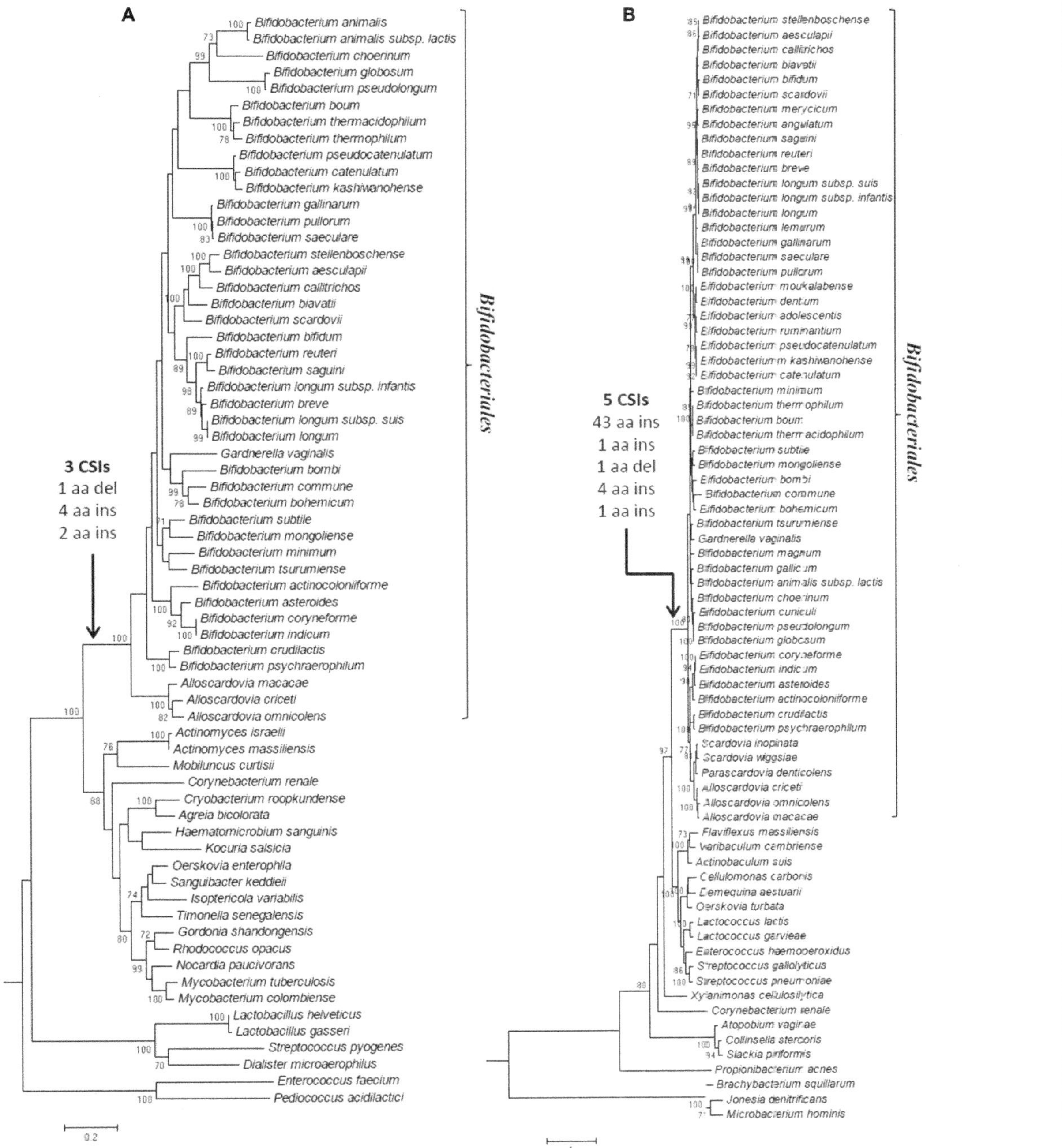

FIGURE 3 | Maximum likelihood trees based on the sequences of **(A)** the large subunit of the class Ib ribonucleotide reductase (NrdE) involving 684 aa residues after the complete deletion of alignment gaps, and **(B)** the large subunit of the class III ribonucleotide reductase (NrdD) involving 549 aa residues after complete deletion. Each tree is based on >40 genome sequenced *Bifidobacteriales* species and ≥20 outgroup members from other Actinobacteria as well as Firmicutes. The trees were constructed using the MEGA6 program and are drawn to scale where horizontal branch length is measured in the inumber of aa substitutions per site. Bootstrap scores that were greater than 70% are shown as a percentage on the nodes. The arrows indicate where the CSI events are likely to have occurred.

from the Firmicutes species were used to root the trees. The sequences from *Bifidobacteriales* species formed strongly supported monophyletic clades in both trees. Because the sequence alignments used for construction of these phylogenetic trees did not contain any sequence gaps, the observed branching pattern was not influenced by the presence of the identified CSIs. Therefore, the distinct branching of bifidobacteria observed in both trees supports the notion that the reported CSIs in the

NrdD and NrdE proteins most likely first occurred in a common ancestor of the order *Bifidobacteriales,* and were inherited by descendants due to incurring an evolutionary advantage. In addition to these trees, we have also constructed a tree based on the sequences of the large subunit from the three main RNR classes (Supplementary Figure 8). The three classes of RNR formed distinct clades in the tree, which were separated from each other by long branches. Based on the midpoint rooting of the tree, the sequences from the class III RNR, which function under strictly anaerobic conditions, were found to form a sister clade to sequences from the classes I and II. The observed branching of the class III RNR in the tree is in agreement with earlier work (Reichard, 1993; Logan et al., 1999; Larsson et al., 2001; Poole et al., 2002; Sintchak et al., 2002; Torrents et al., 2002) suggesting that this class of RNR represents the ancestral form of the reductase. Although phylogenetic analysis can shed light on the evolutionary history of the three types of RNR, it does not explain the variable distribution of these classes in different organisms. As an important protein, there is at least one type of RNR in every organism. However, the combination of different RNRs which are found in various organisms is unpredictable and it does not show any correlation with the evolutionary histories of the organisms (Torrents et al., 2000; Lundin et al., 2009, 2010; Torrents, 2014).

Locations of the CSIs in the Ribonucleotide Reductase Homologs Structures

To gain insights into the possible significance of the identified CSIs, homology models for the class Ib and III RNRs from *B. longum* were constructed (see Materials and Methods section) based on previously crystallized template structures of class Ib and III RNR proteins (Larsson et al., 2001; Uppsten et al., 2003). After the validation of the homology models using a variety of tools described under section "Materials and Methods," a superimposition of the final selected models with the template structures was carried out using PyMOL to determine the locations of the identified CSIs in the structures of the class Ib and III proteins. The locations of the three CSIs identified in class Ib RNR in the modeled structure of the NrdE protein is shown in **Figure 4A**. As seen, all three CSIs in the NrdE homolog were located within the surface loops of the protein (**Figure 4A**). However, of these CSIs, the 2 aa insert also appears to extend a helix. The current model is in agreement with secondary structure analyses (PSIPRED/CONCORD), and yielded reassuring measurements by ERRAT, Verify3D, RAMPAGE and PROSA. The locations of these CSIs in the structure of the *B. longum* NrdE subunit indicate that they are topologically distant from the dimer interface/allosteric regulatory site, as well as the active site as seen in Supplementary Figure 9A (Uppsten et al., 2003, 2006). However, the locations of these CSIs within surface loops in the NrdE structure indicate that they could be involved in mediating novel protein–protein interactions (Cherkasov et al., 2006; Singh and Gupta, 2009; Gupta, 2016; Zhang et al., 2016; Khadka and Gupta, 2017).

Of the five CSIs found in the NrdD homologs (Class III) of bifidobacteria, the structural locations of four indels could be determined and are illustrated on the modeled structure (**Figure 4B**). Structural location of one of the CSIs present near the C-terminal end could not be determined as the structural information for the corresponding region was absent from the template structure (PDB ID: 1H7B) used for homology modeling (Larsson et al., 2001). Similar to the indels found in the NrdE structure, most of the CSIs in the NrdD structure are also be found on surface loops that are structurally distant from the active site (**Figure 4B** and Supplementary Figure 9B). The 4 aa insert, also located on the surface exposed region, appears to form a loop and elongate a helix. The large 43 aa insert in NrdD exists as an elongation of the allosteric regulatory region, in between two helices that form the 4-helix bundle in the NrdD dimer (Uhlin and Eklund, 1994; Logan et al., 1999; Larsson et al., 2001). Since the 43 aa insert did not correspond to a characterized domain or motif, elucidating its structure was a challenging task. We present a model that is in agreement with secondary structure predictions from the *B. longum* primary sequence (PSIPRED/CONCORD) (**Figure 4B**), and an additional model according to the prediction made *in silico* by the I-TASSER server (Supplementary Figure 10B). As seen in Supplementary Figure 10B, in the I-TASSER model, the 43 aa insert appears to form two helices that are connected to one another by a loop, and each are also connected to the two existing helices by loops. The orientation of the insert is such that it folds back toward the bulk of the protein, and the dimer interface is relatively uninfluenced. In the case of the model generated based on the secondary structure predictions, the CSI forms an extension of the two helices, along with two small helices connected by loops in between (**Figure 4B**). In order to be inclusive of all reasonable possibilities, we also modeled the NrdD homolog based on the hypothesis that, instead of small helices with breaks induced by loops according to the PSIPRED/CONCORD model, perhaps the loops connecting the main helices to the small helices may be extended helices without loop-induced breaks (Supplementary Figure 10A). This extended helix hypothesis is a corollary to the observation that the elongated helix in class III compared to class I has important functional significance regarding allosteric binding, and also influences dimer packing (Larsson et al., 2001). All three models were refined and validation scores were maximized in these CSI-containing regions.

Analyzing the Possible Functional Significance of the Large Conserved Insert

To determine the possible role of the large 43 aa insert in dimer formation or complex stability, we have performed a series of docking studies to reveal dimerization potentials of the models compared to an additional model of bifidobacterial NrdD that lacks the CSI. These four models (I-TASSER, PSIPRED/CONCORD, extended helix hypothesis and CSI-lacking) were submitted to two online servers viz. ClusPro and PatchDock. The complexes obtained from ClusPro and PatchDock were refined and scored using ROSIE, and an

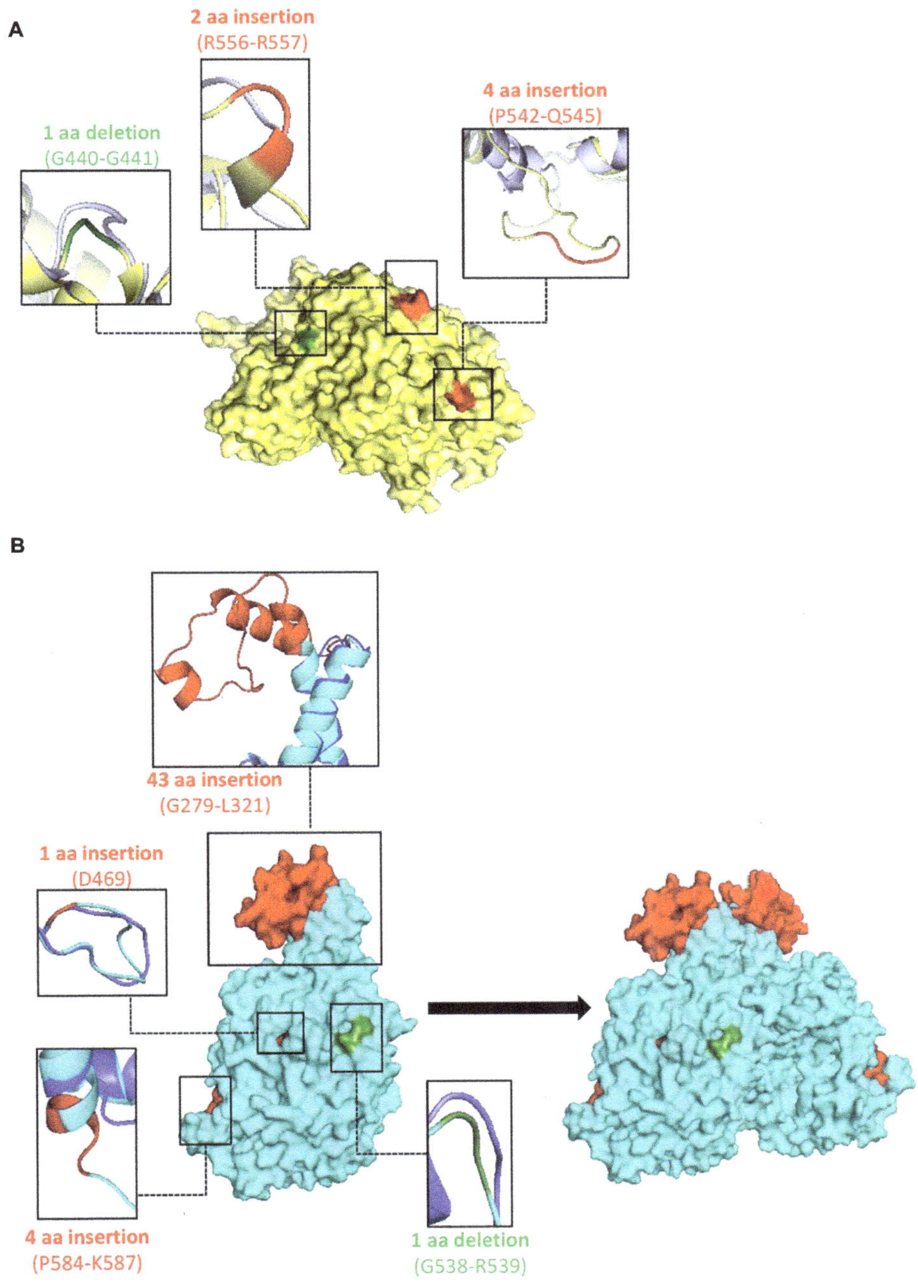

FIGURE 4 | Surface representations of homology models of **(A)** the class Ib ribonucleotide reductase monomer from *Bifidobacterium longum*, modeled from the 1PEQ template (Uppsten et al., 2003); the cartoon representations are of the model (yellow) superimposed on the template (purple) and **(B)** the class III ribonucleotide reductase monomer (NrdD) from *B. longum*, modeled from the 1H7B template (Larsson et al., 2001), with secondary structure in agreement with PSIPRED and CONCORD analyses; the cartoon representations are of the model (cyan) superimposed on the template (dark blue). The observed dimer on the right of **(B)** is a result of docking analysis from ClusPro, refined by ROSIE. The conserved signature insertions identified in these proteins are highlighted in red, and the positions of the deletions are shown in green.

additional dimerization measurement was performed by superimposing the models with the available experimentally solved structure of the NrdD dimer (Larsson et al., 2001). The docking scores of the dimer complexes from the protein–protein docking studies are summarized in **Table 1**. The PatchDock server did not yield any dimer complex for the modeled proteins containing extended helix that was plausible with the known biological assembly. However, the docking scores for the CSI-containing protein model based on PSIPRED/CONCORD were consistently improved (or superior) for all servers in comparison to the protein model lacking the CSI. Thus, the model generated according to the PSIPRED/CONCORD prediction is more likely to approximate the true structure of the class III RNR in *B. longum*. For all docking servers, the I-TASSER generated protein model did not form a plausible dimer complex and hence its results are not shown. This may be due to the fact that in the I-TASSER model of NrdD, the CSI was found to protrude away from the dimeric interface (Supplementary Figure 10B).

To examine if any of the residues from the CSI are involved in dimer interaction, the structural coordinate file for the dimeric form of the PSIPRED/CONCORD model was submitted to the PDBePISA server (Krissinel and Henrick, 2007). Analysis of the results obtained suggests that two residues from the large CSI (viz. Gly279, Met280 for *B. longum* NrdD) are present at the protein dimer interface. However, both of these residues, which are present at the N-terminal end of this large CSI, are not conserved in other *Bifidobacteriales* homologs. Thus, it is difficult to infer with any degree of confidence the possible role of these two residues in protein dimerization. Asides from these two residues, the remaining CSI residues are partly or fully solvent accessible near the interface.

DISCUSSION

The *Bifidobacteriales* are an important group of bacteria that consist of both pathogenic species and health-promoting commensal microorganisms that are frequently exploited in the food industry as probiotics (Gibson et al., 1995; Sanders, 1998; Leahy et al., 2005; Masco et al., 2005; Ventura et al., 2009; Cronin et al., 2011). However, the use of these bacteria as probiotics faces several challenges as very little is understood about the mechanism responsible for the beneficial effects exerted by bifidobacteria (Oberg et al., 2011). In the present work, we have identified several novel signatures in the form of CSIs in the sequences of both classes I and III RNR homologs that differentiate the RNR homologs from *Bifidobacteriales* from all those found in all other organisms. Earlier work on CSIs (including 1–2 amino acid indels) in several important proteins (e.g., GroEL, DnaK, GyrB, PIP5K, etc.) provides evidence that the CSIs play important functional roles in CSI-containing organisms, and deletion or other changes in the CSIs adversely impact cell growth or other critical functions (Chatterji et al., 2000; Singh and Gupta, 2009; Schoeffler et al., 2010; Clarke and Irvine, 2013; Gupta et al., 2017). In this context, the results reported here that both classes I and III RNR homologs harbor multiple CSIs that are uniquely present in all bifidobacteria is of much interest. These CSIs serve to clearly differentiate the bifidobacterial RNR homologs from those found in all other organisms, and they could function in conjunction with each other to impart certain novel functional characteristic(s) that is only shared by the classes I and III RNR homologs from bifidobacteria.

The distribution of RNR homologs in the *Bifidobacteriales* reveals that while all bifidobacteria species contain a class III anaerobic RNR homolog (NrdDG), the class Ib aerobic RNR homologs (NrdEF) were not found (or detected) in a number of *Bifidobacterium* species (viz. *B. adolescentis, B. angulatum, B. dentium*, and *B. gallicum, B. cuniculi, B. lemurum, B. merycicum, B. moukalabense, B. ruminantium*) as well as in members of the genera *Parascardovia* and *Scardovia* (Lundin et al., 2009). The bifidobacteria species lacking the class Ib RNR homologs do not show any specific branching pattern, or belong to any specific clade(s) of bifidobacteria, but they appear to be distributed sporadically within the order *Bifidobacteriales* (**Figure 3**). Thus, it is likely that the genes for both classes I and III RNR were present in the common ancestor of all *Bifidobacteriales* and subsequently some species have lost the genes for the class I RNR. Since the different identified CSIs in the classes I and III RNR homologs are present in all bifidobacteria, the most likely explanation for this fact is that the genetic changes responsible for the observed CSIs occurred in a common ancestor of the order *Bifidobacteriales*, presumably at the time of divergence of this group of bacteria from other organisms, similar to the CSIs in many other proteins that are uniquely found in the members of this order (Zhang et al., 2016).

The biological significance of the wide-spread presence of an aerobic RNR (class Ib) in bifidobacterial species, which are generally regarded as anaerobic organisms, remains to be understood. It is known that bifidobacterial species exhibit varying levels of aerotolerance which affects their viability outside of their natural habitats (e.g., gastrointestinal tract, mouth and vagina of mammals) upon exposure to an oxidative environment (Biavati and Mattarelli, 2006; Oberg et al., 2011). Under these conditions, RNRs are pivotal in maintaining a pool of dNDPs/dNTPs to compensate for DNA and protein damage by reactive oxygen species (ROS), ultimately alleviating the detrimental effects of superoxide stress. Thus, a reasonable assumption is that the class I aerobic RNR may play an important role in the tolerance of bifidobacteria to oxidative environment. Although how bifidobacteria manage oxidative exposure remains to be understood, in few studies that have explored the gene expression of both NrdEFHI and NrdDG systems in *Bifidobacterium* species, including data available from the Gene Expression Omnibus[1], reveal that, under oxidative stress, the expression of both gene systems was rapidly induced by the altered environment; first, the class Ib system was upregulated, followed by the class III system and other proteins, including proteolytic enzymes (Larsson et al., 2001; Edgar et al., 2002; Xiao et al., 2011; Zuo and Chen, 2016). However, further work is necessary to understand the biological and physiological roles of the class I RNR in bifidobacteria under normal conditions and during oxidative stress.

Earlier work on CSIs in protein structures provides evidence that most of the studied CSIs are located in the surface loops of proteins (Geszvain et al., 2004; Akiva et al., 2008; Singh and Gupta, 2009; Gupta et al., 2017; Khadka and Gupta, 2017). The results of our modeling of the identified CSIs in the NrdE and NrdD protein structures also show that all of described

[1]http://www.ncbi.nlm.nih.gov/geo/

CSIs in the classes I and III RNR protein subunits are present on the surface loops of these proteins (**Figure 4**). Based on extensive earlier work, the surface loops in protein structures often serve as platforms for facilitating novel protein–protein or protein–ligand interactions that are specific for the CSI-containing organisms, without affecting the core functions of the target proteins (Geszvain et al., 2004; Akiva et al., 2008; Singh and Gupta, 2009; Schoeffler et al., 2010; Gupta et al., 2017; Khadka and Gupta, 2017). In a number of cases, the surface loops formed by the CSIs have also been shown to play important role in determining the oligomeric state of the proteins (Itzhaki et al., 2006; Akiva et al., 2008; Hashimoto and Panchenko, 2010). Based on these studies, it is expected that the identified CSIs in the RNR homologs of bifidobacteria will also play novel and functionally important roles that are specific for bifidobacteria. There is no information available at present regarding the biochemical properties of the RNR homologs from bifidobacteria or whether they exhibit any novel functional characteristics. However, the identified CSIs provide highly specific tools for the genetic and biochemical exploration of novel functional characteristics of the RNRs from bifidobacteria.

Of the five different CSIs identified within the NrdD homologs, one of these CSIs is a large 43 aa insert, located proximal to the allosteric regulatory site in the protein monomer (**Figure 4B** and Supplementary Figure 9B) (Larsson et al., 2001). This regulatory site consists of a 4-helix bundle where each monomer contributes two helices. These helices are significantly longer in class III RNRs compared to those of class I (~21 aa), resulting in an altered dimer packing, mode of effector binding and consequential conformational changes to the active site (Larsson et al., 2001). Due to the presence of the 43 aa CSI in the class III bifidobacterial RNRs in this region, this helical region is further elongated in bifidobacteria, which is suggestive of a feature uniquely shared by these bacteria. In the absence of any information regarding the structure of the NrdD protein from bifidobacteria, or the conformation of this large CSI in the protein structure, it is difficult to predict the precise role that this CSI may play in the RNR structure and/or function. However, the results of our preliminary *in silico* protein–protein docking using three separate docking servers suggest that when the CSI is oriented in such a way that it can interact with the other monomer, the stability of the dimer complex is improved in the presence of the 43 aa CSI, in comparison to the similar docking studies carried out with the protein lacking the CSI (**Figure 4B** and **Table 1**). The results from our *in silico* analyses are broadly suggestive of one possible function of this large CSI. However, a clearer understanding of the functional significances of the identified CSIs should emerge from future biochemical and structural studies on classes I and III RNRs. Nevertheless, the results presented here highlight the many unique sequence features of the bifidobacterial RNRs, whose further investigations could provide important insights into novel functional aspects of these enzymes in bifidobacteria.

AUTHOR CONTRIBUTIONS

SA identified some of the conserved inserts, carried out phylogenetic and structural work and wrote the draft manuscript. BK was involved in the structural analysis of the identified conserved inserts and read and commented on the draft manuscript. RG identified the initial CSIs and conceived and directed the entire project and edited the final version of the manuscript.

REFERENCES

Akiva, E., Itzhaki, Z., and Margalit, H. (2008). Built-in loops allow versatility in domain-domain interactions: lessons from self-interacting domains. *Proc. Natl. Acad. Sci. U.S.A.* 105, 13292–13297. doi: 10.1073/pnas.0801207105

Altschul, S. F., Madden, T. L., Schäffer, A. A., Zhang, J., Zhang, Z., Miller, W., et al. (1997). Gapped BLAST and PSI-BLAST: a new generation of protein database search programs. *Nucleic Acids Res.* 25, 3389–3402. doi: 10.1093/nar/25.17.3389

Alves, P., Castro, J., Sousa, C., Cereija, T. B., and Cerca, N. (2014). *Gardnerella vaginalis* outcompetes 29 other bacterial species isolated from patients with bacterial vaginosis, using in an *in vitro* biofilm formation model. *J. Infect. Dis.* 210, 593–596. doi: 10.1093/infdis/jiu131

Aurelius, O., Johansson, R., Bagenholm, V., Lundin, D., Tholander, F., Balhuizen, A., et al. (2015). The crystal structure of *Thermotoga maritima* class III ribonucleotide reductase lacks a radical cysteine pre-positioned in the active site. *PLoS ONE* 10:e0128199. doi: 10.1371/journal.pone.0128199

Benson, D. A., Cavanaugh, M., Clark, K., Karsch-Mizrachi, I., Lipman, D. J., Ostell, J., et al. (2017). GenBank. *Nucleic Acids Res.* 45, D37–D42. doi: 10.1093/nar/gkw1070

Biavati, B., and Mattarelli, P. (2006). "The family *Bifidobacteriaceae*," in *The Prokaryotes*, eds M. Dworkin, S. Falkow, E. Rosenberg, K.-H. Schleifer, and E. Stackebrandt (New York, NY: Springer), 322–382.

Biavati, B., Vescovo, M., Torriani, S., and Bottazzi, V. (2000). Bifidobacteria: history, ecology, physiology and applications. *Ann. Microbiol.* 50, 117–132.

Bollinger, J. M. Jr., Jiang, W., Green, M. T., and Krebs, C. (2008). The manganese(IV)/iron(III) cofactor of *Chlamydia trachomatis* ribonucleotide reductase: structure, assembly, radical initiation, and evolution. *Curr. Opin. Struct. Biol.* 18, 650–657. doi: 10.1016/j.sbi.2008.11.007

Bowie, J. U., Luthy, R., and Eisenberg, D. (1991). A method to identify protein sequences that fold into a known three-dimensional structure. *Science* 253, 164–170. doi: 10.1126/science.1853201

Bradshaw, C. S., Tabrizi, S. N., Fairley, C. K., Morton, A. N., Rudland, E., and Garland, S. M. (2006). The association of *Atopobium vaginae* and *Gardnerella vaginalis* with bacterial vaginosis and recurrence after oral metronidazole therapy. *J. Infect. Dis.* 194, 828–836. doi: 10.1086/506621

Brown, N. C., and Reichard, P. (1969). Role of effector binding in allosteric control of ribonucleoside diphosphate reductase. *J. Mol. Biol.* 46, 39–55. doi: 10.1016/0022-2836(69)90056-4

Buchan, D. W., Minneci, F., Nugent, T. C., Bryson, K., and Jones, D. T. (2013). Scalable web services for the PSIPRED protein analysis workbench. *Nucleic Acids Res.* 41, W349–W357. doi: 10.1093/nar/gkt381

Chatterji, M., Unniraman, S., Maxwell, A., and Nagaraja, V. (2000). The additional 165 amino acids in the B protein of *Escherichia coli* DNA gyrase have an important role in DNA binding. *J. Biol. Chem.* 275, 22888–22894. doi: 10.1074/jbc.M001047200

Chaudhury, S., Berrondo, M., Weitzner, B. D., Muthu, P., Bergman, H., and Gray, J. J. (2011). Benchmarking and analysis of protein docking performance in Rosetta v3.2. *PLoS ONE* 6:e22477. doi: 10.1371/journal.pone.0022477

Cherkasov, A., Lee, S. J., Nandan, D., and Reiner, N. E. (2006). Large-scale survey for potentially targetable indels in bacterial and protozoan proteins. *Proteins* 62, 371–380. doi: 10.1002/prot.20631

Clarke, J. H., and Irvine, R. F. (2013). Evolutionarily conserved structural changes in phosphatidylinositol 5-phosphate 4-kinase (PI5P4K) isoforms are responsible for differences in enzyme activity and localization. *Biochem. J.* 454, 49–57. doi: 10.1042/BJ20130488

Colovos, C., and Yeates, T. O. (1993). Verification of protein structures: patterns of nonbonded atomic interactions. *Protein Sci.* 2, 1511–1519. doi: 10.1002/pro.5560020916

Comeau, S. R., Gatchell, D. W., Vajda, S., and Camacho, C. J. (2004). ClusPro: an automated docking and discrimination method for the prediction of protein complexes. *Bioinformatics* 20, 45–50. doi: 10.1093/bioinformatics/btg371

Cotruvo, J. A. Jr., and Stubbe, J. (2008). NrdI, a flavodoxin involved in maintenance of the diferric-tyrosyl radical cofactor in *Escherichia coli* class Ib ribonucleotide reductase. *Proc. Natl. Acad. Sci. U.S.A.* 105, 14383–14388. doi: 10.1073/pnas.0807348105

Crona, M., Torrents, E., Rohr, A. K., Hofer, A., Furrer, E., Tomterm, A. B., et al. (2011). NrdH-redoxin protein mediates high enzyme activity in manganese-reconstituted ribonucleotide reductase from *Bacillus anthracis*. *J. Biol. Chem.* 286, 33053–33060. doi: 10.1074/jbc.M111.278119

Cronin, M., Ventura, M., Fitzgerald, G. F., and van Sinderen, D. (2011). Progress in genomics, metabolism and biotechnology of bifidobacteria. *Int. J. Food Microbiol.* 149, 4–18. doi: 10.1016/j.ijfoodmicro.2011.01.019

Duhovny, D., Nussinov, R., and Wolfson, H. J. (2002). "Efficient unbound docking of rigid molecules," in *Proceedings of the Second International Workshop on Algorithms in Bioinformatics (WABI)* 2452 (Berlin: Springer-Verlag), 185–200. doi: 10.1007/3-540-45784-4_14

Edgar, R., Domrachev, M., and Lash, A. E. (2002). Gene expression omnibus: NCBI gene expression and hybridization array data repository. *Nucleic Acids Res.* 30, 207–210. doi: 10.1093/nar/30.1.207

Eklund, H., Uhlin, U., Farnegardh, M., Logan, D. T., and Nordlund, P. (2001). Structure and function of the radical enzyme ribonucleotide reductase. *Prog. Biophys. Mol. Biol.* 77, 177–268. doi: 10.1016/S0079-6107(01)00014-1

Eliasson, R., Pontis, E., Fontecave, M., Gerez, C., Harder, J., Jörnvall, H., et al. (1992). Characterization of components of the anaerobic ribonucleotide reductase system from *Escherichia coli*. *J. Biol. Chem.* 267, 25541–25547.

Eliasson, R., Pontis, E., Jordan, A., and Reichard, P. (1999). Allosteric control of three B12-dependent (class II) ribonucleotide reductases. Implications for the evolution of ribonucleotide reduction. *J. Biol. Chem.* 274, 7182–7189. doi: 10.1074/jbc.274.11.7182

Eriksson, M., Uhlin, U., Ramaswamy, S., Ekberg, M., Regnström, K., and Sjöberg, B. M. (1997). Binding of allosteric effectors to ribonucleotide reductase protein R1: reduction of active-site cysteines promotes substrate binding. *Structure* 5, 1077–1092. doi: 10.1016/S0969-2126(97)00259-1

Eswar, N., Webb, B., Marti-Renom, M. A., Madhusudhan, M. S., Eramian, D., and Shen, M. Y. (2007). Comparative protein structure modeling using MODELLER. *Curr. Protoc. Protein Sci.* 15:5.6.1–5.6.30. doi: 10.1002/0471140864.ps0209s50

Gao, B., and Gupta, R. S. (2012). Phylogenetic framework and molecular signatures for the main clades of the phylum *Actinobacteria*. *Microbiol. Mol. Biol. Rev.* 76, 66–112. doi: 10.1128/MMBR.05011-11

Garriga, X., Eliasson, R., Torrents, E., Jordan, A., Gibert, I., and Reichard, P. (1996). nrdD and nrdG genes are essential for strict anaerobic growth of *Escherichia coli*. *Biochem. Biophys. Res. Commun.* 229, 189–192. doi: 10.1006/bbrc.1996.1778

Geszvain, K., Gruber, T. M., Mooney, R. A., Gross, C. A., and Landick, R. (2004). A hydrophobic patch on the flap-tip helix of *E.coli* RNA polymerase mediates sigma(70) region 4 function. *J. Mol. Biol.* 343, 569–587. doi: 10.1016/j.jmb.2004.08.063

Gibson, G. R., Beatty, E. R., Wang, X., and Cummings, J. H. (1995). Selective stimulation of bifidobacteria in the human colon by oligofructose and inulin. *Gastroenterology* 108, 975–982. doi: 10.1016/0016-5085(95)90192-2

Goujon, M., McWilliam, H., Li, W., Valentin, F., Squizzato, S., Paern, J., et al. (2010). A new bioinformatics analysis tools framework at EMBL-EBI. *Nucleic Acids Res.* 38, W695–W699. doi: 10.1093/nar/gkq313

Gupta, R. S. (2014). "Chapter 8 - Identification of conserved indels that are useful for classification and evolutionary studies," in *Methods in Microbiology: New Approaches to Prokaryotic Systematics*, eds M. Goodfellow, I. Sutcliffe, and J. Chun (Cambridge, MA: Academic Press), 153–182.

Gupta, R. S. (2016). Impact of genomics on the understanding of microbial evolution and classification: the importance of Darwin's views on classification. *FEMS Microbiol. Rev.* 40, 520–553. doi: 10.1093/femsre/fuw011

Gupta, R. S., Nanda, A., and Khadka, B. (2017). Novel molecular, structural and evolutionary characteristics of the phosphoketolases from bifidobacteria and *Coriobacteriales*. *PLoS ONE* 12:e0172176. doi: 10.1371/journal.pone.0172176

Hashimoto, K., and Panchenko, A. R. (2010). Mechanisms of protein oligomerization, the critical role of insertions and deletions in maintaining different oligomeric states. *Proc. Natl. Acad. Sci. U.S.A.* 107, 20352–20357. doi: 10.1073/pnas.1012999107

Heo, L., Park, H., and Seok, C. (2013). GalaxyRefine: protein structure refinement driven by side-chain repacking. *Nucleic Acids Res.* 41, W384–W388. doi: 10.1093/nar/gkt458

Itzhaki, Z., Akiva, E., Altuvia, Y., and Margalit, H. (2006). Evolutionary conservation of domain-domain interactions. *Genome Biol.* 7:R125. doi: 10.1186/gb-2006-7-12-r125

Jiang, W., Yun, D., Saleh, L., Barr, E. W., Xing, G., Hoffart, L. M., et al. (2007). A manganese(IV)/iron(III) cofactor in *Chlamydia trachomatis* ribonucleotide reductase. *Science* 316, 1188–1191. doi: 10.1126/science.1141179

Jones, D. T. (1999). Protein secondary structure prediction based on position-specific scoring matrices. *J. Mol. Biol.* 292, 195–202. doi: 10.1006/jmbi.1999.3091

Jones, D. T., Taylor, W. R., and Thornton, J. M. (1992). The rapid generation of mutation data matrices from protein sequences. *Comput. Appl. Biosci.* 8, 275–282. doi: 10.1093/bioinformatics/8.3.275

Jordan, A., Aslund, F., Pontis, E., Reichard, P., and Holmgren, A. (1997). Characterization of *Escherichia coli* NrdH. A glutaredoxin-like protein with a thioredoxin-like activity profile. *J. Biol. Chem.* 272, 18044–18050. doi: 10.1074/jbc.272.29.18044

Jordan, A., Pontis, E., Aslund, F., Hellman, U., Gibert, I., and Reichard, P. (1996). The ribonucleotide reductase system of *Lactococcus lactis*. Characterization of an NrdEF enzyme and a new electron transport protein. *J. Biol. Chem.* 271, 8779–8785. doi: 10.1074/jbc.271.15.8779

Kenyon, C. R., and Osbak, K. (2014). Recent progress in understanding the epidemiology of bacterial vaginosis. *Curr. Opin. Obstet. Gynecol.* 26, 448–454. doi: 10.1097/GCO.0000000000000112

Khadka, B., and Gupta, R. S. (2017). Identification of a conserved 8 aa insert in the PIP5K protein in the *Saccharomycetaceae* family of fungi and the molecular dynamics simulations and structural analysis to investigate its potential functional role. *Proteins* doi: 10.1002/prot.25306 [Epub ahead of print].

Kozakov, D., Brenke, R., Comeau, S. R., and Vajda, S. (2006). PIPER: an FFT-based protein docking program with pairwise potentials. *Proteins* 65, 392–406. doi: 10.1002/prot.21117

Krissinel, E., and Henrick, K. (2007). Inference of macromolecular assemblies from crystalline state. *J. Mol. Biol.* 372, 774–797. doi: 10.1016/j.jmb.2007.05.022

Larkin, M. A., Blackshields, G., Brown, N. P., Chenna, R., McGettigan, P. A., McWilliam, H., et al. (2007). Clustal W and Clustal X version 2.0. *Bioinformatics* 23, 2947–2948. doi: 10.1093/bioinformatics/btm404

Larsson, K. M., Andersson, J., Sjoberg, B. M., Nordlund, P., and Logan, D. T. (2001). Structural basis for allosteric substrate specificity regulation in anaerobic ribonucleotide reductases. *Structure* 9, 739–750. doi: 10.1016/S0969-2126(01)00627-X

Larsson, K. M., Logan, D. T., and Nordlund, P. (2010). Structural basis for adenosylcobalamin activation in AdoCbl-dependent ribonucleotide reductases. *ACS Chem. Biol.* 5, 933–942. doi: 10.1021/cb1000845

Leahy, S. C., Higgins, D. G., Fitzgerald, G. F., and van Sinderen, D. (2005). Getting better with bifidobacteria. *J. Appl. Microbiol.* 98, 1303–1315. doi: 10.1111/j.1365-2672.2005.02600.x

Lee, G. R., Heo, L., and Seok, C. (2016). Effective protein model structure refinement by loop modeling and overall relaxation. *Proteins* 84(Suppl. 1), 293–301. doi: 10.1002/prot.24858

Logan, D. T., Andersson, J., Sjoberg, B. M., and Nordlund, P. (1999). A glycyl radical site in the crystal structure of a class III ribonucleotide reductase. *Science* 283, 1499–1504. doi: 10.1126/science.283.5407.1499

Lovell, S. C., Davis, I. W., Arendall, W. B. III, de Bakker, P. I., Word, J. M., Prisant, M. G., et al. (2003). Structure validation by Calpha geometry: phi,psi and Cbeta deviation. *Proteins* 50, 437–450. doi: 10.1002/prot.10286

Lundin, D., Gribaldo, S., Torrents, E., Sjoberg, B. M., and Poole, A. M. (2010). Ribonucleotide reduction - horizontal transfer of a required function spans all three domains. *BMC Evol. Biol.* 10:383. doi: 10.1186/1471-2148-10-383

Lundin, D., Torrents, E., Poole, A. M., and Sjoberg, B. M. (2009). RNRdb, a curated database of the universal enzyme family ribonucleotide reductase, reveals a high level of misannotation in sequences deposited to Genbank. *BMC Genomics* 10:589. doi: 10.1186/1471-2164-10-589

Luthy, R., Bowie, J. U., and Eisenberg, D. (1992). Assessment of protein models with three-dimensional profiles. *Nature* 356, 83–85. doi: 10.1038/356083a0

Lyskov, S., Chou, F. C., Conchuir, S. O., Der, B. S., Drew, K., Kuroda, D., et al. (2013). Serverification of molecular modeling applications: the Rosetta Online Server that Includes Everyone (ROSIE). *PLoS ONE* 8:e63906. doi: 10.1371/journal.pone.0063906

Lyskov, S., and Gray, J. J. (2008). The RosettaDock server for local protein-protein docking. *Nucleic Acids Res.* 36, W233–W238. doi: 10.1093/nar/gkn216

Marchler-Bauer, A., and Bryant, S. H. (2004). CD-Search: protein domain annotations on the fly. *Nucleic Acids Res.* 32, W327–W331. doi: 10.1093/nar/gkh454

Masco, L., Huys, G., De Brandt, E., Temmerman, R., and Swings, J. (2005). Culture-dependent and culture-independent qualitative analysis of probiotic products claimed to contain bifidobacteria. *Int. J. Food Microbiol.* 102, 221–230. doi: 10.1016/j.ijfoodmicro.2004.11.018

Milani, C., Lugli, G. A., Duranti, S., Turroni, F., Bottacini, F., Mangifesta, M., et al. (2014). Genomic encyclopedia of type strains of the genus *Bifidobacterium*. *Appl. Environ. Microbiol.* 80, 6290–6302. doi: 10.1128/AEM.02308-14

Milani, C., Turroni, F., Duranti, S., Lugli, G. A., Mancabelli, L., Ferrario, C., et al. (2015). Genomics of the genus *Bifidobacterium* reveals species-specific adaptation to the glycan-rich gut environment. *Appl. Environ. Microbiol.* 82, 980–991. doi: 10.1128/AEM.03500-15

Mills, S., Stanton, C., Fitzgerald, G. F., and Ross, R. P. (2011). Enhancing the stress responses of probiotics for a lifestyle from gut to product and back again. *Microb. Cell Fact.* 10(Suppl. 1):S19. doi: 10.1186/1475-2859-10-S1-S19

Nordlund, P., and Reichard, P. (2006). Ribonucleotide reductases. *Annu. Rev. Biochem.* 75, 681–706. doi: 10.1146/annurev.biochem.75.103004.142443

Oberg, T. S., Steele, J. L., Ingham, S. C., Smeianov, V. V., Briczinski, E. P., Abdalla, A., et al. (2011). Intrinsic and inducible resistance to hydrogen peroxide in *Bifidobacterium* species. *J. Ind. Microbiol. Biotechnol.* 38, 1947–1953. doi: 10.1007/s10295-011-0983-y

Ollagnier, S., Mulliez, E., Gaillard, J., Eliasson, R., Fontecave, M., and Reichard, P. (1996). The anaerobic *Escherichia coli* ribonucleotide reductase. Subunit structure and iron sulfur center. *J. Biol. Chem.* 271, 9410–9416. doi: 10.1074/jbc.271.16.9410

Palframan, R. J., Gibson, G. R., and Rastall, R. A. (2003). Carbohydrate preferences of *Bifidobacterium* species isolated from the human gut. *Curr. Issues Intest. Microbiol.* 4, 71–75.

Petersson, L., Graslund, A., Ehrenberg, A., Sjoberg, B. M., and Reichard, P. (1980). The iron center in ribonucleotide reductase from *Escherichia coli*. *J. Biol. Chem.* 255, 6706–6712.

Poole, A. M., Logan, D. T., and Sjoberg, B. M. (2002). The evolution of the ribonucleotide reductases: much ado about oxygen. *J. Mol. Evol.* 55, 180–196. doi: 10.1007/s00239-002-2315-3

Reichard, P. (1993). From RNA to DNA, why so many ribonucleotide reductases? *Science* 260, 1773–1777.

Reichard, P. (2010). Ribonucleotide reductases: substrate specificity by allostery. *Biochem. Biophys. Res. Commun.* 396, 19–23. doi: 10.1016/j.bbrc.2010.02.108

Roca, I., Torrents, E., Sahlin, M., Gibert, I., and Sjöberg, B. M. (2008). NrdI essentiality for class Ib ribonucleotide reduction in *Streptococcus pyogenes*. *J. Bacteriol.* 190, 4849–4858. doi: 10.1128/JB.00185-08

Roy, A., Kucukural, A., and Zhang, Y. (2010). I-TASSER: a unified platform for automated protein structure and function prediction. *Nat. Protoc.* 5, 725–738. doi: 10.1038/nprot.2010.5

Sanders, M. E. (1998). Development of consumer probiotics for the US market. *Br. J. Nutr.* 80, S213–S218.

Schneidman-Duhovny, D., Inbar, Y., Nussinov, R., and Wolfson, H. J. (2005). PatchDock and SymmDock: servers for rigid and symmetric docking. *Nucleic Acids Res.* 33, W363–W367. doi: 10.1093/nar/gki481

Schoeffler, A. J., May, A. P., and Berger, J. M. (2010). A domain insertion in *Escherichia coli* GyrB adopts a novel fold that plays a critical role in gyrase function. *Nucleic Acids Res.* 38, 7830–7844. doi: 10.1093/nar/gkq665

Schrödinger, L. (2016). *The PyMOL Molecular Graphics System, Version 1.8.* Available at: https://www.pymol.org/

Shen, M. Y., and Sali, A. (2006). Statistical potential for assessment and prediction of protein structures. *Protein Sci.* 15, 2507–2524. doi: 10.1110/ps.062416606

Singh, B., and Gupta, R. S. (2009). Conserved inserts in the Hsp60 (GroEL) and Hsp70 (DnaK) proteins are essential for cellular growth. *Mol. Genet. Genomics* 281, 361–373. doi: 10.1007/s00438-008-0417-3

Sintchak, M. D., Arjara, G., Kellogg, B. A., Stubbe, J., and Drennan, C. L. (2002). The crystal structure of class II ribonucleotide reductase reveals how an allosterically regulated monomer mimics a dimer. *Nat. Struct. Biol.* 9, 293–300. doi: 10.1038/nsb774

Sippl, M. J., Lackner, P., Domingues, F. S., and Koppensteiner, W. A. (1999). An attempt to analyse progress in fold recognition from CASP1 to CASP3. *Proteins* Suppl. 3, 226–230. doi: 10.1002/(SICI)1097-0134(1999)37:3+<226::AID-PROT29>3.0.CO;2-Z

Smith, S. M., Ogbara, T., and Eng, R. H. (1992). Involvement of *Gardnerella vaginalis* in urinary tract infections in men. *J. Clin. Microbiol.* 30, 1575–1577.

Sun, X., Eliasson, R., Pontis, E., Andersson, J., Buist, G., Sjöberg, B. M., et al. (1995). Generation of the glycyl radical of the anaerobic *Escherichia coli* ribonucleotide reductase requires a specific activating enzyme. *J. Biol. Chem.* 270, 2443–2446. doi: 10.1074/jbc.270.6.2443

Tamao, Y., and Blakley, R. L. (1973). Direct spectrophotometric observation of an intermediate formed from deoxyadenosylcobalamin in ribonucleotide reduction. *Biochemistry* 12, 24–34. doi: 10.1021/bi00725a005

Tamura, K., Stecher, G., Peterson, D., Filipski, A., and Kumar, S. (2013). MEGA6: molecular evolutionary genetics analysis version 6.0. *Mol. Biol. Evol.* 30, 2725–2729. doi: 10.1093/molbev/mst197

Thelander, L., and Reichard, P. (1979). Reduction of ribonucleotides. *Annu. Rev. Biochem.* 48, 133–158. doi: 10.1146/annurev.bi.48.070179.001025

Torrents, E. (2014). Ribonucleotide reductases: essential enzymes for bacterial life. *Front. Cell Infect. Microbiol.* 4:52. doi: 10.3389/fcimb.2014.00052

Torrents, E., Aloy, P., Gibert, I., and Rodriguez-Trelles, F. (2002). Ribonucleotide reductases: divergent evolution of an ancient enzyme. *J. Mol. Evol.* 55, 138–152. doi: 10.1007/s00239-002-2311-7

Torrents, E., Eliasson, R., Wolpher, H., Graslund, A., and Reichard, P. (2001). The anaerobic ribonucleotide reductase from *Lactococcus lactis*. Interactions between the two proteins NrdD and NrdG. *J. Biol. Chem.* 276, 33488–33494. doi: 10.1074/jbc.M103743200

Torrents, E., Jordan, A., Karlsson, M., and Gibert, I. (2000). Occurrence of multiple ribonucleotide reductase classes in gamma-*proteobacteria* species. *Curr. Microbiol.* 41, 346–351. doi: 10.1007/s002840010147

Turroni, F., Ribbera, A., Foroni, E., van Sinderen, D., and Ventura, M. (2008). Human gut microbiota and bifidobacteria: from composition to functionality. *Antonie Van Leeuwenhoek* 94, 35–50. doi: 10.1007/s10482-008-9232-4

Turroni, F., van Sinderen, D., and Ventura, M. (2009). Bifidobacteria: from ecology to genomics. *Front. Biosci.* 14:4673–4684. doi: 10.2741/3559

Turroni, F., Ventura, M., Butto, L. F., Duranti, S., O'Toole, P. W., Motherway, M. O., et al. (2014). Molecular dialogue between the human gut microbiota and the host: a *Lactobacillus* and *Bifidobacterium* perspective. *Cell Mol. Life Sci.* 71, 183–203. doi: 10.1007/s00018-013-1318-0

Uhlin, U., and Eklund, H. (1994). Structure of ribonucleotide reductase protein R1. *Nature* 370, 533–539. doi: 10.1038/370533a0

Uppsten, M., Farnegardh, M., Domkin, V., and Uhlin, U. (2006). The first holocomplex structure of ribonucleotide reductase gives new insight into its mechanism of action. *J. Mol. Biol.* 359, 365–377. doi: 10.1016/j.jmb.2006.03.035

Uppsten, M., Farnegardh, M., Jordan, A., Eliasson, R., Eklund, H., and Uhlin, U. (2003). Structure of the large subunit of class Ib ribonucleotide reductase from *Salmonella* typhimurium and its complexes with allosteric effectors. *J. Mol. Biol.* 330, 87–97. doi: 10.1016/S0022-2836(03)00538-2

Ventura, M., Canchaya, C., Tauch, A., Chandra, G., Fitzgerald, G. F., Chater, K. F., et al. (2007). Genomics of *Actinobacteria*: tracing the evolutionary history of an ancient phylum. *Microbiol. Mol. Biol. Rev.* 71, 495–548. doi: 10.1128/MMBR.00005-07

Ventura, M., O'Flaherty, S., Claesson, M. J., Turroni, F., Klaenhammer, T. R., van Sinderen, D., et al. (2009). Genome-scale analyses of health-promoting bacteria: probiogenomics. *Nat. Rev. Microbiol.* 7, 61–71. doi: 10.1038/nrmicro2047

Ventura, M., Turroni, F., Lugli, G. A., and van Sinderen, D. (2014). Bifidobacteria and humans: our special friends, from ecological to genomics perspectives. *J. Sci. Food Agric.* 94, 163–168. doi: 10.1002/jsfa.6356

Wei, Y., Thompson, J., and Floudas, C. A. (2011). CONCORD: a consensus method for protein secondary structure prediction via mixed integer linear optimization. *Proc. R. Soc. A Math. Phys. Eng. Sci.* 468, 831–850. doi: 10.1098/rspa.2011.0514

Whelan, S., and Goldman, N. (2001). A general empirical model of protein evolution derived from multiple protein families using a maximum-likelihood approach. *Mol. Biol. Evol.* 18, 691–699. doi: 10.1093/oxfordjournals.molbev.a003851

Wiederstein, M., and Sippl, M. J. (2007). ProSA-web: interactive web service for the recognition of errors in three-dimensional structures of proteins. *Nucleic Acids Res.* 35, W407–W410. doi: 10.1093/nar/gkm290

Xiao, M., Xu, P., Zhao, J., Wang, Z., Zuo, F., Zhang, J., et al. (2011). Oxidative stress-related responses of *Bifidobacterium longum* subsp. *longum* BBMN68 at the proteomic level after exposure to oxygen. *Microbiology* 157, 1573–1588. doi: 10.1099/mic.0.044297-0

Yang, J., Yan, R., Roy, A., Xu, D., Poisson, J., and Zhang, Y. (2015). The I-TASSER Suite: protein structure and function prediction. *Nat. Methods* 12, 7–8. doi: 10.1038/nmeth.3213

Zhang, G., Gao, B., Adeolu, M., Khadka, B., and Gupta, R. S. (2016). Phylogenomic analyses and comparative studies on genomes of the *Bifidobacteriales*: identification of molecular signatures specific for the order *Bifidobacteriales* and its different subclades. *Front. Microbiol.* 7:978. doi: 10.3389/fmicb.2016.00978

Zhang, Y. (2008). I-TASSER server for protein 3D structure prediction. *BMC Bioinformatics* 9:40. doi: 10.1186/1471-2105-9-40

Zhi, X. Y., Li, W. J., and Stackebrandt, E. (2009). An update of the structure and 16S rRNA gene sequence-based definition of higher ranks of the class Actinobacteria, with the proposal of two new suborders and four new families and emended descriptions of the existing higher taxa. *Int. J. Syst. Evol. Microbiol.* 59, 589–608. doi: 10.1099/ijs.0.65780-0

Zuo, F., and Chen, S. (2016). Recently performed Next generation sequencing based transcriptomic analysis of oxidative stress response in *Bifidobacterium longum* BBMN68 (data accessible at NCBI GEO database (Edgar et al., 2002), accession GSE65320). NCBI GEO database. Available at: http://www.ncbi.nlm.nih.gov/geo/

10

The Importance of the Expendable: Toxin–Antitoxin Genes in Plasmids and Chromosomes

Ramón Díaz-Orejas[1], Manuel Espinosa[1] and Chew Chieng Yeo[2]**

[1] *Centro de Investigaciones Biológicas, Consejo Superior de Investigaciones Científicas (CSIC), Madrid, Spain,* [2] *Faculty of Medicine, Biomedical Research Centre, Universiti Sultan Zainal Abidin, Kuala Terengganu, Malaysia*

Edited by:
Johann Peter Gogarten, University of Connecticut, United States

Reviewed by:
Francis Repoila, Institut National de la Recherche Agronomique (INRA), France
Nikolai Ravin, Research Center for Biotechnology (RAS), Russia

Correspondence:
Manuel Espinosa
mespinosa@cib.csic.es
Chew Chieng Yeo
chewchieng@gmail.com

Toxin–antitoxin (TA) genes were first reported in plasmids and were considered expendable genetic cassettes involved in the stable maintenance of the plasmid replicon by interfering with growth and/or viability of bacteria in which the plasmid was lost. TAs were later found in bacterial chromosomes and also in integrated mobile genetic elements; they were proposed to be involved in the bacterial response to stressful situations. At present, 100s of TAs have been identified and classified in up to six families (I to VI), with those belonging to the type II (constituted by two protein components) being the most studied. Based on well-characterized examples of several type II TAs, we discuss in this review that irrespective of their locations in plasmids or chromosomes, TAs functionally overlap as indicated by: (i) in both locations they can mediate the maintenance of genetic elements to which they are physical linked, and (ii) they can induce persistence or virulence in response to stress situations. Examples of functional confluences in homologous TA systems with different locations are also given. We also consider whether the physiological role of TAs is due to their genetic organization as operons or to their inherent properties, like the short lifespan of the antitoxin component.

Keywords: toxin–antitoxin, plasmids, post-segregational killing, genomic islands, chromosome, bacterial virulence, persistence

INTRODUCTION

Toxin–antitoxin (TA) genes were initially discovered in two conjugative plasmids of *Escherichia coli*, F (Ogura and Hiraga, 1983) and R1 (Gerdes et al., 1986a; Bravo et al., 1988; Tsuchimoto et al., 1988), as cassette of two genes involved in stable maintenance of these plasmids; they were shown to participate in stable plasmid inheritance because they reduced either the viability or the growth of the cells that had lost the plasmid at the time of cell division (Bravo et al., 1988; Tsuchimoto et al., 1988). Killing of plasmid free segregants was termed post-segregational killing (PSK) (Gerdes et al., 1986b) and shown to be due to the decay of the more unstable antitoxin in plasmid-free cells and to the subsequent activation of the toxin in these cells. A role of plasmidic TAs in outcompeting compatible plasmids was later proposed (Cooper and Heinemann, 2000). TAs were also subsequently shown to be encoded by bacterial chromosomes and some of them integrated within mobile genetic elements (MGEs). One of the hypotheses to explain the presence and function of these TAs was that they participate in the response to stressful conditions

(Christensen and Gerdes, 2003). At present, 100s of TAs have been identified and classified, depending on the nature and activity of antitoxins, in up to six types (type I to type VI), reviewed in Page and Peti (2016). To these toxin–antitoxin pairs can also be added Type II restriction enzymes and their cognate methylases: similar to conventional TAs, restriction-modification systems can also be encoded in plasmids and enforce their maintenance by promoting PSK of plasmid-free cells (Yarmolinsky, 1995), although it must be emphasized that TA and RM complexes that have been characterized so far do not share an evolutionary origin (Mruk and Kobayashi, 2014). Type II TAs, which are constituted by two protein components, are the best characterized (Chan et al., 2016; Kędzierska and Hayes, 2016). Activation of TAs in response to stress was thought to be a way to either eliminate part of the population in benefit of the rest [i.e., "altruistic" cell death (Engelberg-Kulka and Glaser, 1999)], or to reduce the metabolic load during adverse conditions by slowing or arresting cell growth (Maisonneuve and Gerdes, 2014). Chromosomal TAs have also been associated with several bacterial processes, like biofilm formation, survival during infection of eukaryotic cells, defense against invading bacteriophages and entrance and exit into persistence (Goeders and Van Melderen, 2014; Kędzierska and Hayes, 2016; Lobato-Márquez et al., 2016a). Since persistence is believed to be a major factor contributing to the chronic state of infections and tolerance to antibiotic treatments (Michiels et al., 2016), it was proposed that one of the roles of TAs was to contribute to dormancy, i.e., making the cells metabolically inactive (Pedersen et al., 2002; Christensen-Dalsgaard et al., 2010), a state that would lead to persistence due to triggering of the TAs (Maisonneuve et al., 2011; Maisonneuve and Gerdes, 2014). This attractive hypothesis was later considered as too simplistic (Ramisetty et al., 2016; Van Melderen and Wood, 2017). Since the initial discovery of TA systems as plasmid maintenance systems and their subsequent identification in the chromosome, many examples underline the functional confluence of these systems irrespective of their location. In the following sections we will try do discuss this confluence based on a few well-characterized examples of type II TA systems (see **Table 1** for a summary of the TA systems covered in this review).

PLASMID AND CHROMOSOMAL TOXIN–ANTITOXINS AS MEDIATORS OF POST-SEGREGATIONAL KILLING

Several observations indicate the existence of functional overlaps of TAs placed on plasmids or on chromosomes. A role of the chromosomal TAs in stabilization of integrated MGE or adjacent chromosomal regions has been demonstrated (Wozniak and Waldor, 2009), a role that is similar to their function in maintaining plasmid stability through PSK (Hayes, 2003). A novel TA pair designated *mosAT*, was shown to be responsible for maintaining the integrity of the ~100 kb SXT integrative and conjugative element (ICE) that confers resistance to multiple antibiotics in clinical isolates of *Vibrio cholerae* (Wozniak and Waldor, 2009). For a large MGE that can integrate, excise, and transfer to other bacteria, the SXT ICE is remarkably stable, with loss of ICE estimated at only 1 in 10^7 cells (Wozniak and Waldor, 2009). The *mosAT* system has low basal transcriptional levels when SXT is integrated; however its expression is derepressed when SXT is in an extrachromosomal state and vulnerable to loss. Interestingly, a homolog of *mosAT* was located on an octopine-type Ti plasmid of *Agrobacterium tumefaciens* suggesting that this TA system may function to maintain the stability of plasmids as well as ICEs (Wozniak and Waldor, 2009). Another TA system designated *sgiTA* was recently shown to promote the maintenance of a multidrug resistant integrative and mobilizable *Salmonella* Genomic Island 1 (SGI1) in *Salmonella enterica* serovar Typhimurium (Huguet et al., 2016) in a manner similar to *mosAT* for SXT. Intriguingly, SGI1 is only transmissible in the presence of conjugative plasmids of the IncA/C group but paradoxically, SGI1 displayed incompatibility with the IncA/C plasmids. The *sgiTA* locus was shown to play an essential role in SGI1 stability particularly in the concomitant presence of a conjugative IncA/C plasmid when SGI1 is in an extrachromosomal state and is more likely to be lost (Huguet et al., 2016).

Further, elimination of cells that lose a chromosome has been demonstrated in the case of *V. cholerae* (Yuan et al., 2011). Like all vibrios, the *V. cholerae* genome consists of two chromosomes (Heidelberg et al., 2000), and the smaller chromosome II (ChrII, 1.07 Mbp) hosts a large 126 kb superintegron (SI) that gathers 100s of diverse gene cassettes, including antibiotic resistance genes, and contains 17 TA systems (Iqbal et al., 2015). Each of these cassettes is associated with a target recombination sequence (the *attC* site) and in a SI, 100s of gene cassettes with their *attC* sites are arranged in direct orientation, alluding to a likely inherent instability in these SIs. Nevertheless, SIs are remarkably stable, and in an earlier paper, it was elegantly demonstrated that two TA loci from the *Vibrio vulnificus* SI (*relBE1* and *parDE1*) stabilize the SI and prevent large scale deletions from occurring when the SI was devoid of TA loci (Szekeres et al., 2007). Chromosome-specific mechanisms exist to ensure the proper segregation of the two *V. cholerae* chromosomes in daughter cells. In the *V. cholerae* ChrII, the *parAB2* locus was essential for the partitioning of ChrII, and in a *parAB2* deletion mutant, ChrII was mislocalized leading to a complete loss of the entire ChrII in a fraction of the population (Yamaichi et al., 2007). Cells that lost ChrII were non-viable and underwent characteristic cytological changes including cell enlargement, nucleoid condensation and degradation (Yamaichi et al., 2007). It was subsequently shown that the three ParE toxins encoded by their respective *parDE* TA loci in the SI of ChrII were responsible for the PSK of cells that lost ChrII, closely mimicking PSK mediated by plasmid-encoded homologs (Yuan et al., 2011). A recent paper showed that all 17 TA loci in the *V. cholerae* SI were functional, expressed from their own native promoters, and were very specific – i.e., there was no cross-interaction between non-cognate toxins and antitoxins (Iqbal et al., 2015). These advocate for a major role of these 17 TA loci in the stabilization of the *V. cholerae* SI and to prevent the emergence of cells that lack ChrII; in other words, much like their plasmid-encoded homologs, the *V. cholerae* chromosomal

TABLE 1 | Summary of the location and functions of toxin–antitoxin (TA) systems covered in this review.

TA	Host	Location of TA	Function	Reference
MosAT	*Vibrio cholerae*	Chromosomal; integrative and conjugative element (ICE)	Stability of ICE	Wozniak and Waldor, 2009
SgiTA	*Salmonella enterica* serovar Typhimurium	Chromosomal; genomic island 1	Stability of genomic island	Huguet et al., 2016
RelBE1 ParDE1	*Vibrio vulnificus*	Chromosomal; superintegron	Superintegron stability	Szekeres et al., 2007
ParDE-1 ParDE-2 ParDE-3 RelBE-1 RelBE-2 RelBE-3 RelBE-4 HigBA-1 HigBA-2 Phd-Doc 0318-0319 0332-0333 0422-0423 0477-0478 0481-0482 0486-0487 0488-9489	*Vibrio cholerae*	Chromosome II, superintegron	Stability of superintegron as well as chromosome II	Iqbal et al., 2015
Epsilon-Zeta (ε-ζ)	*Streptococcus pyogenes*	Plasmid pSM19035	Plasmid stability	Camacho et al., 2002
PezAT	*Streptococcus pneumoniae*	Chromosomal, pathogenicity island 1	Stability of pathogenicity island, virulence factor	Chan et al., 2014; Iannelli et al., 2014
SezAT	*Streptococcus suis*	Chromosomal; pathogenicity island	Stability of pathogenicity island	Yao et al., 2015
AvrRxo1-Arc1	*Xanthomonas oryzae* pv. *oryzicola*	Unknown	Virulence factor (type III secreted effector)	Triplett et al., 2016
$CcdAB_F$	*Escherichia coli*	Plasmid F	Plasmid stability, persistence	Ogura and Hiraga, 1983; Tripathi et al., 2012
$CcdAB_{O157}$	*Escherichia coli* O157:H7	Chromosomal	Persistence	Gupta et al., 2017
$VapBC2_{ST}$ $CcdAB_{ST}$	*Salmonella enterica* serovar Typhimurium	Plasmid pLST	Plasmid stability, virulence Plasmid stability	Lobato-Márquez et al., 2016b

TA loci are also agents of PSK (Yuan et al., 2011; Iqbal et al., 2015).

Conversely, it has been shown that plasmid-encoded TAs can contribute to overcome stress, induce persistence, and could increase survival of bacterial cells during infection (Helaine et al., 2014), functions that were initially attributed to chromosomal TAs (Lobato-Márquez et al., 2016a). PSK mediated by TAs following the loss of genetic information associated to a plasmid can be considered as a situation of stress, to which the cells react by toxin activation. Transient activation of the *E. coli* F-plasmid-encoded CcdB toxin enhance the generation of drug-tolerant persister cells, and this process was found to be dependent on Lon protease and RecA (Tripathi et al., 2012). The F-plasmid-encoded *ccdAB*$_F$ locus has been well-established as a plasmid maintenance system (Jaffe et al., 1985) and the finding that it plays a role in persistence expands its function as a transmissible persistence factor (Tripathi et al., 2012) (see below).

PezAT AND ITS POTENTIAL ROLE IN VIRULENCE

Two recent articles published in *Frontiers* (Chan and Espinosa, 2016; Lobato-Márquez et al., 2016b) underline the concept that phenotypes associated to plasmid- or to chromosomally encoded TAs do overlap because independent of their location, toxins target similar functions and the TA operons are regulated and induced by similar conditions. The first example is provided by the pneumococcal *pezAT* operon (Khoo et al., 2007; Chan and Espinosa, 2016). The two genes constituting it are placed in the putative mobilizable pathogenicity island 1 (pneumococcal pathogenicity island 1, PPI1) and that is found in nearly half of capsulated (virulent) *Streptococcus pneumoniae* strains (Chan et al., 2012). In some strains there is a second copy of the operon, located on the putative ICE Tn*5253*. A close homolog of the PezAT pair is the Epsilon-Zeta TA, which was discovered in the broad host-range plasmid pSM19035 of *Streptococcus pyogenes* (Camacho et al., 2002). Epsilon-Zeta differs from PezAT as the Epsilon antitoxin does not perform the transcriptional regulation of the operon, but rather by a third component, Omega, which also regulates the transcription of other genes encoded by the pSM19025 plasmid. In the *pezAT* operon, the PezA antitoxin performs the autoregulatory role (Khoo et al., 2007). Epsilon-Zeta plays an essential role in maintaining the stability of plasmid pSM19035 via PSK.

The Zeta/PezT toxins target the cell wall synthesis machinery by phosphorylating the peptidoglycan precursor, UDP-*N*-acetylglucosamine (UNAG) at the 3′-OH group of the *N*-acetylglycosamine moiety. The phosphorylated product, UDP-*N*-acetylglucosamine-3-phosphate (UNAG-3P), accumulates in the cytosol and inhibits MurA, which is the essential enzyme that catalyzes the initial step in peptidoglycan synthesis (Mutschler and Meinhart, 2011; Mutschler et al., 2011). Nevertheless, it was proposed that reduction in the UNAG levels is just one of several responses that is triggered by Zeta/PezT expression in response to stress (Tabone et al., 2014). The *pezAT* operon may play a role in stabilizing the MGE within the pneumococcal host (Chan et al., 2014; Iannelli et al., 2014). Further, a close homolog of *pezA*T was discovered in *Streptococcus suis* (designated *sezAT*), which was shown to be important for the stable inheritance of the Pathogenicity Island 1 (SsPI-1) (Yao et al., 2015). Pneumococcal strains that carry *pezAT* exhibit increased virulence; further, deletion of the operon led to pneumococcal cells exhibiting increased resistance to β-lactam antibiotics and to increased ability to take up homologous DNA by enhancing genetic competence (Chan and Espinosa, 2016). How PezT functions to increase pneumococcal virulence is currently unknown but it was postulated that activation of PezT during environmental stresses or the course of infection would result in inhibition of cell wall synthesis and subsequent lysis of a subpopulation of pneumococcal cells (Mutschler and Meinhart, 2011, 2013). The lysis of these cells would lead to the release of cellular components detrimental to the infected host such as pneumolysin. Interestingly, recent papers showed that a PezT/Zeta homolog, designated AvrRxo1 from the plant pathogen *Xanthomonas oryzae* pv. *oryzicola* functions as a type III-secreted virulence factor which is toxic in plants and is bacteriostatic when expressed in *E. coli* (Han et al., 2015; Triplett et al., 2016). An AvrRxo1 homolog from myxobacterium was also shown to trigger rapid cell death response in tobacco (Triplett et al., 2016). Intriguingly, although the AvrRxo1 toxin was found to be a nucleotide kinase, its target is not UNAG like PezT/Zeta but rather, the coenzyme nicotinamide adenine dinucleotide (NAD) and its biochemical precursor, nicotinic acid adenine dinucleotide (NAAD) leading to the formation of unusual 3′-phosphorylated products, 3′-NADP and 3′-nicotinic acid adenine dinucleotide phosphate (3′-NADDP) (Schuebel et al., 2016). A recent paper showed that 3′-NADP accumulates upon expression of AvrRxo1 in tobacco and rice leaves infected with AvrRxo1-expressing strains of *Xanthomonas oryzae*, thus indicating that the AvrRxo1 effector/toxin targets the coenzyme and redox carrier essential for central metabolic function of the host. However, the actual mechanism of 3′-NADP accumulation *in planta* is currently unknown as NAD and the conventional cofactor, 2′-NADP, are needed in 100s of essential reactions in the cell (Shidore et al., 2017). Hence, it could be possible that PezT/Zeta not only help in triggering the lysis of pneumococcal cells, the toxin itself may also be detrimental to the infected host cells. Indeed, it was recently shown that expression of the pneumococcal *pezT* toxin in the eukaryotic microalgae *Chlorella vulgaris* is lethal, leading to cellular damage and lysis (Ng et al., 2016). We await experimental results that would indicate if expression of PezT/Zeta in mammalian cells would be equally detrimental.

FUNCTIONAL OVERLAPS BETWEEN CHROMOSOMAL AND PLASMID-ENCODED TA SYSTEMS

The second example is related to the role of TAs of *Salmonella enterica* serovar Typhimurium carrying the virulence plasmid pLST during bacterial infection (Lobato-Márquez et al., 2016b). One of the two TAs encoded by plasmid pLST is *vapBC*$_{ST}$. This particular TA contributes to the successful colonization of recipient cells during infection, in conjunction with other type I and type II TAs encoded by the *Salmonella* chromosome (Lobato-Márquez et al., 2015). In addition to its role during infection, the plasmidic copy of *vapBC*$_{ST}$ contributes to the maintenance of the plasmid (Lobato-Márquez et al., 2016b). Interestingly, the chromosomal copy of this particular TA seems to be inactive, so that the role in infection was taken up by the plasmid-encoded copy. Curiously enough, the VapC toxin of vapBC2$_{ST}$ is active as a toxin, indicating that stabilization of pLST could be due to PSK (Lobato-Márquez et al., 2015).

An interesting example showing that location is compatible with different functions is provided by the first type II locus described, the *ccdAB*$_F$ operon encoded by plasmid F (Jaffe et al., 1985). This operon was reported to contribute to plasmid maintenance by killing plasmid free-segregants; PSK was the result of Lon protease-mediated degradation of the CcdA antitoxin and the subsequent activation of the anti-topoisomerase activity of the toxin CcdB (Ogura and Hiraga, 1983). In addition to its role in plasmid maintenance, *ccdAB*$_F$ was shown to contribute to bacterial persistence (Tripathi et al., 2012), a role that was also proposed for several chromosomal TA systems (Maisonneuve et al., 2011) and that has been questioned as reductionist recently (Ramisetty et al., 2016; Van Melderen and Wood, 2017). Furthermore, the *ccdAB*$_{ST}$ system of plasmid pLST seems to participate in plasmid maintenance beyond PSK because, in spite of carrying a single point mutation that inactivates the anti-topoisomerase activity of CcdB$_{ST}$, it contributes significantly to the stabilization of the virulence plasmid by a yet to be identified mechanism

(Lobato-Márquez et al., 2016b). Functional interactions between co-existing *ccd* systems in plasmid and chromosome reported to be present in the pathogenic *E. coli* strain O157:H7 added further versatility to this system (Wilbaux et al., 2007). The chromosomally encoded *ccdAB* genes, like the plasmidic ones, has a toxin that target DNA gyrase and an antitoxin that is degraded by the Lon protease; however, both TAs seem to have evolved to achieve different functions: only the plasmidic antitoxin is able to neutralize the chromosomal toxin but not *vice versa*, and only the plasmidic TA is able to promote plasmid maintenance by PSK (Wilbaux et al., 2007). Nevertheless, a recent paper has shown that the chromosomal $ccdAB_{O157}$ system from *E. coli* O157:H7 also function in the formation of persister cells much like its F-plasmid-encoded counterpart even though the $CcdB_{O157}$ toxin displayed lower toxicity and has fivefold lower affinity for DNA gyrase compared to $CcdB_F$ (Gupta et al., 2017).

Plasmid maintenance linked to the coordination of TAs with plasmid replication was initially reported by the Diaz-Orejas laboratory on the *kis-kid* TA encoded by plasmid R1 (Ruiz-Echevarría et al., 1995a,b), and further analyzed (Pimentel et al., 2005; López-Villarejo et al., 2012, 2015). Additional work revealed a further coordination of the *kis-kid* TA with cell cycle functions (Pimentel et al., 2014). On the whole the above work supports that, failures in plasmid R1 replication reduces the levels of the Kis antitoxin and increases the activity of the Kid toxin. This results in: (i) the rescue of plasmid replication mediated by Kid-dependent decrease in the expression levels of CopB, a secondary inhibitor of plasmid replication, and (ii) the decrease in the levels of key cell division proteins that allows the rescue of plasmid replication before cell division can occur.

Targeting by RNase toxins of mRNAs, tRNAs and rRNAs impact and remodel protein synthesis and are key to the stress response mediated by TAs (reviewed by Moll and Engelberg-Kulka, 2012; Cruz and Woychik, 2015). Targeting tRNA and remodeling of protein synthesis profile seem to be a general mechanism of stress response. Indeed, a recent publication (Chionh et al., 2016) reveals a mechanism related to response to oxidative stress and induction of persistence in *Mycobacterium bovis*. The mechanism implies modification of the tRNA anticodons for threonine or leucine. Due to these modifications, the tRNA will enable the efficient translation of particular proteins related to the oxidative stress response. These will lead, in turn, to: (i) remodeling the protein synthesis potential of the cell to respond to oxidative stress; (ii) preferential synthesis of a set of stress response proteins, and (iii) induction of persistence. Most interestingly these stress response mechanisms seem to be universal and are shared by prokaryotes and eukaryotes. Induction of persistence by TAs in response to stress also involves reduction of the protein synthesis potential and selective synthesis of proteins required to achieve survival to the stress-inducing agent (reviewed by Moll and Engelberg-Kulka, 2012).

CONCLUDING REMARKS

TAs were proposed to be part of the accessory genome but as more and more details of their biological function are uncovered, the importance of these (apparently) expendable genetic entities to the lifestyle of their hosts are becoming clearer. As we have shown in this review, the biological functions of TAs do overlap irrespective of their location in their host genome – i.e., whether they are chromosomally encoded or plasmid-borne. Initially implicated in maintaining the stability of plasmids via PSK, TAs have since been shown to mediate the stability of genomic islands and even chromosome II of *V. cholerae* by PSK. Both plasmid- and chromosomally encoded TAs have also been implicated in persistence and virulence of several pathogens. It is thus clear that these hitherto "expendable" genetic loci have successfully integrated into their hosts' cellular regulatory network, enabling their hosts to better adapt to their distinctive environmental niches.

AUTHOR CONTRIBUTIONS

RD-O, ME, and CCY designed the outline of the review. RD-O wrote the first draft, and all authors worked on it until the production of the final version.

ACKNOWLEDGMENT

Thanks are due to Damián Lobato-Márquez for his critical reading of the manuscript.

REFERENCES

Bravo, A., Ortega, S., De Torrontegui, G., and Díaz, R. (1988). Killing of *Escherichia coli* cells modulated by components of the stability system ParD of plasmid R1. *Mol. Gen. Genet.* 215, 146–151. doi: 10.1007/BF00331316

Camacho, A. G., Misselwitz, R., Behlke, J., Ayora, S., Welfle, K., Meinhart, A., et al. (2002). *In vitro* and *in vivo* stability of the $epsilon_2zeta_2$ protein complex of the broad host-range *Streptococcus pyogenes* pSM19035 addiction system. *Biol. Chem.* 383, 1701–1713. doi: 10.1515/BC.2002.191

Chan, W. T., and Espinosa, M. (2016). The *Streptococcus pneumoniae pezAT* toxin-antitoxin system reduces β-lactam resistance and genetic competence. *Front. Microbiol.* 7:1322. doi: 10.3389/fmicb.2016.01322

Chan, W. T., Espinosa, M., and Yeo, C. C. (2016). Keeping the wolves at bay: antitoxins of prokaryotic type II toxin-antitoxin systems. *Front. Mol. Biosci.* 3:9. doi: 10.3389/fmolb.2016.00009

Chan, W. T., Moreno-Córdoba, I., Yeo, C. C., and Espinosa, M. (2012). Toxin-antitoxin genes of the gram-positive pathogen *Streptococcus pneumoniae*: so few and yet so many. *Microbiol. Mol. Biol. Rev.* 76, 773–791. doi: 10.1128/MMBR.00030-12

Chan, W. T., Yeo, C. C., Sadowy, E., and Espinosa, M. (2014). Functional validation of putative toxin-antitoxin genes from the Gram-positive pathogen *Streptococcus pneumoniae*: *phd-doc* is the fourth bona-fide operon. *Front. Microbiol.* 5:677. doi: 10.3389/fmicb.2014.00677

Chionh, Y. H., Mcbee, M., Babu, I. R., Hia, F., Lin, W., Zhao, W., et al. (2016). tRNA-mediated codon-biased translation in mycobacterial hypoxic persistence. *Nat. Commun.* 7:13302. doi: 10.1038/ncomms13302

Christensen, S. K., and Gerdes, K. (2003). RelE toxins from bacteria and archaea cleave mRNAs on translating ribosomes, which are rescued by tmRNA. *Mol. Microbiol.* 48, 1389–1400. doi: 10.1046/j.1365-2958.2003.03512.x

Christensen-Dalsgaard, M., Jørgensen, M. G., and Gerdes, K. (2010). Three new RelE-homologous mRNA interferases of *Escherichia coli* differentially induced by environmental stresses. *Mol. Microbiol.* 75, 333–348. doi: 10.1111/j.1365-2958.2009.06969.x

Cooper, T. F., and Heinemann, J. A. (2000). Postsegregational killing does not increase plasmid stability but acts to mediate the exclusion of competing plasmids. *Proc. Natl. Acad. Sci. U.S.A.* 97, 12643–12648. doi: 10.1073/pnas.220077897

Cruz, J. W., and Woychik, N. A. (2015). tRNAs taking charge. *Pathog. Dis.* 74:ftv117. doi: 10.1093/femspd/ftv117

Engelberg-Kulka, H., and Glaser, G. (1999). Addiction modules and programmed cell death and antideath in bacterial cultures. *Annu. Rev. Microbiol.* 53, 43–70. doi: 10.1146/annurev.micro.53.1.43

Gerdes, K., Bech, F. W., Jørgensen, S. T., Løbner-Olesen, A., Rasmussen, P. B., Atlung, T., et al. (1986a). Mechanism of postsegregational killing by the hok gene product of the parB system of plasmid R1 and its homology with the relF gene product of the *Escherichia coli relB* operon. *EMBO J.* 5, 2023–2029.

Gerdes, K., Rasmussen, P. B., and Molin, S. (1986b). Unique type of plasmid maintenance function: postsegregational killing of plasmid-free cells. *Proc. Natl. Acad. Sci. U.S.A.* 83, 3116–3120.

Goeders, N., and Van Melderen, L. (2014). Toxin-antitoxin systems as multilevel interaction systems. *Toxins* 6, 304–324. doi: 10.3390/toxins6010304

Gupta, K., Tripathi, A., Sahu, A., and Varadarajan, R. (2017). Contribution of the chromosomal *ccdAB* operon to bacterial drug tolerance. *J. Bacteriol.* doi: 10.1128/jb.00397-17 [Epub ahead of print].

Han, Q., Zhou, C., Wu, S., Liu, Y., Triplett, L., Miao, J., et al. (2015). Crystal structure of *Xanthomonas* AvrRxo1-ORF1, a Type III effector with a polynucleotide kinase domain, and its interactor AvrRxo1-ORF2. *Structure* 23, 1900–1909. doi: 10.1016/j.str.2015.06.030

Hayes, F. (2003). Toxins-antitoxins: plasmid maintenance, programmed cell death, and cell cycle arrest. *Science* 301, 1496–1499. doi: 10.1126/science.1088157

Heidelberg, J. F., Eisen, J. A., Nelson, W. C., Clayton, R. A., Gwinn, M. L., Dodson, R. J., et al. (2000). DNA sequence of both chromosomes of the cholera pathogen *Vibrio cholerae*. *Nature* 406, 477–483. doi: 10.1038/35020000

Helaine, S., Cheverton, A. M., Watson, K. G., Faure, L. M., Matthews, S. A., and Holden, D. W. (2014). Internalization of *Salmonella* by macrophages induces formation of nonreplicating persisters. *Science* 343, 204–208. doi: 10.1126/science.1244705

Huguet, K. T., Gonnet, M., Doublet, B., and Cloeckaert, A. (2016). A toxin antitoxin system promotes the maintenance of the IncA/C-mobilizable *Salmonella* Genomic Island 1. *Sci. Rep.* 6:32285. doi: 10.1038/srep32285

Iannelli, F., Santoro, F., Oggioni, M. R., and Pozzi, G. (2014). Nucleotide sequence analysis of integrative conjugative element *Tn5253* of *Streptococcus pneumoniae*. *Antimicrob. Agents Chemother.* 58, 1235–1239. doi: 10.1128/AAC.01764-13

Iqbal, N., Guérout, A.-M., Krin, E., Le Roux, F., and Mazel, D. (2015). Comprehensive functional analysis of the 18 *Vibrio cholerae* N16961 toxin-antitoxin systems substantiates their role in stabilizing the superintegron. *J. Bacteriol.* 197, 2150–2159. doi: 10.1128/jb.00108-15

Jaffe, A., Oruga, T., and Hiraga, S. (1985). Effects of the *ccd* function of the F plasmid on bacterial growth. *J. Bacteriol.* 163, 841–849.

Kędzierska, B., and Hayes, F. (2016). Emerging roles of Toxin-Antitoxin modules in bacterial pathogenesis. *Molecules* 21:E790. doi: 10.3390/molecules21060790

Khoo, S. K., Loll, B., Chan, W. T., Shoeman, R. L., Ngoo, L., Yeo, C. C., et al. (2007). Molecular and structural characterization of the PezAT chromosomal toxin-antitoxin system of the human pathogen *Streptococcus pneumoniae*. *J. Biol. Chem.* 282, 19606–19618. doi: 10.1074/jbc.M701703200

Lobato-Márquez, D., Díaz-Orejas, R., and García-Del Portillo, F. (2016a). Toxin-antitoxins and bacterial virulence. *FEMS Microbiol. Rev.* 40, 592–609. doi: 10.1093/femsre/fuw022

Lobato-Márquez, D., Molina-García, L., Moreno-Córdoba, I., García-Del Portillo, F., and Díaz-Orejas, R. (2016b). Stabilization of the virulence plasmid pSLT of *Salmonella* Typhimurium by three maintenance systems and its evaluation by using a new stability test. *Front. Mol. Biosci.* 3:66. doi: 10.3389/fmolb.2016.00066

Lobato-Márquez, D., Moreno-Córdoba, I., Figueroa, V., Díaz-Orejas, R., and García-Del Portillo, F. (2015). Distinct type I and type II toxin-antitoxin modules control *Salmonella* lifestyle inside eukaryotic cells. *Sci. Rep.* 5:9374. doi: 10.1038/srep09374

López-Villarejo, J., Diago-Navarro, E. E., Hernández-Arriaga, A. M., and Díaz-Orejas, R. (2012). Kis antitoxin couples plasmid R1 replication and *parD* (*kis, kid*) maintenance modules. *Plasmid* 67, 118–127. doi: 10.1016/j.plasmid.2011.12.015

López-Villarejo, J., Lobato-Márquez, D., and Díaz-Orejas, R. (2015). Coupling between the basic replicon and the Kis-Kid maintenance system of plasmid R1: modulation by Kis antitoxin levels and involvement in control of plasmid replication. *Toxins* 7, 478–492. doi: 10.3390/toxins7020478

Maisonneuve, E., and Gerdes, K. (2014). Molecular mechanisms underlying bacterial persisters. *Cell* 157, 539–548. doi: 10.1016/j.cell.2014.02.050

Maisonneuve, E., Shakespeare, L. J., Jørgensen, M. G., and Gerdes, K. (2011). Bacterial persistence by RNA endonucleases. *Proc. Natl. Acad. Sci. U.S.A.* 108, 13206–13211. doi: 10.1073/pnas.1100186108

Michiels, J. E., Van Den Bergh, B., Verstraeten, N., and Michiels, J. (2016). Molecular mechanisms and clinical implications of bacterial persistence. *Drug Resist. Updat.* 29, 76–89. doi: 10.1016/j.drup.2016.10.002

Moll, I., and Engelberg-Kulka, H. (2012). Selective translation during stress in *Escherichia coli*. *Trends Biochem. Sci.* 11, 493–498. doi: 10.1016/j.tibs.2012.07.007

Mruk, I., and Kobayashi, I. (2014). To be or not to be: regulation of restriction–modification systems and other toxin–antitoxin systems. *Nucleic Acids Res.* 42, 70–86. doi: 10.1093/nar/gkt711

Mutschler, H., Gebhardt, M., Shoeman, R. L., and Meinhart, A. (2011). A novel mechanism of programmed cell death in bacteria by toxin-antitoxin systems corrupts peptidoglycan synthesis. *PLoS Biol.* 9:e1001033. doi: 10.1371/journal.pbio.1001033

Mutschler, H., and Meinhart, A. (2011). ε/ζ systems: their role in resistance, virulence, and their potential for antibiotic development. *J. Mol. Med.* 89, 1183–1194. doi: 10.1007/s00109-011-0797-4

Mutschler, H., and Meinhart, A. (2013). "Type II toxin-antitoxin loci: the epsilon/zeta family," in *Prokaryotic Toxin-Antitoxins*, ed. K. Gerdes (Berlin: Springer), 205–223.

Ng, S. L., Harikrishna, J. A., Abu Bakar, F., Yeo, C. C., and Cha, T. S. (2016). Heterologous expression of the *Streptococcus pneumoniae yoeB* and *pezT* toxin genes is lethal in *Chlorella vulgaris*. *Algal Res.* 19, 21–29. doi: 10.1016/j.algal.2016.07.011

Ogura, T., and Hiraga, S. (1983). Mini-F plasmid genes that couple host cell division to plasmid proliferation. *Proc. Natl. Acad. Sci. U.S.A.* 80, 4784–4788. doi: 10.1073/pnas.80.15.4784

Page, R., and Peti, W. (2016). Toxin-antitoxin systems in bacterial growth arrest and persistence. *Nat. Chem. Biol.* 12, 208–214. doi: 10.1038/nchembio.2044

Pedersen, K., Christensen, K. S., and Gerdes, K. (2002). Rapid induction and reversal of a bacteriostatic condition by controlled expression of toxins and antitoxins. *Mol. Microbiol.* 45, 501–510. doi: 10.1046/j.1365-2958.2002.03027.x

Pimentel, B., Madine, M. A., and De La Cueva-Mendez, G. (2005). Kid cleaves specific mRNAs at UUACU sites to rescue the copy number of plasmid R1. *EMBO J.* 24, 3459–3469. doi: 10.1038/sj.emboj.7600815

Pimentel, B., Nair, R., Bermejo-Rodríguez, C., Preston, M. A., Agu, C. A., Wang, X., et al. (2014). Toxin Kid uncouples DNA replication and cell division to enforce retention of plasmid R1 in *Escherichia coli* cells. *Proc. Natl. Acad. Sci. U.S.A.* 111, 2734–2739. doi: 10.1073/pnas.1308241111

Ramisetty, B. C. M., Ghosh, D., Roy Chowdhury, M., and Santhosh, R. S. (2016). What Is the link between stringent response, endoribonuclease encoding Type II Toxin–Antitoxin systems and persistence? *Front. Microbiol.* 7:1882. doi: 10.3389/fmicb.2016.01882

Ruiz-Echevarría, M. J., De La Torre, M. A., and Díaz-Orejas, R. (1995a). A mutation that decreases the efficiency of plasmid R1 replication leads to the activation of *parD*, a killer stability system of the plasmid. *FEMS Microbiol. Lett.* 130, 129–135. doi: 10.1111/j.1574-6968.1995.tb07709.x

Ruiz-Echevarría, M. J., Giménez-Gallego, G., Sabariegos-Jareño, R., and Díaz-Orejas, R. (1995b). Kid, a small protein of the *parD* stability system of plasmid

R1 is an inhibitor of DNA replication acting at the initiation of DNA synthesis. *J. Mol. Biol.* 247, 568–577. doi: 10.1016/S0022-2836(05)80138-X

Schuebel, F., Rocker, A., Edelmann, D., Schessner, J., Brieke, C., and Meinhart, A. (2016). 3′-NADP and 3′-NAADP, two metabolites formed by the bacterial Type III effector AvrRxo1. *J. Biol. Chem.* 291, 22868–22880. doi: 10.1074/jbc.M116.751297

Shidore, T., Broeckling, C. D., Kirkwood, J. S., Long, J. J., Miao, J., Zhao, B., et al. (2017). The effector AvrRxo1 phosphorylates NAD *in planta*. *PLOS Pathog.* 13:e1006442. doi: 10.1371/journal.ppat.1006442

Szekeres, S., Dauti, M., Wilde, C., Mazel, D., and Rowe-Magnus, D. A. (2007). Chromosomal toxin-antitoxin loci can diminish large-scale genome reductions in the absence of selection. *Mol. Microbiol.* 63, 1588–1605. doi: 10.1111/j.1365-2958.2007.05613.x

Tabone, M., Lioy, V. S., Ayora, S., Machón, C., and Alonso, J. C. (2014). Role of Toxin ζ and starvation responses in the sensitivity to antimicrobials. *PLoS ONE* 9:e86615. doi: 10.1371/journal.pone.0086615

Tripathi, A., Dewan, P. C., Barua, B., and Varadarajan, R. (2012). Additional role for the ccd operon of F-plasmid as a transmissible persistence factor. *Proc. Natl. Acad. Sci. U.S.A.* 109, 12497–12502. doi: 10.1073/pnas.1121217109

Triplett, L. R., Shidore, T., Long, J., Miao, J., Wu, S., Han, Q., et al. (2016). AvrRxo1 is a bifunctional Type III secreted effector and Toxin-Antitoxin system component with homologs in diverse environmental contexts. *PLoS ONE* 11:e0158856. doi: 10.1371/journal.pone.0158856

Tsuchimoto, S., Ohtsubo, H., and Ohtsubo, E. (1988). Two genes, *pemK* and *pemI*, responsible for stable maintenance of resistance plasmid R100. *J. Bacteriol.* 170, 1461–1466. doi: 10.1128/jb.170.4.1461-1466.1988

Van Melderen, L., and Wood, T. K. (2017). Commentary: What is the link between stringent response, endoribonuclease encoding Type II Toxin-Antitoxin systems and persistence? *Front. Microbiol.* 8:191. doi: 10.3389/fmicb.2017.00191

Wilbaux, M., Mine, N., Guerout, A.-M., Mazel, D., and Van Melderen, L. (2007). Functional interactions between coexisting toxin-antitoxin systems of the *ccd* family in *Escherichia coli* O157:H7. *J. Bacteriol.* 189, 2712–2719. doi: 10.1128/jb.01679-06

Wozniak, R. A. F., and Waldor, M. K. (2009). A toxin–antitoxin system promotes the maintenance of an integrative conjugative element. *PLoS Genet.* 5:e1000439. doi: 10.1371/journal.pgen.1000439

Yamaichi, Y., Fogel, M. A., and Waldor, M. K. (2007). *par* genes and the pathology of chromosome loss in *Vibrio cholerae*. *Proc. Natl. Acad. Sci. U.S.A.* 104, 630–635. doi: 10.1073/pnas.0608341104

Yao, X., Chen, T., Shen, X., Zhao, Y., Wang, M., Rao, X., et al. (2015). The chromosomal SezAT toxin-antitoxin system promotes the maintenance of the SsPI-1 pathogenicity island in epidemic *Streptococcus suis*. *Mol. Microbiol.* 98, 243–257. doi: 10.1111/mmi.13116

Yarmolinsky, M. (1995). Programmed cell death in bacterial populations. *Science* 267, 836–838. doi: 10.1126/science.7846528

Yuan, J., Yamaichi, Y., and Waldor, M. K. (2011). The three *Vibrio cholerae* chromosome II-encoded ParE toxins degrade chromosome I following loss of chromosome II. *J. Bacteriol.* 193, 611–619. doi: 10.1128/jb.01185-10

11

Small, Enigmatic Plasmids of the Nosocomial Pathogen, *Acinetobacter baumannii*

*Soo Sum Lean[1] and Chew Chieng Yeo[2]**

[1] Saw Swee Hock School of Public Health, National University of Singapore, Singapore, Singapore, [2] Faculty of Medicine, Biomedical Research Centre, Universiti Sultan Zainal Abidin, Kuala Terengganu, Malaysia

Edited by:
Feng Gao,
Tianjin University, China

Reviewed by:
Christopher Morton Thomas,
University of Birmingham,
United Kingdom
Jordi Vila Estapé,
Hospital Clinic of Barcelona, Spain

***Correspondence:**
Chew Chieng Yeo
chewchieng@gmail.com

Acinetobacter baumannii is a Gram-negative nosocomial pathogen that has become a serious healthcare concern within a span of two decades due to its ability to rapidly acquire resistance to all classes of antimicrobial compounds. One of the key features of the *A. baumannii* genome is an open pan genome with a plethora of plasmids, transposons, integrons, and genomic islands, all of which play important roles in the evolution and success of this clinical pathogen, particularly in the acquisition of multidrug resistance determinants. An interesting genetic feature seen in majority of *A. baumannii* genomes analyzed is the presence of small plasmids that usually ranged from 2 to 10 kb in size, some of which harbor antibiotic resistance genes and homologs of plasmid mobilization genes. These plasmids are often overlooked when compared to their larger, conjugative counterparts that harbor multiple antibiotic resistance genes and transposable elements. In this mini-review, we will examine our current knowledge of these small *A. baumannii* plasmids and look into their genetic diversity and phylogenetic relationships. Some of these plasmids, such as the Rep-3 superfamily group and the pRAY-type, which has no recognizable replicase genes, are quite widespread among diverse *A. baumannii* clinical isolates worldwide, hinting at their usefulness to the lifestyle of this pathogen. Other small plasmids especially those from the Rep-1 superfamily are truly enigmatic, encoding only hypothetical proteins of unknown function, leading to the question of whether these small plasmids are "good" or "bad" to their host *A. baumannii*.

Keywords: *Acinetobacter baumannii*, small plasmids, antibiotic resistance genes, mobilizable plasmids, Rep-1 superfamily, Rep-3 superfamily, pRAY plasmids, toxin–antitoxin

INTRODUCTION

Acinetobacter baumannii is a Gram-negative nosocomial pathogen that has become a serious healthcare concern especially in the last two decades due to its rapid ability to acquire antimicrobial resistance leading to the development of pandrug resistant (PDR) isolates that are resistant to all classes of antimicrobial compounds (Magiorakos et al., 2012; Göttig et al., 2014; Lean et al., 2014). Advances in genome sequencing and their increasing affordability have led to the availability of a plethora of *A. baumannii* genomes in the public databases (Peleg et al., 2012; Liu et al., 2013; Lean et al., 2015, 2016; Wallace et al., 2016). One of the key features of the *A. baumannii* genome is an open pan genome with a wide variety of mobile genetic elements, particularly integrons and transposons in genomic islands, some of which are known as resistance islands due to the presence of multiple antibiotic resistance genes (Fournier et al., 2006; Bonnin et al., 2012; Ramírez et al., 2013). Resistance genes are also plasmid-borne and in *A. baumannii*, plasmids range from as small as 2 kb to more than 100 kb in size (Gallagher et al., 2015; Hamidian et al., 2016a,b). The large plasmids of *A. baumannii* are often the focus of analyses due mainly to the presence of multiple antibiotic resistance genes and the self-transmissible nature of these plasmids (Hamidian et al., 2014a,b, 2016a; Hamidian and Hall, 2014) although small plasmids have been highlighted especially those that harbor antibiotic resistance genes (D'Andrea et al., 2009; Merino et al., 2010; Grosso et al., 2012; Hamidian et al., 2012, 2016b). Despite the importance of plasmids in the potential transmission of resistance and virulence genes in *A. baumannii*, there has been surprisingly very little experimental work done on the basic biology of these plasmids. We know next-to-nothing with regards to the basic replicons of these plasmids, their replication mechanisms and transmissibility. The rapidly increasing volume of *Acinetobacter* plasmid sequences in the databases from numerous whole genome sequencing projects has led to often conflicting and chaotic annotations, complicating their *in silico* analyses, a fact that was recently highlighted for all plasmid sequences in an excellent review paper by Thomas et al. (2017). So far, *A. baumannii* plasmids have been classified according to their replicase (Rep) proteins with Bertini et al. (2010) showing that there are 19 homology groups (GR1–GR19) and developing a plasmid-based replicon typing scheme based on their *rep* genes. In this mini-review, we shall examine our current knowledge of the small plasmids of *A. baumannii* (for this purpose, we shall define "small" as any plasmid that is around 10 kb and less) and present their genetic diversity and phylogenetic relationships. We will also discuss the importance of these small plasmids to their host *A. baumannii*.

THE REP-3 SUPERFAMILY PLASMIDS

Majority of plasmids from *A. baumannii* encode replicase proteins belonging to the Rep-3 superfamily (identified by the pfam0151 conserved domain) with the larger plasmids usually harboring more than one replicon type (Bertini et al., 2010). In most of the Rep-3 superfamily replicons, the *rep* gene, which is usually annotated as *repB*, is preceded by three to six direct repeats (19–22 nucleotides in length and mainly located between 10 and 200 bp upstream of the *repB* start codon; majority are four direct repeats) that could be considered as the iterons for the RepB basic replicon (please see **Supplementary Table S1** and Data Sheet 1 for further details). In enterobacterial plasmids, these iterons serve as the origin of replication whereby the replication initiation protein binds and interacts with other host proteins (such as DnaA and the DnaBC helicase complex) required for replication initiation (Bertini et al., 2010; Konieczny et al., 2014). To the best of our knowledge, there has only been one experimental demonstration of the functionality of the *Acinetobacter* basic replicon. Dorsey et al. (2006) showed that the minimal replicon for the 9,540-bp plasmid pMAC from *A. baumannii* 19,606 was the *repB* gene [denoted as open reading frame-1 (ORF1)] and the four direct repeats that preceded the gene in experiments using the *Escherichia coli* cloning vector pCR-Blunt II-TOPO and *Acinetobacter calcoaceticus* BD413 as host.

Phylogenetic analysis using the RepB protein sequences of 50 of these Rep-3 superfamily plasmids (**Figure 1**) was largely in agreement with the plasmid homology groups proposed by Bertini et al. (2010). However, we are of the opinion that pABVA01 which was categorized under the GR2 group by Bertini et al. (2010) warrants a separate grouping along with similar plasmids such as pMMCU3 and pAbATCC329, which we designate GR20, as the phylogenetic tree clearly showed that this group of plasmids belonged to a separate clade (**Figure 1**).

Interestingly, in a majority of these small *A. baumannii* plasmids that belonged to the Rep-3 superfamily, the reading frame immediately downstream of the *repB* gene is highly conserved and is usually annotated as "*repA*" (**Supplementary Figure S1**). We could not find any homology to known replicase proteins for the translated "*repA*" gene and we are uncertain as to why this reading frame was designated *repA* in the absence of homology and/or experimental evidence. The translated protein contains a DNA-binding helix-turn-helix motif at its N-terminus and is usually annotated as a "conserved hypothetical protein" or a "DNA-binding protein" in the various database entries. The pMAC plasmid harbors this gene, which was designated ORF2, and which was shown by RT-PCR to be actively transcribed (Dorsey et al., 2006). Although for the pMAC plasmid, ORF2 was shown not to be part of the minimal replicon (Dorsey et al., 2006), its conservation in a vast majority of the small Rep-3 superfamily plasmids is suggestive of its importance. We have not found any evidence so far of the existence of any *Acinetobacter* plasmid that harbors only this "*repA*" reading frame without the *repB* gene. Nevertheless, a small number of *repB*-only plasmids do exist (such as p1ABAYE and the pABUH2a plasmids) and they form a distinct clade in the RepB phylogenetic tree (grouped under GR11; **Figure 1**) with their own unique iteron sequences (**Supplementary Table S1** and Data Sheet 1). Hence, in the absence of further experimental evidence, we could neither confirm nor completely rule out the involvement of this "*repA*" gene in the replication function of this group of plasmids. It is possible that some of these plasmids do require two replication genes, similar to IncQ plasmids such as RSF1010 which contained three replication genes with RepA functioning as the helicase, RepB as the primase, and RepC as the iteron-binding *oriV*-activator (Meyer, 2009).

Another key feature found in majority of the small Rep-3 superfamily plasmids is XerC/XerD recombination sites flanking various gene modules (**Supplementary Figure S1** and **Table S1**). Some of these gene modules include antibiotic resistance determinants such as *bla*$_{OXA-24}$/*bla*$_{OXA-40}$ (in pABVA01, pMMCU3, pAbATCC329, pABUH3a-8.2, and pABUH2a-5.6), *bla*$_{OXA-72}$ (in p2ABST25 and pAB-NCGM253), and the *tet(39)* tetracycline-resistance gene (in pRCH52-1). XerC and XerD recombinases usually function to convert plasmid and chromosomal dimers to monomers during cell division with

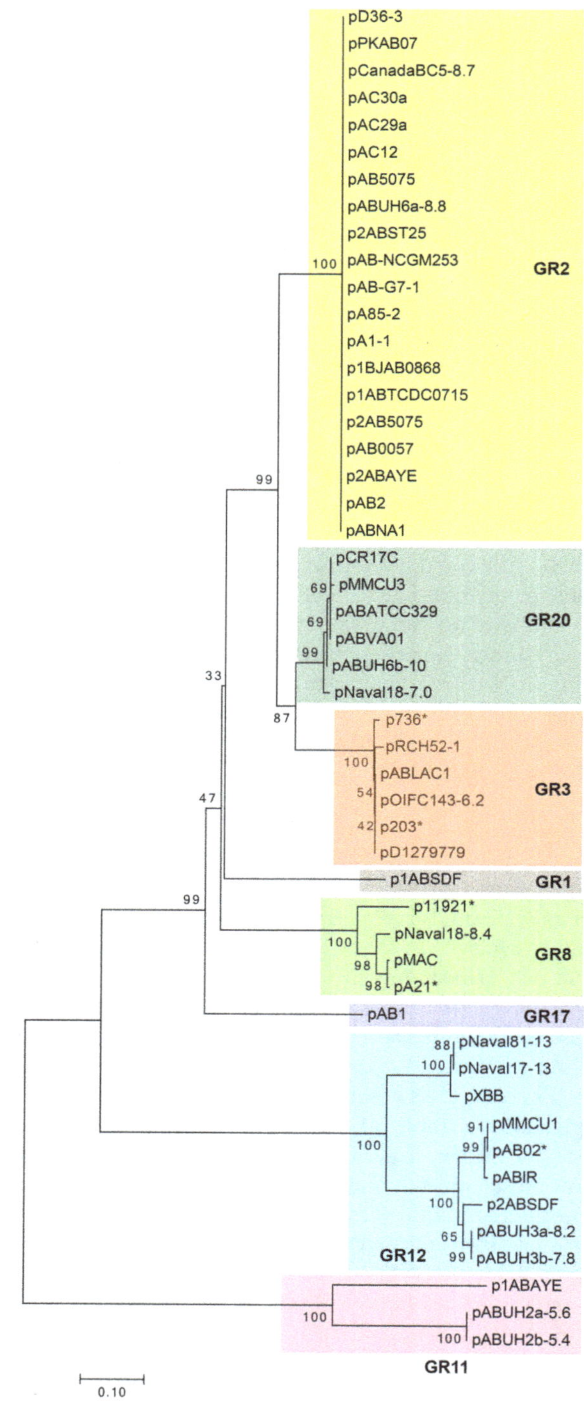

FIGURE 1 | Phylogenetic tree of the small *Acinetobacter* plasmids of the Rep-3 superfamily based on the RepB replicase protein sequences as analyzed and drawn using MEGA7 (Kumar et al., 2016). Alignment of the RepB protein sequences was carried out using MUSCLE (Edgar, 2004) and the evolutionary history was inferred using the Neighbor-Joining method. The optimal tree with the sum of branch length = 3.32974406 is shown. The percentage of replicate trees in which the associated taxa clustered together in the bootstrap test (1000 replicates) is shown next to the branches.

(Continued)

FIGURE 1 | Continued
The tree is drawn to scale, with branch lengths in the same units as those of the evolutionary distances used to infer the phylogenetic tree. The evolutionary distances were computed using the Poisson correction method and are in the units of the number of amino acid substitutions per site. The analysis involved 50 RepB amino acid sequences with the GenBank accession numbers of the plasmids as listed in **Supplementary Table S1**. Each clade of the tree corresponded with the plasmid homology grouping (GR classification) as proposed by Bertini et al. (2010) and indicated by different colored boxes. Plasmid names marked with an asterisk (*) indicate partial plasmid sequences that covered only the *oriV–repB* sequences and were included in the analysis to validate the plasmid groupings as they were used by Bertini et al. (2010) in their classification scheme.

each recombinase catalyzing the exchange of a specific pair of strands between the recombining sites via a Holliday Junction, which is an essential reaction intermediate (Midonet and Barre, 2014). These recombinases are also involved in the integration of phage CTX-Φ in the *Vibrio cholerae* genome (Val et al., 2005) and transposition of certain conjugative transposons (Bui et al., 2006; Midonet and Barre, 2014). The DNA sequence of these small plasmids strongly infer the involvement of the XerC/XerD recombination system in the mobilization of discrete DNA modules, including antibiotic resistance genes, in *A. baumannii* although experimental proof of this has yet to be demonstrated.

Type II toxin–antitoxin (TA) systems are also found in most of the Rep-3 superfamily group of small plasmids. Type II toxin–antitoxin systems are known to mediate the stable maintenance of plasmids which harbor them through the post-segregational killing of any plasmid-free daughter cells that developed, making it difficult for the host cells to lose these plasmids (Hayes, 2003). Their presence may partly explain the widespread prevalence of this group of plasmids among *A. baumannii*. The AbkB/AbkA TA system (also known as SplT/SplA) has been shown to be a functional TA system with the AbkB (or SplT) toxin as an endoribonuclease and translational inhibitor, and AbkA (or SplA) as its cognate antitoxin (Jurenaite et al., 2013; Mosqueda et al., 2014). Other TA pairs found in these plasmids in place of AbkB/AbkA include RelE/Cro-CI (in pMAC, pABLAC1, and pD36-3), *phd-yoeB* (in p1ABAYE), *dinJ-yafQ* (in pABUH2 plasmids), and *rnlA–rnlB* (found flanked by XerC/XerD sites in pNaval18-7.0) (**Supplementary Table S1**). The functionality of these putative TA systems has yet to be experimentally verified.

Some of these Rep-3 superfamily small plasmids also harbor putative virulence factors in the form of a TonB-dependent receptor, septicolysin (Lean et al., 2016) and Sel1-repeat protein. TonB-dependent receptors are known to play a role in iron acquisition (Zimbler et al., 2013) whereas septicolysins are thiol-activated cytolysins with cytolytic activity toward eukaryotic cells and have been implicated in the pathogenesis of bacteria such as *Clostridium perfringens*, *Listeria monocytogenes*, and *Streptococcus pneumoniae* (Billington et al., 2000). Sel1-repeat proteins have diverse biological roles, often as adaptor proteins for the assembly of macromolecular complexes (Mittl and Schneider-Brachert, 2007). Bacterial Sel1-repeat proteins mediate interactions between the pathogen and its eukaryotic host cells and have been described in

Helicobacter pylori, *Legionella pneumophila*, and *Pseudomonas aeruginosa* as important virulence factors, as reviewed in Mittl and Schneider-Brachert (2007). In *Neisseria meningitidis*, a Sel1-repeat protein, NMB0419, was shown to be involved in meningococcal interactions with epithelial cells (Li et al., 2003) and in a recent paper, it was intriguingly shown that the expression of NMB0419 led to transcriptional changes in genes involved in iron uptake, energy metabolism, and virulence functions in a manner counteracting the global regulator, Fur (Li et al., 2017). It would therefore be of interest to experimentally investigate if these genes encoded by some of the small plasmids of the Rep-3 superfamily truly function as virulence factors for *A. baumannii*, thereby contributing to the pathogenicity of the bacterium.

Some of these small Rep-3 superfamily plasmids also encode orthologs of the MobL or MobA mobilization proteins identified by the pfam03389 conserved domain found in the MobA/MobL protein family. Plasmids that encode genes for these proteins are mobilizable by other self-transmissible plasmids. Nevertheless, the only experimental evidence for the mobilization potential of these plasmids was for pMAC of *A. baumannii* 19606 with the experiment carried out using the cloned *mobA/mobL* gene in an *E. coli* DH5α host and an *E. coli* HB101 recipient (Dorsey et al., 2006). Until now, the mobilization potential of this group of plasmids from an *Acinetobacter* donor to an *Acinetobacter* recipient has yet to be shown.

THE REP-1 SUPERFAMILY

There is a group of small cryptic plasmids from *A. baumannii* that usually comprise of a single *rep* gene and between two and five hypothetical genes. The *rep* gene of this group of plasmids encodes a replicase of the Rep-1 superfamily. Phylogenetic analysis of the Rep proteins from this group of plasmids showed that they could be divided into two subgroups: the p4ABAYE subgroup and the Rep63 subgroup (**Supplementary Figure S2**). The 2,726 bp p4ABAYE from *A. baumannii* AYE encodes a *rep* gene and four hypothetical ORFs (Fournier et al., 2006) and was categorized under the GR14 group of *Acinetobacter* plasmids (Bertini et al., 2010). The second subgroup contained two of the smallest reported *Acinetobacter* plasmids, the 1,967 bp p3AB5075 from *A. baumannii* AB5075 (Gallagher et al., 2015) and the 1,958 bp pM131-10 plasmid from *Acinetobacter* sp. M131 (accession no. JX101639). The small size of p3AB5075 has been validated by plasmid extraction and agarose gel electrophoresis (Gallagher et al., 2015), and the plasmid consisted of the *rep* gene and two other reading frames of unknown function. Although Gallagher et al. (2015) stated that the *rep* gene of p3AB5075 was of undefined plasmid replication group, our phylogenetic analysis indicated that it is grouped with an unpublished 2,343 bp *A. baumannii* plasmid pAB49 (accession no. L77992.1) (**Supplementary Figure S2**), which was previously categorized by Bertini et al. (2010) under the GR16 group. Furthermore, the *rep*-encoded protein of pAB49 had been previously shown to be homologous to the Rep63 replication initiation protein encoded by pBL63.1 of *Bacillus licheniformis* and orthologs in rolling-circle replicating (RCR) plasmids from various other bacterial species (Guglielmetti et al., 2005).

PLASMID pRAY AND ITS DERIVATIVES

The 6,076 bp plasmid pRAY was first isolated from a South African clinical *Acinetobacter* strain designated SUN, which is of unknown clonal origin, through its carriage of the *aadB* gene which conferred resistance to the aminoglycosides gentamicin, kanamycin, and tobramycin (Segal and Elisha, 1999). The *aadB* gene is usually associated with class I integrons (Recchia and Hall, 1995), but in *Acinetobacter* sp. SUN and subsequently, in other *Acinetobacter* spp. isolated worldwide, *aadB* is found in pRAY and its closely related derivatives (Segal and Elisha, 1999; Adams et al., 2010; Nigro et al., 2011; Hamidian et al., 2012; Gifford et al., 2014; Ou et al., 2015; Kurakov et al., 2016). The *aadB* gene is likely acquired as its G+C content of 58% is higher than the G+C content of 37% for the rest of pRAY (Segal and Elisha, 1999) and the presence of an *attC* site immediately downstream of *aadB* is indicative of its gene cassette origin (Nigro et al., 2011).

A total of 10 ORFs, including *aadB*, was identified from the pRAY sequence, with two ORFs (designated ORF3 and ORF6) encoding proteins that were homologous to mobilization proteins (Segal and Elisha, 1999). A putative origin of transfer (*oriT*) was also identified upstream of ORF3 (Segal and Elisha, 1999), inferring the potential transmissibility of pRAY.

Derivatives of pRAY have been characterized from Australian *A. baumannii* clinical strains with a plasmid designated pRAY* isolated from strain D36 and pRAY*-v1 from strain C2 (Hamidian et al., 2012). The *mobA* gene from pRAY* is larger than ORF3 of pRAY but is still categorized within the ColE1 superfamily of MobA proteins (MOB_{HEN} family) with the putative *oriT* located upstream of *mobC* (Hamidian et al., 2012). Plasmid pRAY*-v1 differed from pRAY* by 66 single nucleotide differences, 65 of which were within the *mobC–mobA* region leading only to amino acid substitutions of MobC and MobA but without any frameshifts. *A. baumannii* E7 harbored pRAY*-v2 which was 2.5 kb larger than pRAY and sequence analysis indicated complete identity with pRAY* but with the insertion of two IS elements, an IS*18*-like element which is found within IS*Aba22* and located upstream of the *aadB* gene (Hamidian et al., 2012) (**Figure 2**). A single nucleotide variant of pRAY*-v1, designated pRAY*-v3, was isolated from a clinical strain of *A. nosocomialis* from Melbourne, Australia (Gifford et al., 2014).

Analysis of a 4,135 bp plasmid designated pALWED1.8 harbored in *A. lwoffii* isolated from the permafrost in Russia indicated conservation of the *oriT–mobC–mobA* region with pRAY and its derivatives (**Figure 2**) (Kurakov et al., 2016). The pALWED1.8 plasmid contained an *aadA27* gene downstream of *mobA* that conferred resistance to streptomycin/spectinomycin but without an *attC* site that was observed for *aadB* in pRAY and its variants (Nigro et al., 2011). Interestingly, the *oriT–mobC–mobA* backbone was identified from the genome sequence of various *Acinetobacter* species with various genes found in the

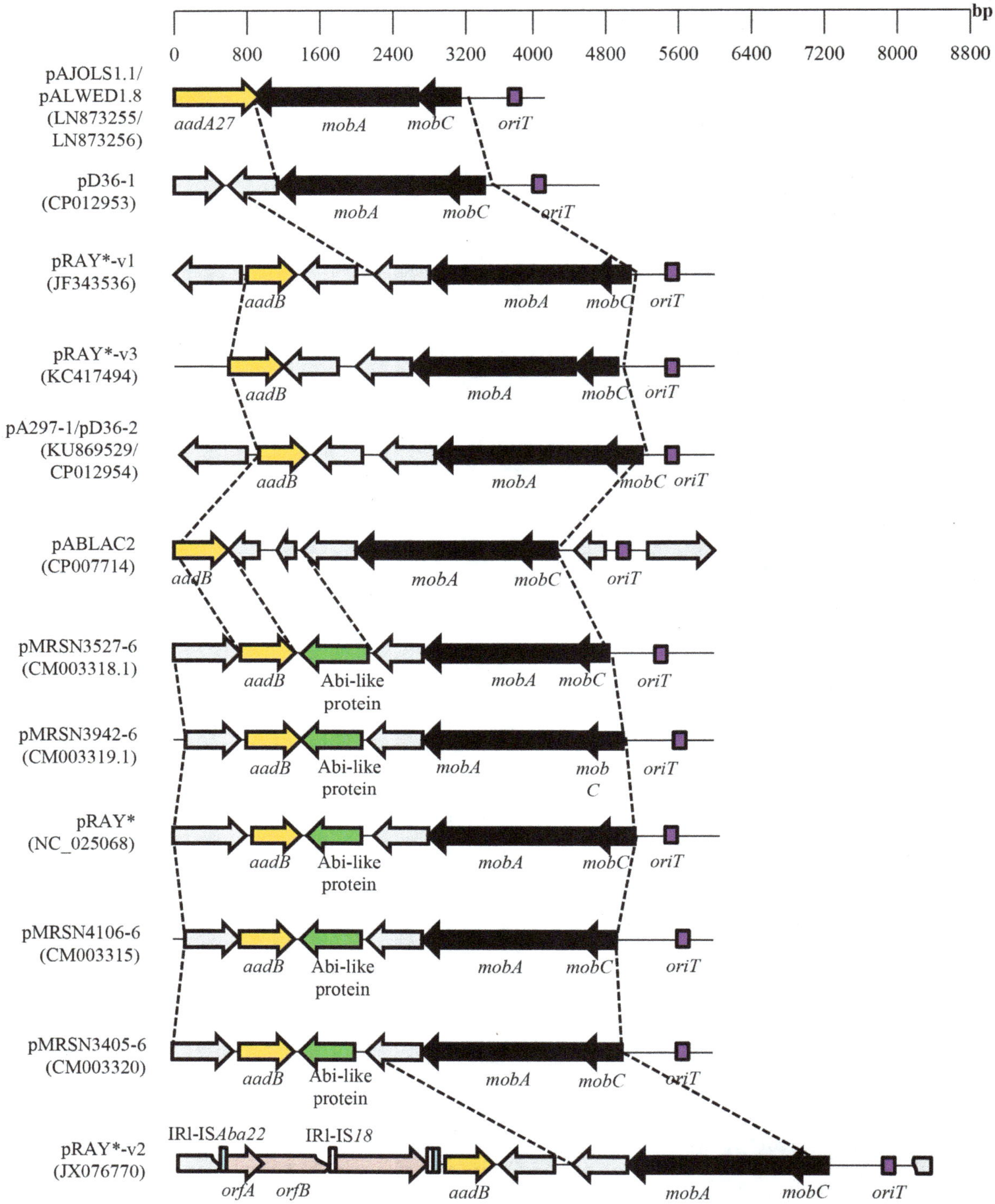

FIGURE 2 | Comparative map of the pRAY plasmid and its derivatives from various *Acinetobacter* spp. The *mobA* and *mobC* genes are indicated as black arrows whereas the AT-rich putative *oriT* sequence is indicated as a purple box. Antimicrobial resistance genes (either *aadA27* for streptomycin/spectinomycin resistance or *aadB* for aminoglycoside resistance) are depicted as a dark yellow arrow while the Abi-like protein (identified by pfam07751) gene is indicated as a green arrow. IS element-encoded transposases are depicted in pink, the inverted repeats (IRs) for IS*Aba22* are shown as blue rectangles whereas the IRs for IS*18* are shown as white rectangles (for pRAY*-v2). Hypothetical open reading frames are depicted as gray arrows. Accession numbers for the plasmids shown are in parentheses following their names.

accessory regions of these plasmids such as an alkyl sulfatase gene (involved in the degradation of surface-active substances such as sodium dodecylsulfate, or SDS) in the plasmid from *A. radioresistens* SK82 (Kurakov et al., 2016). Thus, members of this group of plasmids, including pRAY and pALWED1.8, might have originated from a common ancestor and independently acquired different genes into the accessory region of the plasmid. The mobilization of pALWED1.8 was demonstrated in conjugation experiments between *A. lwoffii* strain ED23-35 which contained pALWED1.8 and a large conjugative plasmid pKLH208 (Kholodii et al., 2004) and *A. baylyi* BD413rif as the recipient.

Intriguingly, until now, no potential replication initiation protein could be identified for pRAY and its derivatives based on sequence homology (Hamidian et al., 2012; Kurakov et al., 2016). Nevertheless, a potential origin of replication was identified for pRAY upstream of *aadB* where eight copies of an AT-rich repeat sequence, AAAAAATAT, were found (Segal and Elisha, 1999). The replication of these plasmids may mirror that of plasmids such as ColE1 which do not encode a *rep* gene since their replicon only consists of an *oriV* with the host RNA polymerase transcriptional machinery taking care of the melting of duplex DNA and synthesis of pre-primer RNA for replication initiation (Brantl, 2014; Thomas et al., 2017). Efforts to transform pRAY into *E. coli* were not successful, implying that pRAY and its derivatives might be specific for *Acinetobacter* (Segal and Elisha, 1999).

CONCLUDING REMARKS

This mini-review has highlighted the small plasmids of *A. baumannii*, whether cryptic, resistance-related, or even mobilizable plasmids, and inferred the likely importance of these plasmids to their host. The potential of these small plasmids in transferring antibiotic resistance and possibly, even virulence genes, among *Acinetobacter* species should not be overlooked as their promiscuity could be comparable to that of larger plasmids and thus, would have a significant impact on the evolution of *A. baumannii*. The dearth of experimental studies with regards to these small *Acinetobacter* plasmids, given the importance of *A. baumannii* in the World Health Organization list of priority pathogens (World Health Organization, 2017), is indeed surprising and needs to be addressed. The PCR-based replicon typing (PBRT) scheme developed by Bertini et al. (2010) would probably need updating in view of an ever increasing amount of *A. baumannii* plasmid sequence data although their Rep-based classification scheme into different GR groupings appeared to be still valid with respect to the small plasmids. Nevertheless, plasmids of the pRAY-type would require another classification scheme due to the lack of a replicase protein. Other plasmid typing schemes such as plasmid multi-locus sequence typing (pMLST) and MOB classification based on plasmid mobility genes (Francia et al., 2004; Garcillán-Barcia et al., 2011) would be difficult to apply for these small *Acinetobacter* plasmids due to their lack of loci used in these typing schemes. There is clearly a need for us to accurately identify individual plasmids especially in this era of big data and whole genome sequencing (Orlek et al., 2017; Thomas et al., 2017), tracking the movement of plasmids and understanding their dynamic evolution, and small plasmids should not escape from our consideration simply because of their size.

AUTHOR CONTRIBUTIONS

SSL and CCY conceived, analyzed the data, wrote, edited, and approved this manuscript.

ACKNOWLEDGMENTS

Our thanks and appreciation to T. Venkova for her help in checking the iteron sequences and for her suggestions. Our gratitude also to A. Brahmavamso for inspiring us with the title of this manuscript.

FIGURE S1 | Comparative map of the small *Acinetobacter* plasmids of the Rep-3 superfamily. The *repB* replicase gene is indicated as a dark blue filled arrow, the putative *repA* gene is depicted in light blue. Hypothetical open reading frames are shown as unfilled arrows whereas black arrows are for the *mobA/mobL* mobilization genes. Red crosses indicate the XerC/XerD recombination sites. Filled blue twin-triangles depict the iterons that make up the putative origin of replication, *oriV*. Accession numbers and further details of the plasmids are as in **Supplementary Table S1** with detailed teron sequences and locations on the respective plasmids in Supplementary Data Sheet 1.

FIGURE S2 | Phylogenetic tree of the small *Acinetobacter* plasmids of the Rep-1 superfamily based on the Rep protein sequences, analyzed and drawn using MEGA7 (Kumar et al., 2016). Protein sequences were aligned using MUSCLE (Edgar, 2004), evolutionary history was inferred using the Neighbor-Joining method and the optimal tree (with the sum of branch length = 3.01508040) is shown. The percentage of replicate trees in which the associated taxa clustered together in the bootstrap test (1000 replicates) is shown next to the branches. The tree is drawn to scale, with branch lengths in the same units as those of the evolutionary distances used to infer the phylogenetic tree. The plasmids were grouped according to the GR classification scheme proposed by Bertini et al. (2010) and indicated here as GR14 and GR16 in different colored boxes. Accession numbers for the plasmids used in the analysis are as follows: p3AB5075 (NZ_CP008709.1), pBL63.1 (NC_006959.1), pM131-10 (NC_025169.1), pAB49 (L77992.1), pMRSN7339-2.3 (NZ_CM003313.1), p4ABAYE (NC_010403.1), pMRSN58-2.7 (NZ_CM003316.1), pA85-1 (NC_025107.1), and pTS236 (NC_016977.1). Note that pBL63.1 was isolated from *Bacillus lichineformis* and was included in the analysis based on the findings of Guglielmetti et al. (2005).

TABLE S1 | Features of the Rep-3 superfamily group of *Acinetobacter* plasmids.

REFERENCES

Adams, M. D., Chan, E. R., Molyneaux, N. D., and Bonomo, R. A. (2010). Genomewide analysis of divergence of antibiotic resistance determinants in closely related isolates of *Acinetobacter baumannii*. *Antimicrob. Agents Chemother.* 54, 3569–3577. doi: 10.1128/AAC.00057-10

Bertini, A., Poirel, L., Mugnier, P. D., Villa, L., Nordmann, P., and Carattoli, A. (2010). Characterization and PCR-based replicon typing of resistance plasmids in *Acinetobacter baumannii*. *Antimicrob. Agents Chemother.* 54, 4168–4177. doi: 10.1128/AAC.00542-10

Billington, S. J., Jost, B. H., and Songer, J. G. (2000). Thiol-activated cytolysins: structure, function and role in pathogenesis. *FEMS Microbiol. Lett.* 182, 197–205. doi: 10.1111/j.1574-6968.2000.tb08895.x

Bonnin, R. A., Poirel, L., and Nordmann, P. (2012). AbaR-type transposon structures in *Acinetobacter baumannii*. *J. Antimicrob. Chemother.* 67, 234–236. doi: 10.1093/jac/dkr413

Brantl, S. (2014). Plasmid replication control by antisense RNAs. *Microbiol. Spectr.* 2:PLAS-0001-2013. doi: 10.1128/microbiolspec.PLAS-0001-2013

Bui, D., Ramiscal, J., Trigueros, S., Newmark, J. S., Do, A., Sherratt, D. J., et al. (2006). Differences in resolution of *mwr*-containing plasmid dimers mediated by the *Klebsiella pneumoniae* and *Escherichia coli* XerC recombinases: potential implications in dissemination of antibiotic resistance genes. *J. Bacteriol.* 188, 2812–2820. doi: 10.1128/JB.188.8.2812-2820.2006

D'Andrea, M. M., Giani, T., D'Arezzo, S., Capone, A., Petrosillo, N., Visca, P., et al. (2009). Characterization of pABVA01, a plasmid encoding the OXA-24 carbapenemase from Italian isolates of *Acinetobacter baumannii*. *Antimicrob. Agents Chemother.* 53, 3528–3533. doi: 10.1128/AAC.00178-09

Dorsey, C. W., Tomaras, A. P., and Actis, L. A. (2006). Sequence and organization of pMAC, an *Acinetobacter baumannii* plasmid harboring genes involved in organic peroxide resistance. *Plasmid* 56, 112–123. doi: 10.1016/j.plasmid.2006.01.004

Edgar, R. C. (2004). MUSCLE: a multiple sequence alignment method with reduced time and space complexity. *BMC Bioinformatics* 5:113. doi: 10.1186/1471-2105-5-113

Fournier, P.-E., Vallenet, D., Barbe, V., Audic, S., Ogata, H., Poirel, L., et al. (2006). Comparative genomics of multidrug resistance in *Acinetobacter baumannii*. *PLoS Genet.* 2:e7. doi: 10.1371/journal.pgen.0020007

Francia, M. V., Varsaki, A., Garcillán-Barcia, M. P., Latorre, A., Drainas, C., and De La Cruz, F. (2004). A classification scheme for mobilization regions of bacterial plasmids. *FEMS Microbiol. Rev.* 28, 79–100. doi: 10.1016/j.femsre.2003.09.001

Gallagher, L. A., Ramage, E., Weiss, E. J., Radey, M., Hayden, H. S., Held, K. G., et al. (2015). Resources for genetic and genomic analysis of emerging pathogen *Acinetobacter baumannii*. *J. Bacteriol.* 197, 2027–2035. doi: 10.1128/JB.00131-15

Garcillán-Barcia, M. P., Alvarado, A., and de la Cruz, F. (2011). Identification of bacterial plasmids based on mobility and plasmid population biology. *FEMS Microbiol. Rev.* 35, 936–956. doi: 10.1111/j.1574-6976.2011.00291.x

Gifford, B., Tucci, J., McIlroy, S. J., and Petrovski, S. (2014). Isolation and characterization of two plasmids in a clinical *Acinetobacter nosocomialis* strain. *BMC Res. Notes* 7:732. doi: 10.1186/1756-0500-7-732

Göttig, S., Gruber, T. M., Higgins, P. G., Wachsmuth, M., Seifert, H., and Kempf, V. A. J. (2014). Detection of pan drug-resistant *Acinetobacter baumannii* in Germany. *J. Antimicrob. Chemother.* 69, 2578–2579. doi: 10.1093/jac/dku170

Grosso, F., Quinteira, S., Poirel, L., Novais, Â., and Peixe, L. (2012). Role of common $bla_{OXA-24/OXA-40}$-carrying platforms, and plasmids in the spread of OXA-24/OXA-40 among *Acinetobacter* species clinical isolates. *Antimicrob. Agents Chemother.* 56, 3969–3972. doi: 10.1128/AAC.06255-11

Guglielmetti, S., Mora, D., Manachini, P. L., and Parini, C. (2005). Genetic relationship among *Bacillus licheniformis* rolling-circle-replicating plasmids and complete nucleotide sequence of pBL63.1, an atypical replicon. *Plasmid* 54, 93–103. doi: 10.1016/j.plasmid.2005.01.002

Hamidian, M., Ambrose, S. J., and Hall, R. M. (2016a). A large conjugative *Acinetobacter baumannii* plasmid carrying the *sul2* sulphonamide and *strAB* streptomycin resistance genes. *Plasmid* 8, 43–50. doi: 10.1016/j.plasmid.2016.09.001

Hamidian, M., and Hall, R. M. (2014). *pACICU2* is a conjugative plasmid of *Acinetobacter* carrying the aminoglycoside resistance transposon *TnaphA6*. *J. Antimicrob. Chemother.* 69, 1146–1148. doi: 10.1093/jac/dkt488

Hamidian, M., Holt, K. E., Pickard, D., Dougan, G., and Hall, R. M. (2014a). A GC1 *Acinetobacter baumannii* isolate carrying AbaR3 and the aminoglycoside resistance transposon Tn*aphA6* in a conjugative plasmid. *J. Antimicrob. Chemother.* 69, 955–958. doi: 10.1093/jac/dkt454

Hamidian, M., Holt, K. E., Pickard, D., and Hall, R. M. (2016b). A small *Acinetobacter* plasmid carrying the *tet39* tetracycline resistance determinant. *J. Antimicrob. Chemother.* 71, 269–271. doi: 10.1093/jac/dkv293

Hamidian, M., Kenyon, J. J., Holt, K. E., Pickard, D., and Hall, R. M. (2014b). A conjugative plasmid carrying the carbapenem resistance gene bla_{OXA-23} in AbaR4 in an extensively resistant GC1 *Acinetobacter baumannii* isolate. *J. Antimicrob. Chemother.* 69, 2625–2628. doi: 10.1093/jac/dku188

Hamidian, M., Nigro, S. J., and Hall, R. M. (2012). Variants of the gentamicin and tobramycin resistance plasmid pRAY are widely distributed in *Acinetobacter*. *J. Antimicrob. Chemother.* 67, 2833–2836. doi: 10.1093/jac/dks318

Hayes, F. (2003). Toxins-antitoxins: plasmid maintenance, programmed cell death, and cell cycle arrest. *Science* 301, 1496–1499. doi: 10.1126/science.1088157

Jurenaite, M., Markuckas, A., and Suziedeliene, E. (2013). Identification and characterization of type II toxin-antitoxin systems in the opportunistic pathogen *Acinetobacter baumannii*. *J. Bacteriol.* 195, 3165–3172. doi: 10.1128/JB.00237-13

Kholodii, G., Mindlin, S., Gorlenko, Z., Petrova, M., Hobman, J., and Nikiforov, V. (2004). Translocation of transposition-deficient (TndPKLH2-like) transposons in the natural environment: mechanistic insights from the study of adjacent DNA sequences. *Microbiology* 150(Pt 4), 979–992. doi: 10.1099/mic.0.26844-0

Konieczny, I., Bury, K., Wawrzycka, A., and Wegrzyn, K. (2014). Iteron plasmids. *Microbiol. Spectr.* 2:PLAS-0026-2014. doi: 10.1128/microbiolspec.PLAS-0026-2014

Kumar, S., Stecher, G., and Tamura, K. (2016). MEGA7: molecular evolutionary genetics analysis version 7.0 for bigger datasets. *Mol. Biol. Evol.* 33, 1870–1874. doi: 10.1093/molbev/msw054

Kurakov, A., Mindlin, S., Beletsky, A., Shcherbatova, N., Rakitin, A., Ermakova, A., et al. (2016). The ancient small mobilizable plasmid pALWED1.8 harboring a new variant of the non-cassette streptomycin/spectinomycin resistance gene *aadA27*. *Plasmid* 84, 36–43. doi: 10.1016/j.plasmid.2016.02.005

Lean, S. S., Suhaili, Z., Ismail, S., Rahman, N. I. A., Othman, N., Abdullah, F. H., et al. (2014). Prevalence and genetic characterization of carbapenem- and polymyxin-resistant *Acinetobacter baumannii* isolated from a tertiary hospital in Terengganu, Malaysia. *ISRN Microbiol.* 2014:953417. doi: 10.1155/2014/953417

Lean, S. S., Yeo, C. C., Suhaili, Z., and Thong, K.-L. (2015). Whole-genome analysis of an extensively drug-resistant clinical isolate of *Acinetobacter baumannii* AC12: insights into the mechanisms of resistance of an ST195 clone from Malaysia. *Int. J. Antimicrob. Agents* 45, 178–182. doi: 10.1016/j.ijantimicag.2014.10.015

Lean, S. S., Yeo, C. C., Suhaili, Z., and Thong, K. L. (2016). Comparative genomics of two ST 195 carbapenem-resistant *Acinetobacter baumannii* with different susceptibility to polymyxin revealed underlying resistance mechanism. *Front. Microbiol.* 6:1445. doi: 10.3389/fmicb.2015.01445

Li, M. S., Farrant, J. L., Langford, P. R., and Kroll, J. S. (2003). Identification and characterization of genomic loci unique to the Brazilian purpuric fever clonal group of *H. influenzae* biogroup aegyptius: functionality explored using meningococcal homology. *Mol. Microbiol.* 47, 1101–1111. doi: 10.1046/j.1365-2958.2003.03359.x

Li, M.-S., Langford, P. R., and Kroll, J. S. (2017). Inactivation of NMB0419 encoding a Sel1-like repeat (SLR) protein in *Neisseria meningitidis* is associated with differential expression of genes belonging to the Fur regulon and reduced intra-epithelial replication. *Infect. Immun.* 85:e00574-16. doi: 10.1128/IAI.00574-16

Liu, C.-C., Tang, C. Y., Kuo, H.-Y., Lu, C.-W., Chang, K.-C., and Liou, M.-L. (2013). The origin of *Acinetobacter baumannii* TYTH-1: a comparative genomics study. *Int. J. Antimicrob. Agents* 41, 318–324. doi: 10.1016/j.ijantimicag.2012.12.010

Magiorakos, A., Srinivasan, A., Carey, R., Carmeli, Y., Falagas, M., Giske, C., et al. (2012). Multidrug-resistant, extensively drug-resistant and pandrug-resistant bacteria: an international expert proposal for interim standard definitions for acquired resistance. *Clin. Microbiol. Infect.* 18, 268–281. doi: 10.1111/j.1469-0691.2011.03570.x

Merino, M., Acosta, J., Poza, M., Sanz, F., Beceiro, A., Chaves, F., et al. (2010). OXA-24 carbapenemase gene flanked by XerC/XerD-like recombination sites in different plasmids from different *Acinetobacter* species isolated during a nosocomial outbreak. *Antimicrob. Agents Chemother.* 54, 2724–2727. doi: 10.1128/AAC.01674-09

Meyer, R. (2009). Replication and conjugative mobilization of broad host-range IncQ plasmids. *Plasmid* 62, 57–70. doi: 10.1016/j.plasmid.2009.05.001

Midonet, C., and Barre, F.-X. (2014). Xer site-specific recombination: promoting vertical and horizontal transmission of genetic information. *Microbiol. Spectr.* 2:MDNA3-0056-2014. doi: 10.1128/microbiolspec.MDNA3-0056-2014

Mittl, P. R. E., and Schneider-Brachert, W. (2007). Sel1-like repeat proteins in signal transduction. *Cell. Signal.* 19, 20–31. doi: 10.1016/j.cellsig.2006.05.034

Mosqueda, N., Gato, E., Roca, I., López, M., de Alegría, C. R., Fernández-Cuenca, F., et al. (2014). Characterization of plasmids carrying the $bla_{\mathrm{OXA-24/40}}$ carbapenemase gene and the genes encoding the AbkA/AbkB proteins of a toxin/antitoxin system. *J. Antimicrob. Chemother.* 69, 2629–2633. doi: 10.1093/jac/dku179

Nigro, S. J., Post, V., and Hall, R. M. (2011). Aminoglycoside resistance in multiply antibiotic-resistant *Acinetobacter baumannii* belonging to global clone 2 from Australian hospitals. *J. Antimicrob. Chemother.* 66, 1504–1509. doi: 10.1093/jac/dkr163

Orlek, A., Stoesser, N., Anjum, M. F., Doumith, M., Ellington, M. J., Peto, T., et al. (2017). Plasmid classification in an era of whole-genome sequencing: application in studies of antibiotic resistance epidemiology. *Front. Microbiol.* 8:182. doi: 10.3389/fmicb.2017.00182

Ou, H.-Y., Kuang, S. N., He, X., Molgora, B. M., Ewing, P. J., Deng, Z., et al. (2015). Complete genome sequence of hypervirulent and outbreak-associated *Acinetobacter baumannii* strain LAC-4: epidemiology, resistance genetic determinants and potential virulence factors. *Sci. Rep.* 5:8643. doi: 10.1038/srep08643

Peleg, A. Y., de Breij, A., Adams, M. D., Cerqueira, G. M., Mocali, S., Galardini, M., et al. (2012). The success of *Acinetobacter* species; genetic, metabolic and virulence attributes. *PLoS ONE* 7:e46984. doi: 10.1371/journal.pone.0046984

Ramírez, M. S., Vilacoba, E., Stietz, M. S., Merkier, A. K., Jeric, P., Limansky, A. S., et al. (2013). Spreading of AbaR-type genomic islands in multidrug resistance *Acinetobacter baumannii* strains belonging to different clonal complexes. *Curr. Microbiol.* 67, 9–14. doi: 10.1007/s00284-013-0326-5

Recchia, G. D., and Hall, R. M. (1995). Gene cassettes: a new class of mobile element. *Microbiology* 141, 3015–3027. doi: 10.1099/13500872-141-12-3015

Segal, H., and Elisha, B. G. (1999). Characterization of the *Acinetobacter* plasmid, pRAY, and the identification of regulatory sequences upstream of an *aadB* gene cassette on this plasmid. *Plasmid* 42, 60–66.

Thomas, C. M., Thomson, N. R., Cerdeño-Tárraga, A. M., Brown, C. J., Top, E. M., and Frost, L. S. (2017). Annotation of plasmid genes. *Plasmid* 91, 61–67. doi: 10.1016/j.plasmid.2017.03.006

Val, M.-E., Bouvier, M., Campos, J., Sherratt, D., Cornet, F., Mazel, D., et al. (2005). The single-stranded genome of phage CTX is the form used for integration into the genome of *Vibrio cholerae*. *Mol. Cell* 19, 559–566. doi: 10.1016/j.molcel.2005.07.002

Wallace, L., Daugherty, S. C., Nagaraj, S., Johnson, J. K., Harris, A. D., and Rasko, D. A. (2016). Use of comparative genomics to characterize the diversity of *Acinetobacter baumannii* surveillance isolates in a health care institution. *Antimicrob. Agents Chemother.* 60, 5933–5941. doi: 10.1128/AAC.00477-16

World Health Organization (2017). *WHO Publishes List of Bacteria for Which New Antibiotics are Urgently Needed.* Geneva: World Health Organization. Available at: http://www.who.int/mediacentre/news/releases/2017/bacteria-antibiotics-needed/en/

Zimbler, D. L., Arivett, B. A., Beckett, A. C., Menke, S. M., and Actis, L. A. (2013). Functional features of TonB energy transduction systems of *Acinetobacter baumannii*. *Infect. Immun.* 81, 3382–3394. doi: 10.1128/IAI.00540-13

12

Characterization of the Sorbitol Utilization Cluster of the Probiotic *Pediococcus parvulus* 2.6: Genetic, Functional and Complementation Studies in Heterologous Hosts

Adrian Pérez-Ramos[1], Maria L. Werning[1,2], Alicia Prieto[1], Pasquale Russo[3], Giuseppe Spano[3], Mari L. Mohedano[1] and Paloma López[1]*

[1] Biological Research Center (CIB), Consejo Superior de Investigaciones Científicas, Madrid, Spain, [2] Center of Research and Transfer of Catamarca (CITCA), Consejo Nacional de Investigaciones Científicas y Técnicas, Catamarca, Argentina, [3] Department of Agricultural, Food and Environmental Sciences, University of Foggia, Foggia, Italy

Edited by:
Tatiana Venkova, Fox Chase Cancer Center, United States

Reviewed by:
Maria Jesus Yebra, Instituto de Agroquímica y Tecnología de Alimentos (CSIC), Spain
Bopda Waffo Alain, Alabama State University, United States
Antonius Suwanto, Bogor Agricultural University, Indonesia

***Correspondence:**
Paloma López
plg@cib.csic.es

Pediococcus parvulus 2.6 secretes a 2-substituted (1,3)-β-D-glucan with prebiotic and immunomodulatory properties. It is synthesized by the GTF glycosyltransferase using UDP-glucose as substrate. Analysis of the *P. parvulus* 2.6 draft genome revealed the existence of a sorbitol utilization cluster of six genes (*gutFRMCBA*), whose products should be involved in sorbitol utilization and could generate substrates for UDP-glucose synthesis. Southern blot hybridization analysis showed that the cluster is located in a plasmid. Analysis of metabolic fluxes and production of the exopolysaccharide revealed that: (i) *P. parvulus* 2.6 is able to metabolize sorbitol, (ii) sorbitol utilization is repressed in the presence of glucose and (iii) sorbitol supports the synthesis of 2-substituted (1,3)-β-D-glucan. The sorbitol cluster encodes two putative regulators, GutR and GutM, in addition to a phosphoenolpyruvate-dependent phosphotransferase transport system and sorbitol-6-phosphate dehydrogenase. Therefore, we investigated the involvement of GutR and GutM in the expression of *gutFRMCBA*. The promoter-probe vector pRCR based on the *mrfp* gene, which encodes the fluorescence protein mCherry, was used to test the potential promoter of the cluster (P_{gut}) and the genes encoding the regulators. This was performed by transferring by electrotransformation the recombinant plasmids into two hosts, which metabolize sorbitol: *Lactobacillus plantarum* and *Lactobacillus casei*. Upon growth in the presence of sorbitol, but not of glucose, only the presence of P_{gut} was required to support expression of *mrfp* in *L. plantarum*. In *L. casei* the presence of sorbitol in the growth medium and the pediococcal *gutR* or *gutR* plus *gutM* in the genome was required *for* P_{gut} functionality. This demonstrates that: (i) P_{gut} is required for expression of the *gut* cluster, (ii) P_{gut} is subjected to catabolic repression in lactobacilli, (iii) GutR is an activator, and (iv) in the presence of sorbitol, *trans*-complementation for activation of P_{gut} exists in *L. plantarum* but not in *L. casei*.

Keywords: *Pediococcus parvulus*, exopolysaccharides, β-glucans, sorbitol, lactic acid bacteria, probiotic

INTRODUCTION

Sorbitol, also named D-glucitol, is a six-carbon sugar polyol widespread in plants, particularly in fruits, such as berries, cherries, plums, pears and apples. However, sorbitol is obtained industrially, by catalytic hydrogenation of glucose or glucose/fructose mixtures. This polyol has a relative sweetness of about 60% compared to that of sucrose, high-water solubility and is largely used as a low calorie sweetener, humectant, texturizer and softener (Zumbé et al., 2001). In addition, sorbitol is used in the production of pharmaceutical compounds, such as sorbose and ascorbic acid, and as a vehicle for drug-suspension (Silveira and Jonas, 2002). Sorbitol has also a potential prebiotic effect *in vivo*, since it does not contribute to the formation of dental caries, is slowly and only partially absorbed in the small intestine and can reach the colon where it can act as substrate for bacterial fermentation. Supplementation with sorbitol resulted in enrichment of lactobacilli in rat colon and cecum (Sarmiento-Rubiano et al., 2007).

Sorbitol absorption is mediated by dose and concentration. Doses greater than 30 g can cause water retention, resulting in osmotic diarrhea, bloating, flatulence, cramping and abdominal pain (Fernández-Bañares et al., 2009). These doses vary depending on the condition of the intestinal absorption surface. In patients with malabsorption, the ingestion of 5–20 g, provoked diarrhea and gastrointestinal complications (Montalto et al., 2013). In the colon, this sugar alcohol is metabolized by some species of *Lactobacillus* and is also a preferred carbon source for human intestinal bifidobacteria (Sarmiento-Rubiano et al., 2007).

Furthermore, utilization of sorbitol as a carbon source has been described in a variety of bacteria within the fila proteobacteria (Yamada and Saier, 1988; Aldridge et al., 1997) and firmicutes (Tangney et al., 1998; Boyd et al., 2000; Yebra and Pérez-Martínez, 2002). Among the firmicutes, there are some lactic acid bacteria (LAB) with catabolic pathways for sorbitol metabolism (Rhodes and Kator, 1999; Sarmiento-Rubiano et al., 2007). These pathways are encoded by genes organized in *gut* operons, and include the sorbitol transport system, sorbitol-6-phosphate dehydrogenase (S6PD) as well as regulatory protein(s), and those of *Lactobacillus casei* and *Lactobacillus plantarum* have been characterized (Nissen et al., 2005; Ladero et al., 2007; Alcantara et al., 2008).

Sorbitol is transported into the cells and phosphorylated to sorbitol-6-phosphate by a phosphopyruvate-dependent phosphotransferase (PTS) sorbitol system (PTS^{gut}). Each PTS is composed of two cytoplasmic enzymes, common to the transport of different compounds (EI and HPr) and of different membrane-associated enzyme complexes (EII), specific for one, or several substrates. The genes *gutC*, *gutB* and *gutA* encode the EII domain of a sorbitol PTS (Alcantara et al., 2008). The *gutF* gene encodes a sorbitol-6-P dehydrogenase, which catalyzes the conversion of sorbitol-6-phosphate to fructose-6-phosphate, a compound that is introduced into the glycolytic pathway with NADH regeneration (Nissen et al., 2005). The *gutR* and *gutM* genes encode two regulatory proteins. The role of the GutM and GutR proteins has been studied in *Escherichia coli*, operating GutM as an activator and GutR as a repressor (Yamada and Saier, 1988). In the firmicutes group, the analyzed *gut* operons contain homologs to the *gutM* and *gutR* genes, but the role of GutR regulator is different from that of *E. coli*. The GutR of *L. casei* has been functionally characterized and it has been shown to be a PTS-controlled transcriptional activator, via a PTS regulation binding domain (PRD) (Stülke et al., 1998). Also, both the GutR binding sequence and the PRD domain are conserved in firmicutes. GutM encodes a highly conserved protein in firmicutes and in *L. casei* plays a regulatory role (Alcantara et al., 2008).

Pediococcus parvulus 2.6 (Werning et al., 2006) (previously named *Pediococcus damnosus*) is a lactic acid bacteria isolated from a ropy cider (Fernández et al., 1995). This LAB produces a 2-substituted (1,3)-β-D-glucan exopolysaccharide (EPS) (Dueñas-Chasco et al., 1997), with high molecular mass ($>10^6$ Da), and whose rheological properties showed its potential utility as a biothickening agent (Velasco et al., 2009). The presence of this EPS improves some probiotic features of *P. parvulus* 2.6, including tolerance to simulated gastrointestinal conditions and adherence to Caco-2 cell lines and reduces inflammation-related cytokine levels produced by polarized macrophages (Fernández de Palencia et al., 2009; Immerstrand et al., 2010). Moreover, the purified EPS improves the growth, viability and adhesion capability of probiotic microorganisms (Russo et al., 2012), also it activates macrophages with anti-inflammatory effects (Notararigo et al., 2014), and decreases the levels of the proinflammatory IL8 in human intestine cultures (Notararigo et al., unpublished data). The draft genome of *P. parvulus* 2.6 has been determined (Pérez-Ramos et al., 2016), and its analysis showed the existence of a putative sorbitol utilization *gut* operon in this bacterium. Thus, this current work focuses on the genomic location, expression and metabolic involvement of the *gut* operon of *P. parvulus* 2.6 in sorbitol catabolism, as well as its interplay with EPS production by this bacterium.

MATERIALS AND METHODS

Bacterial Strains and Growth Conditions

The bacteria used in this work are listed in **Table 1**. *Pediococcus* and *Lactobacillus* strains were routinely grown in de Man Rogosa Sharpe (MRS) broth (Pronadisa, Madrid, Spain) at 30°C and 37°C, respectively. *Lactococcus lactis* strains were grown in ESTY broth (Pronadisa) supplemented with 0.5% glucose at 30°C. When bacteria carried the pRCR plasmid or its derivatives the medium was supplemented with chloramphenicol (Cm) at 5 $\mu g\ mL^{-1}$ for *L. lactis* and at 10 $\mu g\ mL^{-1}$ for lactobacilli. *E. coli* V517 was grown in LB broth and incubated at 37°C.

For evaluation of sorbitol utilization, *P. parvulus* strains were grown in a MRS broth made by components (de Man et al., 1960) without glucose, pH was adjusted to 5.2 and the medium supplemented with 10 mM glucose (MRSG), 30 mM sorbitol (MRSS) or 10 mM glucose plus 30 mM sorbitol (MRSGS) at 30°C. Prior selection of conditions for growth in presence of sorbitol several tests were performed. First various carbon sources were tested (10 mM glucose, 10 mM fructose or 10 mM maltose) and

TABLE 1 | Bacteria used in this work.

Bacteria	Plasmid	Resistance	Characteristics	Reference
Pediococcus parvulus 2.6	pPP1, pPP2, and pPP3	–	2-substituted (1,3)-β-D-glucan producer	Pérez-Ramos et al., 2016
P. parvulus 2.6NR	pPP1 and pPP3	–	Non-EPS-producing strain. Derivative of 2.6 strain by pPP2 plasmid curing	Fernández et al., 1995
Lactococcus lactis subsp. *cremoris* MG1363	–	–	Plasmid free type strain used for plasmid cloning	Wegmann et al., 2007
L. lactis subsp. *cremoris* MG1363[pRCR]	pRCR	Cm^R	Source of promoter probe pRCR containing the *mrfp* gene, which encodes the fluorescent mCherry protein	Mohedano et al., 2015
Escherichia coli V517	8 plasmids pVA517A through pVA517H	ND	Source of plasmids used as references in agarose gel analysis	Macrina et al., 1978
Lactobacillus casei BL23	–	–	Bacteria used for heterologous gene expression	Mazé et al., 2010
L. casei BL23[pRCR16]	[pRCR16]	Cm^R	Derivative of pRCR by cloning of P_{gut} upstream of *mrfp*	This study
L. casei BL23[pRCR17]	[pRCR17]	Cm^R	Derivative of pRCR16 by cloning of *gutR* upstream of *mrfp*	This study
L. casei BL23[pRCR18]	[pRCR18]	Cm^R	Derivative of pRCR16 by cloning of *gutM* upstream of *mrfp*	This study
L. casei BL23[pRCR19]	[pRCR19]	Cm^R	Derivative of pRCR16 by cloning of *gutMR* upstream of *mrfp*	This study
Lactobacillus plantarum 90	1 uncharacterized plasmid	ND	Bacteria used for heterologous gene expression	Lamontanara et al., 2015
L. plantarum 90[pRCR16]	[pRCR16]	Cm^R	Derivative of pRCR by cloning of P_{gut} upstream of *mrfp*	This study
L. plantarum 90[pRCR17]	[pRCR17]	Cm^R	Derivative of pRCR16 by cloning of *gutR* upstream of *mrfp*	This study
L. plantarum 90[pRCR18]	[pRCR18]	Cm^R	Derivative of pRCR16 by cloning of *gutM* upstream of *mrfp*	This study
L. plantarum 90[pRCR19]	[pRCR19]	Cm^R	Derivative of pRCR16 by cloning of *gutMR* upstream of *mrfp*	This study

ND, no determined; Cm^R, resistance to chloramphenicol.

pH at 6.8, 5.2 or 4.0 and then influence of aeration was evaluated in presence of 10 mM glucose at either pH 6.8 and 5.2 (results not show).

For evaluation of mCherry expression, *Lactobacillus* strains were grown in a MRSG containing 55 mM (1% w/v) glucose or in a MRSS containing 55 mM (1% w/v) sorbitol at 37°C.

Plasmidic DNA Preparations

Total plasmidic DNA preparations of *P. parvulus* 2.6 and 2.6NR strains were prepared as follows. Bacterial cultures were grown to an optical density at 600 nm ($OD_{600\ nm}$) of 2.5, and 100 mL of each culture were sedimented by centrifugation at 10,000 × *g* for 20 min at 4°C. The cells were resuspended in 4 mL of a solution containing 50 mM Tris/HCl pH 8.0, 10 mM EDTA, lysozyme (30 mg mL^{-1}) and RNasa A (10 μg mL^{-1}), and incubated for 30 min at 37°C. Then, 4 mL of a solution containing 220 mM NaOH and 1.33% sodium dodecyl sulfate) were added and samples were incubated for 5 min at room temperature. Upon addition of 5 M potassium acetate pH 5.0 (4 mL), samples were centrifugated at 10,000 × *g* for 15 min at 21°C. The DNA present in the supernatants was precipitated, concentrated by addition of 8.7 mL of isopropanol, sedimented by centrifugation at 10,000 × *g* for 15 min at 4°C, and resuspended in 10 mM Tris, 1 mM EDTA buffer (4.3 mL). The DNA preparation was deproteinated by treatment with 7.5 M ammonium acetate (2.7 mL) and phenol (4.3 mL) during 5 min at room temperature and then sedimented at 10,000 × *g* for 5 min at 21°C. The aqueous phase containing total plasmidic DNA was further purified by isopycnic CsCl density gradient centrifugation and dialysis as previously described (López et al., 1989). The final recovery was 54 μg and 58 μg for 2.6 and 2.6NR DNA preparations, respectively.

The recombinant plasmids from the lactococcal and lactobacilli strains were isolated using the High pure plasmid isolation kit (Roche) as follows. Bacteria were grown until stationary phase (10^9 colony forming units mL^{-1}) and 1 mL of each culture were sedimented by centrifugation at 10,000 × *g* for 10 min at 4°C. Cells were resuspended in solution I of the kit supplemented with lysozyme (30 mg mL^{-1}) and were incubated for 30 min at 37°C. Then, plasmid isolation were performed as described in the kit protocol, eluting the plasmidic DNA in 100 μL at approximately 100 ng μL^{-1}.

Sequencing

DNA sequencing was performed by the dideoxy method at Secugen (Madrid, Spain). The sequencing of the sorbitol utilization cluster and the flanking regions of pPP1 of *P. parvulus* 2.6 was performed using total plasmidic DNA preparations of the bacterium (see above) with the walking strategy and the sequence has been deposited in GenBank (accession No MF766019). The lack of sorbitol cluster in the 2.6NR strain was confirmed by sequencing of its pPP1 plasmid by using as substrates a total plasmidic preparation of 2.6NR strain and either pPP1*F or pPP1*R primers (see **Table 2**) In addition, in the case of sequencing with pPP1*F, it was also used as substrate the product of a polymerization reaction catalyzed by the bacteriophage Φ29 DNA polymerase with plasmidic DNA of *P. parvulus* 2.6NR and hexamers containing random sequences.

Construction of pRCR16, pRCR17, pRCR18, and pRCR19

A region located upstream of the *P. parvulus* 2.6 *gut* operon carrying the putative P_{gut} promoter and the *gutR* and *gutM* genes was cloned into the promoter probe pRCR vector. To this end, three DNA regions of pPP1 plasmid were amplified with Phusion High Fidelity Polymerase (PHFP, ThermoFisher Scientific) by using a plasmidic DNA preparation of *P. parvulus* 2.6 and the primers depicted in **Table 2**, which have homology with pPP1 DNA and carry restriction sites suitable for cloning. Plasmid pRCR16 (**Figure 1**) was generated by ligation of the P_{gut} promoter to the pRCR promoter probe vector (Mohedano et al., 2015), after double digestion of both DNAs with BglII and XmaI (New England Biolabs, Ipswich, MA, United States), with the T4 DNA ligase (New England Biolabs). Then, between the XmaI and XbaI restriction sites of pRCR16 three amplicons were independently cloned, containing *gutR*, *gutM* or *gutRM*, generating plasmids pRCR17, pRCR18 and pRCR19, respectively. The clonings were performed in *L. lactis* MG1363, the ligations mixtures were used to transform the bacteria by electroporation (25 μF, 2.5 kV and 200 Ω in 0.2 cm cuvettes), as previously described (Dornan and Collins, 1987) and transformants were selected in ESTY-agar plates supplemented with Cm at 5 $\mu g\ mL^{-1}$. The inserts present in the new four recombinant plasmids were confirmed by automated sequencing. Then, DNA preparations of pRCR17, pRCR18 and pRCR19 obtained from *L. lactis* MG1363 (0.5 μg) were used for transfer to lactobacilli by electroporation (25 μF, 1.3 kV and 200 Ω in 0.1 cm cuvettes) as previously described (Berthier et al., 1996) and transformants were selected in MRSG-agar plates supplemented with Cm at 10 $\mu g\ mL^{-1}$.

Southern Hybridization

Plasmid samples were fractionated by electrophoresis in a 0.7% agarose gel and DNA molecules were revealed by staining with ethidium bromide at 0.5 $\mu g\ mL^{-1}$. The image of the gels was obtained with GelDoc 200 (BioRad) and the bands were quantitated with the Quantity One 4.5.2 software (BioRad). The DNA fragments were transferred to a nylon membrane Biodyne A (PALL Gelman Laboratory, AnnArbor, MI, United States) by 5 inches Hg of vacuum for 2 h using the Vacuum Blotter model 785 (Bio-Rad). Internal regions of *gutF*, *gutR* and *gutB* genes were amplified by PCR generating amplicons 1, 2, and 3, respectively, in reactions catalyzed by PHFP, and by using as substrate total plasmidic DNA preparation of *P. parvulus* 2.6 and the primer pairs shown in **Table 2**. Then, the amplicons were labeled with digoxigenin-dUTP by using the DIG high prime DNA labeling and detection starter kit II (Roche, Mannheim, Germany). Each DIG-labeled DNA probe (25 ng mL^{-1}) was used for hybridization at 45°C following the specifications of the kit's supplier. The hybridization bands were revealed with the chemiluminescent substrate CSPD, and the signals were detected

TABLE 2 | Oligonucleotides used in this work.

Primers	Sequence (5′- 3′)	Utilization	Amplicon size (bp)
pts1F	TGCGGAAGCGGTTAATCGGCT	*gutF* DNA probe[a]	618
pts1R	CCACGACTCTTGCCTCCCGCA		
ptsRF	CGAACTGGAAGCAACCTGGGA	*gutR* DNA probe[a]	652
ptsRR	CCGATGAATAATTGGCGCTGC		
ptsBF	GGAATGGAAGCTGTTGATGGC	*gutB* DNA probe[a]	643
ptsBR	CAACGCCAATCAAGGTCCCGA		
pgutBglIIF	GAAGATCTACCATATGGCGATAATGAAAA	Cloning of P_{gut} in pRCR[a]	186
pgutXmaIR	TCGCTCCCGGGTCATTTCTTTTC		
gutRXmaIF	CGTGGTTAACCCGGGAATTTTAGTTG	Cloning of *gutR* in pRCR16[a]	1952
gutRXbaIR	GCTCTAGAAACGCACTGACTAGGATCA		
gutMXmaIF	TCCCCCCGGGTTAAATCAGTTGATGGA	Cloning of *gutM* in pRCR16[a]	597
gutMXbaIR	GCTCTAGAACAGCCCATAAAGCCC		
gutRXmaIF	CGTGGTTAACCCGGGAATTTTAGTTG	Cloning of *gutRM* in pRCR16[a]	2472
gutMXbaIR	GCTCTAGAACAGCCCATAAAGCCC		
pPP1*F	CATAGTTCACTGGGCTACCA	Sequencing of pPP1*[b]	–
pPP1*R	TAGCGGTGCCTCCCTTTAAT		–

[a]*Plasmidic DNA preparations of P. parvulus 2.6 was used as substrate for the PCR reactions.* [b]*Plasmidic DNA preparations of P. pervulus 2.6NR was used as substrate for the DNA sequencing.*

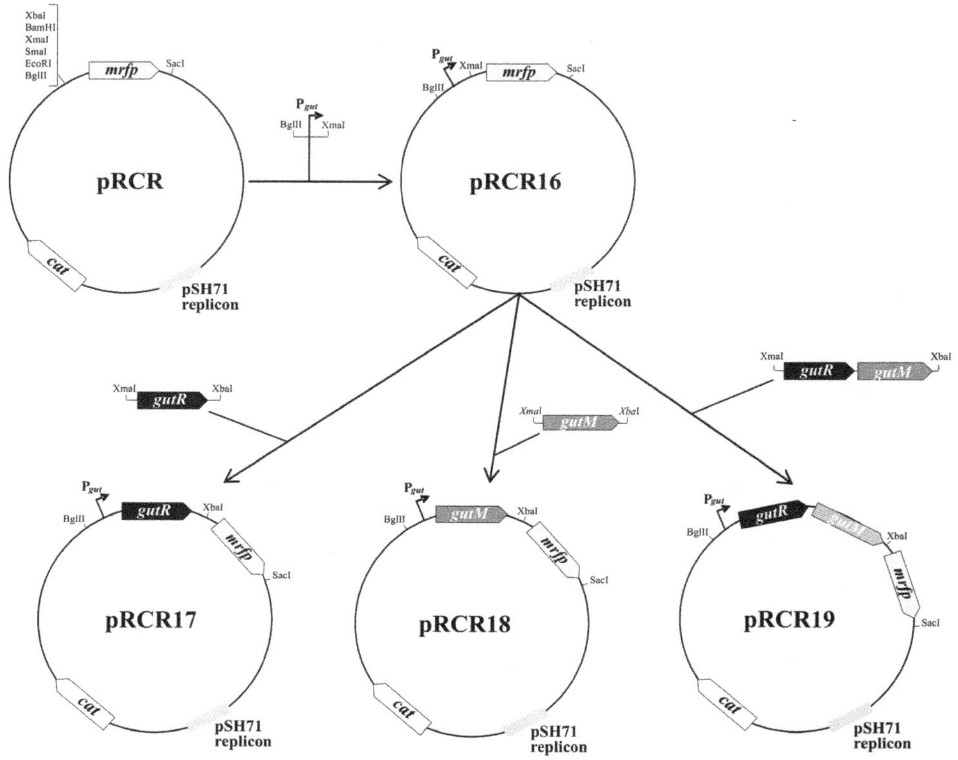

FIGURE 1 | Scheme of construction of plasmids pRCR16, pRCR17, pRCR18, and pRCR19. Maps of these plasmids and of the parental pRCR are depicted. The pertinent restriction sites are shown, as well as the putative promoter of the *gut* operon (P_{gut}) and the following genes: *mrfp, cat, gutR* and *gutM* encoding the mCherry, a chloramphenicol acetyltransferase, GutR and GutM, respectively.

with the LAS-3000 imaging system (Fujifilm, Stamford, CT, United States).

Analysis of the Metabolic Fluxes of *P. parvulus* and Its EPS Production

P. parvulus 2.6 and 2.6NR strains were grown in either MRSG or MRSGS under aerobic conditions (shaking at 180 rpm), at 30°C during 66 h, and samples were taken at the times indicated in **Figure 2** to monitor growth by determination of optical density at 600 nm and of acidification of the media by measuring pH. Also, samples were centrifuged at 16,000 × *g* for 30 min at 4°C, and the levels of glucose, sorbitol, lactic acid and EPS in the supernatants were analyzed. The experiments were performed in triplicate for each strain and in each condition of growth.

Analysis of Culture Supernatants by Gas Chromatography-Mass Spectrometry (GC-MS)

The concentration of glucose, sorbitol and lactic acid was determined by GC–MS using myo-inositol as internal standard. For this analysis, *myo*-inositol (100 μg) was first added to aliquots of the bacterial culture supernatants. The mixture was lyophilized and derivatized with 2.5% hydroxylamine chloride in pyridine for 30 min at 70°C, to form the sugar oximes. Afterward, *bis*-trimethylsilyl trifluoroacetamide (BSTFA) was added and samples were incubated for 45 min at 80°C, to form the trimethylsilylated derivatives. Identification and quantification of the compounds were performed by GC–MS on a 7980A-5975C instrument (Agilent, Santa Clara, CA, United States) equipped with a HP-5MS column (30 m × 0.25 mm I.D. × 0.2 μm film thickness) with helium as the carrier gas. Injector and detector were set at 275°C. Samples (1 μL) were injected with a split ratio of 1:50 with a temperature program: 80°C for 4 min, then 15°C min^{-1} to 270°C and finally 30°C min^{-1} to 310°C (2 min). The peaks in the chromatograms corresponding to sugars and lactic acid were identified by their retention times. Quantifications were calculated using the peak areas and the calibration standard curve for each compound.

Quantification of the 2-Substituted (1,3)-β-D-Glucan Produced by *P. parvulus*

A competition (ELISA) method for the specific detection of the EPS synthesized by *P. parvulus* 2.6, based on *Streptococcus pneumoniae* serotype 37 antibodies, was performed as previously described (Werning et al., 2014). Briefly, the ELISA assay was carried out in 96-Well Nunc-Immuno MicroWell MaxiSorp plates (Thermo Fisher Scientific), and the EPS of *P. parvulus* 2.6, purified as previously described (Notararigo et al., 2013), was immobilized in each well (62.5 ng per well).

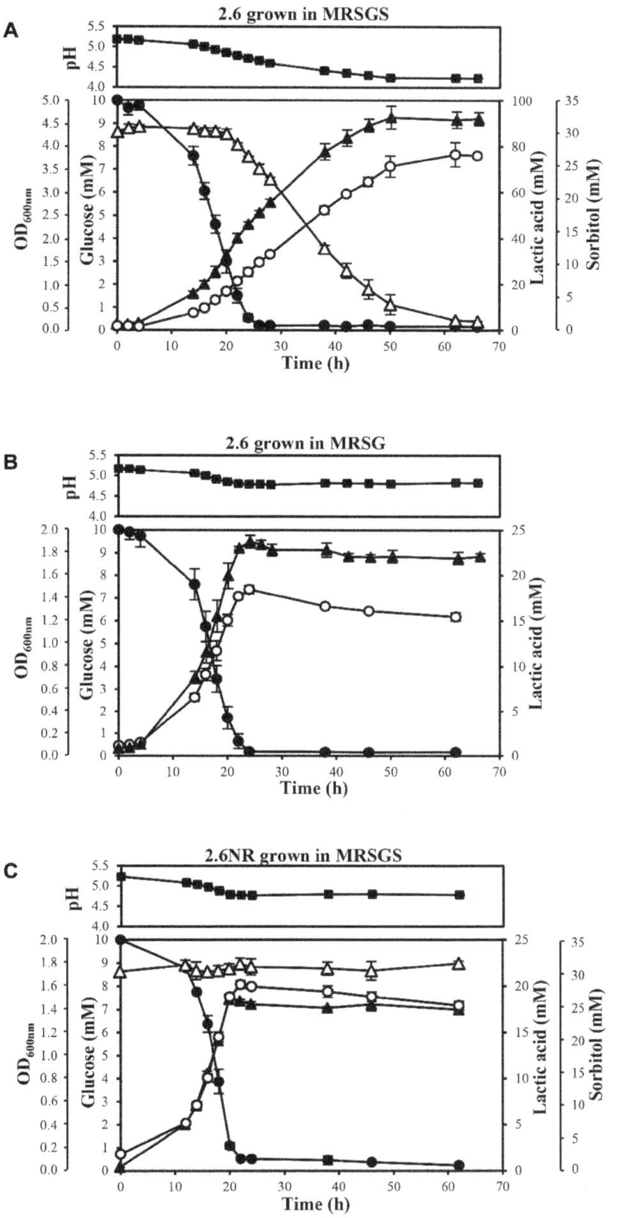

FIGURE 2 | Analysis of metabolism of *P. parvulus* 2.6 **(A,B)** and 2.6NR **(C)** strains. Bacteria were grown in the indicated media. Symbols: $OD_{600\ nm}$ (▲), glucose (●), sorbitol (△), lactic acid (○) and pH (■). The experiments were performed in triplicate and the mean value and standard deviation is depicted.

Culture supernatants [diluted with phosphate-buffered saline (PBS) pH 7.2 when necessary] were used as competitor for binding to the primary antibody (dilution 1:800 of anti-serotype 37, Statens Serum Institut, Copenhagen, Denmark). Then, primary antibody was conjugated with a secondary antibody, polyclonal Anti-Rabbit IgG alkaline phosphatase (Sigma–Aldrich, Saint Louis, MO, United States) diluted 1:25,000, and finally was revealed with *p*-nitrophenylphosphate in diethanolamine buffer (Sigma–Aldrich). Reaction signals were detected with a microtiter plate reader model 680 (Bio-Rad, Hercules, CA, United States), measuring the OD at 415 nm. Quantification was performed using a standard curve generated by the competition for the primary antibody of serial dilutions of the purified *P. parvulus* 2.6 EPS dissolved in PBS.

Detection of mCherry Fluorescence in LAB Carrying pRCR16, pRCR17, pRCR18, or pRCR19

To detect the expression levels of the mCherry fluorescent protein, *L. plantarum* Lp90 strains carrying the pRCR derivatives were diluted 1:100 and grown in MRS supplemented with 1% glucose in static mode at 37°C, until mid-exponential phase. Then the cultures were centrifuged at 9,000 × *g* for 10 min at room temperature, and the cells were washed with one volume of PBS pH 7.2 prewarmed at 37°C. Then, the bacteria were resuspended in the same volume of MRS broth supplemented with 1% sorbitol or 1% glucose prewarmed at 37°C. Cultures were incubated at 37°C with agitation of 180 rpm, and samples were taken each hour. Two hundred microliter of all chilled samples were centrifuged at 9,000 × *g* for 10 min at 4 °C and cells were washed once with chilled PBS buffer pH 7.2. Samples were resuspended in 200 μL of PBS buffer pH 7.2 and used to measure the fluorescence levels of mCherry protein in a 96-Well Nunc U96 MicroWell plate (Thermo Fisher Scientific) in a Varioskan Flash equipment (Thermo Fisher Scientific), using 587 and 610 nm wavelengths for excitation and detection of emission, respectively. In addition, appropriate dilutions were prepared to estimate culture biomass by measuring the $OD_{600\ nm}$. Three independent trials were performed and the same fresh suspensions, without fixing, were used for phase contrast and fluorescent microscopy analysis with a Leica DM1000 model microscope (Leica Microsystems, Mannheim, Germany) with a light source EL6000 and a filter system TX2 ET for detection of red fluorescence. The microscope was connected to a DFC3000G camera (Leica Microsystems) with a CCD sensor. Image analysis was performed using Leica Application Suite X Software (Leica Microsystems).

To detect the expression of the mCherry fluorescent protein, *L. casei* BL23 strains carrying the pRCR derivatives were grown and processed in the same manner as the *L. plantarum* cultures, except that preinoculum cultures were diluted in MRS supplemented with 1% glucose or 1% sorbitol to an $OD_{600\ nm} = 0.1$ and then were incubated at 37°C with agitation of 180 rpm for 16 h, until they reached early stationary phase. Then, 1 mL of each culture was centrifuged and washed with PBS as above. Samples were concentrated five-fold and used to measure the fluorescence levels and to take fluorescence images as described above.

Bioinformatic Analysis

The DNA sequence of plasmid pPP1 was analyzed with the programs included in the DNASTAR Lasergene 12 (DNAstar Inc. Madison, WI, United States). Homologies of pPP1 DNA

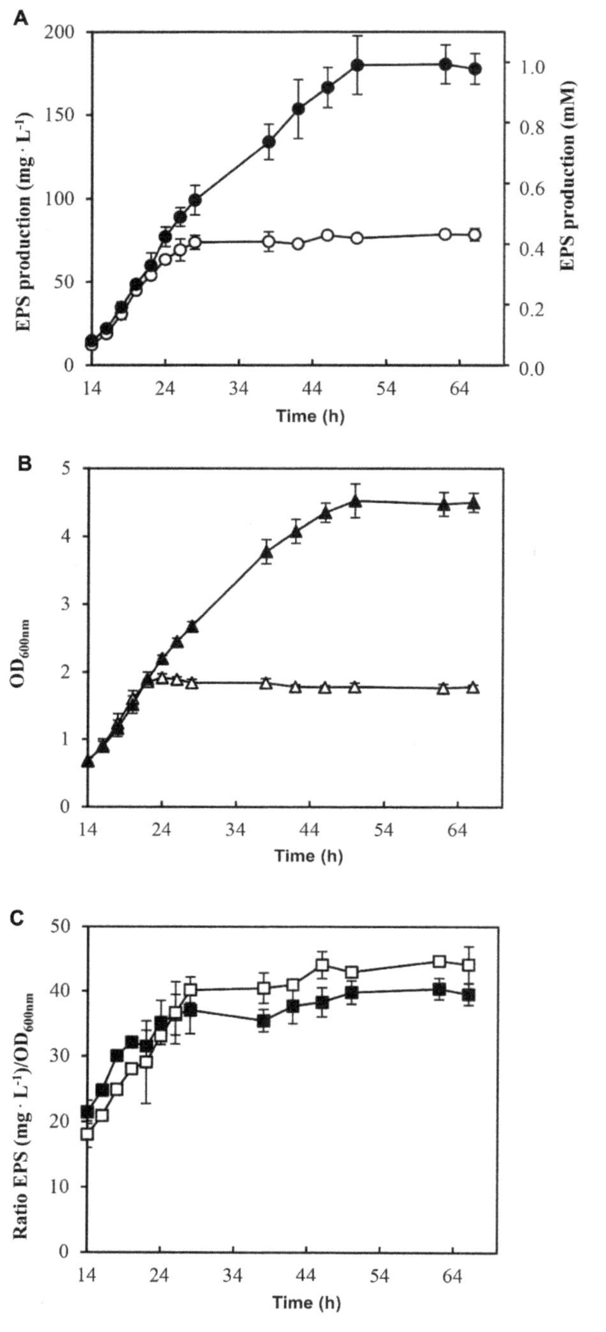

FIGURE 3 | Analysis of EPS production by *P. parvulus* 2.6 in MRSGS (●), (▲) and (■) and in MRSG (○), (△) and (□). **(A)** Concentration of EPS present in cultures supernatants is depicted. The results were expressed in mg of EPS per L or in mmol of glucose per L. **(B)** The $OD_{600\ nm}$ of the cultures is depicted. **(C)** Specific EPS concentration is shown and it was calculated as the ratio EPS concentration/$OD_{600\ nm}$. The experiments were performed in triplicate and the mean value and standard deviation is depicted

sequences and of its inferred translated products with the NCBI data bases of the National Center for Biotechnology Information (NCBI) were analyzed with the Basic Local Alignment Search Tool (BLAST)[1]. Multiple sequence alignment of genes and proteins were performed with Clustalx 2.1[2] programs.

Transmembrane helices in GutM were predicted using TMHMM 2.0[3] (TMpred[4]) programs. Prediction of secondary structures in the gut mRNA was accomplished with the mfold 2.3 program[5].

RESULTS

Analysis of *P. parvulus* Sorbitol Metabolism

Sorbitol could be a substrate for the synthesis of *P. parvulus* 2.6 EPS and analysis of the DNA sequence of the draft genome of this bacterium (Pérez-Ramos et al., 2016) with the BLAST program revealed a putative *gut* operon, that could be involved in transport and catabolism of this compound. Therefore, growth of *P. parvulus* 2.6 and its isogenic EPS-non-producing (non-ropy) 2.6NR strain in MRS (without glucose) and MRSS (medium containing sorbitol) was tested. The 2.6NR strain showed the same poor growth in both media (Supplementary Figure S1). However, the presence of sorbitol in the medium significantly improved the growth of the 2.6 strain (Supplementary Figure S1), reaching a final $OD_{600\ nm}$ of 3.0 in MRSS *versus* 0.45 in MRS, indicating that this bacterium was able to utilize sorbitol. Nevertheless, the growth of *P. parvulus* 2.6 in MRSS was very slow and took more than 12 days to reach the final optical density (Supplementary Figure S1). Therefore, in order to improve the growth rate of 2.6 strain, the influence of modifying various parameters in bacterial growth in MRSS was investigated. The best inferred conditions were the usage of a MRSGS containing as carbon sources 10 mM glucose plus 30 mM sorbitol, pH = 5.2, and growth with aeration at 30°C. Thus, these conditions were used to investigate a potential interplay between sorbitol utilization and EPS production by *P. parvulus* 2.6.

A comparative study of the metabolic fluxes of *P. parvulus* strains by analysis of culture supernatants during growth in MRSG or MRSGS corroborated that 2.6, but not 2.6NR, was able to ferment sorbitol (**Figure 2**). Co-metabolism of sorbitol and glucose by the 2.6 strain resulted in an increase of 2.5-fold in the final biomass estimated by the $OD_{600\ nm}$ of the cultures. Values of 4.48 ± 0.18 in MRGS (**Figure 2A**) compared to 1.77 ± 0.06 reached in MRSG (**Figure 2B**), the latter being similar to 1.36 ± 0.05 observed for the 2.6NR strain in MRGS (**Figure 2C**). In addition, a prolonged exponential growth phase of the 2.6 strain was observed in the MRSGS medium (50 h *versus* 20 h, **Figures 2A,B**). In the 2.6NR culture supernatants, the initial sorbitol levels (30 mM) remained constant during the entire time period of the assays, revealing that this bacterium was unable to transport sorbitol to the cytosol (**Figure 2C**). Moreover, the analysis of the carbon source consumption by the 2.6 strain

[1]https://blast.ncbi.nlm.nih.gov/Blast.cgi

[2]http://www.ebi.ac.uk/Tools/msa/clustalw2/

[3]http://www.cbs.dtu.dk/services/TMHMM-2.0/

[4]http://embnet.vital-it.ch/software/TMPRED_form.html

[5]http://unafold.rna.albany.edu/?q=mfold/RNA-Folding-Form

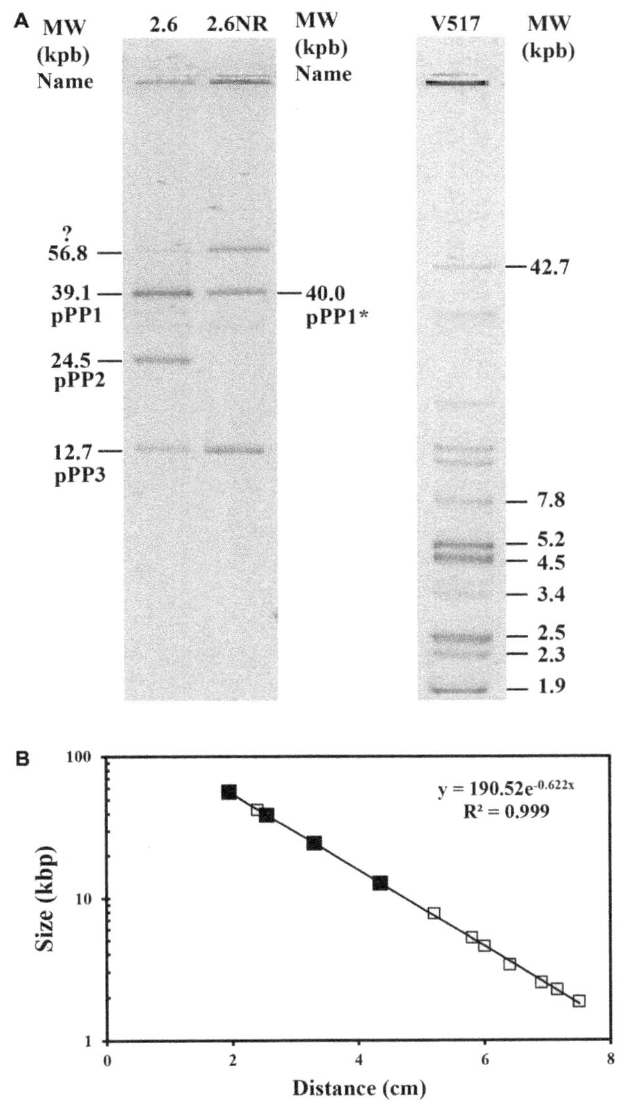

FIGURE 4 | Detection of plasmids of *P. parvulus* 2.6 and 2.6NR strains. Plasmids preparations of *P. parvulus* strains and of *E. coli* V517 were analyzed in a 0.7% agarose gel. **(A)** The image of the gel. The three samples were run in the same gel. In **(B)** the calibration curve for plasmid size determination is depicted. Symbols: plasmids from *E. coli* V517 (□), plasmids from *P. parvulus* strains (■).

showed that glucose started to be transported to the cytosol after 2 h of growth, and upon 26 h of incubation the monosaccharide was undetectable in the culture supernatants (**Figures 2A,B**). Furthermore, only after 20 h of incubation did the 2.6 strain start to internalize the sorbitol and presumably to metabolize it, because the bacterium did not enter into the stationary phase until the sorbitol was consumed (**Figure 2A**). The metabolic activity of the two strains was monitored by detecting the lactic acid production, since it is the main metabolic end-product because pediococci are homofermentative bacteria. The results showed that the 2.6 strain grown in MRSG (**Figure 2B**) and the 2.6NR strain grown in MRSGS (**Figure 2C**) released to the culture media similar amounts of lactic acid, the maximum levels being 18.45 ± 0.45 mM and 20.19 ± 0.42 mM, respectively. By contrast, the 2.6 strain grown in the presence of both carbon sources showed a higher lactic acid production, up to 76.03 ± 0.43 mM (**Figure 2A**). Correlating with these results, the final pH of the 2.6 cultures in MRSG and of the 2.6NR cultures in MRSGS was similar (4.81 ± 0.02 *versus* 4.78 ± 0.02), and higher than that of the 2.6 cultures in MRSGS (4.23 ± 0.02).

Furthermore, the EPS production by *P. parvulus* 2.6 in the presence or absence of sorbitol was investigated. Significant EPS levels were detected after 14 h of growth in MRSG and MRSGS media. Therefore, the data depicted in **Figure 3** correspond to those obtained within the 14–62 h incubation period. The results revealed that the bacterium produced EPS during the growth in MRSGS and synthesized higher levels of the polymer in this medium than in MRSG (**Figure 3A**). Thus, after 62 h of growth in MRSG, the 2.6 strain produced 78.6 ± 3.7 mg L^{-1} of EPS, while in MRSGS synthesized 180.5 ± 11.8 mg L^{-1}. Additionally, in order to evaluate the specific efficiency of the EPS production depending on the carbon source used, the ratio between EPS concentration and the biomass estimated from the $OD_{600\ nm}$ (**Figure 3B**) was calculated (**Figure 3C**). The results showed that irrespectively of the carbon source, the bacteria had almost identical efficiency of EPS production, which increased during the exponential and stationary phases of growth (**Figure 3C**).

Determination of Genomic Location of the *gut* Operon

P. parvulus 2.6 probably carries three natural plasmids, which were previously named pPP1, pPP2 and pPP3 (Werning et al., 2006), and we have identified only three plasmid replication machineries in the *P. parvulus* draft genome (Pérez-Ramos et al., 2016). In addition, the *P. parvulus* 2.6 EPS is synthesized by the GTF glycosyltransferase encoded by the *gtf* gene, which is located in the pPP2 plasmid (Werning et al., 2006). Thus, the 2.6NR strain was generated from 2.6 by pPP2 plasmid curing after treatment with the DNA intercalating agent ethidium bromide and the gyrase inhibitor novobiocin (Fernández et al., 1995).

Consequently, given that 2.6NR does not utilize sorbitol, it was feasible that the *gut* operon was encoded by pPP2 and this hypothesis was investigated. First, total plasmidic DNA preparations of the two *Pediococcus* strains were purified by fractionation in a CsCl gradient to eliminate non-supercoiled (open circles and linear) forms of the plasmids. Then, the purified plasmidic DNA preparations were analyzed in an agarose gel (**Figure 4**). Four and three bands were detected, respectively, in preparations of the 2.6 and 2.6NR strains. The sizes of the bands were inferred from their migration using a calibration curve (**Figure 4B**) generated with the plasmids of the *E. coli* V517 strain and are shown in **Figure 4A**. Two of the bands apparently were shared by 2.6 and 2.6NR, and were initially ascribed to the monomeric forms of pPP1 (39.1 kpb in 2.6 and 40.0 kpb in 2.6NR) and pPP3 (12.7 kpb). As expected, pPP2 (24.5 kpb) was not detected in 2.6NR DNA preparations. Moreover, we could not ascribe to any plasmid the band with less mobility and a

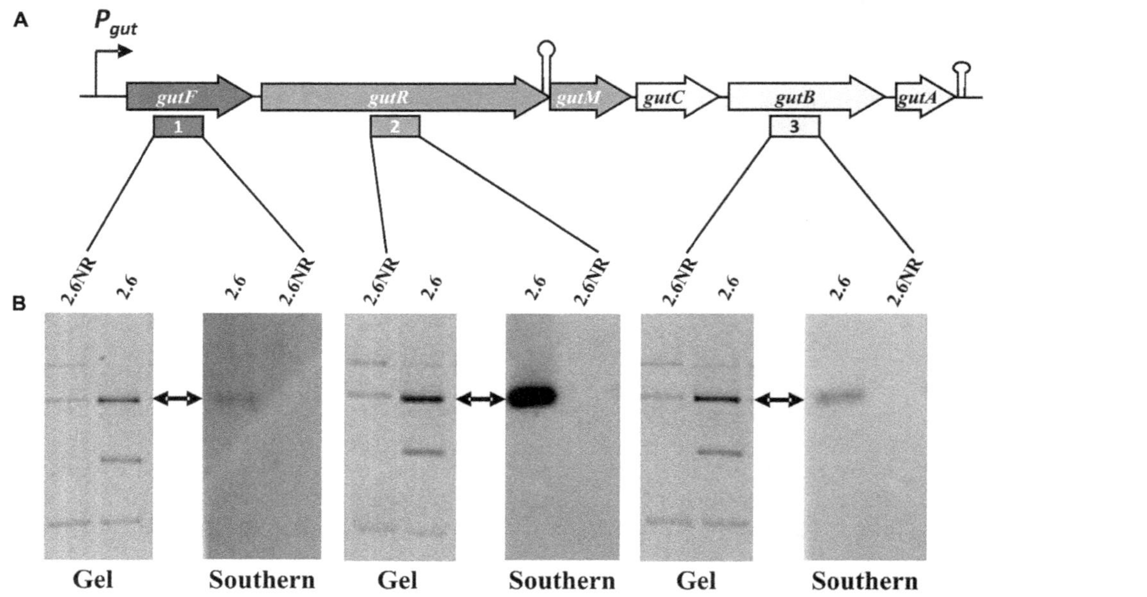

FIGURE 5 | Detection of the *P. parvulus* 2.6 *gut* operon by Southern blot hybridization. Plasmids preparations of the *P. parvulus* strains were fractionated by electrophoresis in agarose gel, transferred to membranes and hybridized for detection of *gutF*, *gutR* and *gutB* with the probes 1, 2, and 3, respectively. **(A)** Physical map of the *gut* operon, including the putative promoter of the *gut* operon (P_{gut}), the operon genes (*gutF, gutR, gutM gutC, gutB* and *gutA*), a secondary structure including the 3′-end of *gutR* and the 5′-end of *gutM* (coordinates 3483-3512 in GenBank accession No MF766019) as well as a transcriptional terminator (coordinates 6103-6027 in GenBank accession No MF766019) downstream of *gutA*. In addition, location of probes 1, 2, and 3 is indicated. **(B)** Images of the corresponding gels and hybridized membranes. The double headed arrows indicate the position of the hybridized bands in the corresponding gel.

theoretical molecular weight of 56.8 kbp that was present in DNA preparations of both strains. Quantification of the bands from agarose gels (**Figures 5**, **6**) revealed different proportions of the plasmidic forms in 2.6 (0.3:5.9:2.5:1.0) and 2.6NR (0.8:1.0:0.0:1.0) samples.

The *gut* operon of *P. parvulus* 2.6 (**Figures 5A**, **6A**) is composed of six genes, of which *gutF* encodes a sorbitol-6-phosphate dehydrogenase; *gutRM* encodes two putative regulators; and *gutCBA* encodes the proteins EIIC, EIIBC and EIIA which are components of a phosphoenolpyruvate-dependent sorbitol phosphotransferase system (PTS^{gut}). Thus, to detect the location of the *gut* operon, Southern blot hybridization of total plasmidic DNA preparations was performed using as a probe internal regions of *gutF*, *gutR* or *gutB*. One hybridization signal was observed with the three probes at the position of the 39.1 kb pPP1 plasmid in the 2.6 DNA sample (**Figure 5B**). Surprisingly, this plasmid was apparently present in both *P. parvulus* strains, but in the 2.6NR DNA sample no signal was observed. Nevertheless, the results demonstrated that the *gut* operon was not located in the pPP2 plasmid, but rather was carried by the pPP1 plasmid of the 2.6 strain and not of the newly designated pPP1* plasmid of 2.6NR strain.

Analysis of Plasmids pPP1 of *P. parvulus* 2.6 and pPP1* of *P. parvulus* 2.6NR

The results obtained by Southern blot analysis prompted us to obtain further information of pPP1 and pPP1* plasmids. Thus, the total plasmidic DNA preparation of the 2.6 strain was used as a substrate to confirm the sequence of the gut operon and to determine the unknown nucleotide sequence of the flanking regions (undetected in the draft genome of the bacterium) by the dideoxynucleotide method and with the walking strategy. The sequence of a DNA segment of 11,746 bp (**Figure 6A** and GenBank accession No MF766019) was obtained and its analysis revealed the existence of nine open reading frames (ORF), in addition to the 6 genes (*gutFRMCBA*) of the *gut* operon (**Figure 6A** and Supplementary Table S1). One open reading frame was detected upstream of the gut operon and was designated *tnp*, since its product has 100% identity with a multispecies transposase (Genbank accession No WP_003606336.1) widely distributed in the Lactobacillaceae family. Downstream of the gut operon were detected four ORF named orf1, orf2, orf3 and orf4, which could encode hypothetical proteins conserved in other LAB. In addition, the product of the named *res* gene belongs to the Ser-recombinase superfamily (cl02788) and specifically to the PinE conserved protein domain family (COG1961), showing more than 90% amino acid identity with proteins from oenococci, lactobacilli and pediococci annotated as Pin-related site-specific recombinases/DNA invertases. Also, two divergent genes named *tauE* and *tetR* seem to encode a TauE sulfite exporter which belongs to the TauE conserved domain family (pfam01925) and a transcriptional regulator belonging to the TetR family (domain architecture ID 11442015), and both proteins have more than 95% amino acid identity with their homologues in *Oenoccocus oeni* and lactobacilli.

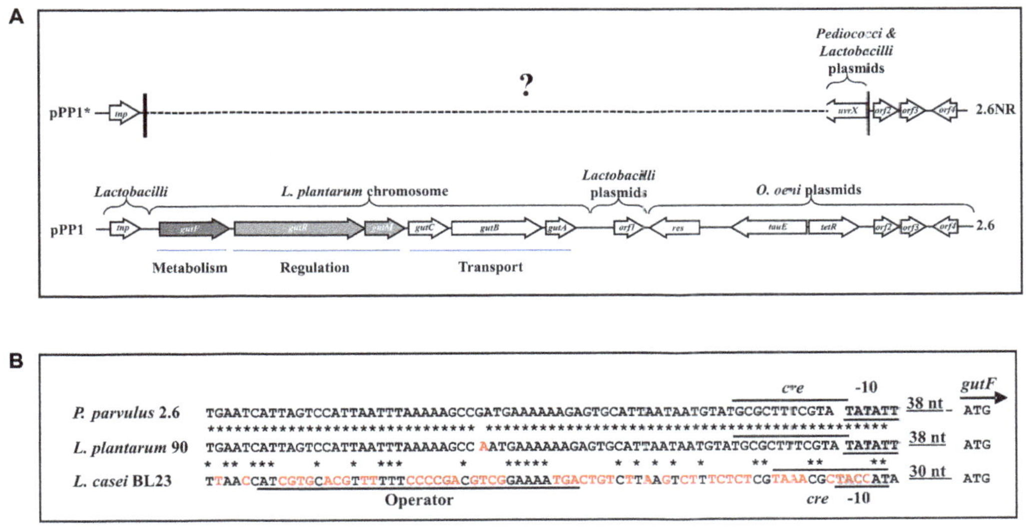

C

Bacteria	GutF	GutR	GutM	GutC	GutB	GutA
P. parvulus 2.6	100%	100%	100%	100%	100%	100%
L. plantarum 90	95%	98%	98%	99%	99%	100%
L. casei BL23	68%	24%	41%	68%	61%	55%

FIGURE 6 | The sorbitol operon and flanking DNA regions of *P. parvulus* 2.6. **(A)** Sequenced regions of pPP1 and pPP1* plasmids are shown. Genes and ORF are depicted and homologies at the nucleotide level are indicated (see details in the text). ? Indicates unknown sequence. Homology of regulatory genetic regions **(B)** and proteins **(C)** involved in sorbitol utilization between *P. parvulus* 2.6, *L. plantarum* 90 and *L. casei* BL23 is shown. **(B)** DNA sequences of: the putative P_{gut} promoters (–10 boxes) and the GutR operator of *L. casei* are underlined; the putative *cre* sequences for catabolic repression are overlined; also the start codon (ATG) of *gutF* gene and the nucleotides between this and the –10 boxes are depicted. Red letters indicate mismatches with the *P. parvulus* 2.6 DNA sequence. *Means identity of nucleotides. **(C)** Identity of the amino acid sequence of the Gut proteins from *L. plantarum* 90 and *L. casei* BL23 with those of *P. parvulus* 2.6 is depicted.

Based on the DNA sequence of the *gut* operon flanking regions in pPP1, and on the lack of the *gut* operon in 2.6NR, primers were designed and used to try to detect if there exists any identity between pPP1 and pPP1* by DNA sequencing. Two of these, pPP1*F and pPP1*R, located respectively upstream and downstream of the *gut* operon, provided the desired information (**Figure 6A**). A good chromatogram of the DNA sequencing of pPP1* using the 2.6NR plasmidic preparation and pPP1*F primer with 100% identity with pPP1 was obtained until nucleotide 156 in the chromatogram (548 nt in Genbank accession No WP_003606336.1), then at least two overlapping sequences were observed (Supplementary Figure S2A), and it was not possible from this point to deduce a further correct DNA sequence. This was not the case when DNA from the 2.6 strain was used as substrate, since a good chromatogram of the pPP1 DNA sequencing was obtained (Supplementary Figure S2B). However, the usage of pPP1*R allowed not only to determine that the homology between pPP1 and pPP1* starts again at nucleotide 10,021 (in Genbank accession No WP_003606336.1), but also that upstream of this position in pPP1* there exists a region including a *uvrX* putative gene identical to those of other pediococci (i.e., in pPC892-2 plasmid, Genbank accession No CP021472.1) and *Lactobacilli* (i.e., in pH10 plasmid, Genbank accession No CP002430.1) plasmids, which do not carry *orf2, orf3 and orf4*.

With regard to the *gut* operon of *P. parvulus* 2.6, the identity of the region including the genes and the upstream regulatory regions with the homologues of *L. plantarum* strains was 99% (**Figures 6A,B**) and nucleotides from 694-to 60020 in GenBank accession No MF766019), consequently the amino acid sequence of the Gut proteins of *P. parvulus* showed an identity ranging from 95 to 100%, with those of *L. plantarum* 90 (**Figure 6C**). No significant homology at the DNA sequence level was detected between the characterized operons of *L. casei* and those of *P. parvulus* (**Figure 6B** and results not shown). However, presumably due to convergent evolution, homology ranging from 68 to 24% amino acid identity was detected between the Gut proteins of *P. parvulus* 2.6 and of *L. casei* BL23 (**Figure 6C**).

Analysis of the *Gut* Operon Regulation

GutR and GutM of *P. parvulus* could be involved in regulation of the *gut* operon expression and upstream of the start codon of *P. parvulus* 2.6 *gutF* gene, a TATAtT sequence was detected that only deviates one nucleotide from the consensus −10 promoter region (**Figure 6B**). Thus, to gain insight into this potential regulation, complementation studies in heterologous LAB hosts able to utilize sorbitol were carried out. First, we cloned

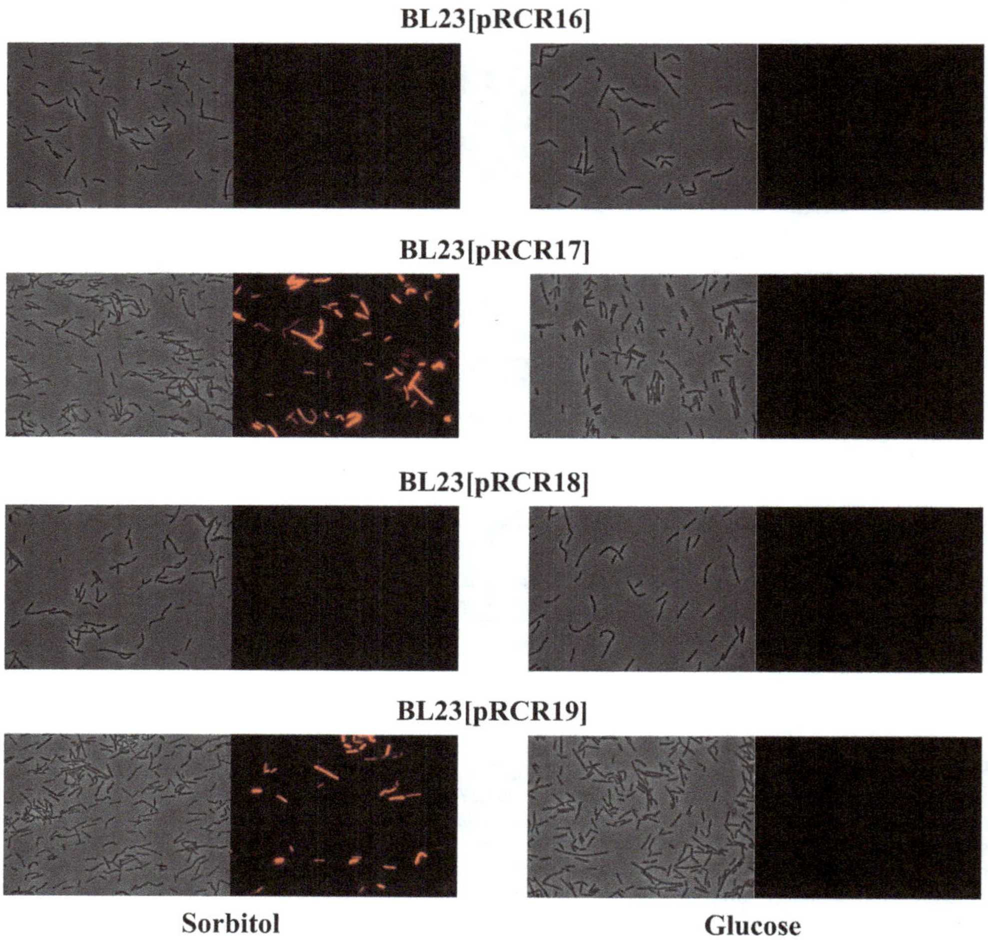

FIGURE 7 | Detection of fluorescence in *L. casei* BL23 strains at stationary phase. Cultures of the indicated strains in MRSS (sorbitol) or MRSG (glucose) were analyzed by phase contrast (left panels) or fluorescence (right panels) microscopy.

independently the putative promoter sequence (designated P_{gut}) and its upstream region (**Figure 6B**), as well as the transcriptional fusions P_{gut}-*gutR*, P_{gut}-*gutM* and P_{gut}-*gutRM* into the pRCR promoter probe vector (Mohedano et al., 2015) upstream of the *mrfp*, generating the pRCR16, pRCR17, pRCR18 and pRCR19 plasmids, respectively (**Figure 1**). Thus, functionality of the promoter and influence of GutR and GutM could be detected by measuring the levels of fluorescence of the mCherry encoded by the *mrfp* gene. As hosts to perform the studies, we chose: (i) the plasmid free *L. casei* BL23, because its sorbitol utilization and the regulation of its *gut* operon is known (Yebra and Pérez-Martínez, 2002; Nissen et al., 2005; Alcantara et al., 2008) and, (ii) *L. plantarum* 90, because we have previously detected in this bacterium efficient functional expression of mCherry from a pRCR derivative, without problems of plasmid incompatibility and that the copy number of the plasmid was 62 ± 2 molecules per bacterial genome (Russo et al., 2015).

The well characterized transcriptional activator GutR of *L. casei* BL23 controls expression of the *gut* operon of this bacteria and its operator site upstream of the P_{gut} has been identified as well as a catabolite repression element (*cre*) overlapping the −10 region of the promoter (Alcantara et al., 2008) (**Figure 6B**). The *P. parvulus* 2.6 GutR has only a low homology of amino acids (24%) with its homologue of *L. casei*, but like its counterpart belongs to the BglG transcriptional antiterminators family, possesses the PRD domain and the DNA helix turn helix binding domain. Therefore, both proteins could have a similar role. Alignment of the *L. casei* and *P. parvulus* -10 regions revealed that the upstream regulatory regions of BL23 strain has no clear homologs in the 2.6 strain (**Figure 6B**). Consequently, cross talk between transcriptional signals of *P. parvulus* and *L. casei* regulators should not take place, and influence of the pediococcal GutR and GutM in expression of P_{gut} from the 2.6 strain could be investigated in the BL23 strain without interferences. Thus, the pRCR derivatives were transferred independently to the BL23 strain and the recombinant bacteria were grown in MRS supplemented with either 1% glucose or 1% sorbitol until stationary phase prior to analysis. Examination of the cultures by fluorescent and phase contrast optical microscopy revealed that only bacteria carrying pRCR17 and pRCR19 and grown in

TABLE 3 | Heterologous expression of components of the *P. parvulus* 2.6 *gut* operon in *L. casei* BL23 carrying pRCR derivatives plasmids grown in either MRSS or MRSG.

L. casei strains	DNA insert in pRCR derivatives	Specific fluorescence[a]	
		MRSS	MRSG
BL23[pRCR16]	P_{gut}	0.14 ± 0.07	0.14 ± 0.01
BL23[pRCR17]	P_{gut}-*gutR*	5.69 ± 0.44	0.07 ± 0.09
BL23[pRCR18]	P_{gut}-*gutM*	0.19 ± 0.06	0.11 ± 0.02
BL23[pRCR19]	P_{gut}-*gutRM*	3.58 ± 0.06	0.05 ± 0.02

[a]*The specific fluorescence is depicted and it was calculated as the ratio of the detected fluorescence (5×) and the bacterial biomass estimated from the $OD_{600\ nm}$ of the culture.*

presence of sorbitol have fluorescence (**Figure 7**). In addition, fluorescence as well as the optical density of the cultures was measured and the specific fluorescence, referred to the biomass, was calculated. The fluorescence quantification confirmed that the P_{gut}-*gutRmrfp*, and P_{gut}-*gutRMmrfp* transcriptional fusions are activated upon growth in the presence of sorbitol (**Table 3**). Thus, these results revealed that expression from the P_{gut} required the activation by GutR and the presence of sorbitol in the growth medium. Moreover, they indicated that activation by GutR decreased, when GutM was present (5.69 ± 0.44 *versus* 3.58 ± 0.06).

Concerning the *L. plantarum* 90 host, its GutR has 98% homology to that of *P. parvulus* 2.6 (**Figure 6C**) and the DNA sequence of the region located upstream of the two P_{gut} promoters only differs in one nucleotide (**Figure 6B**). Consequently, both operons must have the same regulatory gene system, which implies that both systems could recognize each other. Thus, a trans-complementation process was expected between the regulatory proteins of Lp90 and the promoter region of 2.6. Therefore, the pRCR derivatives were transferred independently to the 90 strain and, since a cross talk is more complex situation, its comprehension required a more detailed analysis. For this reason, the recombinant bacteria, after growth in MRS supplemented with 1% glucose, were transferred to MRS fresh medium supplemented with either 1% sorbitol or 1% glucose and a time course assay of fluorescence and growth of the cultures was performed. The results revealed that all recombinant strains became fluorescent, when grown in the presence of sorbitol and, with the time of incubation the fluorescence increased (**Figure 8** and **Table 4**). In addition, analysis of the bacterial growth showed that all cultures in MRSG have very similar exponential growth rates (ranging from 0.889 ± 0.059 to 0.803 ± 0.049) and all entered slowly into stationary phase after 2 h of incubation (**Figure 8F** and **Table 4**). Initial transfer of the cultures to MRSS resulted in a similar decrease (around 50%) of the growth rate (values from 0.416 ± 0.045 to 0.495 ± 0.011) during the first 2 h of induction. Then, probably after consumption of the residual intracellular glucose or due to the induction process, bacteria decreased their growth rate to levels ranging from 0.259 ± 0.020 to 0.251 ± 0.034, besides the 90[pRCR18] (GutM overexpressor), that after stalling its growth from 2 h to 3 h incubation time decreased its growth rate to 0.239 ± 0.048, indicating that overexpression of GutM in absence of high levels of GutR has a negative impact for the cells. Furthermore, analysis of the specific levels of fluorescence of the cultures referred to their biomass (**Table 4**) showed different levels for the different fusions (P_{gut}-*mrfp* < P_{gut}-*gutRMmrfp* < P_{gut}-*gutRmrfp* < P_{gut}-*gutMmrfp*), showing that overexpression of GutM provokes the highest induction of expression from P_{gut}. In addition, the highest levels were observed after 4 h of induction for cells carrying either pRCR17 (22.38 ± 2.02) or pRCR19 (19.26 ± 2.10) *versus* the end of the incubation (6 h) for cells carrying pRCR16 (14.81 ± 0.66) and pRCR18 (26.83 ± 1.83).

Thus, the results revealed a trans-complementation of the *L. plantarum* regulatory proteins on expression driven from the *P. parvulus* P_{gut} promoter. Moreover, the results confirmed the role of inducer of GutR as well as requirement of sorbitol for expression from P_{gut} and support that co-expression of GutR and GutM decrease the activation mediated by GutR.

DISCUSSION

The overall metabolic results obtained here support that *P. parvulus* is able to synthesize EPS in MRS medium using either glucose or sorbitol as carbon sources. We have previously demonstrated that the 2-substituted (1,3)-β-D-glucan of *P. parvulus* 2.6 is synthesized by the GTF glycosyltransferase utilizing UDP-glucose as substrate (Werning et al., 2014). In addition, Velasco et al. (2007) determined that the 2.6 strain transport the glucose by a PMF-permease and possesses the α-phosphoglucomutase and the UDP-glucose pyrophosphorylase activities responsible for the conversion of glucose-6-P to glucose-1-P and further conversion of this compound to UDP-glucose. Thus, Velasco et al. (2007) showed how the 2.6 strain uses the glucose, not only for the central metabolism, *via* the glycolytic pathway, but also for the secondary metabolism involving a biosynthetic pathway for its EPS synthesis. In addition, the detection of the genetic determinants of sorbitol utilization by the 2.6 strain obtained in this work supports that the bacterium transports sorbitol by a PTS^{gut} system and converts sorbitol-6-P into fructose-6-P by the action of sorbitol-6-P dehydrogenase. Fructose-6-P can be converted to glucose-6-P by a reaction catalyzed by phosphoglucose isomerase, enzymatic activity that was also previously detected in the 2.6 strain (Velasco et al., 2007). Therefore, the 2.6 strain possesses the transport and enzymatic machineries for synthesis of the EPS from sorbitol. In addition, we have detected that aeration of the cultures during the growth improves sorbitol consumption (results not shown). Accordingly, the conversion of sorbitol-6-P into fructose-6-P requires NAD^- as an oxidative co-factor to produce NADH (Zarour et al., 2017). Analysis of the draft genome of the 2.6 strain showed the existence of a putative NADH oxidase coding gene. If this enzyme exists, it could unbalance the NAD^+/NADH equilibrium toward the oxidized form NAD^+.

The *P. parvulus* 2.6 2-substituted (1,3)-β-D-glucan is composed of molecules of glucose and consequently the

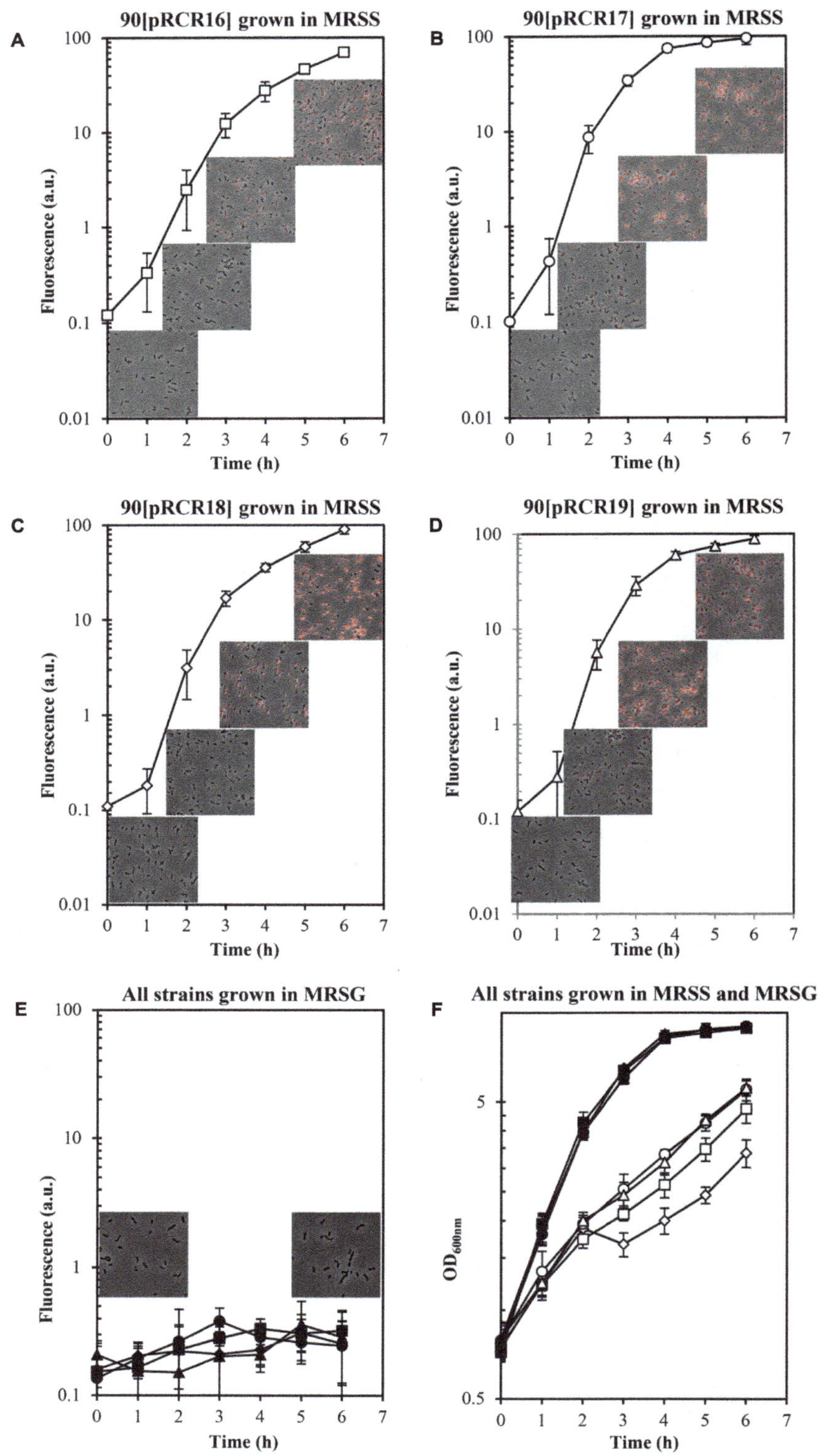

FIGURE 8 | Trans complementation of the sorbitol utilization regulatory machinery in *L. plantarum* 90. Bacteria carrying **(A,E,F)** pRCR16 (□, ■), **(B,E,F)** pRCR17 (○, ●), **(C,E,F)** pRCR18 (◇, ◆) or **(D,E,F)** pRCR19 (△, ▲) were grown in MRSG and at time 0 were transferred to fresh MRSS (open symbols) or to MRSG (closed symbols). OD_{600nm} and fluorescence of the cultures were monitored every hour. Overlays of images of the cultures taken at time 0 and after 2, 4, and 6 h incubation by phase contrast and fluorescent microscopy are depicted. The experiments were performed in triplicate and the mean value and standard deviation is depicted.

TABLE 4 | Heterologous expression of components of *P. parvulus* 2.6 *gut* operon in *L. plantarum* 90 carrying pCRC derivatives plasmids.

L. plantarum strains	DNA insert in pRCR derivatives	Medium	Initial growth rate[a] (μ)	Sorbitol induced growth rate[b] (μ)	Specific fluorescence[c]				
					2 h	3 h	4 h	5 h	6 h
90[pRCR16]	P_{gut}	MRSS	0.416 ± 0.045	0.251 ± 0.034	1.48 ± 1.03	5.86 ± 1.43	10.42 ± 1.56	13.44 ± 0.51	14.81 ± 0.66
90[pRCR17]	P_{gut}-*gutR*	MRSS	0.455 ± 0.065	0.259 ± 0.020	4.42 ± 1.22	13.41 ± 0.90	22.38 ± 2.02	20.30 ± 1.62	17.28 ± 1.65
90[pRCR18]	P_{gut}-*gutM*	MRSS	0.429 ± 0.025	0.239 ± 0.048	1.62 ± 0.69	10.26 ± 2.30	18.05 ± 2.85	24.26 ± 1.51	26.83 ± 1.83
90[pRCR19]	P_{gut}-*gutRM*	MRSS	0.495 ± 0.011	0.256 ± 0.016	2.80 ± 0.77	11.74 ± 1.51	19.26 ± 2.10	17.20 ± 1.71	15.98 ± 1.21
90[pRCR16]	P_{gut}	MRSG	0.889 ± 0.059	*ND*	0.06 ± 0.03	0.04 ± 0.00	0.04 ± 0.00	0.04 ± 0.01	0.04 ± 0.01
90[pRCR17]	P_{gut}-*gutR*	MRSG	0.825 ± 0.029	*ND*	0.07 ± 0.06	0.06 ± 0.02	0.04 ± 0.01	0.03 ± 0.00	0.03 ± 0.02
90[pRCR18]	P_{gut}-*gutM*	MRSG	0.803 ± 0.049	*ND*	0.06 ± 0.03	0.03 ± 0.03	0.03 ± 0.01	0.04 ± 0.01	0.03 ± 0.01
90[pRCR19]	P_{gut}-*gutRM*	MRSG	0.832 ± 0.034	*ND*	0.04 ± 0.02	0.03 ± 0.02	0.03 ± 0.01	0.04 ± 0.02	0.03 ± 0.02

[a]Growth rate of the cultures grown in MRSS was calculated from the $OD_{600\ nm}$ obtained during the first 2 h of induction. [b]Growth rate of the cultures grown in MRSS was calculated from the $OD_{600\ nm}$ obtained during the 2–6 h of induction, besides for 90[pRCR18], that due to the stalling of growth from 2 h to 3 h of incubation the growth rate was calculated from the data obtained from 3 h to 6 h of incubation. ND, the growth rate was not determined because the cultures have entered in the stationary phase of growth. [c]Specific fluorescence was calculated as the ratio of the detected fluorescence (5x) and the bacterial biomass estimated from the OD_{600} nm of the culture.

EPS concentration can be calculated as molarity of this monosaccharide (see secondary Y axis in **Figure 3A**). This calculation revealed that in both media this bacterium only used a small percentage of the substrate molecules (10 mM glucose plus 30 mM sorbitol in MRSGS or 10 mM glucose in MRSG) for synthesis of EPS (0.99 or 0.45 mM, respectively), whereas more than 90% was utilized in the glycolytic pathway to synthesize pyruvic acid (2 molecules per 1 molecule of substrate) and by action of the lactate dehydrogenase to finally generate lactic acid (1 molecule per 1 molecule of pyruvate, 79 mM or 18 mM). Moreover, the specific quantification method for 2-substituted (1,3)-β-D-glucan used here and the estimation of the specific concentration of EPS synthesized (**Figure 3C**) showed that, using as substrate either glucose or sorbitol, the bacterium synthesizes the same polymer and suggests that with the same efficiency. This was not the case when synthesis of this EPS utilizing fructose was tested, since levels were low compared with that obtained from glucose (Velasco et al., 2007). We have also observed a temporal delay of 2.6 to start to consume sorbitol in MRSGS (**Figure 3A**). This could be due to the existence of a catabolite repression of sorbitol utilization by glucose. Supporting this hypothesis, we have detected a potential *cre* operator (**Figure 6**) for the CcpA, which mediates with HPR this regulation in firmicutes (Deutscher, 2008).

In *P. parvulus* 2.6, the *gut* operon, as in other LAB, constitutes the genetic determinant for sorbitol transport and conversion into fructose-6-P. In addition, we have established here that it is located in a plasmid named pPP1 (**Figure 6**) which is unusual, since the almost identical operon of *L. plantarum* and that of *Lactobacillus pentosus* strain SLC13 (82% homologous, Genbank accession No CP022130.1) as well as the unrelated one from *L. casei* are located in the chromosome. As far as we know, only the location of an unrelated *gut* operon in the megaplasmid pMP118 from *L. salivarius* UCC118 has been previously described (Claesson et al., 2006). A search of the protein data banks revealed that *L. salivarius* 5713 and JCM1046 strains possess, respectively, the pHN3 and pMP1046A megaplasmids which carry *gut* operons homologous to that of pMP118 (Jiménez et al., 2010; Raftis et al., 2014). Flanking the operon two inverted repeat sequences (nucleotides 604-627 and 6612-6635 of Genbank accession No WP_003606336.1) were identified, which are also present at the same relative location in *L. plantarum* strains and at various locations in lactobacilli chromosomes and plasmids (even more than one copy per genome). The upstream region is preceded by a *tnp* gene encoding a putative transposase, which could be responsible for mobilization of the *gut* operon from plasmid to chromosome or *vice versa*.

Lactic acid bacteria are prone to carry more than one compatible plasmid and this facilitates exchange of different regions with physiological significance, that later on can be transferred to other bacteria by plasmid conjugation or mobilization (Cui et al., 2015). Thus, downstream of the *gut* operon of pPP1 there are DNA regions almost identical to that present in plasmids of lactobacilli, which along with *P. parvulus* can be contaminants of alcoholic beverages. Furthermore, the

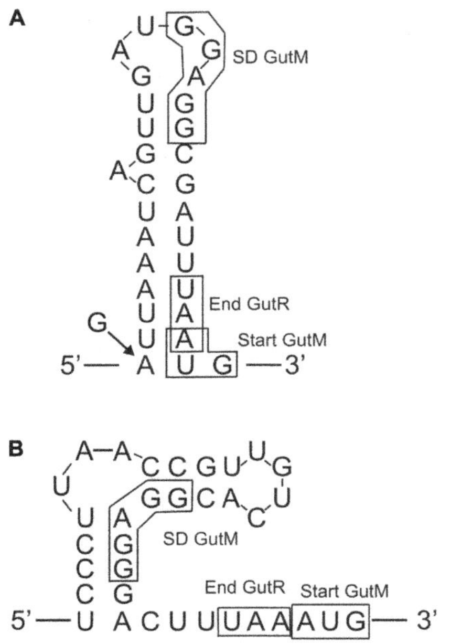

FIGURE 9 | Predicted folding of the *gut* mRNA region including the 3′-end of *gutR* and the 5′-end of *gutM*. **(A)** Secondary structure of the wild-type *P. parvulus* 2.6 and *L. plantarum* 90 *gut* operons. The arrow and the G indicate that in the insert of pRCR18 G substitute to the A in the indicated position. **(B)** Secondary structure of the *L. casei* BL23 *gut* operon.

Oenococcus oeni pOENI-1 and pOENI-1v2 plasmids (Favier et al., 2012) and pPP1 carry a region containing among others the *res, tauE* and *tetR* genes. The putative TauE sulfite exporter is possibly involved in adaptation to stress conditions during alcoholic beverage production (Favier et al., 2012). Thus, the recombinase or invertase site specific Res could be responsible for a mobilization of an element composed of a truncated *res, tauE* and *tetR* to a stable location, since at the 3′-end region of *res* and downstream of *tetR* unit exist inverted complementary sequences 5′-TTTTAAAGC-3′ and 5′-GCTTTAAAA-3′ (nucleotides 7774–7778 and 10021–10029 of Genbank accession No WP_003606336.1).

Another instance of plasmids rearrangement in *P. parvulus* is that which generated the profile and DNA sequence of 2.6NR strain plasmids (**Figure 6**). The initial isolate of 2.6NR strain generated in the Basque country University (BCU, Spain) and described in Fernández et al. (1995), kindly provided by Dr. Maria Teresa Dueñas (BCU) was studied in this work. Thus, the changes in plasmid cassettes were not produced in our laboratory, and presumably they took place upon treatment of 2.6 strain with ethidium bromide and novobiocin and was selected for the loss of the ropy phenotype. Thus, it is feasible that a formation of a co-integrate of pPP1 with other plasmid, may be pPP2, took place and convergent replication from two origins prompted to a deletion of one of the replicons, may be the pPP2, since its loss was envisaged, to generate pPP1*.

Concerning the regulation of expression of the *gut* operon, the overall results showed that it is repressed in the absence of sorbitol in the growth medium and that the *P. parvulus* GutR is an activator like the *L. casei* BL23 regulator (Alcantara et al., 2008). In this system, it has been proposed that GutM is involved in the activation, since a decreased expression of the *gut* operon was detected in a GutM deficient mutant (Alcantara et al., 2008). Furthermore, the complementation studies in *L. plantarum* 90 performed here showed a heterologous regulation of gene expression from the pediococcal P_{gut} promoter by the GutR from *Lactobacillus*, and a positive effect when only the pediococcal GutM was overexpressed (**Figure 8** and **Table 4**). Thus, these results suggest that a protein-protein interaction between the *P. parvulus* GutM and the *L. plantarum* GutR could potentiate the activation of the P_{gut} promoter, since, *P. parvulus* 2.6 GutR and GutM have 98% identity with their homologues of *L. plantarum* 90. In addition, either in *L. casei* BL23 and in *L. plantarum* a decrease of expression from P_{gut} was observed when GutM was overexpressed in combination with GutR (**Figure 8** and **Tables 3, 4**). This prompted us to analyze the genetic environment of *gutR* and *gutM*. An overlapping of the last nucleotide of the termination codon (TAA) of *gutR* and the first nucleotide of the start codon (ATG) of *gutM* was detected in *P. parvulus* 2.6 and *L. plantarum* 90 genomes. This indicated that post-transcriptional regulation of the *gut* operon could exist in this bacterium. For this reason, the secondary structure of the region surrounding the overlapping in the *gut* mRNA was folded with the Mfol program (**Figure 9A**). A secondary stem-loop structure was predicted with a $\Delta G = -5.6$ kcal mol^{-1}, the ribosomal binding site (RBS) of *gutM* (5′-GGAGG-3′) was located at the loop and partially blocked in the stem of the structure. Thus, even though the sequence of the RBS of *gutM* indicates a high efficiency of utilization for the ribosome, the initiation of translation of *gutM* could be partially impaired by the partial RBS blockage, which would be released by the opening provoked by the passage of the ribosomes translating *gutR*. In addition, the overlapping of *gutR* and *gutM* is located at the end of the stem of the structure. Thus, two post-transcriptional regulations could take place: (i) translation of *gutR* can act by favoring translation of *gutM* by exposition of its RBS and (ii) a -1 frameshift (Atkins et al., 2016) could happen at the TAA termination codon of *gutR* and ribosomes translating this could step back one nucleotide and upon charging the corresponding tRNA read the Leu (TTA) codon and continue translating *gutM*. In this way a fused peptide GutR-M could be synthesized. The same structure could be formed in the transcript encoded by the plasmid pRCR19 with a $\Delta G = -5.9$ kcal mol^{-1}, containing *gutRM* and which could be a substrate for the two proposed post-transcriptional regulations. Furthermore, the DNA fragment cloned in pRCR18, lacks most of the *gutR* gene but still retains some of the 3′-end region of this gene and the encoded mRNA can form a secondary structure almost identical to the wild-type structure (with only a change of A-U by G-U pairing at the end of the stem, **Figure 9**). Thus in bacteria carrying pRCR18 partial blockage of the RBS could take place, but synthesis of GutR-M could not occur. This could explain the antagonistic effect of overexpression of *gutM* from pRCR18 (increase of expression from P_{gut}) and pRCR19 (decrease of expression from P_{gut}), if GutR-M exists and has a role.

Prediction of transmembrane regions in the regulatory proteins with the TM-Pred revealed that GutR is a soluble protein and that the first amino acids from 1 to 21 of GutM constituted a transmembrane region also predicted for the GutR-M fused polypeptide. This fused polypeptide could provide an efficient anchoring of the regulator to the membrane bringing it close to the PTS^{gut} system facilitating the phosphorylation of GutR and resulting in the physiological optimal expression of the operon. This generation of a fused polypeptide could also take place in *L. plantarum* but does not seems to occur in *L. casei,* since in this bacterium the TAA translational stop codon of GutR and the ATG start codon of GutM are adjacent and not overlapped (**Figure 9**). However, the *L. casei gut* transcript can form a secondary structure with a $\Delta G = -9.8$ kcal mol^{-1} which could block the RBS of *gutM* gene, couple translation of GutR and GutM could take place, and protein-protein interaction could be responsible for higher activation of the system at the beginning of the induction process. Our results indicate that high levels of GutM synthesized from a multicopy plasmid have a deleterious effect for the bacteria (**Figure 8**) and probably the proposed models of posttranscriptional regulation are designed to have the right concentration of regulatory proteins. Nevertheless, further experiments are required to pinpoint the role of GutM and of the putative GutR-M polypeptide of *P. parvulus*.

AUTHOR CONTRIBUTIONS

AP-R contributed to all parts of the experimental work and wrote a draft of the manuscript. MW performed the initial detection of sorbitol utilization and characterization of *gut* genes. AP contributed to the characterization of the sorbitol metabolism. PR participated in the elaboration of the manuscript and analysis of the DNA sequences. GS contributed to the design and analysis of the experimental work involving characterization of regulation of *gut* operon expression. MM contributed to the design of strategies to determine trans complementation of the *gut* operon and corrected the manuscript. PL participated in study conception, data interpretation and generated the final version of the manuscript. All authors have read and approved the final manuscript.

ACKNOWLEDGMENTS

The authors thank Dr. Stephen W. Elson for critical reading of the manuscript and to Dr. Mª Teresa Dueñas for providing the original *P. parvulus* 2.6NR isolate to perform this study.

REFERENCES

Alcantara, C., Sarmiento-Rubiano, L. A., Monedero, V., Deutscher, J., Perez-Martinez, G., and Yebra, M. J. (2008). Regulation of *Lactobacillus casei* sorbitol utilization genes requires DNA-binding transcriptional activator GutR and the conserved protein GutM. *Appl. Environ. Microbiol.* 74, 5731–5740. doi: 10.1128/AEM.00230-08

Aldridge, P., Metzger, M., and Geider, K. (1997). Genetics of sorbitol metabolism in *Erwinia amylovora* and its influence on bacterial virulence. *Mol. Gen. Genet.* 256, 611–619.

Atkins, J. F., Loughran, G., Bhatt, P. R., Firth, A. E., and Baranov, P. V. (2016). Ribosomal frameshifting and transcriptional slippage: from genetic steganography and cryptography to adventitious use. *Nucleic Acids Res.* 44, 7007–7078. doi: 10.1093/nar/gkw530

Berthier, F., Zagorec, M., Champomier-Vergès, M., Ehrlich, S. D., and Morel-Deville, F. (1996). Efficient transformation of *Lactobacillus sake* by electroporation. *Microbiology* 142, 1273–1279. doi: 10.1099/13500872-142-5-1273

Boyd, D. A., Thevenot, T., Gumbmann, M., Honeyman, A. L., and Hamilton, I. R. (2000). Identification of the operon for the sorbitol (glucitol) phosphoenolpyruvate: sugar phosphotransferase system in *Streptococcus mutans Infect. Immun.* 68, 925–930.

Claesson, M. J., Li, Y., Leahy, S., Canchaya, C., van Pijkeren, J. P., Cerdeño-Tárraga, A. M., et al. (2006). Multireplicon genome architecture of *Lactobacillus salivarius. Proc. Natl. Acad. Sci. U.S.A.* 103, 6718–6723. doi: 10.1073/pnas.0511060103

Cui, Y., Hu, T., Qu, X., Zhang, L., Ding, Z., Dong, A., et al. (2015). Plasmids from food lactic acid bacteria: diversity, similarity, and new developments. *Int. J. Mol. Sci.* 16, 13172–13202.

de Man, J. C., Rogosa, M., and Sharpe, M. E. (1960). A medium for the cultivation of lactobacilli. *J. Appl. Microbiol.* 23, 130–135. doi: 10.1111/j.1365-2672.1960.tb00188.x

Deutscher, J. (2008). The mechanisms of carbon catabolite repression in bacteria. *Curr. Opin. Microbiol.* 11, 87–93. doi: 10.1016/j.mib.2008.02.007

Dornan, S., and Collins, M. (1987). High efficiency electroporation of *Lactococcus lactis* subsp. *lactis* LM0230. *Lett. Appl. Microbiol.* 11, 62–64. doi: 10.1111/j.1472-765X.1990.tb01275.x

Dueñas-Chasco, M. T., Rodríguez-Carvajal, M. A., Mateo, P. T., Franco-Rodríguez, G., Espartero, J. L., Irastorza-Iribas, A., et al. (1997). Structural analysis of the exopolysaccharide produced by *Pediococcus damnosus* 2.6. *Carbohydr Res.* 303, 453–458. doi: 10.1016/S0008-6215(97)00192-4

Favier, M., Bilhère, E., Lonvaud-Funel, A., Moine, V., and Lucas, P. M. (2012). Identification of pOENI-1 and related plasmids in *Oenococcus oeni* strains performing the malolactic fermentation in wine. *PLOS ONE* 7:e49082. doi: 10.1371/journal.pone.0049082

Fernández-Bañares, F., Esteve, M., and Viver, J. M. (2009). Fructose-sorbitol malabsorption. *Curr. Gastroenterol. Rep.* 11, 368–374.

Fernández, K., Dueñas, M. T., Irastorza, A., Bilbao, A., and del Campo, G. (1995). Characterization and DNA plasmid analysis of ropy *Pediococcus* spp. strains isolated from Basque Country ciders. *J. Food Prot.* 59, 35–40.

Fernández de Palencia, P., Werning, M. L., Sierra-Filardi, E., Dueñas, M. T., Irastorza, A., Corbí, A. L. et al. (2009). Probiotic properties of the 2-substituted (1,3)-β-d-Glucan-producing bacterium *Pediococcus parvulus* 2.6. *Appl. Environ. Microbiol.* 75, 4887–4891. doi: 10.1128/AEM.00394-09

Immerstrand, T., Paul, C. J., Rosenquist, A., Deraz, S., Mårtensson, O. B., Ljungh, A., et al. (2010). Characterization of the properties of *Pediococcus parvulus* for probiotic or protective culture use. *J. Food Prot.* 73, 960–966.

Jiménez, E., Martín, R., Maldonado, A., Martín, V., Gómez de Segura, A., Fernández, L., et al. (2010). Complete genome sequence of *Lactobacillus salivarius* CECT 5713, a probiotic strain isolated from human milk and infant feces. *J. Bacteriol.* 192, 5266–5267. doi: 10.1128/jb.00703-10

Ladero, V., Ramos, A., Wiersma, A., Goffin, P., Schanck, A., Kleerebezem, M., et al. (2007). High-level production of the low-calorie sugar sorbitol by *Lactobacillus plantarum* through metabolic engineering. *Appl. Environ. Microbiol.* 73, 1864–1872. doi: 10.1128/AEM.02304-06

Lamontanara, A., Caggianiello, G., Orrù, L., Capozzi, V., Michelotti, V., Bayjanov, J. R., et al. (2015). Draft genome sequence of *Lactobacillus plantarum* Lp90 isolated from wine. *Genome Announc.* 3:e00097-15. doi: 10.1128/genomeA.00097-15

López, P., Martínez, S., Diaz, A., Espinosa, M., and Lacks, S. A. (1989). Characterization of the *polA* gene of *Streptococcus pneumoniae* and comparison of the DNA polymerase I it encodes to homologous enzymes from *Escherichia coli* and phage T7. *J. Biol. Chem.* 264, 4255–4263.

Macrina, F. L., Kopecko, D. J., Jones, K. R., Ayers, D. J., and McCowen, S. M. (1978). A multiple plasmid-containing *Escherichia coli* strain: convenient source of size reference plasmid molecules. *Plasmid* 1, 417–420. doi: 10.1016/0147-619X(78)90056-2

Mazé, A., Boël, G., Zúñiga, M., Bourand, A., Loux, V., Yebra, M. J., et al. (2010). Complete genome sequence of the probiotic *Lactobacillus casei* strain BL23. *J. Bacteriol.* 192, 2647–2648. doi: 10.1128/JB.00076-10

Mohedano, M. L., García-Cayuela, T., Pérez-Ramos, A., Gaiser, R., Requena, T., and López, P. (2015). Construction and validation of a mCherry protein vector for promoter analysis in *Lactobacillus acidophilus*. *J. Ind. Microbiol. Biotechnol.* 42, 247–253. doi: 10.1007/s10295-014-1567-4

Montalto, M., Gallo, A., Ojetti, V., and Gasbarrini, A. (2013). Fructose, trehalose and sorbitol malabsorption. *Eur. Rev. Med. Pharmacol. Sci.* 17, 26–29.

Nissen, L., Perez-Martinez, G., and Yebra, M. J. (2005). Sorbitol synthesis by an engineered *Lactobacillus casei* strain expressing a sorbitol-6-phosphate dehydrogenase gene within the lactose operon. *FEMS Microbiol. Lett.* 249, 177–183. doi: 10.1016/j.femsle.2005.06.010

Notararigo, S., Casas-Engel, M., Fernández de Palencia, P., Corbí, A. L., and López, P. (2014). Immunomodulation of human macrophages and myeloid cells by 2-substituted (1-3)-β-D-glucan from *P. Parvulus* 2.6. *Carbohydr. Polym.* 112, 109–113. doi: 10.1016/j.carbpol.2014.05.073

Notararigo, S., Nacher-Vazquez, M., Ibarburu, I., Werning, M. L., de Palencia, P. F., Duenas, M. T., et al. (2013). Comparative analysis of production and purification of homo- and hetero-polysaccharides produced by lactic acid bacteria. *Carbohydr. Polym.* 93, 57–64. doi: 10.1016/j.carbpol.2012.05.016

Pérez-Ramos, A., Mohedano, M. L., Puertas, A., Lamontanara, A., Orru, L., Spano, G., et al. (2016). Draft genome sequence of *Pediococcus parvulus* 2.6, a probiotic β-glucan producer strain. *Genome Announc.* 4:e01381-16. doi: 10.1128/genomeA.01381-16

Raftis, E. J., Forde, B. M., Claesson, M. J., and O'Toole, P. W. (2014). Unusual genome complexity in *Lactobacillus salivarius* JCM1046. *BMC Genomics* 15:771. doi: 10.1186/1471-2164-15-771

Rhodes, M. W., and Kator, H. (1999). Sorbitol-fermenting bifidobacteria as indicators of diffuse human faecal pollution in estuarine watersheds. *J. Appl. Microbiol.* 87, 528–535.

Russo, P., Iturria, I., Mohedano, M. L., Caggianiello, G., Rainieri, S., Fiocco, D., et al. (2015). Zebrafish gut colonization by mCherry-labelled lactic acid bacteria. *Appl. Microbiol. Biotechnol.* 99, 3479–3490. doi: 10.1007/s00253-014-6351-x

Russo, P., López, P., Capozzi, V., Férnandez de Palencia, P., Dueñas, M. T., Spano, G., et al. (2012). Beta-glucans improve growth, viability and colonization of probiotic microorganisms. *Int. J. Mol. Sci.* 13, 6026–6039. doi: 10.3390/ijms13056026

Sarmiento-Rubiano, L. A., Zuniga, M., Perez-Martinez, G., and Yebra, M. J. (2007). Dietary supplementation with sorbitol results in selective enrichment of lactobacilli in rat intestine. *Res. Microbiol.* 158, 694–701. doi: 10.1016/j.resmic.2007.07.007

Silveira, M. M., and Jonas, R. (2002). The biotechnological production of sorbitol. *Appl. Microbiol. Biotechnol.* 59, 400–408. doi: 10.1007/s00253-002-1046-0

Stülke, J., Arnaud, M., Rapaport, G., and Martin-Verstraete, I. (1998). PRD — a protein domain involved in PTS-dependent induction and carbon catabolite repression of catabolic operons in bacteria. *Mol. Microbiol.* 28, 865–874.

Tangney, M., Brehm, J. K., Minton, N. P., and Mitchell, W. J. (1998). A gene system for glucitol transport and metabolism in *Clostridium beijerinckii* NCIMB 8052. *Appl. Environ. Microbiol.* 64, 1612–1619.

Velasco, S. E., Areizaga, J., Irastorza, A., Dueñas, M. T., Santamaria, A., and Muñoz, M. E. (2009). Chemical and rheological properties of the β- glucan produced by *Pediococcus parvulus* 2.6. *J. Agric. Food Chem.* 57, 1827–1834. doi: 10.1021/jf803065w

Velasco, S. E., Yebra, M. J., Monedero, V., Ibarburu, I., Dueñas, M. T., and Irastorza, A. (2007). Influence of the carbohydrate source on β-glucan production and enzyme activities involved in sugar metabolism in *Pediococcus parvulus* 2.6. *Int. J. Food Microbiol.* 115, 325–334. doi: 10.1016/j.ijfoodmicro.2006.12.023

Wegmann, U., O'connell-Motherway, M., Zomer, A., Buist, G., Shearman, C., Canchaya, C., et al. (2007). Complete genome sequence of the prototype lactic acid bacterium *Lactococcus lactis* subsp. *cremoris* MG1363. *J. Bacteriol.* 189, 3256–3270. doi: 10.1128/JB.01768-06

Werning, M. L., Ibarburu, I., Dueñas, M. T., Irastorza, A., Navas, J., and López, P. (2006). *Pediococcus parvulus gtf* gene encoding the GTF glycosyltransferase and its application for specific PCR detection of beta-D-glucan-producing bacteria in foods and beverages. *J. Food Prot.* 69, 161–169.

Werning, M. L., Pérez-Ramos, A., Fernández de Palencia, P., Mohedano, M. L., Dueñas, M. T., Prieto, A., et al. (2014). A specific immunological method to detect and quantify bacterial 2-substituted (1,3)-β-d-glucan. *Carbohydr. Polym.* 113, 39–45. doi: 10.1016/j.carbpol.2014.06.072

Yamada, M., and Saier, M. H. J. (1988). Positive and negative regulators for glucitol (gut) operon expression in *Escherichia coli*. *J. Mol. Biol.* 203, 569–583.

Yebra, M. J., and Pérez-Martínez, G. (2002). Cross-talk between the L-sorbose and D-sorbitol (D-glucitol) metabolic pathways in *Lactobacillus casei*. *Microbiology* 148, 2351–2359.

Zarour, K., Vieco, N., Pérez-Ramos, A., Nácher-Vázquez, M., Mohedano, M. L., and López, P. (2017). "Food ingredients synthesised by lactic acid bacteria," in *Microbial Production of Ingredients and Additives*, eds A. M. Holban and A. M. Grumezescu (Cambridge: Elsevier), 89–124.

Zumbé, A., Lee, A., and Storey, D. (2001). Polyols in confectionery: the route to sugar-free, reduced sugar and reduced calorie confectionery. *Br. J. Nutr.* 85, S31–S45.

13

The Transcriptome of *Streptococcus pneumoniae* Induced by Local and Global Changes in Supercoiling

Adela G. de la Campa[1,2]*, María J. Ferrándiz[1], Antonio J. Martín-Galiano[1], María T. García[3] and Jose M. Tirado-Vélez[1]

[1] Unidad de Genética Bacteriana, Centro Nacional de Microbiología, Instituto de Salud Carlos III, Madrid, Spain,
[2] Presidencia, Consejo Superior de Investigaciones Científicas, Madrid, Spain, [3] Departamento de Microbiología, Facultad de Ciencias Biológicas, Universidad Complutense, Madrid, Spain

Edited by:
Manuel Espinosa,
Molecular Microbiology and Infection Biology (CIB, CSIC), Spain

Reviewed by:
Andrea M. Mitchell,
University of Birmingham, United Kingdom
Jorge Bernardo Schvartzman,
Cellular and Molecular Biology (CIB, CSIC), Spain

***Correspondence:**
Adela G. de la Campa
agcampa@isciii.es

The bacterial chromosome is compacted in a manner optimal for DNA transactions to occur. The degree of compaction results from the level of DNA-supercoiling and the presence of nucleoid-binding proteins. DNA-supercoiling is homeostatically maintained by the opposing activities of relaxing DNA topoisomerases and negative supercoil-inducing DNA gyrase. DNA-supercoiling acts as a general *cis* regulator of transcription, which can be superimposed upon other types of more specific *trans* regulatory mechanism. Transcriptomic studies on the human pathogen *Streptococcus pneumoniae*, which has a relatively small genome (~2 Mb) and few nucleoid-binding proteins, have been performed under conditions of local and global changes in supercoiling. The response to local changes induced by fluoroquinolone antibiotics, which target DNA gyrase subunit A and/or topoisomerase IV, involves an increase in oxygen radicals which reduces cell viability, while the induction of global supercoiling changes by novobiocin (a DNA gyrase subunit B inhibitor), or by seconeolitsine (a topoisomerase I inhibitor), has revealed the existence of topological domains that specifically respond to such changes. The control of DNA-supercoiling in *S. pneumoniae* occurs mainly via the regulation of topoisomerase gene transcription: relaxation triggers the up-regulation of gyrase and the down-regulation of topoisomerases I and IV, while hypernegative supercoiling down-regulates the expression of topoisomerase I. Relaxation affects 13% of the genome, with the majority of the genes affected located in 15 domains. Hypernegative supercoiling affects 10% of the genome, with one quarter of the genes affected located in 12 domains. However, all the above domains overlap, suggesting that the chromosome is organized into topological domains with fixed locations. Based on its response to relaxation, the pneumococcal chromosome can be said to be organized into five types of domain: up-regulated, down-regulated, position-conserved non-regulated, position-variable non-regulated, and AT-rich. The AT content is higher in the up-regulated than in the down-regulated domains. Genes within the different domains share structural and functional characteristics. It would seem that a topology-driven selection pressure has defined the chromosomal location of the metabolism, virulence and competence genes, which suggests the existence of topological rules that aim to improve bacterial fitness.

Keywords: DNA supercoiling, DNA topoisomerases, fluoroquinolones, global transcription, interactome, novobiocin, seconeolitsine, topological domains

INTRODUCTION

The compaction of DNA by up to 1000-fold (Holmes and Cozzarelli, 2000) in the bacterial chromosome, or nucleoid, achieves the optimal condition under which its essential functions – replication, segregation and gene expression (reviewed by Dorman, 2013) – can be reconciled. This compaction is mediated by both the natural supercoiling of the DNA, and by the binding of nucleoid-associated proteins (NAPs) (Wang et al., 2013). NAPs form a functional network that maintains DNA topology by bending, wrapping, bridging and constraining supercoils. Although several NAPs have been characterized in the Gram-negative bacterium *Escherichia coli*, very few have been detected in Gram-positive bacteria, including the human pathogen *Streptococcus pneumoniae* (Dillon and Dorman, 2010). In bacteria, gene transcription is regulated by DNA-supercoiling. This functions as a general *cis* regulator of transcription, and can be superimposed upon other types of more specific *trans* regulatory mechanisms. *cis* regulation can also occur via promoter DNA sequences. Factors acting in *trans* include structural and regulatory proteins. NAPs (structural proteins) target a number of genes (Dillon and Dorman, 2010), while specific regulatory proteins facilitate or inhibit the interaction of RNA polymerase with specific promoter regions (Browning and Busby, 2004). The precision balance of DNA supercoiling is thus modulated by a network of self-regulating factors.

DNA topoisomerases, which are present in all bacteria, are responsible for the maintenance of DNA-supercoiling. These enzymes are classified into two types based on their DNA cleavage pattern: type I, which cleaves only one DNA strand, and type II, which cleaves both. The type II topoisomerases, gyrase and topoisomerase IV (Topo IV), are tetrameric proteins with two subunits: $GyrA_2GyrB_2$ in gyrase, and $ParC_2ParE_2$ in Topo IV. Supercoiling homeostasis is achieved by the competing activities of gyrase and topoisomerase I (Topo I, a type I isomerase) plus IV (Champoux, 2001); gyrase introduces negative supercoils into DNA (Gellert et al., 1976), Topo I relaxes DNA, and Topo IV both relaxes DNA and participates in chromosome partitioning (Kato et al., 1990). *S. pneumoniae* (the pneumococcus) has a relatively small genome (~2 Mb compared to ~4.6 Mb for *E. coli*) rich in AT (60%), that carries genes for all three of the above enzymes. These characteristics are shared by other pathogens of the genus *Streptococcus*, including *S. pyogenes* and *S. suis*.

Streptococcus pneumoniae is the primary cause of community-acquired pneumonia, meningitis, bacteremia, and otitis media in children. Worldwide, 1 million children under 5 years of age die every year of pneumococcal infections (World Health Organization, 2007). The use of the pneumococcal 7-valent conjugate vaccine, which covers the serotypes most often associated with resistance to antibiotics, has achieved a decline in the incidence of invasive pneumococcal disease (Whitney et al., 2003; Kyaw et al., 2006) and a reduction in penicillin resistance rates (Kyaw et al., 2006; Pilishvili et al., 2010). However, serotypes not included in the vaccine soon emerged, highlighting the limitations of anti-pneumococcal prophylaxis (Moore et al., 2008; Fenoll et al., 2009).

The post-genomic age is beginning to provide answers to questions regarding how chromosomes are topologically organized, and how this organization influences bacterial evolution. Several degrees of organization in bacterial chromosomes have been observed, based on size (for a recent review see Badrinarayanan et al., 2015). Macrodomains are found at the megabase-size range. *E. coli*, for example, has four macrodomains: Ori (origin of replication), Ter (terminus of replication), Left, and Right, plus two less-structured regions flanking the Ori macrodomain (Espeli et al., 2008). Macrodomains may be maintained by specific proteins, such as the macrodomain Ter proteins (MatPs) that bind, as the name suggests, to specific sites in the Ter macrodomain (Dupaigne et al., 2012). However, no such proteins stabilizing the other macrodomains have been identified, and MatP proteins are found only in enteric bacteria. Non-homologous proteins may therefore take on similar roles in other bacteria. Supercoiling domains are found at the kilobase range. These are isolated loops that coil up around themselves; proteins at their bases help to topologically isolate the looped DNA. These loops were initially detected in electron micrographs of lysed *E. coli* cells (Kavenoff and Bowen, 1976). Later studies estimated the number of supercoil domains by assessing the numbers of nicks required to fully relax the chromosome. From these experiments it was estimated that the *E. coli* chromosome contains about 40 domains of around 100 kb (Worcel and Burgi, 1972; Sinden and Pettijohn, 1981). Studies in *Caulobacter crescentus* suggested domains ranging in length from 30 to 420 kb (Le et al., 2013). In *Salmonella enterica*, these domains were estimated to be 20 kb long by taking into account the site-specific recombination events that occurred between chromosomal sites distant from one another (Higgins et al., 1996). Later, transcriptional data predicted sizes of ~10 kb for *E. coli* (Postow et al., 2004). Controversy regarding the size and definition of domains remains, perhaps as a consequence of the different methods being used in their calculation.

The availability of drugs against all the topoisomerases of *S. pneumoniae* (**Figure 1**) has helped in determining the existence of chromosomal domains. This review summarizes the transcriptomic alterations induced by these agents, and how these changes can be interpreted to provide definitions of the chromosome domains in this bacterium. Changes induced by the clinically used fluoroquinolones (FQs) levofloxacin (LVX), and moxifloxacin (MOX) are first considered, followed by those that occur concomitantly with a global change in supercoiling, as induced by novobiocin (NOV, an inhibitor of the gyrase B subunit) and seconeolitsine (SCN, an inhibitor of Topo I). Overall, these studies reveal the *S. pneumoniae* genome to be organized into topology-reacting gene clusters, or supercoiling domains. The conservation of the location of these domains in the *Streptococcus* genus, and their enrichment for specific functions, suggests the existence of topological rules that aim to improve fitness via tight physiological feedback.

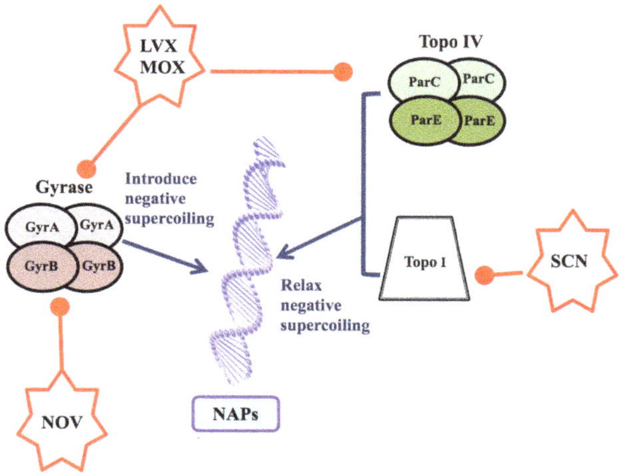

FIGURE 1 | Factors determining the topology of the *Streptococcus pneumoniae* chromosome. The level of supercoiling is controlled by three DNA topoisomerases. Gyrase ($GyrA_2GyrB_2$) is inhibited by LVX, MOX (which inhibit GyrA) and NOV (which inhibits GyrB). Topo IV ($ParC_2ParE_2$) is inhibited by LVX and MOX (which inhibit ParC). Topo I, a monomer, is inhibited by SCN. The topological organization of the chromosome depends on the level of DNA supercoiling and on the presence of NAPs.

CONTROL OF TRANSCRIPTION BY LOCAL CHANGES IN SUPERCOILING

Strains of *S. pneumoniae* resistant to antibiotics that act on the cell wall (beta-lactams) and on protein synthesis (macrolides) have proliferated in the last 30 years (Jacobs et al., 2003; Liñares et al., 2010). Consequently, pneumococcal infections are nowadays fought with LVX and MOX, which inhibit DNA topoisomerases. FQs target the type II DNA topoisomerases gyrase and Topo IV. Their mechanism of action involves the formation of DNA-FQ-topoisomerase complexes, which sterically inhibit replication and transcription and the subsequent generation of detrimental double-stranded DNA breaks (Drlica et al., 2008). Bacterial survival depends on the resolution of these breaks. Reactive oxygen species (ROS), such as superoxide anions, hydrogen peroxide and hydroxyl radicals contribute to FQ-mediated cell death via a protein synthesis-dependent pathway (Wang et al., 2010). This observation is consistent with the general model explaining the lethality of bactericidal antibiotics, which attributes a role to ROS generated via the Fenton reaction. The original reports supporting this model based their conclusions on the use of microarrays to study the transcriptional response to the inhibition of *E. coli* GyrA by an FQ or the peptide toxin CcdB. Under these conditions, global transcription was altered. In addition to the up-regulation of SOS damage response genes, genes related to superoxide stress, iron-sulfur cluster synthesis and iron uptake were up-regulated too (Dwyer et al., 2007). ROS production was also observed with a variety of bactericidal antibiotic families, in addition to FQs, each with a different intracellular target (reviewed by Dwyer et al., 2015). However, the intervening pathways lying between the initial antibiotic-target interaction and ROS formation have yet to be fully characterized.

The treatment of *S. pneumoniae* with FQs involves causing double-stranded breaks in the bacterial chromosome (Ferrándiz et al., 2016b), and as in other bacteria this requires active protein synthesis (Brito et al., 2017). Treatment with LVX or MOX (Ferrándiz and de la Campa, 2014; Ferrándiz et al., 2016b) is reported not to alter the level of global supercoiling. Nor are changes in supercoiling observed in *E. coli* exposed to oxolinic acid (Snyder and Drlica, 1979), although changes have been observed in the latter after treatment with the FQ norfloxacin (Peter et al., 2004). These differences might be attributable to species-dependent affinities of each drug for Topo IV or gyrase. For instance, Topo IV is the primary target of most FQs in Gram-positive bacteria, including *S. pneumoniae*, with gyrase a secondary target (Janoir et al., 1996; Muñoz and de la Campa, 1996; Tankovic et al., 1996; Fernández-Moreira et al., 2000). In contrast, in Gram-negative bacteria, including *E. coli*, gyrase is the primary target. At the LVX concentrations used in *S. pneumoniae* experiments, only Topo IV would have been inhibited, and no global change in supercoiling would be expected. However, at the MOX concentrations used, both gyrase and Topo IV would have been inhibited, suggesting that the inhibition of their opposing activities preserved the net level of supercoiling. Nevertheless, local topological changes are predictable in both cases and these would produce alterations in the transcriptome. Indeed, FQs induce a transcriptional response in *S. pneumoniae*, in which the differentially expressed genes (DEGs) account for 5.2 and 6.5% of the genome for LVX and MOX, respectively. In this bacterium, which lacks a proper SOS-like system, activation of the competence regulon has been reported with both FQs (Ferrándiz and de la Campa, 2014; Ferrándiz et al., 2016b), supporting the idea that competence is a general stress response in *S. pneumoniae* (Prudhomme et al., 2006). In addition, both LVX and MOX induce transcriptional alterations, which, although different, ultimately stimulate the Fenton reaction, increasing ROS accumulation and contributing to cell death (Ferrándiz and de la Campa, 2014; Ferrándiz et al., 2016b). Although *S. pneumoniae* is a facultative anaerobe, the increased lethality of FQs mediated by an increase in ROS fits with the antibiotic lethality model proposed for aerobic bacteria (Dwyer et al., 2007, 2014; Kohanski et al., 2007; Wang and Zhao, 2009). Via local supercoiling changes, the response to LVX specifically triggers the up-regulation of the *fatDCEB* operon. This causes an increase in intracellular iron, and in turn, a shift in the Fenton reaction toward the production of hydroxyl radicals. With MOX, the response leads to the up-regulation of the glycolytic pathway, with a noticeable increase in pyruvate and a subsequent increase in hydrogen peroxide (**Figure 2**). The different alterations in the patterns of gene expression induced by LVX and MOX are due to local changes in supercoiling, which are dependent on whether Topo IV (LVX) or both Topo IV and gyrase (MOX) are inhibited.

Since both Topo IV and gyrase produce double-stranded breaks in the DNA when bound to FQs, the differential transcriptional alterations caused by these drugs might also

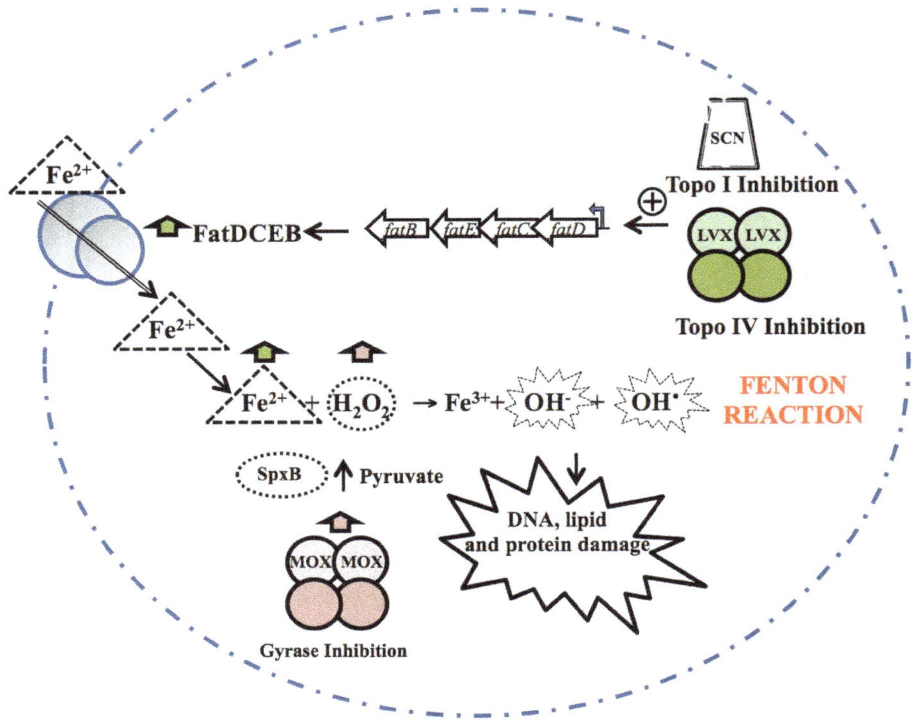

FIGURE 2 | Oxidative damage cell death pathway. The inhibition of Topo IV by LVX, or of Topo I by SCN, causes a local increase in supercoiling, resulting in the up-regulation of the *fatDCEB* operon. The consequent increase in the iron transporter it encodes causes an increase in intracellular Fe^{2+}. MOX alters the transcriptome, up-regulating genes from several metabolic pathways involved in the production of pyruvate. Pyruvate can then be converted by pyruvate oxidase (SpxB) into hydrogen peroxide (H_2O_2). Fe^{2+} and H_2O_2 are the substrates of the Fenton reaction. The Fenton reaction renders hydroxyl radicals, which oxidatively damage DNA, proteins and lipids. Taken from Ferrándiz and de la Campa (2014), with modifications.

be related to subtle, yet important, differences in sequence recognition (Leo et al., 2005), which are themselves affected by DNA supercoiling and bending (Arnoldi et al., 2013). Sequence recognition mediated by local supercoiling levels might explain the unique distribution of genes affected by LVX or MOX. In addition, the location of FQ-topoisomerase complexes relative to the replication forks, which is different for gyrase and Topo IV (Postow et al., 2001), may be involved in their different transcriptional outcomes.

CONTROL OF TRANSCRIPTION BY GLOBAL CHANGES IN SUPERCOILING

Response to Relaxation Caused by the Inhibition of Gyrase

The homeostatic control of supercoiling was first described in *E. coli.* In this bacterium, the transcription of *topA* (which codes for Topo I) was found to decrease under DNA relaxation (Tse-Dinh, 1985), while that of *gyrA*, and *gyrB* (which code for the two gyrase subunits) were found to increase (Menzel and Gellert, 1983, 1987a,b). An increase in gyrase expression in response to relaxation has also been observed in *Streptomyces* and *Mycobacterium* (Thiara and Cundliffe, 1989; Unniraman et al., 2002). However, in *Staphylococcus aureus*, treatment with NOV affects the transcription of the gyrase genes but not of *topA* (Schroder et al., 2014). In *S. pneumoniae*, treatment with NOV was also found to increase the transcription of gyrase genes, and diminish the expression of Topo I and Topo IV. In addition, global relaxation followed by a recovery of the native level of supercoiling was observed at low drug concentrations (Ferrándiz et al., 2010). The distribution of topoisomers in plasmid pLS1 (Stassi et al., 1981) was used to estimate the chromosomal superhelical density (σ), and returned a mean value of about -0.06 (**Figure 3**), which is within the range reported for the *E. coli* chromosome (Deng et al., 2005). At subinhibitory NOV concentrations ($0.5\times$ MIC), a transcriptomic response allowed the restoration of the native level of supercoiling after an initial relaxation causing a σ variation of 23%. A similar effect was observed at $1\times$ MIC. However, higher concentrations of NOV increased the degree of relaxation with no further restoration of supercoiling, compatible with the saturation of the homeostatic capacity that results in the inhibition of cell division. The range of σ variation permitting homeostatic recovery of the supercoiling observed in *S. pneumoniae* is in agreement with the estimated $\pm 20\%$ variation compatible with normal cell growth in *E. coli* (Drlica, 1992). Supercoiling recovery in the pneumococcus occurred after the up-regulation of the gyrase genes *gyrA* and *gyrB* and the down-regulation of the Topo I

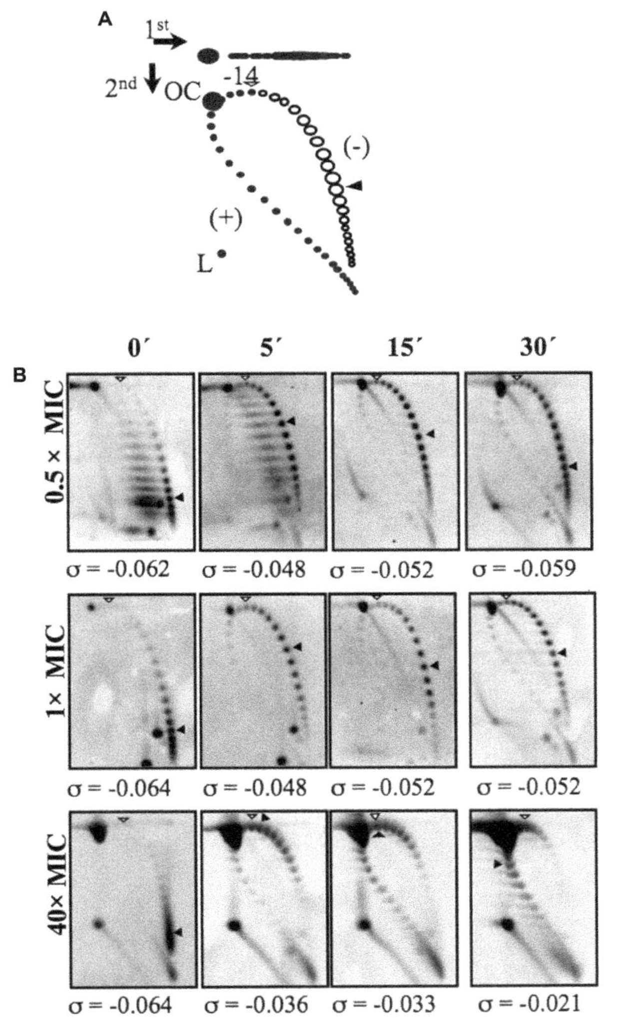

FIGURE 3 | Treatment with the GyrB inhibitor NOV causes relaxation and subsequent recovery of supercoiling levels. **(A)** Diagram showing plasmid pLS1 topoisomer distribution after two-dimensional electrophoresis in agarose gels run in the presence of 1 and 2 μg/ml chloroquine in the first and second dimensions, respectively. Arrows at the top left corner indicate the running direction of the first and second dimensions, respectively. OC, open circle; L, linear form. Negative supercoiled topoisomers are in white and positive supercoiled topoisomers in black. 2 μg/ml chloroquine introduces 14 positive supercoils. A white arrowhead indicates the topoisomer that migrated with ΔLk of 0 in the second dimension; it migrated with a ΔWr of −14 in the first dimension. A black arrowhead indicates the most abundant topoisomer. **(B)** pLS1 topoisomer distribution after different NOV treatments. Samples were taken before the addition of the drug (time 0 min) and at the times indicated. The corresponding supercoiling density (σ) value is indicated below each autoradiogram. Taken from Ferrándiz et al. (2010), with modifications.

(*topA*) and Topo IV (*parEC*) genes (Ferrándiz et al., 2010). In *E. coli*, the expression of the gyrase and Topo I genes is also mediated by NAPs, which affect DNA supercoiling (Travers and Muskhelishvili, 2005; Vora et al., 2009). However, these regulatory mechanisms may not function in *S. pneumoniae* for which NAP scarcity is predicted, and which certainly lacks most of the NAPs found in *E. coli*. Thus, supercoiling maintenance in *S. pneumoniae* appears to depend mainly on the regulation of topoisomerase transcription.

The Transcriptional Response to DNA Relaxation Involves Topology-Reactive Gene Clusters

The modulation of the expression of topoisomerase genes in *S. pneumoniae* is part of a global genome response (Ferrándiz et al., 2010). At subinhibitory concentrations, i.e., under physiological conditions, and short treatment times (5 and 15 min), DEGs were found to account for about 13% of the genome. An attenuation in the response at 30 min was observed, the number of DEGs being reduced to account for just 5.7% of the genome (**Figure 4A**), reflecting the recovery of supercoiling (**Figure 3**). Some 13% of the pneumococcal genome was therefore involved in the cellular response to moderate relaxation, allowing the recovery of the initial level of supercoiling. At fully inhibitory concentrations, the proportion of the genome covered by DEGs increased with time, from 14.4% at 5 min to 24% over longer periods (**Figure 4**). This agrees with the inhibition of cell division and with the continuous relaxation of the DNA (**Figure 3**). This proportion of the genome covered by DEGs upon relaxation is larger than in other bacteria. In Gram-negative bacteria, DEGs were found to account for 7% of the genome in *E. coli* [as determined using both gyrase inhibitors and gyrase thermosensitive mutants (Peter et al., 2004)], and for 8% in *Haemophilus influenzae* [as determined using NOV (Gmüender et al., 2001)]. In Gram-positive *Staphylococcus aureus*, treatment with NOV affected the transcription of 11% of the genome (Schroder et al., 2014).

It should be noted that the transcriptomic response to relaxation in *S. pneumoniae* involves topology-reactive gene clusters, or domains, that show coordinated up- or down-regulation. A total of 15 clusters have been detected, corresponding to 37% of the genome (**Figure 4**) (Ferrándiz et al., 2010). The sizes of these clusters varies from 14.6 to 85.6 kb (mean ± SD: 51.8 ± 21.8) and they contain 15–43 responsive genes (mean ± SD: 28 ± 9). They also include more than 68% of the DEGs. This has allowed topological clusters to be identified in which gene co-regulation is clearly more complex than would be expected simply from the number of genes in operons. In addition, the direction of transcription of the DEGs showed no preference for leading or lagging strands, providing additional evidence that topological control is structurally dependent.

The AT content over the genome correlates with domain location, and is higher in up-regulated (UP) than in down-regulated (DOWN) domains. These results suggest that the relaxation of DNA in AT-rich (ATr) regions favors the access of RNA polymerase to their promoters. On the contrary, a low AT content in DOWN clusters obstructs the access of RNA polymerase. Enrichment in the AT content of the region from positions −800 to +200 of genes up-regulated under relaxation has been reported in *E. coli* (Peter et al., 2004).

The organization of the *S. pneumoniae* chromosome into domains was further confirmed by the introduction of a *cat* heterologous gene cassette into the different types of domain (**Figure 5A**) (Ferrándiz et al., 2014). In response to

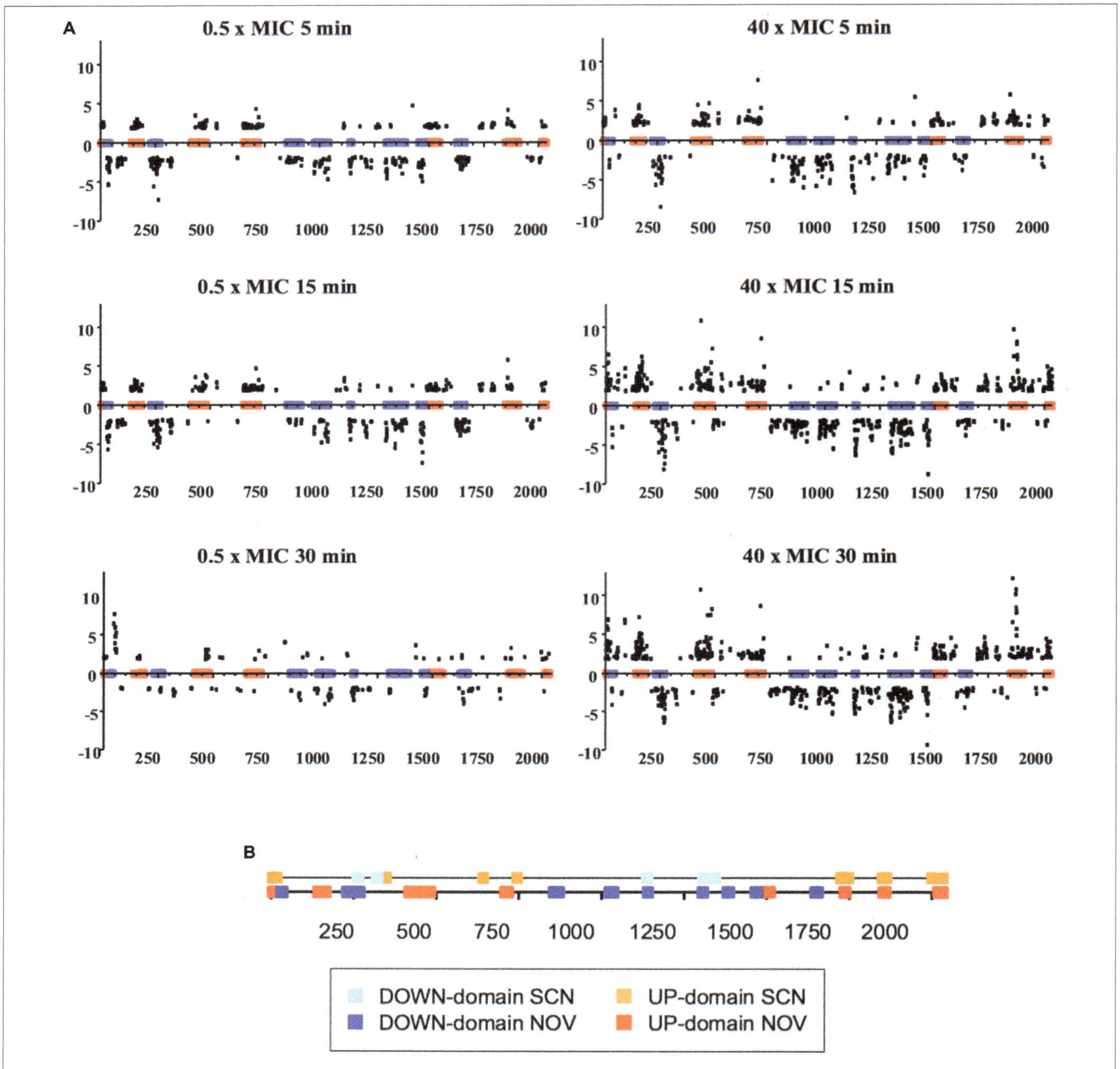

FIGURE 4 | Global transcriptomic responses of *S. pneumoniae* to relaxation with NOV. **(A)** The relative fold variation of each gene is represented against the 3′ location of each open reading frame in the *S. pneumoniae* R6 chromosome (bases 1 to 2,038,615). Boxes indicate the transcriptional clusters: up-regulated in red, down-regulated in blue. **(B)** Localization of topological clusters detected as a result of treatment with either SCN or NOV. Taken from Ferrándiz et al. (2010, 2016a), with modifications.

relaxation with NOV, the transcription of *cat* was dependent on its chromosomal location, being up-regulated when located in UP domains, down-regulated when located in DOWN domains, and showing almost no changes when located in the non-regulated (NR) domains (**Figure 5B**). This all supports the idea that the chromosome is organized into topological domains that are reactive to interference in the supercoiling status. These results contrast, however, with those obtained in *E. coli,* in which the 306 DEGs were not only functionally diverse but widely dispersed throughout the chromosome (Peter et al., 2004), and with results obtained for *Staphylococcus aureus,* in which NOV-responsive genes were randomly distributed throughout the chromosome (Schroder et al., 2014).

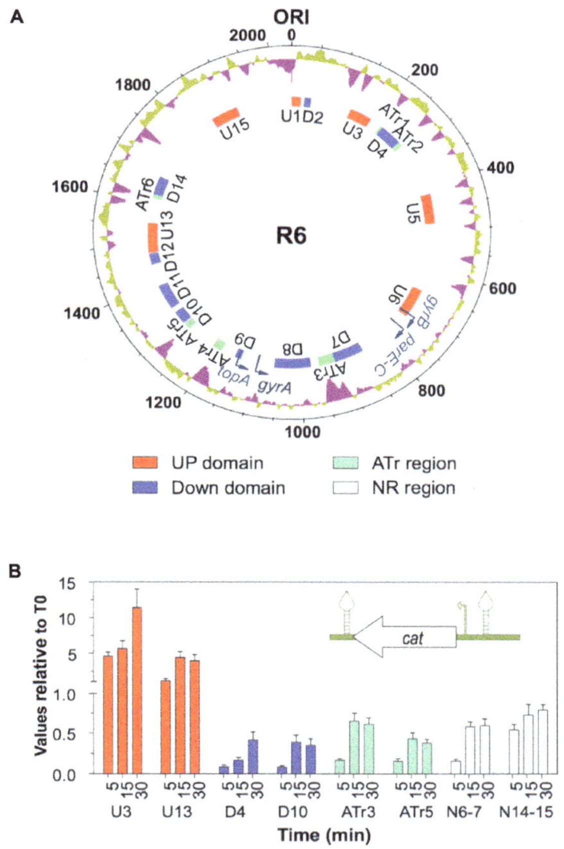

FIGURE 5 | The topology-dependent transcription of P_{C}*cat* is dependent on its chromosomal location. **(A)** Organization of the *S. pneumoniae* R6 chromosome in topological domains. Circles, from outside to inside, represent: % GC (values above the average in purple); DNA topoisomerase genes (dark blue curved arrows); topology-responsive domains. The chromosome is organized into domains up-regulated (U, red boxes) or down-regulated (D, blue boxes) in response to DNA relaxation, and ATr domains (green boxes). **(B)** Transcriptional response to DNA relaxation by NOV measured by qRT–PCR. A P_{tc}*cat* cassette, coding for chloramphenicol-acetyl-transferase, which carries its own promoter (curved arrow) and is flanked by two transcriptional terminators (stem and loop structures), was inserted into different supercoiling domains. Cultures of the R6-CAT strains were treated with NOV and the transcription of *cat* analyzed by qRT–PCR. Taken from Ferrándiz et al. (2014), with modifications.

Response to Hypernegative Supercoiling Caused by the Inhibition of Topo I

The negative supercoiled state is the natural state of DNA homeostatic equilibrium in many bacteria. However, hypernegative supercoiling has been reported in *E. coli topA* mutants. With the exception of the *topA10* mutant, all have acquired compensatory mutations in the gyrase genes (DiNardo et al., 1982). The *topA10* mutant shows a notable 22% increase in negative supercoiling (Pruss et al., 1982), which probably represents the limit viable cells can afford in the long term. The inhibition of Topo I would produce greater hyper-supercoiling. Topo I plays an essential role in transcription, given its physical interaction with RNA polymerase (Cheng et al., 2003). During transcription, hypernegative supercoiling occurs behind the RNA polymerase, leading to RNA-DNA hybrid (R-loop) stabilization (Drolet, 2006). Topo I relaxes this supercoiling and prevents R-loop formation (Drolet et al., 1994; Phoenix et al., 1997; Masse and Drolet, 1999), allowing transcription to continue. Thus, the effects of hypernegative supercoiling in transcription depend directly on the activity of Topo I.

However, Topo I-targeting compounds are extremely scarce. Cheng et al. (2007) identified an alkaloid, which, although it inhibits the activity of *E. coli* Topo I, did not inhibit cell growth significantly. Our group discovered a new inhibitor of *S. pneumoniae* Topo I, SCN, which inhibits its relaxation activity at concentrations equivalent to those that inhibit cell growth. The modeling of pneumococcal Topo I, based on the crystal structure of the *E. coli* enzyme (**Figure 6**), and docking to SCN, revealed strong interactions between the drug and the DNA-binding site of Topo I to correlate with the inhibitory effect observed (García et al., 2011).

Our group was the first to use SCN in studies of the transcriptomic response to hypernegative supercoiling in bacteria (Ferrándiz et al., 2016a). The viability of *S. pneumoniae* and the increase in supercoiling is affected by SCN in a concentration-dependent manner (**Figure 7**). Treatment with 6 μM SCN produced a peak σ increase of 41% at 5 min, which later recovered. Treatment with 8 μM SCN resulted in higher and longer lasting increases in the σ value, with partial recovery after 120 min. These results show that treatment with subinhibitory SCN concentrations permit the recovery of peak σ increases of up to 41% without affecting cell viability. This tolerance to increases in supercoiling levels is greater than the 25% observed for DNA relaxation upon NOV treatment (**Figure 8A**) (Ferrándiz et al., 2010), and indicates that *S. pneumoniae*, and very likely genetically related bacteria, are naturally more tolerant to hyper-negative supercoiling than to hyper-relaxation. Similarly, the results of experimental evolution assays with *E. coli* revealed increasing supercoiling (associated with mutations in *topA*) to increase bacterial fitness (Crozat et al., 2005). A similar homeostatic mechanism allowing increased negative supercoiling might also exist in bacteria with reverse gyrase. These bacteria keep DNA in a slightly overwound state to protect their genome from heat damage (Ogawa et al., 2015).

The transcription levels of *topA* in *S. pneumoniae* at subinhibitory concentrations of SCN or NOV (which allow for cell growth and the recovery of supercoiling) show a good correlation with the induced variation in σ (**Figure 8B**). The regulation of *topA* therefore plays a fundamental role in the recovery of supercoiling levels. The variations seen in *topA* expression were, however, only part of a global transcriptomic response. Treatment with subinhibitory concentrations of SCN (8 μM, 0.5× MIC) generated a two-stage transcriptomic response: (i) early response and (ii) recovery. The former, which represents an active response against sharply increased supercoiling, was observed at 5 and 15 min of treatment, and involved about 11% of the genome. During recovery, only about 2% of the genome was involved at 30 min. In the early response, transcriptional variations also occurred in clusters, with DEGs

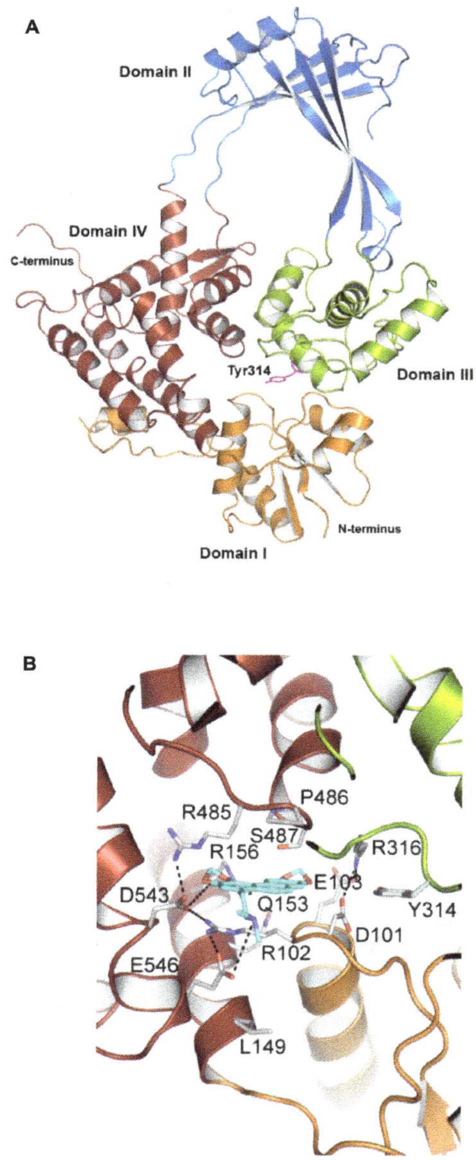

FIGURE 6 | Structural modeling of the interaction of *N*-methyl SCN with *S. pneumoniae* topoisomerase I. **(A)** Modeling of the 67 kDa fragment of Topo I, showing domains I–IV and the catalytic Tyr314. **(B)** *N*-methyl-SCN (in blue) bound to the nucleotide-binding site of Topo I. Hydrogen bonds and salt-bridge interactions are indicated by dashed lines. Taken from García et al. (2011), with modifications.

grouping into topologically sensitive domains. The average size of a SCN cluster is 14.0 ± 7.6, similar to the 10 kb *E. coli* domains predicted using transcriptional data (Postow et al., 2004). Although the NOV and SCN clusters are not identical, their position in the chromosome nearly overlap (**Figure 4B**) – an unexpected finding given the opposing nature of DNA relaxation and supercoiling. These results support the idea that the chromosome is divided into topological domains with fixed locations.

Regulation of DNA Topoisomerase Gene Transcription

In *E. coli*, several NAPs are involved in the regulation of topoisomerases. One such NAP is the FIS protein, which regulates the expression of genes coding for the subunits of gyrase (Schneider et al., 1999), Topo I (Weinstein-Fischer and Altuvia, 2007), and the genes coding for other NAPs involved in DNA supercoiling (Claret and Rouviere-Yaniv, 1996; Falconi et al., 1996; Grainger et al., 2008). In addition, two further NAPs, FIS, and H-NS proteins control both the level of supercoiling and global transcription (Blot et al., 2006; Marr et al., 2008). The corresponding situation in *S. pneumoniae*, which lacks these NAPs, seems to be much simpler.

The transcription of *gyrB* and *topA* in *S. pneumoniae* is regulated by their strategic chromosomal location in topological domains, since the expression driven by their promoters differs whether they are located in their natural chromosomal locations or in a replicating plasmid (Ferrándiz et al., 2014). Transcriptional fusions of these promoters to a reporter gene in plasmid pLS1 have been measured after DNA relaxation induced by NOV. As expected, relaxation caused down-regulation of *topA* and up-regulation of *gyrB* when the genes were located in their native chromosomal sites (DOWN9 for *topA* and UP6 for *gyrB* in **Figure 5A**). However, transcription from both promoters in the plasmid fusions was down-regulated. These results indicate that both *topA* and *gyrB* are under supercoil-mediated regulation, and that the plasmid behaves as a DOWN domain. This may serve to neutralize the high copy number of the plasmid genes and/or favor their replication.

In contrast, the Topo IV genes (*parE* and *parC*) and *gyrA* are located in NR domains, and their expression depends on specific regulatory signals located in the promoter region. The expression of the Topo IV genes from their common promoter (Balsalobre and de la Campa, 2008) is equivalent in their natural chromosomal location and in plasmids (Ferrándiz et al., 2014). With respect to the *gyrA* gene, its upstream region ($P_{gyrA126}$, nt −126 to +1 in **Figure 9A**) shows an intrinsic DNA curvature (Balas et al., 1998). This was fused to *cat* and cloned into plasmid pLS1, and the curvature either eliminated by a 5 bp insertion ($P_{gyrA126Pae}$) or by a 5 bp deletion ($P_{gyrA121Pae}$), and a direct correlation observed between *cat* expression and the curvature under basal conditions (the specific activity of the $P_{gyrA126}$ fusion was ~3-fold higher than that recorded for plasmids lacking the curvature). This shows that the curvature behaves as an activator *per se*, providing better recruitment of either the RNA polymerase complex or specific regulatory proteins. The role of curvatures as regulators of transcription has previously been established in bacteria (Pérez-Martín et al., 1994), including *S. pneumoniae* (Pérez-Martín and Espinosa, 1991). In addition, the transcription levels from the chromosomal P_{gyrA} and the P_{gyrA}*cat* fusions in plasmids in the presence of NOV have been determined. While in the plasmid carrying the wild-type promoter ($P_{gyrA126}$) the up-regulation of *cat* was similar to that of the chromosomal *gyrA*, down-regulation of *cat* was observed in the plasmids lacking the curvature (**Figure 9B**). These results suggest that the signals regulating *gyrA* transcription are included within the above-mentioned 126 nt region, and that bending is a key element

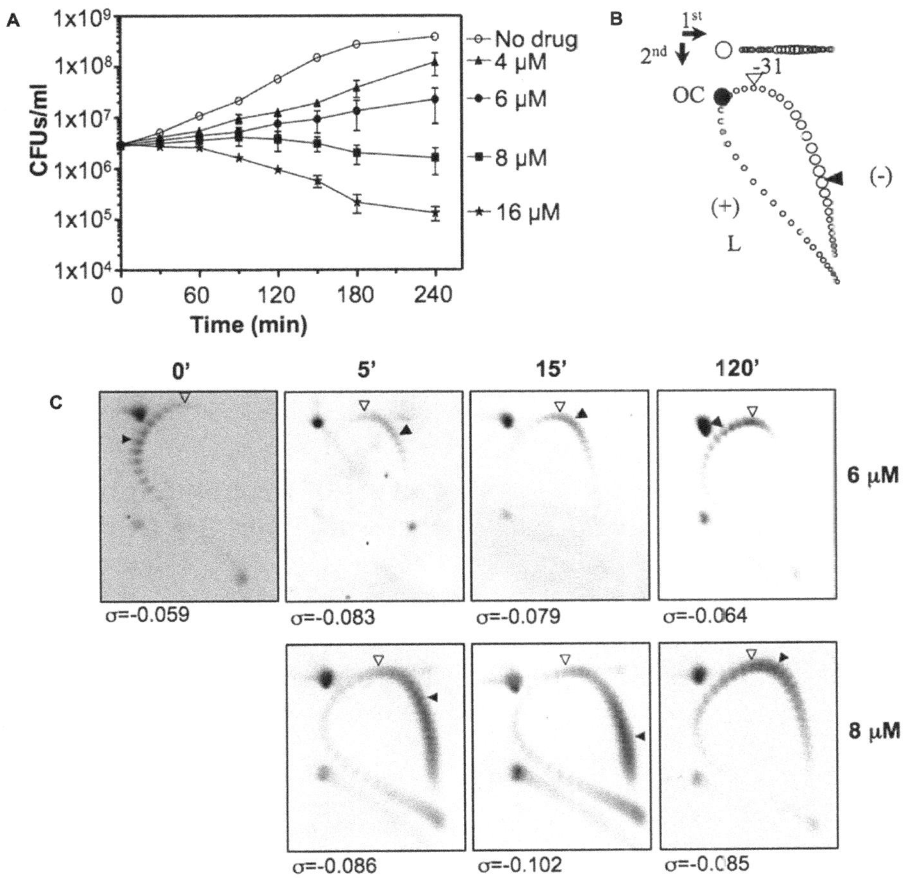

FIGURE 7 | Seconeolitsine affects cell viability and induces hypersupercoiling. **(A)** Viability of R6 (pLS1) in a medium containing different SCN concentrations. Samples were taken before the addition of the drug (zero time), and at the indicated times were plated on drug-free agar medium. **(B)** Diagram showing topoisomer distribution in plasmid pLS1 subjected to 2D-agarose gel electrophoresis run in the presence of 5 and 15 μg/ ml chloroquine in the first and second dimensions, respectively. OC, open circle; L, linear form. A white arrowhead points to the topoisomer migrating with a ΔLk of 0 in the second dimension; it had a ΔWr of −31 in the first dimension. A black arrowhead points to the more abundant topoisomer. **(C)** Distribution of pLS1 topoisomers in 2D-gels after treatment with the indicated SCN concentrations. The supercoiling density (σ) values are indicated. Taken from Ferrándiz et al (2016a), with modifications

for its regulation under relaxation by acting as a sensor of the supercoiling level.

Chromatin immunoprecipitation experiments using antibodies directed against the pneumococcal GyrA subunit and Topo I (Ferrándiz et al., 2016a) have shown P_{gyrA} to recruit Topo I, but not gyrase (**Figure 9C**). The region to which Topo I binds includes the −35 and extended −10 boxes on P_{gyrA}, plus the DNA curvature (Balas et al., 1998). Thus, Topo I, the transcription of which is regulated by supercoiling levels, appears to be the key factor regulating *gyrA* expression.

EVOLUTIONARY PRESSURE DRIVES THE ORGANIZATION OF THE CHROMOSOME INTO DOMAINS

Domain Conservation in Streptococci

Gene order in bacterial chromosomes surpasses the level of the operon (Lathe et al., 2000; Reams and Neidle, 2004). As explained above, and based on its transcriptome under DNA relaxation, the chromosome of *S. pneumoniae* R6 appears to be organized into four types of topological domains: UP, DOWN, NR, and ATr. The analysis of 12 *S. pneumoniae* complete genome sequences has revealed the conservation of the UP and DOWN domains (**Figure 10**). The gene-lack index (number of genomes in which a gene is absent divided by the total number of genomes) revealed lower values for the UP (1.51) and DOWN (1.65) domains than the genome average (1.91). However, ATr domains have high gene-lack indices (average 4.66), suggesting extensive gene interchange in these domains. To study the conservation of domains, normalized location dispersion indices (nLDI: values that quantify the position deviation of a given gene with respect to the Ori, and relative to homologs in several genomes (Martín-Galiano et al., 2017)) were calculated across *S. pneumoniae* genomes; the values returned were very small since synteny is highly conserved in this species. The same was then calculated for representative strains of 25 species of *Streptococcus* in order to detect distinguishing differences. The conservation of

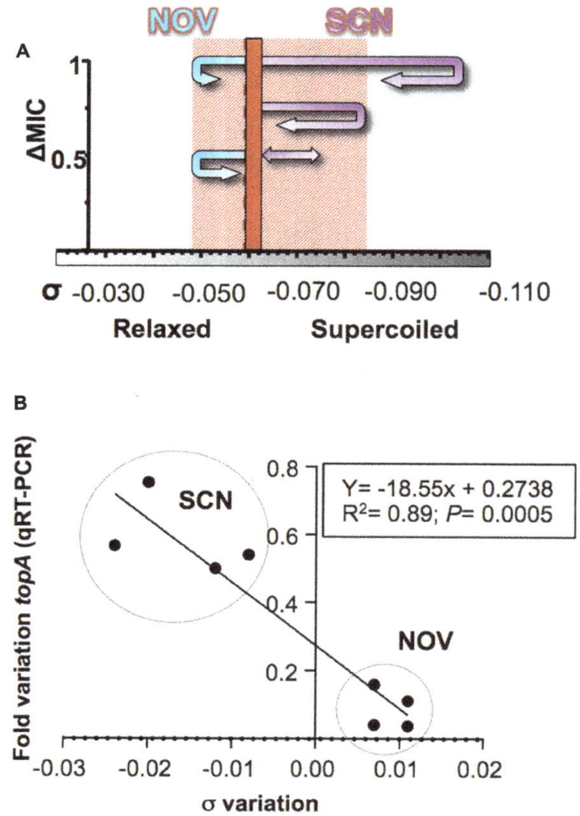

FIGURE 8 | Responses to changes in supercoiling are conducted by Topo I. **(A)** Diagram showing the trajectory of the supercoiling density at the indicated NOV and SCN concentrations. A dashed line indicates the supercoiling density of DNA of non-treated cells. The shaded area represents the supercoiling density interval in which cells can survive. **(B)** Correlation between changes in supercoiling level and the transcription of *topA*. The data correspond to samples treated with either SCN or NOV at concentrations that allowed cell growth and the recovery of DNA supercoiling. Taken from Ferrándiz et al. (2016a), with modifications.

S. pneumoniae domains across these *Streptococcus* representatives was then determined. Two assumptions were made: (i) that the gene order is relatively conserved, as seen in gamma-proteobacteria (Sobetzko et al., 2012), and (ii) that chromosomal topology is conserved, given that species share core gene pools (Lefebure and Stanhope, 2007), similar genome lengths, and a similar AT content. Similar approaches have been followed to examine chromosomal patterning in other bacteria (Wright et al., 2007; Khedkar and Seshasayee, 2016). In *S. pneumoniae*, 571 genes (28.0%) had nLDI values of <1, which indicates they tend to locate to positions more stable than the average for maintained homologs (Martín-Galiano et al., 2017). Several genes from the UP and DOWN domains were present in most streptococci at equivalent positions. The greatest position conservation was observed in 40 genes near the Ori, indicating strong topological pressure to maintain functionalities in this region. Genes near the Ori have high copy numbers (Slager and Veening, 2016) and show a peculiar pattern of NAP binding (Sobetzko et al., 2012). Moreover, seven clusters with conserved positions were detected for NR genes, and named pcNR domains (position-conserved Non-Regulated domains). Most of the remaining NR genes were organized into 14 domains (≥10 genes) termed pvNR domains (position-variable Non-Regulated). ATr regions accounted for 13 domains (**Figure 11**). Strikingly, the pcNR domains appeared symmetrically located at regular intervals (~200, 400, and 800 kb) on both sides of the Ori and were interleaved between UP, DOWN, and pvNR domains (**Figure 11A**). The size of these domains appeared compatible with the 100 kb lengths estimated for them using different techniques (Worcel and Burgi, 1972; Sinden and Pettijohn, 1981; Le et al., 2013). This suggests a potential higher-order macrostructural unit above the domain level controlling the genetic stability and plasticity required to face new environments (Rocha, 2004a).

Levels of Protein Expression and Essentiality of the Domains

The transcriptomes of exponentially growing cultures (Ferrándiz et al., 2016a,b) showed the pcNR domain transcription level to be higher than that of the ATr domains (**Figure 11B**). Two factors contribute to these transcriptional differences. First, long repeat sequences (BOX, RUP, and SPRITE) (Croucher et al., 2011), which are associated with the repression of transcription, are few in pcNR domains, and second, the codon adaptation index (CAI), which is related to the translation rate and mRNA levels (Martín-Galiano et al., 2004), is high in pcNR domains (Martín-Galiano et al., 2017). Gene location also affects protein levels (Ochman et al., 2000; Rocha, 2004b), a pattern associated with the distance to the Ori. Genes at the Ori are doubly represented with respect to genes at the Ter in *E. coli* during exponential growth (Chandler and Pritchard, 1975). Accordingly, the relocation of genes coding for ribosomal proteins and the RNA polymerase alpha subunit to positions distant to the Ori, reduces their transcription rates, which was associated with slower growth in *Vibrio cholerae* (Soler-Bistue et al., 2015). Similarly, in *Salmonella typhimurium*, genes relocated near the Ori are expressed more strongly than those relocated near the Ter (Schmid and Roth, 1987). The regular positioning of strongly expressed genes may mark the limits of domains, as reported for *Caulobacter crescentus* (Le et al., 2013).

The fraction of essential genes, as determined by Tn-seq (van Opijnen and Camilli, 2012), is notably higher in pcNR domains than in the other domains (**Figure 12A**). The co-localization of essential genes beyond randomness has also been reported for *Bacillus subtilis* and *E. coli* (Fang et al., 2005), perhaps because clustering makes genomes more resistant to deletions (Fang et al., 2008). The number of pcNR genes in the lagging strand was 15.6%, significantly lower than the average in the remaining *S. pneumoniae* genome (22.3%). This would reduce the chances of collision between DNA and RNA polymerases, resulting in the discontinuation of transcription (French, 1992). Essential genes also tend to be more strongly expressed (Rocha and Danchin, 2003), as confirmed for pneumococcal pcNR genes. Essential gene clustering at regular intervals, and not affected by topological stress as defined for pcNR, appears

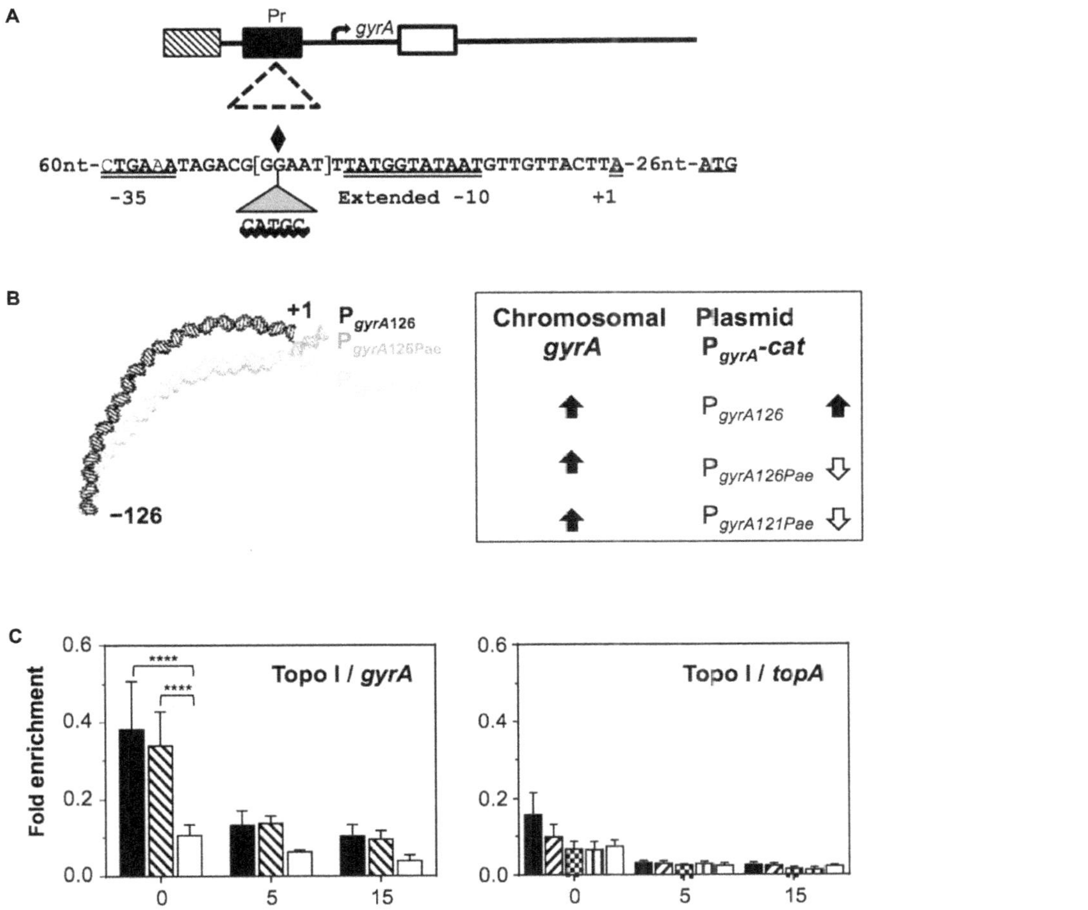

FIGURE 9 | Control of the transcription of DNA topoisomerase genes by supercoiling. **(A)** Representation of the *gyrA* coding region and of the regions tested in chromatin immunoprecipitation, showing the sequences of the wild-type $P_{gyrA126}$, $P_{gyrA126Pae}$, and $P_{gyrA121Pae}$ derivatives. The −35 and extended −10 boxes, the nucleotide at which transcription is initiated (+1), the center of the intrinsic DNA curvature (diamond), and the location of the inserted CATGC sequence that creates a *Pae*I restriction site, are all indicated. The five nucleotides deleted in $P_{gyrA121Pae}$ are in brackets. **(B)** The relaxation-induced up-regulation of P_{gyrA} depends on intrinsic bending: curvature prediction and results obtained from qRT-PCR analysis. **(C)** Recruitment of Topo I to *topA* and *gyrA* upstream sequences. Exponentially growing cells were subjected to chromatin immunoprecipitation using anti-Topo I antibodies; the pulled-down DNA was subsequently analyzed by qPCR. The graphs show the pulldown efficiency (ChIP-DNA/input DNA) for each primer pair. Values are the average ± SD of three independent replicates. ****$P < 0.0001$. Taken from Ferrándiz et al. (2014, 2016a), with modifications.

to reflect a favorable "supercoiling environment" for protein expression.

The Different Domains Contain Genes with Different Functions

Importance of the Protein Interaction Network

A significant fraction of the pcNR genes codes for proteins with important roles in central metabolism and that have a high number of protein–protein interactions (PPIs). PPIs provide a rough estimate of a protein's importance in cell physiology. The estimated amounts of protein produced, and their functions, support the idea that the genes of pcNR domains are more involved in the central metabolic network than are those of the pvNR domains. In stark contrast, ATr genes appear to play little or no role in central metabolism; their PPI values are at most only about one third of the average for the remaining genome. As mentioned above, changes in the location of genes could lead to alterations in cell physiology, which holds true for both central metabolic (Soler-Bistue et al., 2015) and regulatory genes (Gerganova et al., 2015). The physical positioning of specific supercoiling-favorable regions in the chromosome is also related to the ability to gain access to cytoplasmic regions rich in ribosomes (Soler-Bistue et al., 2015).

Overall, the evidence supports the idea that the function, expression, essentiality and stability of genomic positions are interconnected, as reported for *Dickeya dadantii* and *E. coli* (Sobetzko et al., 2012; Jiang et al., 2015). Altogether, the pcNR genes reflect a multistep adaptation in the transcription-translation-interaction cascade that facilitates the activity of these genes' products, thereby increasing bacterial fitness.

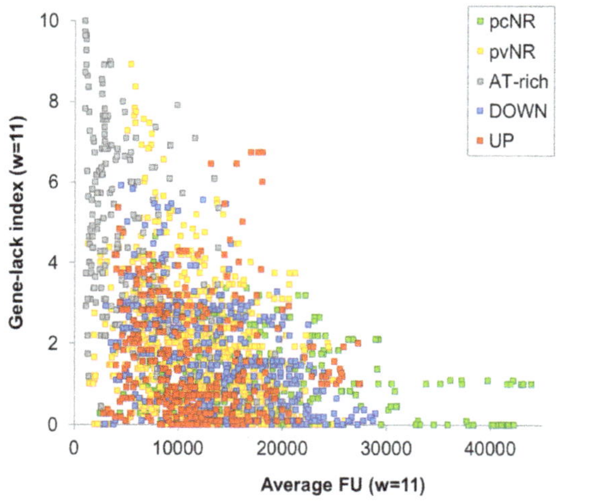

FIGURE 10 | Evidence for conservation of UP and DOWN domains in *S. pneumoniae*. Relationship between the gene-lack index and gene expression (fluorescent units, FU) in *S. pneumoniae* R6 as detected in high density microarrays. A total of 12 genome sequences (from 11 clinical isolates and R6) were analyzed. Genes of clinical isolates were considered equivalent to those of R6 when their products shared ≥80% similarity over ≥80% of the sequence length. An 11-gene window (about 10 kb) was contemplated.

Pathogenesis and Immunogenicity

DNA topology regulates the expression of virulence factors in several bacteria (Dorman and Porter, 1998; Cameron and Dorman, 2012; Reverchon and Nasser, 2013; Jiang et al., 2015). In *S. pneumoniae*, three types of virulence genes show differences in their distribution among domains. Widely accepted virulence factors are more abundant in pvNR domains (**Figure 12B**), while genes contributing [as estimated by signature-tagged mutagenesis (Hensel et al., 1995)] to intranasal colonization, meningitis or otitis (Chen et al., 2008; Molzen et al., 2011) are more abundant in DOWN domains (**Figure 12C**). Finally, genes coding for proteins that trigger an immune response in humans (Giefing et al., 2009), and which are therefore candidate targets for a serotype-independent protein-based vaccine against pneumococcus, are predominant in the pvNR domains. The pvNR domains also contain more genes coding for extracellular proteins or proteins anchored in the cell wall than do pcNR domains. All in all, pvNR domains show strong allelic variation by being subjected to selective pressure during adhesion, cytotoxic challenge and immune system evasion. This variation also increases the genome pool of the species via gene duplication/paralogs in which one copy is not subject to immediate pressure (Mira et al., 2010). The link between supercoiling stress and virulence enhancement does not seem to be the rule for *S. pneumoniae*, the canonical virulence and accessory factors of which are preferentially encoded in the pvNR or DOWN domains.

Genes Involved in Competence

Gene transfer is a primary driver of evolution in bacteria, but the introduction of new genetic material at random can perturb chromosomal topology. *S. pneumoniae* is a naturally transformable bacterium (Claverys et al., 2006; Martin et al., 2006), the evolution of which (including its antibiotic-resistance and virulence factors) depends on both intra-species and inter-species chromosomal transformation (Dowson et al., 1989, 1990; Balsalobre et al., 2003; Ferrándiz et al., 2005). Competence involves the transient transcriptional modulation of ~10% of the genome with strict timing (Peterson et al., 2004). When under stress (the X state), the competence system -which bears some resemblance to the SOS repair system of *E. coli* and other bacteria – is activated (Claverys et al., 2006). In fact, FQs induce the SOS response since they cause double-strand breaks in chromosomes (Drlica et al., 2008). As described above, local supercoiling changes triggered by FQs activate pneumococcal competence, but global supercoiling changes do so too. The early and delayed up-regulated competence genes (those activated during stress) are mainly located in UP domains. Many pcNR genes are, however, down-regulated, indicating that during the X-state the topology of the chromosome is perturbed to a degree that threatens cell viability via effects on the central metabolic machinery. This explains why growth is slowed during competence (Oggioni et al., 2004) and why several mechanisms have been acquired, including the use of small untranslated RNAs and proteases to actively terminate the X-state and promptly recover the normal topological situation (Echenique et al., 2000; Cassone et al., 2012).

Horizontally Acquired Genes

In *S. pneumoniae* R6, up to 12.1% of the genome is thought to have been acquired by horizontal gene transfer. The distribution of these acquired genes among domains is uneven, with a clear bias toward ATr domains (**Figure 12D**). This suggests that these domains act as structural or parasitic DNA hotspots, which agrees with their low transcriptional level and annotated functions (Ferrándiz et al., 2010, 2014, 2016a). It remains open the possibility that the ATr regions influence the organization of topological dynamics, or that they are involved in the acquisition of foreign genes.

CONCLUSION AND PERSPECTIVES

The transcriptome of *S. pneumoniae* alters with local or global changes in supercoiling. Local changes induced by the clinically used FQs LVX, and MOX, which target GyrA and/or Topo IV, trigger a transcriptional response. Both FQs up-regulate the competence regulon in response to stress, and, respectively, cause an increase in intracellular ROS by increasing the uptake of iron (through up-regulation of the *fatDCEB* transporter) and hydrogen peroxide (through up-regulation of the glycolytic pathway), both of which are involved in the Fenton reaction.

Changes in global supercoiling induced by NOV (which targets GyrB), or by SCN (which targets Topo I), have revealed the existence of topological domains that react in a coordinated fashion. In *S. pneumoniae*, the control of DNA-supercoiling occurs mainly via the regulation of transcription of the topoisomerase genes: relaxation triggers the up-regulation of

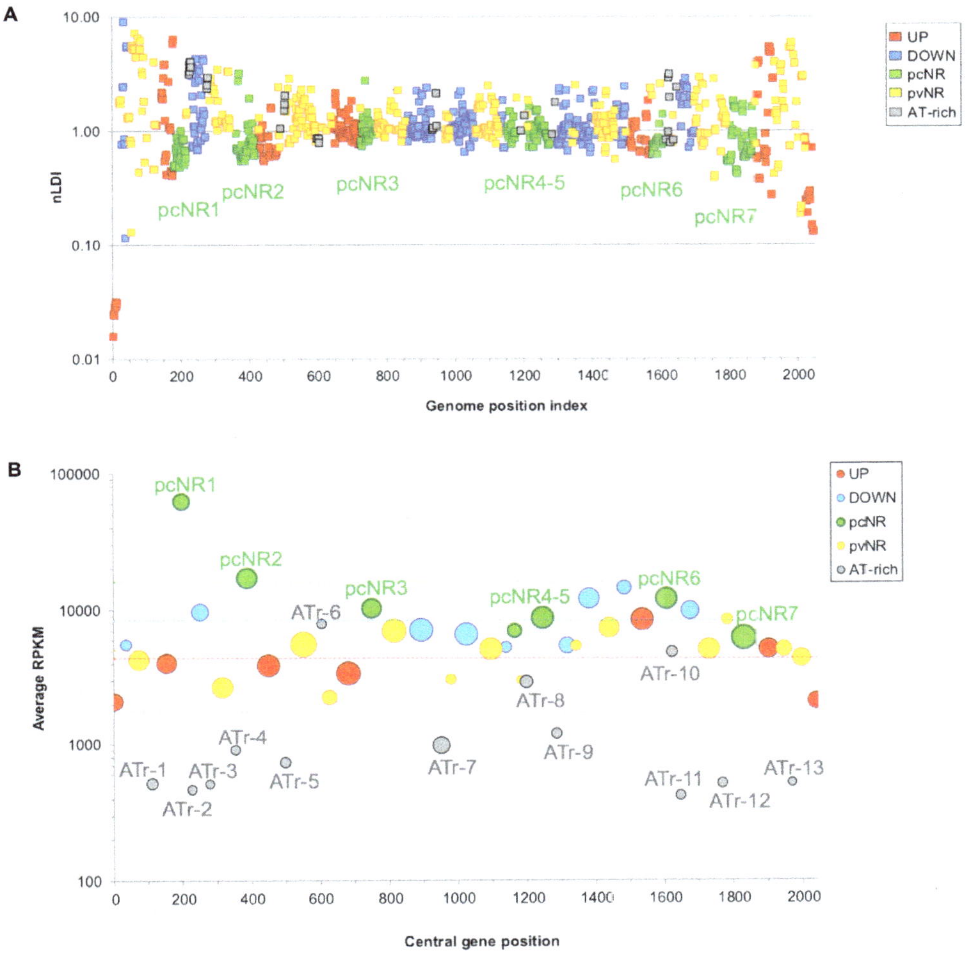

FIGURE 11 | Different domains have different locations and transcription levels. **(A)** Location of NOV-reactive and pcNR domains. Dots indicate nLDI values for *S. pneumoniae* R6 genes for the whole *Streptococcus* genus. A total of 1216 genes coding for proteins in *S. pneumoniae* R6, with homologs in ≥10 species, were contemplated in the analysis. Gene indexing starts from the replication origin. **(B)** Abundance of RNA transcripts according to domain type. RNA-Seq data were normalized by gene size (kb) using the reads per kb per million mapped reads (RPKM). The positions of the domains in the genome are represented by the position of the central gene (central gene position, *X*-axis). The size of the circles is proportional to the number of genes in the domain. Only pvNR clusters with at least 10 members are shown. Horizontal lines in the corresponding color indicate the average RPKM value for each domain class. Taken from Martín-Galiano et al. (2017).

gyrA and *gyrB* and the down-regulation of the Topo I (*topA*) and Topo IV (*parEC*) genes, while hypernegative supercoiling triggers the down-regulation of *topA*. The transcription of *gyrB* and *topA* is regulated by their strategic chromosomal location in the topological domains, while the expression of *parEC* and *gyrA* depends on the specific regulation of their promoters. Although the regulators of *parEC* are unknown, the promoter of *gyrA* shows an intrinsic curvature that acts as a sensor of the supercoiling level. In addition, chromatin immunoprecipitation experiments have revealed Topo I to bind to the *gyrA* promoter. Therefore, Topo I, the transcription of which is regulated by the supercoiling level, appears to regulate *gyrA* expression.

The regulation of topoisomerase genes is part of a global response to changes in supercoiling. Relaxation affects >13% of the genome (from 13 to 24%), while hypernegative supercoiling affects 10%. In both cases, responsive genes are grouped into domains that essentially overlap, suggesting that they have a fixed chromosomal location. Based on their structural and functional characteristics, and the change in the domains detected under relaxation, the following types can be defined: UP, DOWN, pcNR, pvNR, and ATr. The genes of the UP, DOWN, and pcNR domains have been found at equivalent positions present in most streptococci, especially near the Ori. pcNR domains are interleaved between UP, DOWN, and pvNR domains, which suggests a higher-order macrostructural unit. The pcNRs genes show the highest level of transcription, and contain most of the essential genes plus those involved in the central metabolic network. In stark contrast, the ATr domains show the lowest transcriptional levels, and the genes they contain appear to have little to do with the central metabolic network. This explains the tropism of pcNR genes for topologically secure areas, helping to maintain the constant provision of central proteins.

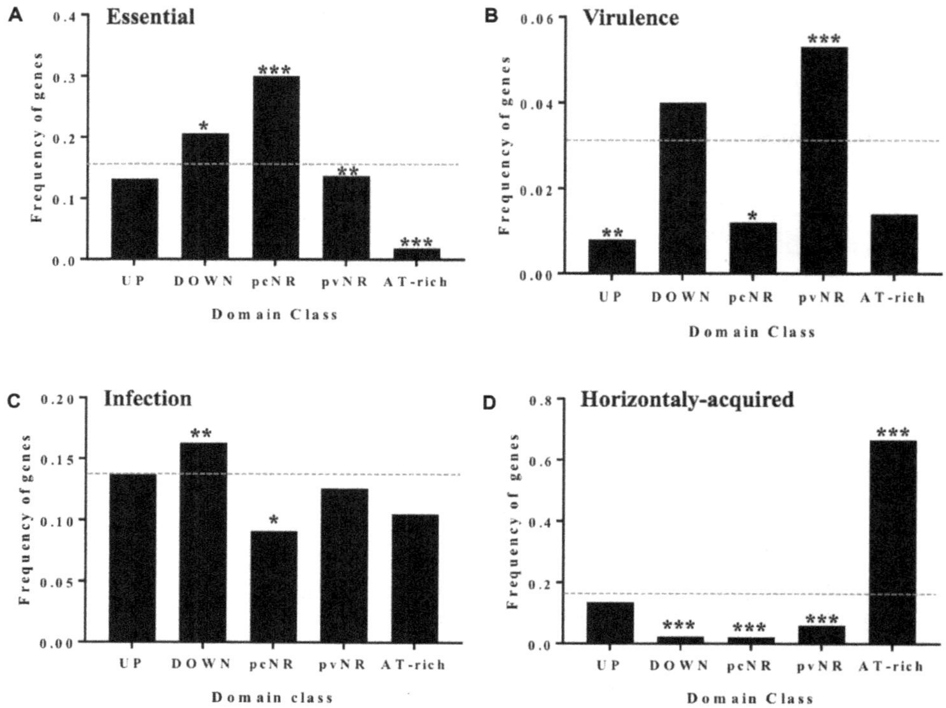

FIGURE 12 | Genes of the different domains have different functionalities. **(A)** Fraction of essential genes in different domains. **(B)** Virulence factors. **(C)** Horizontally acquired genes abundance and domain class. **(D)** Essential genes for infection by STM. Statistical significance with respect to the genome average (dashed line): $^{*}P \leq 0.05$, $^{**}P \leq 0.01$, $^{***}P \leq 0.001$. Taken from Martín-Galiano et al. (2017), with modifications.

The genes coding for the classical virulence factors, plus those coding for immunogenic proteins, are more common in the pvNR domains, while genes contributing toward the establishment of infection are more common in the DOWN domains. The distribution of horizontally acquired genes is clearly biased toward ATr domains, suggesting these to be hotspots for the acquisition of foreign genes.

In general, UP gene expression is favored by topological stress; DOWN genes are highly expressed under favorable conditions and less so during such stress. ATr domains may sense topological stress and modify supercoiling in their area to reduce the transcription of adjacent genes, preferentially those in the DOWN domains. The chromosome supercoiling structure may act as a multi-sensor with homeostatic capacity, adapted to react to unfavorable conditions.

Pneumococcal genes appear to be subject to topology-driven selection that defines the chromosomal location of genes involved in metabolism, virulence and competence. Together, these organizational features reveal the genome of *S. pneumoniae* to be influenced by physiology-related topological rules. A global topology theory might be envisaged in which gene positioning is far from random. Many aspects of the importance of gene location – such as the idiosyncrasy of the domains and how this affects fundamental aspects of bacterial biology – are only now becoming understood.

Topological genomics – topogenomics – provides an alternative paradigm of genome analysis. Certainly, genome architecture plays an important role in the pathobiology and evolution of *S. pneumoniae*, and it is tempting to speculate that in other species too, the genes are subjected to topology-driven selection pressure that defines their chromosomal locations. Data from many species will, however, be needed before a full understanding of all the rules underlying topogenomics are known and understood.

AUTHOR CONTRIBUTIONS

All authors made intellectual contributions to the work and approved it for publication. AdC supervised all the studies and wrote the manuscript. MF performed most of the experiments related to determinations of supercoiling densities and transcriptomic studies. AM-G performed the bioinformatic studies. MG performed the characterization of topoisomerase I and its inhibition by seconeolitsine. JT-V contributed to the experiments of chromatin immunoprecipitation.

ACKNOWLEDGMENTS

We thank Monica Amblar (*Centro Nacional de Microbiología*, ISCIII, Madrid, Spain), Pablo Hernández (*Centro de Investigaciones Biológicas* CSIC, Madrid, Spain), and Pedro A. Lazo-Zbikowski (*Instituto de Biología Molecular y Celular del Cáncer*, CSIC, Salamanca, Spain), and for their critical reading of the manuscript.

REFERENCES

Arnoldi, E., Pan, X. S., and Fisher, L. M. (2013). Functional determinants of gate-DNA selection and cleavage by bacterial type II topoisomerases. *Nucleic Acids Res.* 41, 9411–9423. doi: 10.1093/nar/gkt696

Badrinarayanan, A., Le, T. B., and Laub, M. T. (2015). Bacterial chromosome organization and segregation. *Annu. Rev. Cell Dev. Biol.* 31, 171–199. doi: 10.1146/annurev-cellbio-100814-125211

Balas, D., Fernández-Moreira, E., and de la Campa, A. G. (1998). Molecular characterization of the gene encoding the DNA gyrase A subunit of *Streptococcus pneumoniae*. *J. Bacteriol.* 180, 2854–2861.

Balsalobre, L., and de la Campa, A. G. (2008). Fitness of *Streptococcus pneumoniae* fluoroquinolone-resistant strains with topoisomerase IV recombinant genes. *Antimicrob. Agents Chemother.* 52, 822–830. doi: 10.1128/AAC.00731-07

Balsalobre, L., Ferrándiz, M. J., Liñares, J., Tubau, F., and de la Campa, A. G. (2003). Viridans group streptococci are donors in horizontal transfer of topoisomerase IV genes to *Streptococcus pneumoniae*. *Antimicrob. Agents Chemother.* 47, 2072–2081. doi: 10.1128/AAC.47.7.2072-2081.2003

Blot, N., Mavathur, R., Geertz, M., Travers, A., and Muskhelishvili, G. (2006). Homeostatic regulation of supercoiling sensitivity coordinates transcription of the bacterial genome. *EMBO Rep.* 7, 710–715. doi: 10.1038/sj.embor.7400729

Brito, L., Wilton, J., Ferrándiz, M. J., Gómez-Sanz, A., de la Campa, A. G., and Amblar, M. (2017). Absence of tmRNA has a protective effect against fluoroquinolones in *Streptococcus pneumoniae*. *Front. Microbiol.* 7:2164. doi: 10.3389/fmicb.2016.02164

Browning, D. F., and Busby, S. J. (2004). The regulation of bacterial transcription initiation. *Nat. Rev. Microbiol.* 2, 57–65. doi: 10.1038/nrmicro787

Cameron, A. D., and Dorman, C. J. (2012). A fundamental regulatory mechanism operating through OmpR and DNA topology controls expression of *Salmonella* pathogenicity islands SPI-1 and SPI-2. *PLoS Genet.* 8:e1002615. doi: 10.1371/journal.pgen.1002615

Cassone, M., Gagne, A. L., Spruce, L. A., Seeholzer, S. H., and Sebert, M. E. (2012). The HtrA protease from *Streptococcus pneumoniae* digests both denatured proteins and the competence-stimulating peptide. *J. Biol. Chem.* 287, 38449–38459. doi: 10.1074/jbc.M112.391482

Champoux, J. J. (2001). DNA topoisomerases: structure, function, and mechanism. *Annu. Rev. Biochem.* 70, 369–413. doi: 10.1146/annurev.biochem.70.1.369

Chandler, M. G., and Pritchard, R. H. (1975). The effect of gene concentration and relative gene dosage on gene output in *Escherichia coli*. *Mol. Gen. Genet.* 138, 127–141. doi: 10.1007/BF02428117

Chen, H., Ma, Y., Yang, J., O'Brien, C. J., Lee, S. L., Mazurkiewicz, J. E., et al. (2008). Genetic requirement for pneumococcal ear infection. *PLoS ONE* 3:e2950. doi: 10.1371/journal.pone.0002950

Cheng, B., Liu, I. F., and Tse-Dinh, Y. C. (2007). Compounds with antibacterial activity that enhance DNA cleavage by bacterial DNA topoisomerase I. *J. Antimicrob. Chemother.* 59, 640–645. doi: 10.1093/jac/dkl556

Cheng, B., Zhu, C. X., Ji, C., Ahumada, A., and Tse-Dinh, Y. C. (2003). Direct interaction between *Escherichia coli* RNA polymerase and the zinc ribbon domains of DNA topoisomerase I. *J. Biol. Chem.* 278, 30705–30710. doi: 10.1074/jbc.M303403200

Claret, L., and Rouviere-Yaniv, J. (1996). Regulation of HU alpha and HU beta by CRP and FIS in *Escherichia coli*. *J. Mol. Biol.* 263, 126–139. doi: 10.1006/jmbi.1996.0564

Claverys, J. P., Prudhomme, M., and Martin, B. (2006). Induction of competence regulons as a general response to stress in gram-positive bacteria. *Annu. Rev. Microbiol.* 60, 451–475. doi: 10.1146/annurev.micro.60.080805.142139

Croucher, N. J., Vernikos, G. S., Parkhill, J., and Bentley, S. D. (2011). Identification, variation and transcription of pneumococcal repeat sequences. *BMC Genomics* 12:120. doi: 10.1186/1471-2164-12-120

Crozat, E., Philippe, N., Lenski, R. E., Geiselmann, J., and Schneider, D. (2005). Long-term experimental evolution in *Escherichia coli*. XII. DNA topology as a key target of selection. *Genetics* 169, 523–532. doi: 10.1534/genetics.104.035717

Deng, S., Stein, R. A., and Higgins, N. P. (2005). Organization of supercoil domains and their reorganization by transcription. *Mol. Microbiol.* 57, 1511–1521. doi: 10.1111/j.1365-2958.2005.04796.x

Dillon, S. C., and Dorman, C. J. (2010). Bacterial nucleoid-associated proteins, nucleoid structure and gene expression. *Nat. Rev. Microbiol.* 8, 185–195. doi: 10.1038/nrmicro2261

DiNardo, S., Voelkel, K. A., Sternglanz, R., Reynolds, A. E., and Wright, A. (1982). *Escherichia coli* DNA topoisomerase I mutants have compensatory mutations in DNA gyrase genes. *Cell* 31 43–51. doi: 10.1016/0092-8674(82)90403-2

Dorman, C. J. (2013). Genome architecture and global gene regulation in bacteria: making progress towards a unified model? *Nat. Rev. Microbiol.* 11, 349–355. doi: 10.1038/nrmicro3007

Dorman, C. J., and Porter, M. E. (1998). The *Shigella* virulence gene regulatory cascade: a paradigm of bacterial gene control mechanisms. *Mol. Microbiol.* 29, 677–684. doi: 10.1046/j.1365-2958.1998.00902.x

Dowson, C. G., Hutchison, A., Brannigan, J. A., George, R. C., Hansman, D., Liñares, J., et al. (1989). Horizontal transfer of penicillin-binding protein genes in penicillin-resistant clinical isolates of *Streptococcus pneumoniae*. *Proc. Natl. Acad. Sci. U.S.A.* 86, 8842–8846. doi: 10.1073/pnas.86.22.8842

Dowson, C. G., Hutchinson, A., Woodford, N., Johnson, A. P., George, R. C., and Spratt, B. G. (1990). Penicillin-resistant viridans streptococci have obtained altered penicillin-binding protein genes from penicillin-resistant strains of *Streptococcus pneumoniae*. *Proc. Natl. Acad. Sci. U.S.A.* 87, 5858–5862. doi: 10.1073/pnas.87.15.5858

Drlica, K. (1992). Control of bacterial DNA supercoiling. *Mol. Microbiol.* 6, 425–433. doi: 10.1111/j.1365-2958.1992.tb01486.x

Drlica, K., Malik, M., Kerns, R. J., and Zhao, X. (2008). Quinolone-mediated bacterial death. *Antimicrob. Agents Chemother.* 52, 385–392. doi: 10.1128/AAC.01617-06

Drolet, M. (2006). Growth inhibition mediated by excess negative supercoiling: the interplay between transcription elongation, R-loop formation and DNA topology. *Mol. Microbiol.* 59, 723–730. doi 10.1111/j.1365-2958.2005.05006.x

Drolet, M., Bi, X., and Liu, L. F. (1994). Hypernegative supercoiling of the DNA template during transcription elongation in vitro. *J. Biol. Chem.* 269, 2068–2074.

Dupaigne, P., Tonthat, N. K., Espeli, O., Whitfill, T., Boccard, F., and Schumacher, M. A. (2012). Molecular basis for a protein-mediated DNA-bridging mechanism that functions in condensation of the E. *coli* chromosome. *Mol. Cell* 48, 560–571. doi: 10.1016/j.molcel.2012.09.009

Dwyer, D. J., Belenky, P. A., Yang, J. H., MacDonald, I. C., Martell, J. D., Takahashi, N., et al. (2014). Antibiotics induce redox-related physiological alterations as part of their lethality. *Proc. Natl. Acad. Sci. U.S.A.* 111, E2100–E2109. doi: 10.1073/pnas.1401876111

Dwyer, D. J., Collins, J. J., and Walker, G. C. (2015). Unraveling the physiological complexities of antibiotic lethality. *Annu. Rev. Pharmacol. Toxicol.* 55, 313–332. doi: 10.1146/annurev-pharmtox-010814-124712

Dwyer, D. J., Kohanski, M. A., Hayete, B., and Collins, J. J. (2007). Gyrase inhibitors induce an oxidative damage cellular death pathway in *Escherichia coli*. *Mol. Syst. Biol.* 3, 91 doi: 10.1038/msb4100135

Echenique, J. R., Chapuy-Regaud, S., and Trombe, M. C. (2000). Competence regulation by oxygen in *Streptococcus pneumoniae*: involvement of ciaRH and comCDE. *Mol. Microbiol.* 36, 688–696. doi: 10.1046/j.1365-2958.2000.01891.x

Espeli, O., Mercier, R., and Boccard, F. (2008). DNA dynamics vary according to macrodomain topography in the E. *coli chromosome*. *Mol. Microbiol.* 68, 1418–1427. doi: 10.1111/j.1365-2958.2008.06239.x

Falconi, M., Brandi, A., La Teana, A., Gualerzi, C. O., and Pon, C. L. (1996). Antagonistic involvement of FIS and H-NS proteins in the transcriptional control of hns expression. *Mol. Microbiol.* 19, 965–975. doi: 10.1046/j.1365-2958.1996.436961.x

Fang, G., Rocha, E., and Danchin, A. (2005). How essential are nonessential genes? *Mol. Biol. Evol.* 22, 2147–2156. doi: 10.1093/molbev/msi211

Fang, G., Rocha, E. P., and Danchin, A. (2008). Persistence drives gene clustering in bacterial genomes. *BMC Genomics* 9:4. doi: 10.1186/1471-2164-9-4

Fenoll, A., Gimenez, M. J., Vicioso, M. D., Granizo, J. J., Robledo, O., and Aguilar, L. (2009). Susceptibility of pneumococci causing meningitis in Spain and prevalence among such isolates of serotypes contained in the 7-valent pneumococcal conjugate vaccine. *J. Antimicrob. Chemother.* 64, 1338–1340. doi: 10.1093/jac/dkp376

Fernández-Moreira, E., Balas, D., González, I., and de la Campa, A. G. (2000). Fluoroquinolones inhibit preferentially *Streptococcus pneumoniae* DNA topoisomerase IV than DNA gyrase native proteins. *Microb. Drug Resist.* 6, 259–267. doi: 10.1089/mdr.2000.6.259

Ferrándiz, M. J., Ardanuy, C., Liñares, J., García-Arenzana, J. M., Cercenado, E., Fleites, A., et al. (2005). New mutations and horizontal transfer of *rpoB* among rifampin-resistant *Streptococcus pneumoniae* from four Spanish hospitals. *Antimicrob. Agents Chemother.* 49, 2237–2245. doi: 10.1128/AAC.49.6.2237-2245.2005

Ferrándiz, M. J., Arnanz, C., Martín-Galiano, A. J., Rodríguez-Martin, C., and de la Campa, A. G. (2014). Role of global and local topology in the regulation of gene expression in *Streptococcus pneumoniae*. *PLoS ONE* 9:e101574. doi: 10.1371/journal.pone.0101574

Ferrándiz, M. J., and de la Campa, A. G. (2014). The fluoroquinolone levofloxacin triggers the transcriptional activation of iron transport genes that contribute to cell death in *Streptococcus pneumoniae*. *Antimicrob. Agents Chemother.* 58, 247–257. doi: 10.1128/AAC.01706-13

Ferrándiz, M. J., Martín-Galiano, A. J., Arnanz, C., Camacho-Soguero, I., Tirado-Vélez, J. M., and de la Campa, A. G. (2016a). An increase in negative supercoiling in bacteria reveals topology-reacting gene clusters and a homeostatic response mediated by the DNA topoisomerase I gene. *Nucleic Acids Res.* 44, 7292–7303. doi: 10.1093/nar/gkw602

Ferrándiz, M. J., Martín-Galiano, A. J., Arnanz, C., Zimmerman, T., and de la Campa, A. G. (2016b). Reactive oxygen species contribute to the bactericidal effects of the fluoroquinolone moxifloxacin in *Streptococcus pneumoniae*. *Antimicrob. Agents Chemother.* 60, 409–417. doi: 10.1128/AAC.02299-15

Ferrándiz, M. J., Martín-Galiano, A. J., Schvartzman, J. B., and de la Campa, A. G. (2010). The genome of *Streptococcus pneumoniae* is organized in topology-reacting gene clusters. *Nucleic Acids Res.* 38, 3570–3581. doi: 10.1093/nar/gkq106

French, S. (1992). Consequences of replication fork movement through transcription units in vivo. *Science* 258, 1362–1365. doi: 10.1126/science.1455232

García, M. T., Blázquez, M. A., Ferrándiz, M. J., Sanz, M. J., Silva-Martín, N., Hermoso, J. A., et al. (2011). New alkaloid antibiotics that target the DNA topoisomerase I of *Streptococcus pneumoniae*. *J. Biol. Chem.* 286, 6402–6413. doi: 10.1074/jbc.M110.148148

Gellert, M., Mizuuchi, K., ODea, H., and Nash, H. A. (1976). DNA gyrase: an enzyme that introduces superhelical turns into DNA. *Proc. Natl. Acad. Sci. U.S.A.* 73, 3872–3876. doi: 10.1073/pnas.73.11.3872

Gerganova, V., Berger, M., Zaldastanishvili, E., Sobetzko, P., Lafon, C., Mourez, M., et al. (2015). Chromosomal position shift of a regulatory gene alters the bacterial phenotype. *Nucleic Acids Res.* 43, 8215–8226. doi: 10.1093/nar/gkv709

Giefing, C., Nagy, E., and von Gabain, A. (2009). The antigenome: from protein subunit vaccines to antibody treatments of bacterial infections? *Adv. Exp. Med. Biol.* 655, 90–117. doi: 10.1007/978-1-4419-1132-2-9

Gmüender, H., Kuratli, K., Di, Padova K, Gray, C. P., Keck, W., and Evers, S. (2001). Gene expression changes triggered by exposure of *Haemophilus influenzae* to novobiocin or ciprofloxacin: combined transcription and translation analysis. *Genome Res.* 11, 28–42. doi: 10.1101/gr.157701

Grainger, D. C., Goldberg, M. D., Lee, D. J., and Busby, S. J. (2008). Selective repression by Fis and H-NS at the *Escherichia coli dps* promoter. *Mol. Microbiol.* 68, 1366–1377. doi: 10.1111/j.1365-2958.2008.06253.x

Hensel, M., Shea, J. E., Gleeson, C., Jones, M. D., Dalton, E., and Holden, D. W. (1995). Simultaneous identification of bacterial virulence genes by negative selection. *Science* 269, 400–403. doi: 10.1126/science.7618105

Higgins, N. P., Yang, X., Fu, Q., and Roth, J. R. (1996). Surveying a supercoil domain by using the gamma delta resolution system in *Salmonella typhimurium*. *J. Bacteriol.* 178, 2825–2835. doi: 10.1128/jb.178.10.2825-2835.1996

Holmes, V. F., and Cozzarelli, N. R. (2000). Closing the ring: links between SMC proteins and chromosome partitioning, condensation, and supercoiling. *Proc. Natl. Acad. Sci. U.S.A.* 97, 1322–1324. doi: 10.1073/pnas.040576797

Jacobs, M. R., Felmingham, D., Appelbaum, P. C., Grüneberg, R. N., and Group, T. A. P. (2003). The Alexander project 1998-200: susceptibility of pathogens isolated from community-acquired respiratory tract infection to commonly used antimicrobial agents. *J. Antimicrob. Chemother.* 52, 229–246. doi: 10.1093/jac/dkg321

Janoir, C., Zeller, V., Kitzis, M.-D., Moreau, N. J., and Gutmann, L. (1996). High-level fluoroquinolone resistance in *Streptococcus pneumoniae* requires mutations in *parC* and *gyrA*. *Antimicrob. Agents Chemother.* 40, 2760–2764.

Jiang, X., Sobetzko, P., Nasser, W., Reverchon, S., and Muskhelishvili, G. (2015). Chromosomal "stress-response" domains govern the spatiotemporal expression of the bacterial virulence program. *MBio* 6, e00353-15. doi: 10.1128/mBio.00353-15

Kato, J., Nishimura, Y., Imamura, R., Niki, H., Hiraga, S., and Suzuki, H. (1990). New topoisomerase essential for chromosome segregation in E. *coli*. *Cell* 63, 393–404. doi: 10.1016/0092-8674(90)90172-B

Kavenoff, R., and Bowen, B. C. (1976). Electron microscopy of membrane-free folded chromosomes from *Escherichia coli*. *Chromosoma* 59, 89–101. doi: 10.1007/BF00328479

Khedkar, S., and Seshasayee, A. S. (2016). Comparative genomics of interreplichore translocations in bacteria: a measure of chromosome topology? *G3 (Bethesda)* 6, 1597–1606. doi: 10.1534/g3.116.028274

Kohanski, M. A., Dwyer, D. J., Hayete, B., Lawrence, C. A., and Collins, J. J. (2007). A common mechanism of cellular death induced by bactericidal antibiotics. *Cell* 130, 797–810. doi: 10.1016/j.cell.2007.06.049

Kyaw, M. H., Lynfield, R., Schaffner, W., Craig, A. S., Hadler, J., Reingold, A., et al. (2006). Effect of introduction of the pneumococcal conjugate vaccine on drug-resistant *Streptococcus pneumoniae*. *N. Engl. J. Med.* 354, 1455–1463. doi: 10.1056/NEJMoa051642

Lathe, W. C. III, Snel, B., and Bork, P. (2000). Gene context conservation of a higher order than operons. *Trends. Biochem. Sci.* 25, 474–479. doi: 10.1016/S0968-0004(00)01663-7

Le, T. B., Imakaev, M. V., Mirny, L. A., and Laub, M. T. (2013). High-resolution mapping of the spatial organization of a bacterial chromosome. *Science* 342, 731–734. doi: 10.1126/science.1242059

Lefebure, T., and Stanhope, M. J. (2007). Evolution of the core and pan-genome of *Streptococcus*: positive selection, recombination, and genome composition. *Genome Biol.* 8:R71. doi: 10.1186/gb-2007-8-5-r71

Leo, E., Gould, K. A., Pan, X. S., Capranico, G., Sanderson, M. R., Palumbo, M., et al. (2005). Novel symmetric and asymmetric DNA scission determinants for *Streptococcus pneumoniae* topoisomerase IV and gyrase are clustered at the DNA breakage site. *J. Biol. Chem.* 280, 14252–14263. doi: 10.1074/jbc.M500156200

Liñares, J., Ardanuy, C., Pallarés, R., and Fenoll, A. (2010). Changes in antimicrobial resistance, serotypes and genotypes in *Streptococcus pneumoniae* over a 30-year period. *Clin. Microbiol. Infect.* 16, 402–410. doi: 10.1111/j.1469-0691.2010.03182.x

Marr, C., Geertz, M., Hutt, M. T., and Muskhelishvili, G. (2008). Dissecting the logical types of network control in gene expression profiles. *BMC Syst. Biol.* 2:18. doi: 10.1186/1752-0509-2-18

Martin, B., Quentin, Y., Fichant, G., and Claverys, J. P. (2006). Independent evolution of competence regulatory cascades in streptococci? *Trends Microbiol.* 14, 339–345. doi: 10.1016/j.tim.2006.06.007

Martín-Galiano, A. J., Ferrándiz, M. J., and de la Campa, A. G. (2017). Bridging chromosomal architecture and pathophysiology of *Streptococcus pneumoniae*. *Genome Biol. Evol.* 9, 350–361. doi: 10.1093/gbe/evw299

Martín-Galiano, A. J., Wells, J. M., and de la Campa, A. G. (2004). Relationship between codon biased genes, microarray expression values and physiological characteristics of *Streptococcus pneumoniae*. *Microbiology* 150, 2313–2325. doi: 10.1099/mic.0.27097-0

Masse, E., and Drolet, M. (1999). *Escherichia coli* DNA topoisomerase I inhibits R-loop formation by relaxing transcription-induced negative supercoiling. *J. Biol. Chem.* 274, 16659–16664. doi: 10.1074/jbc.274.23.16659

Menzel, R., and Gellert, M. (1983). Regulation of the genes for E. *coli* DNA gyrase: homeostatic control of DNA supercoiling. *Cell* 34, 105–113. doi: 10.1016/0092-8674(83)90140-X

Menzel, R., and Gellert, M. (1987a). Fusions of the *Escherichia coli* gyrA and *gyrB* control regions to the galactokinase gene are inducible by coumermycin treatment. *J. Bacteriol.* 169, 1272–1278.

Menzel, R., and Gellert, M. (1987b). Modulation of transcription by DNA supercoiling: a deletion analysis of the *Escherichia coli gyrA* and *gyrB* promoters. *Proc. Natl. Acad. Sci. U.S.A.* 84, 4185–4189.

Mira, A., Martín-Cuadrado, A. B., D'Auria, G., and Rodríguez-Valera, F. (2010). The bacterial pan-genome:a new paradigm in microbiology. *Int. Microbiol.* 13, 45–57. doi: 10.2436/20.1501.01.110

Molzen, T. E., Burghout, P., Bootsma, H. J., Brandt, C. T., van, der Gaast-de Jongh, C. E., Eleveld, M. J., et al. (2011). Genome-wide identification of *Streptococcus pneumoniae* genes essential for bacterial replication during experimental meningitis. *Infect. Immun.* 79, 288–297. doi: 10.1128/IAI.00631-10

Moore, M. R., Gertz, J. R. E., Woodbury, R. L., Barkocy-Gallagher, G. A., Schaffner, W., Lexau, C., et al. (2008). Population snapshot of emergent *Streptococcus pneumoniae* serotype 19A in the United States, 2005. *J. Infect. Dis.* 197, 1016–1027. doi: 10.1086/528996

Muñoz, R., and de la Campa, A. G. (1996). ParC subunit of DNA topoisomerase IV of *Streptococcus pneumoniae* is a primary target of fluoroquinolones and cooperates with DNA gyrase A subunit in forming resistance phenotype. *Antimicrob. Agents Chemother.* 40, 2252–2257.

Ochman, H., Lawrence, J. G., and Groisman, E. A. (2000). Lateral gene transfer and the nature of bacterial innovation. *Nature* 405, 299–304. doi: 10.1038/35012500

Ogawa, T., Yogo, K., Furuike, S., Sutoh, K., Kikuchi, A., and Kinosita, K. Jr. (2015). Direct observation of DNA overwinding by reverse gyrase. *Proc. Natl. Acad. Sci. U.S.A.* 112, 7495–7500. doi: 10.1073/pnas.1422203112

Oggioni, M. R., Iannelli, F., Ricci, S., Chiavolini, D., Parigi, R., Trappetti, C., et al. (2004). Antibacterial activity of a competence-stimulating peptide in experimental sepsis caused by *Streptococcus pneumoniae*. *Antimicrob. Agents Chemother.* 48, 4725–4732. doi: 10.1128/AAC.48.12.4725-4732.2004

Pérez-Martín, J., and Espinosa, M. (1991). The RepA repressor can act as a transcriptional activator by inducing DNA bends. *EMBO J.* 10, 1375–1382.

Pérez-Martín, J., Rojo, F., and de Lorenzo, V. (1994). Promoters responsive to DNA bending: a common theme in prokaryotic gene expression. *Microbiol. Rev.* 58, 268–290.

Peter, B. J., Arsuaga, J., Breier, A. M., Khodursky, A. B., Brown, P. O., and Cozzarelli, N. R. (2004). Genomic transcriptional response to loss of chromosomal supercoiling in *Escherichia coli*. *Genome Biol.* 5:R87. doi: 10.1186/gb-2004-5-11-r87

Peterson, S. N., Sung, C. K., Cline, R., Desai, B. V., Snesrud, E. C., Luo, P., et al. (2004). Identification of competence pheromone responsive genes in *Streptococcus pneumoniae* by use of DNA microarrays. *Mol. Microbiol.* 51, 1051–1070. doi: 10.1046/j.1365-2958.2003.03907.x

Phoenix, P., Raymond, M. A., Masse, E., and Drolet, M. (1997). Roles of DNA topoisomerases in the regulation of R-loop formation in vitro. *J. Biol. Chem.* 272, 1473–1479. doi: 10.1074/jbc.272.3.1473

Pilishvili, T., Lexau, C., Farley, M. M., Hadler, J., Harrison, L. H., Bennett, N. M., et al. (2010). Sustained reductions in invasive pneumococcal disease in the era of conjugate vaccine. *J. Infect. Dis.* 201, 32–41. doi: 10.1086/648593

Postow, L., Crisona, N. J., Peter, B. J., Hardy, C. D., and Cozzarelli, N. R. (2001). Topological challenges to DNA replication: conformations at the fork. *Proc. Natl. Acad. Sci. U.S.A.* 98, 8219–8226. doi: 10.1073/pnas.111006998

Postow, L., Hardy, C. D., Arsuaga, J., and Cozzarelli, N. R. (2004). Topological domain structure of the *Escherichia coli* chromosome. *Genes Dev.* 18, 1766–1779. doi: 10.1101/gad.1207504

Prudhomme, M., Attaiech, L., Sanchez, G., Martin, B., and Claverys, J. P. (2006). Antibiotic stress induces genetic transformability in the human pathogen *Streptococcus pneumoniae*. *Science* 313, 89–92. doi: 10.1126/science.1127912

Pruss, G. J., Manes, S. H., and Drlica, K. (1982). *Escherichia coli* DNA topoisomerase I mutants: increased supercoiling is corrected by mutations near gyrase genes. *Cell* 31, 35–42. doi: 10.1016/0092-8674(82)90402-0

Reams, A. B., and Neidle, E. L. (2004). Selection for gene clustering by tandem duplication. *Annu. Rev. Microbiol.* 58, 119–142. doi: 10.1146/annurev.micro.58.030603.123806

Reverchon, S., and Nasser, W. (2013). Dickeya ecology, environment sensing and regulation of virulence programme. *Environ. Microbiol. Rep.* 5, 622–636. doi: 10.1111/1758-2229.12073

Rocha, E. P. (2004a). Order and disorder in bacterial genomes. *Curr. Opin. Microbiol.* 7, 519–527. doi: 10.1016/j.mib.2004.08.006

Rocha, E. P. (2004b). The replication-related organization of bacterial genomes. *Microbiology* 150, 1609–1627. doi: 10.1099/mic.0.26974-0

Rocha, E. P., and Danchin, A. (2003). Essentiality, not expressiveness, drives gene-strand bias in bacteria. *Nat. Genet.* 34, 377–378. doi: 10.1038/ng1209

Schmid, M. B., and Roth, J. R (1987). Gene location affects expression level in *Salmonella typhimurium*. *J. Bacteriol.* 169, 2872–2875. doi: 10.1128/jb.169.6.2872-2875.1987

Schneider, R., Travers, A., Kutateladze, T., and Muskhelishvili, G. (1999). A DNA architectural protein couples cellular physiology and DNA topology in *Escherichia coli*. *Mol. Microbiol.* 34, 953–964. doi: 10.1046/j.1365-2958.1999.01656.x

Schroder, W., Bernhardt, J., Marincola, G., Klein-Hitpass, L., Herbig, A., Krupp, G., et al. (2014). Altering gene expression by aminocoumarins: the role of DNA supercoiling in Staphylococcus aureus. *BMC Genomics* 15:291. doi: 10.1186/1471-2164-15-291

Sinden, R. R., and Pettijohn, D. E. (1981). Chromosomes in living *Escherichia coli* cells are segregated into domains of supercoiling. *Proc. Natl. Acad. Sci. U.S.A.* 78, 224–228. doi: 10.1073/pnas.78.1.224

Slager, J., and Veening, J. W. (2016). Hard-wired control of bacterial processes by chromosomal gene location. *Trends Microbiol.* 24, 788–800. doi: 10.1016/j.tim.2016.06.003

Snyder, M., and Drlica, K. (1979). DNA gyrase on the bacterial chromosome: DNA cleavage induced by oxolinic acid. *J. Mol. Biol.* 131, 287–302. doi: 10.1016/0022-2836(79)90077-9

Sobetzko, P., Travers, A., and Muskhelishvili, G. (2012). Gene order and chromosome dynamics coordinate spatiotemporal gene expression during the bacterial growth cycle. *Proc. Natl. Acad. Sci. U.S.A.* 109, E42–E50. doi: 10.1073/pnas.1108229109

Soler-Bistue, A., Mondotte, J. A., Bland, M. J., Val, M. E., Saleh, M. C., and Mazel, D. (2015). Genomic location of the major ribosomal protein gene locus determines *Vibrio cholerae* global growth and infectivity. *PLoS Genet.* 11:e1005156. doi: 10.1371/journal.pgen.1005156

Stassi, D., López, P., Espinosa, M., and Lacks, S. A. (1981). Cloning of chromosomal genes in *Streptococcus pneumoniae*. *Proc. Natl. Acad. Sci. U.S.A.* 78, 7028–7032. doi: 10.1073/pnas.78.11.7028

Tankovic, J., Perichon, B., Duval, J., and Courvalin, P. (1996). Contribution of mutations in *gyrA* and *parC* genes to fluoroquinolone resistance of mutants of *Streptococcus pneumoniae* obtained in vivo and in vitro. *Antimicrob. Agents Chemother.* 40, 2505–2510.

Thiara, A. S., and Cundliffe, E. (1989). Interplay of novobiocinresistant and –sensitive DNA gyrase activities in selfprotection of the novobiocin producer *Streptomyces sphaeroides*. *Gene* 81, 65–72 doi: 10.1016/0378-1119(89)90337-5

Travers, A., and Muskhelishvili, G. (2005). DNA supercoiling - a global transcriptional regulator for enterobacterial growth? *Nat. Rev. Microbiol.* 3, 157–169. doi: 10.1038/nrmicro1088

Tse-Dinh, Y.-C. (1985). Regulation of the *Escherichia coli* DNA topoisomerase I gene by DNA supercoiling. *Nucleic Acids Res.* 13, 4751–4763. doi: 10.1093/nar/13.13.4751

Unniraman, S., Chatterji, M., and Nagaraja, V. (2002). DNA gyrase genes in Mycobacterium tuberculosis: a single operon driven by multiple promoters. *J. Bacteriol.* 184, 5449–5456. doi: 10.1128/JB.184.19.5449-5456.2002

van Opijnen, T., and Camilli, A. (2012). A fine scale phenotype-genotype virulence map of a bacterial pathogen. *Genome Res.* 22, 2541–2551. doi: 10.1101/gr.137430.112

Vora, T., Hottes, A. K., and Tavazoie, S. (2009). Protein occupancy landscape of a bacterial genome. *Mol. Cell* 35, 247–253. doi: 10.1016/j.molcel.2009.06.035

Wang, X., Montero Llopis, P., and Rudner, D. Z. (2013). Organization and segregation of bacterial chromosomes. *Nat. Rev. Genet.* 14, 191–203. doi: 10.1038/nrg3375

Wang, X., and Zhao, X. (2009). Contribution of oxidative damage to antimicrobial lethality. *Antimicrob. Agents Chemother.* 53, 1395–1402. doi: 10.1128/AAC.01087-08

Wang, X., Zhao, X., Malik, M., and Drlica, K. (2010). Contribution of reactive oxygen species to pathways of quinolone-mediated bacterial cell death. *J. Antimicrob. Chemother.* 65, 520–524. doi: 10.1093/jac/dkp486

Weinstein-Fischer, D., and Altuvia, S. (2007). Differential regulation of *Escherichia coli* topoisomerase I by Fis. *Mol. Microbiol.* 63, 1131–1144. doi: 10.1111/j.1365-2958.2006.05569.x

Whitney, C. G., Farley, M. M., Hadler, J., Harrison, L. H., Bennett, N. M., Lynfield, R., et al. (2003). Decline in invasive pneumococcal disease after the introduction of protein-polysaccharide conjugate vaccine. *N. Engl. J. Med.* 348, 1737–1746. doi: 10.1056/NEJMoa022823

Worcel, A., and Burgi, E. (1972). On the structure of the folded chromosome of *Escherichia coli*. *J. Mol. Biol.* 71, 127–147. doi: 10.1016/0022-2836(72)90342-7

World Health Organization (2007). Pneumococcal conjugate vaccine for childhood immunization-WHO position paper. *Wkly. Epidemiol. Rec.* 82, 93–104.

Wright, M. A., Kharchenko, P., Church, G. M., and Segre, D. (2007). Chromosomal periodicity of evolutionarily conserved gene pairs. *Proc. Natl. Acad. Sci. U.S.A.* 104, 10559–10564. doi: 10.1073/pnas.0610776104

14

Dextransucrase Expression is Concomitant with that of Replication and Maintenance Functions of the pMN1 Plasmid in *Lactobacillus sakei* MN1

Montserrat Nácher-Vázquez[1], *José A. Ruiz-Masó*[1], *María L. Mohedano*[1], *Gloria del Solar*[1], *Rosa Aznar*[2,3] *and Paloma López*[1*]

[1] Department of Molecular Microbiology and Infection Biology, Biological Research Center, Spanish National Research Council (CSIC), Madrid, Spain, [2] Department of Food Safety and Preservation, Institute of Agrochemistry and Food Technology, CSIC, Paterna, Spain, [3] Department of Microbiology and Ecology, University of Valencia, Burjassot, Spain

Edited by:
Tatiana Venkova,
Fox Chase Cancer Center,
United States

Reviewed by:
Preeti Srivastava,
Indian Institute of Technology Delhi,
India
Stephen M. Kwong,
University of Sydney, Australia
Reinaldo Fraga,
Cuban Research Institute on
Sugarcane By-Products (ICIDCA),
Cuba

***Correspondence:**
Paloma López
plg@cib.csic.es

The exopolysaccharide synthesized by *Lactobacillus sakei* MN1 is a dextran with antiviral and immunomodulatory properties of potential utility in aquaculture. In this work we have investigated the genetic basis of dextran production by this bacterium. Southern blot hybridization experiments demonstrated the plasmidic location of the *dsrLS* gene, which encodes the dextransucrase involved in dextran synthesis. DNA sequencing of the 11,126 kbp plasmid (pMN1) revealed that it belongs to a family which replicates by the theta mechanism, whose prototype is pUCL287. The plasmid comprises the origin of replication, *repA*, *repB,* and *dsrLS* genes, as well as seven open reading frames of uncharacterized function. *Lb. sakei* MN1 produces dextran when sucrose, but not glucose, is present in the growth medium. Therefore, plasmid copy number and stability, as well as *dsrLS* expression, were investigated in cultures grown in the presence of either sucrose or glucose. The results revealed that pMN1 is a stable low-copy-number plasmid in both conditions. Gene expression studies showed that *dsrLS* is constitutively expressed, irrespective of the carbon source present in the medium. Moreover, *dsrLS* is expressed from a monocistronic transcript as well as from a polycistronic *repA-repB-orf1-dsrLS* mRNA. To our knowledge, this is the first report of a plasmid-borne dextransucrase-encoding gene, as well as the first time that co-transcription of genes involved in plasmid maintenance and replication with a gene encoding an enzyme has been established.

Keywords: dextran, dextransucrase, lactic acid bacteria, *Lactobacillus sakei*, plasmid, probiotics

INTRODUCTION

Lactic acid bacteria (LAB) play an important role in the production of fermented foods based on milk, meat, and vegetables as well as alcoholic beverages, due to their metabolic pathways, whose products contribute to food safety (e.g., lactic acid or hydrogen peroxide) and to the organoleptic characteristics [e.g., the diacetyl and other aroma compounds or texturizing exopolysaccharides (EPS)]. In addition, some LAB have beneficial health characteristics (probiotic properties) or metabolic capacities such as the production of enzymes (amylases, phytases), vitamins (folates, riboflavin), or EPS, which are of particular interest for the agro-food industry and for the formulation of new functional foods (Anastasio et al., 2010; Badel et al., 2011; Capozzi et al., 2012).

The EPS produced by LAB can be classified by (i) composition, because they include different types of bonds and monosaccharide subunits; (ii) types and degrees of branching; (iii) molecular mass; and (iv) three-dimensional (3D) structural conformation. EPS are classified as homopolysaccharides (HoPS), consisting of a single type of monosaccharide, or heteropolysaccharides (HePS), composed of two or more types of monosaccharides. HoPS are glucans, fructans, or galactans, made up of repeating units of glucose, fructose, or galactose, respectively (Pérez-Ramos et al., 2015).

The α-D-glucans are the most widely produced HoPS and according to the linkage in the main chain, they are subdivided in dextrans α-(1,6), mutans α-(1,3), reuterans α-(1,4), and alternans α-(1,3) and α-(1,6) and may present different types and degrees of branching (Monsan et al., 2001). Of these, dextrans are currently used in the food and pharmaceutical industries (Aman et al., 2012). The viscosity and rheological properties of the dextran solutions are influenced by their molecular masses and consequently define their applications. Thus, low-molecular mass dextrans are used in the photographic and pharmaceutical industries, whereas the high-molecular mass dextrans are utilized in the chemical industry. Among various medical applications, dextrans are used for anticoagulant therapy as heparin substitutes and blood plasma replacers/expanders. In addition, there is evidence that dextran sulfate has an antiviral effect against human immunodeficiency virus (Piret et al., 2000) and we have recently demonstrated that dextrans synthesized by LAB have potential as antivirals and immunomodulatory agents in trout (Nácher-Vázquez et al., 2015). In the food industry HoPS are added to bakery products and confectionery to improve softness or moisture retention, to prevent crystallization, and to increase viscosity, rheology, texture, and volume (Pérez-Ramos et al., 2015). They are also used as films to protect surfaces of frozen fish, meat, vegetables, or cheese from oxidation and other chemical changes.

Dextrans are synthesized by dextransucrases (Dsr), which catalyze the transfer of D-glucopyranosyl residues from sucrose to the growing polymer, accompanied by fructose release (Werning et al., 2012). The Dsr-encoding genes are carried by strains belonging to the genera *Lactobacillus*, *Leuconostoc*, *Oenococcus*, *Streptococcus*, *Weissella*, and *Pediococcus* (Kralj et al., 2004a; Naessens et al., 2005; Bounaix et al., 2010; Werning et al., 2012; Amari et al., 2013; Rühmkorf et al., 2013; Dimopoulou et al., 2015; Yanping et al., 2015). Many Dsr have been characterized, but despite the interest of dextrans in various applications, little is known about the regulation of the expression of Dsr. By determining the levels or the activity of the Dsr in bacterial cultures grown in presence of sucrose or other sugars it has been inferred that their expression may be constitutive or sucrose-inducible. However, transcriptional analysis of *dsr* genes expression to validate sucrose inducibility has been performed in a few cases only (e.g., in *Leuconostoc mesenteroides* NRRL B-512F, Quirasco et al., 1999).

We have previously demonstrated that *Lb. sakei* MN1 isolated from a fermented meat product synthesizes an α-(1-6) glucan with ~6% substitution, at positions O-3, by side chains composed of a single residue of glucose and with a molecular mass of 1.7 × 10^8 Da (Nácher-Vázquez et al., 2015; Zarour et al., 2017). We have performed *in vitro* and *in vivo* experiments that support that this dextran has antiviral and immunomodulatory properties of interest in aquaculture (Nácher-Vázquez et al., 2015). Moreover, we have demonstrated that the purified HoPS is able to efficiently immunomodulate *in vitro* human macrophages (Zarour et al., 2017). In addition, we have provided evidences that *Lb. sakei* MN1 has probiotic properties and we have shown that the production of dextran influences, *in vitro*, the bacterial capability for aggregation, biofilm formation, and adhesion to enterocytes as well as *in vivo* bacterial colonization and competition with pathogens in gnotobiotic zebrafish models (Nácher-Vázquez et al., 2017). Therefore this bacterium and its dextran seem to have potential for development of functional synbiotic food and feed. Thus, in this work, we have characterized the genetic basis of dextran production in this bacterium with the aim of having a better knowledge of the practical utility of *Lb. sakei* MN1.

MATERIALS AND METHODS

Bacterial Strains and Growth Media

The bacterial strains used in this work are shown in **Table 1**. The *Latococcus lactis* strains were grown in M17 broth (Oxoid) supplemented with 0.5% glucose (M17G) or 0.5% glucose plus 0.8% sucrose (M17GS) and *Lb. sakei* strains were grown in Man Rogosa Sharpe broth (de Man et al., 1960) with 2% glucose (MRSG) or 2% sucrose (MRSS) or in defined medium with 0.8% glucose (CDMG) or 0.8% sucrose (CDMS) (Sánchez et al., 2008) and incubated at 30°C. *Escherichia coli* strains were grown in LB medium containing 10 g L^{-1} of tryptone, 5 g L^{-1} of yeast extract, and 10 g L^{-1} of NaCl (pH 7.0) and incubated at 37°C. When the bacteria carried pRCR or pRCR-based plasmid derivatives conferring resistance to chloramphenicol (Cm) the medium was supplemented with Cm at 5 μg mL^{-1} for *L. lactis* and *Lb. sakei* strains or at 10 μg mL^{-1} for *E. coli* DH5α.

Genomic and Plasmidic DNA Preparations

To isolate plasmidic DNA, bacterial cultures were grown to an absorbance at 600 nm (A_{600}) of 2 at 30°C in MRSG and 10 mL of each culture were sedimented by centrifugation (15,700 × g, 10 min, 4°C). For cellular lysis, cells were washed with phosphate buffer saline (PBS, pH 7.4), resuspended in 2 mL of a solution containing 25% sucrose, 30 mg mL^{-1} lysozyme, 120 U mL^{-1} mutanolysin, and 40 μg mL^{-1} RNase A, and incubated for 15 min at 37°C. Then, cell debris and chromosomal DNA were removed from the extracts by: (i) treatment for 7 min at 21°C with 4 mL of a solution containing 0.13 N NaOH and 2% SDS, (ii) incubation for 15 min at 0°C with 3 mL of 1 M potassium acetate pH 4.8, and (iii) sedimentation (centrifugation at 15,700 × g, 15 min, 4°C). The plasmidic DNA present in the supernatants was precipitated and concentrated by

TABLE 1 | Description of bacteria used in this work.

Strain	Plasmids	[a]Resistance	Information	References
Lb. sakei MN1 (CECT 8329)	pMN1 pMN2	–	Isolated from a fermented meat product	Chenoll et al., 2007
Lb. sakei MN1[pRCR12]	pMN1 pMN2 pRCR12	Cm^R	Derivative of pRCR. It contains the promoter Px of *malXCD* operon of *Streptococcus pneumoniae* JNR7/87 fused to the *mrfp* gene	Nácher-Vázquez et al., 2017
Lb. sakei MN1[pRCR13]	pMN1 pMN2 pRCR13	Cm^R	Derivative of pRCR by cloning of region A (GenBank MF590088 position: 38–363) upstream of *mrfp*	This study
Lb. sakei MN1[pRCR14]	pMN1 pMN2 pRCR14	Cm^R	Derivative of pRCR by cloning of region B (GenBank MF590088 position: 1,476–1,819) upstream of *mrfp*	This study
Lb. sakei MN1[pRCR15]	pMN1 pMN2 pRCR15	Cm^R	Derivative of pRCR by cloning of region C (GenBank MF590088 position: 2,392–2,749) upstream of *mrfp*	This study
L. lactis MG1363	–	–	Strain derived from *L. lactis* 712 by plasmids curing	Gasson, 1983
L. lactis MG1363[pRCR13]	pRCR13	Cm^R	Derivative of pRCR by cloning of region A upstream of *mrfp*	This study
L. lactis MG1363[pRCR14]	pRCR14	Cm^R	Derivative of pRCR by cloning of region B upstream of *mrfp*	This study
L. lactis MG1363[pRCR15]	pRCR15	Cm^R	Derivative of pRCR by cloning of region C upstream of *mrfp*	This study
E. coli DH5α[pRCR]	pRCR	Cm^R	Promoters probe vector containing the *mrfp* gene, which encodes the fluorescent mCherry protein	Mohedano et al., 2015
E. coli V517	Eight plasmids pVA517A through pVA517H	ND	Source of plasmids used as references in agarose gel analysis	Macrina et al., 1978

[a]*ND, No determined; Cm^R, resistance to chloramphenicol.*

addition of 42% (final concentration) isopropanol, centrifugation as above and dissolution in 3.2 mL of ultrapure water. The DNA preparation was purified and deproteinated by treatment with 2 mL of a solution containing 7.5 M ammonium acetate and ethidium bromide at 0.5 mg mL^{-1}, addition of 1:1 (v/v) of a mixture of phenol, chloroform, and isoamyl alcohol (50:48:2, vol/vol/vol) and centrifugation (15,700 × g, 10 min, 21°C). Plasmidic DNA was precipitated from the aqueous phase with 69.5% (final concentration) ethanol for 12 h at −20°C and recovered by centrifugation (11,269 × g, 45 min, −10°C). The precipitated DNA was washed with 1 mL of 70% ethanol, sedimented by centrifugation (11,269 × g, 30 min, −10°C), and dissolved in 10 mM Tris buffer pH 8.0 (100 μL).

For the isolation of genomic DNA, bacterial cultures were grown to an $A_{600} = 2$ at 30°C in MRSG and 1 mL of each culture were sedimented by centrifugation (15,700 × g; 10 min, 4°C). For cellular lysis, cells were washed with phosphate buffer saline (PBS, pH 7.4), resuspended in a solution (100 μL) containing 25% sucrose, 50 mM Tris pH 8.0, 0.1 M NaCl, 30 mg mL^{-1} lysozyme, 240 U mL^{-1} mutanolysin, and 80 μg mL^{-1} RNase A, and incubated for 15 min at 37°C. Then, lysed cells were treated with 1% (final concentration) SDS for 2 min at 21°C and passed through a 25 GA needle (0.5 × 16 mm) three times. The extracts were deproteinated by treatment with an equal volume of a mixture of phenol, chloroform, and isoamyl alcohol (50:48:2, vol/vol/vol) for 5 min at 21°C and centrifugation (15,700 × g, 10 min, 21°C). The DNA contained in the aqueous phase was precipitated with 69.5% (final concentration) ethanol and 83 mM (final concentration) sodium acetate pH 7.0 for 12 h at −20°C and recovered by centrifugation (11,269 × g, 45 min, −10°C). The precipitated DNA was washed with 1 mL of 70% ethanol, sedimented by centrifugation (11,269 × g, 30 min, −10°C), and dissolved in 10 mM Tris buffer pH 8.0 (100 μL).

Total RNA Preparations

For the isolation of total RNA, bacterial cultures were grown to $A_{600} = 2$ at 30°C in CDMG or CDMS for RT-PCR. Total RNAs were isolated using the kit "FastRNA Pro Blue" (QBIOgene) and subjected to electrophoresis in 0.8% agarose gels at a constant voltage of 135 V for 20 min to check the integrity of the rRNAs. The total RNA concentration was determined with a Qubit 2.0 fluorimeter (Invitrogen) following the instructions of the supplier. To ensure absence of DNA, the RNA preparations were incubated for 1 h at 37°C with 1 μg mL^{-1} DNase I (Sigma-Aldrich) and then purified following the "RNA Cleanup" protocol of the "RNeasy Midi" kit from QIAGEN. In addition, the samples were subjected to three cycles of deproteinization, which involved incubation with acid phenol for 5 min at 70°C with shaking and subsequent centrifugation (15,700 × g, 21°C, 5 min). The collected supernatants were treated with one volume of phenol:chloroform-isoamyl alcohol (50:48:2) and centrifuged at 15,700 × g, 21°C, 10 min). The RNA present in the aqueous phase was precipitated by the addition of 1/10 volume of 3 M sodium acetate (pH 7.0) and three volumes of absolute ethanol, followed by storage at −20°C for 12 h. The precipitated RNA was recovered by centrifugation (11,269 × g, −10°C, 45 min), washed with 1 mL of 75% ethanol and finally dissolved in 200 μL of diethylpyrocarbonate-treated water.

Detection of the Genes

Oligonucleotides, PCR, RT-PCR, and Sequencing

The oligonucleotides used are summarized in **Table 2**. The nucleotide sequences of 14 bacterial genes encoding dextransucrase enzymes, obtained from GenBank of the National Centre for Biotechnology (NCBI, USA), were analyzed with the BLAST program in order to design the primers dsrF and dsrR for further amplification of a conserved dextransucrase coding region, located at the catalytic site of these enzymes. Genomic DNA and plasmidic DNA of *Lb. sakei* MN1 were used as templates in a reaction with primers dsrF and dsrR with Phusion Hot Start High Fidelity Polymerase (HSHFP, ThermoFisher Scientific) following the instructions of the enzyme supplier.

For RT-PCR, total RNA preparations were used. For the synthesis of the cDNAs, 400 ng of RNA, 0.75 μM of each of the oligonucleotides complementary to different regions of pMN1 and 500 μM of each of the 4 dNTPs (dATP, dGTP, dCTP, and dTTP) were used. The mixtures were incubated for 5 min at 65°C, transferred to 4°C and to each were added 1X cDNA synthesis buffer (Invitrogen), 5 mM DTT (Invitrogen), 2 U μL^{-1} of RNaseOUT (Invitrogen), and 0.75 U μL^{-1} of avian reverse transcriptase from ThermoScript (Invitrogen). The reaction was allowed to proceed for 1 h at 50°C and samples were next incubated 5 min at 85°C to stop the reaction. To remove RNA residues, 0.2 μg μL^{-1} of RNase A (Sigma-Aldrich) was added and the mixture was incubated for 20 min at 37°C. Finally, the samples were dialyzed for 30 min against 10 mM Tris pH 8.0 using membranes of the Millipore V series (Merck). PCRs were carried out according to the instructions of the Thermo Fisher Scientific Phusion DNA polymerase using the primers described in **Table 2**.

DNA sequencing was performed by the dideoxynucleotide method at Secugen (Madrid, Spain). For the determination of the nucleotide sequence of the pMN1 plasmid, the walking strategy was followed, after detection of the *dsrLS* gene by PCR with the dsrF and dsrR primers and further sequencing of the amplicon. To determine the sequence of pMN1 by the dideoxynucleotide method, it was used as substrate the product of a polymerization reaction catalyzed by the bacteriophage Φ29 DNA polymerase with plasmidic DNA of *Lb. sakei* MN1 and hexamers containing random sequences. The DNA sequence of plasmid pMN1 was deposited in GenBank (accession No MF590088).

Southern Blot Hybridization

Genomic and plasmidic DNA preparations were subjected to electrophoresis in a 0.7% agarose gel with 40 mM Tris, 20 mM acetate, and 1 mM EDTA buffer at constant amperage of 40 mA for ~4 h, and DNA molecules were revealed by staining with ethidium bromide at 0.5 μg mL^{-1}. Then, DNA was transferred to a 0.45 μm nylon membrane (Biodyne A, Pall Corporation) and hybridized with the probe. The temperature of hybridization was 45°C. Probe labeling and detection procedures were performed with the NEBlot® Phototope® Kit (New England BioLabs) and the Phototope®-Stars Detection Kit (New England Biolabs). Substrate for probe labeling was the 695-bp amplicon synthesized with primers dsrF and dsrR (**Table 2**) and *Lb. sakei* MN1 genomic DNA following the indications of the kit supplier.

Determination of pMN1 Plasmid Copy Number

Preparation of Template DNA for Real Time-qPCR

Exponential cultures (A_{600} = 1.0) of *Lb. sakei* MN1 grown in either MRSG or MRSS at 30°C were generated by inoculation of the media (dilution 1/1,000) with a bacterial stock culture previously grown in MRSG. These two cultures (designated generation 0) were always maintained in the exponential phase and sub-cultured in the corresponding medium by dilutions 1/1,000 for 60 more generations. Then genomic DNA was isolated from ~0.5 × 10^9 bacteria of the 0 and 60 generations cultures by using the Wizard® Genomic DNA Purification Kit (Promega). At the cell lysis step, 30 mg mL^{-1} of lysozyme and 30 U of mutanolysin were added. Concentration of the genomic DNA was determined with a Qubit fluorometer by using the Qubit HS dsDNA Assay Kit (Molecular Probes). The DNA extracts were digested with EcoRI, a restriction enzyme that linearizes the pMN1 plasmid leaving intact the *repA* and *pcrA* amplicons. This method was developed to obtain accurate qPCR-based plasmid copy numbers (Providenti et al., 2006).

Real Time-qPCR Analysis

Two primer sets were designed, based on the pMN1 sequence (this work) and on the *Lb. sakei* 23K (NC_007576.1) *pcrA* sequence, specific for either the pMN1 replication protein coding gene (*repA*) or the PcrA helicase chromosomal reference gene (*pcrA*). The criteria used during primer design was that the primers had a predicted Tm of ~59°C and that they generated amplicons ~140-bp long.

The qPCR were conducted in a total volume of 20 μL using an iQ5 real-time detection system (BIO-RAD) and the IQ™ SYBR® Green Supermix (Bio-Rad Laboratories), following the manufacturer's recommendations. Decimally diluted EcoRI-digested total DNA preparations (15, 1.5, 0.15, and 0.015 ng per reaction) were analyzed using 0.5 μM (final concentration) of the specific forward and reverse primers. To prepare the reactions and minimize pipetting errors 2 μl of template DNA were added to individual qPCRs. Thermal cycling conditions were as follows:

TABLE 2 | Description of primers used in this work.

Target	Primer[a]	Sequence (5′-3′)	Product (bp)
pcrA	pcrF	ACAAACATGGCGCATCAACG	146
	pcrR	GCGAAGGTGCTCAAGATGTTT	
repA	repF	GGCAAGCCGGTTATTGGTTAC	140
	repR	TTTTCCTGCTCTGTTAATTCACCAT	
dsrLS	dsrF	GATGATGGTCAATATATGGCAA	695
	dsrR	CTTGAACGATATTGTGGTGCCAA	
Region A	P1F	GAAGATCTTCTTTTAGACCCCTCTTGAGGCT	343
	P1R	GCTCTAGAGCAGTATCATCACCTTTATCGCGC	
Region B	P2F	GGAAGATCTTCCTCAGCAACAACGGTTAGCCT	366
	P2R	GCTCTAGAGCCGCCAGTGATCATATAACCGA	
Region C	P3F	GGAAGATCTTCCAAATTAACCAGAGACCGC	376
	P3R	GCTCTAGAGCTGGCTGGCTGGTAACTAGCA	
RT-PCR Amplicon 1	rt1F	AGCTGGGTTCGATATGCTTTA	1,150
	rt1R	CCCACCCCTCGCTCTTTA	
RT-PCR Amplicon 2	P2F	GGAAGATCTTCCTCAGCAACAACGGTTAGCCT	1,952
	rt2R	CGGTTGGCAAAGACGTTTTG	
RT-PCR Amplicon 3	P3F	GGAAGATCTTCCAAATTAACCAGAGACCGC	1,033
	rt2R	CGGTTGGCAAAGACGTTTTG	
RT-PCR Amplicon 4	rt2F	ACGGCTGCGATCACTACTG	496
	rt2R	CGGTTGGCAAAGACGTTTTG	
RT-PCR Amplicon 5	rt3F	AGCTTACGCTGCTACCAAGGC	862
	rt3R	ATGGCTGGAGTAAAATGGATCAGCT	

[a]F, Forward; R, reverse.

initial denaturation at 95°C for 5 min, followed by 40 cycles of 95°C for 10 s (denaturation), 59°C for 30 s (primer annealing), and 72°C for 20 s (elongation). A melting curve analysis of the PCR products, with a temperature gradient of 0.1°C/s from 59 to 95°C, was performed to confirm the purity and specificity of the PCR products. Two independent qPCR trials were conducted for each template source. In each trial, triplicate samples of the four different amounts of template were analyzed.

Relative copy number of pMN1 was calculated using equation (1):

$$PCN = \left(1 + E_{pcrA}\right)^{Ct_{pcrA}} \Big/ \left(1 + E_{repA}\right)^{Ct_{repA}}, \quad (1)$$

where E_{pcrA} and E_{repA} are, respectively, the PCR amplification efficiencies of the chromosomal and plasmidic amplicons, and Ct_{pcrA} and Ct_{repA} are the mean threshold cycle values obtained for the corresponding amplicons. A PCN value was calculated for each of the four template concentrations analyzed, and the average and standard deviation of the four values was estimated.

E values of target (E_{repA}) and reference (E_{pcrA}) sequences were empirically calculated for each qPCR trial. For that purpose, mean Ct values were plotted against the logarithm of the amount of total DNA template in the assay (**Figure 2C**). From the slope of the curve generated by linear regression of the plotted points, the PCR amplification efficiency was determined according to the equation:

$$E = 10^{-1/slope} - 1 \quad (2)$$

Although the *E* values for both amplicons was higher than 0.9, we have chosen Equation (1) to calculate the relative plasmid copy number as it allows taking into account the slight differences between E_{target} and $E_{reference}$ that we have observed.

Construction of pRCR-pMN1 Derivative Plasmids to Detect Promoter Regions Driving Transcription of *dsrLS*

The pRCR vector containing the *mrfp* gene, which encodes the fluorescent mCherry protein, was used to detect and evaluate the performance of pMN1 promoter(s) of *dsrLS*. For the construction of pRCR13, pRCR14, and pRCR15 (pRCR derivatives, Figure S1), amplicons of 343-, 366-, and 376-bp were synthesized by PCR with HSHFP following the instructions of the DNA polymerase supplier. The substrate for the reaction was pMN1 present in a plasmidic DNA preparation of *Lb. sakei* MN1 and the following oligonucleotide pairs, P1F and P1R for pRCR13; P2F and P2R for pRCR14; P3F and P3R for pRCR15 (**Table 2**). The plasmid vector pRCR, obtained from *E. coli* DH5α[pRCR], and the amplicons generated by PCR were subjected to digestion with BglII and XbaI (New England Biolabs) and ligated into the pRCR vector with the T4 DNA ligase (New England Biolabs) to obtain the recombinant plasmids. The ligations mixtures were used to transform *L. lactis* MG1363 by electroporation (25 μF, 2.5 kV and 200 Ω in 0.2 cm cuvettes), as previously described (Dornan and Collins, 1987), and transformants were selected in M17G agar plates supplemented with Cm at 5 μg mL^{-1}. The three new plasmid constructs were confirmed by automated sequencing. DNA preparations of pRCR13, pRCR14, and pRCR15 obtained from *L. lactis* MG1363 were then used to transform *Lb. sakei* MN1 by electroporation (25 μF, 1.8 kV and 600 Ω in 0.2 cm cuvettes) as previously described (Berthier et al., 1996) and transformants were selected in MRSG-agar plates supplemented with Cm at 5 μg mL^{-1}.

Detection of mCherry Fluorescence in LAB Carrying pRCR12, pRCR13, pRCR14, or pRCR15

To detect the expression of the mCherry fluorescent protein, *L. lactis* strains containing pRCR12, pRCR13, pRCR14, or pRCR15 were grown in 10 mL of M17G or M17GS till the initial stationary phase (A_{660} = 2.5–2.6) or to late stationary phase (A_{660} = 3.0–3.2). *Lb. sakei* strains containing the same plasmid constructs were grown in 10 mL of MRSG or MRSS medium to middle exponential phase (A_{600} = 2) or to late exponential phase (A_{600} = 5 or A_{600} = 10 for cultures grown in MRSG or in MRSS, respectively). All cultures were centrifuged (16,000 × g, 15 min, 4°C), resuspended in PBS buffer (pH 7.4), sedimented by centrifugation as above and finally resuspended in 500 μL of PBS buffer (pH 7.4). Suspensions (200 μL of each) were used to measure the fluorescence levels of the mCherry protein in a Varioskan Flash (Thermo Fisher Scientific) equipment, using 587 and 610 nm wavelengths for excitation and emission detection, respectively. In addition, appropriate dilutions were prepared to estimate culture biomass by measuring the absorbance at A_{600} or A_{480} for *L. lactis* or *Lb. sakei*, respectively. Three independent trials were performed and the same fresh suspensions (8 μL of each), without fixing, were used for phase contrast and fluorescent microscopy analysis as previously described (Nácher-Vázquez et al., 2017). A Leica AF6000LX-DMI6000B model microscope (Leica Microsystems, Mannheim,Germany) was used. Illumination was provided with a 100× objective. For detection of mCherry BP 620/60 excitation and BP 700/75 emission filters were used. Image analysis was performed using LAS AF CoreSoftware (Leica Microsystems).

Bioinformatic Analysis of DNA Sequences and Modeling of Dsr

The DNA sequence of plasmid pMN1 was analyzed with the programs included in the DNASTAR Lasergene 12 (DNAstar Inc.). Homologies of pMN1 DNA sequences and of its inferred translated products with the NCBI data bases of the National Center for Biotechnology Information (NCBI) were analyzed with the Basic Local Alignment Search Tool (BLAST) (https://blast.ncbi.nlm.nih.gov/Blast.cgi). Multiple sequence alignment of genes and proteins were performed with the Megaling (DNASTAR laser gene 12) and Clustalx 2.1 (http://www.ebi.ac.uk/Tools/msa/clustalw2/) programs.

The primers for the qPCR experiments were designed with Primer3 v0.4.0 (Koressaar and Remm, 2007; Untergasser et al., 2012).

The modeling of the DsrLS was generated based on its homology with the glucansucrase of *Lactobacillus reuteri* 180, using the 3D-structure of the GTF180-ΔN glucansucrase and the I-TASSER program (Roy et al., 2010). The superposition of the DsrLS model and the 3D-structure of GTF180-ΔN was performed with the CE program at the http://source.rcsb.org.

RESULTS

Detection and Genomic Localization of *dsrLS*

Previous characterization of the EPS produced by *Lb. sakei* MN1 revealed that it is a dextran (Nácher-Vázquez et al., 2015) and indicated that this bacterium produces a Dsr (named DsrLS) responsible for the polymer synthesis. Thus, based on the known sequences of *dsr* genes from other bacteria, primers dsrF and dsrR were designed to amplify by PCR a DNA fragment located at the coding sequence of the dextransucrase catalytic domain. The expected amplicon (695-bp) was obtained using either genomic or plasmidic DNA preparations from *Lb. sakei* MN1 as template (data not shown). The determination of the nucleotide sequence of this amplicon (GenBank accession No KJ161305) and its BLAST analysis against the nucleotide (nr/nt) data base of NCBI revealed only high homologies (100, 99, and 71% identity, with 40 gaps) with genes encoding dextransucrases of *Lactobacillus curvatus* TMW1624 (GenBank accession No HE972512), *Lb. sakei* Kg15 (GenBank accession No AY697434) and *Weisella confusa* strain Cab3 (GenBank accession No KP729387.1), respectively. The overall results supported that the *dsrLS* gene has been detected, and, generation of the expected amplicon using the plasmidic DNA preparation strongly suggested that it was plasmid-borne. In addition, they revealed that the dsrF and dsrR primers pair designed by us are useful for the detection of *dsr* genes in a similar way to those degenerate oligonucleotides previously used for the detection of genes encoding glucosyltransferases of *Lactobacilli* (Kralj et al., 2003).

Analysis of a plasmidic DNA preparation from *Lb. sakei* MN1 in agarose gel revealed the presence of two groups of two bands each (**Figure 1A**). In addition, it was detected that exposure of the DNA preparations to repeating freezing (at −80°C) and thawing (at 4°C) cycles resulted in alteration of the intensity of the bands: an increase in the two upper bands accompanied by a decrease in the two faster migrating ones (results not shown). The sizes of the bands were inferred from their migration using a calibration curve (**Figure 1B**) generated with the plasmids of the *E. coli*

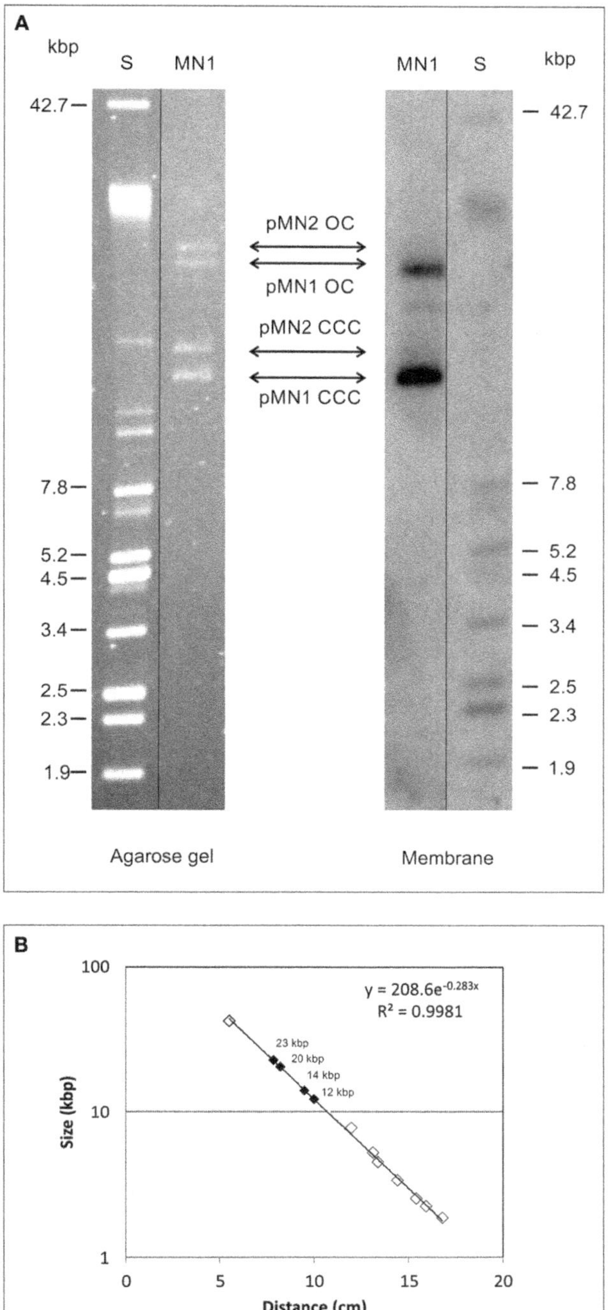

FIGURE 1 | Detection of *dsrLS* gene by Southern blot hybridization. Plasmids preparations of *Lb. sakei* MN1 (MN1) and of *E. coli* V517 (S) were analyzed in 0.7% agarose gel (**A** left) transferred to a membrane and hybridized for detection of *dsrLS* (**A** right). In **(B)**, the calibration curve for plasmid size determination is depicted.

V517 strain (**Figure 1A**, lane S), and the overall results indicated that bands with higher mobility could be the covalently closed circles of two plasmids of ~12 and 14 kbp named respectively pMN1 and pMN2, whereas bands with less mobility could be the open circle forms of those plasmids. To disclose the location of *dsrLS*, the 695-bp amplicon was used as a probe for Southern blot hybridization and the blot revealed the presence of two hybridization bands corresponding to the two proposed pMN1 forms (**Figure 1A**).

Characterization of Plasmid pMN1

Following DNA sequencing, the size of pMN1 plasmid was estimated as 11,126 kbp. A genetic map of pMN1 is depicted in **Figure 2A**. Blast alignment of its DNA sequence with those deposited in Genbank lead to confirm the existence of the 5,304-bp *dsrLS* gene. Moreover, in pMN1, a 1,674-bp replicon homologous to the replicons of the pUCL287 plasmid family was identified that replicates bidirectionally by theta mechanism (Benachour et al., 1995). This region contains the origin of replication and the *repA* and *repB* genes. In addition, in pMN1, seven other open reading frames were identified and designated ORFs 1-7, which could encode hypothetical proteins. Blast analysis of inferred amino acid sequence of the ORFs with those deposited in the Non-redundant protein sequences database reveled that *orf1*, *orf3*, *orf4*, and *orf5* could encode respectively a truncated type I restriction endonuclease subunit R, a transcriptional regulator belonging to the XRE family, a RelE type toxin addiction module and a site-specific integrase.

Furthermore, the copy number of the plasmid was investigated by real time-qPCR by analysis of the *repA* gene vs. that of the chromosomal housekeeping and monocopy *pcrA* gene. *Lb. sakei* MN1 cultures were maintained in exponential growth phase in either MRSG or MRSS and DNA preparations of cultures grown to exponential phase once (from the glycerol stock) or for 60 successive generations (by six subsequent 1/1,000 dilutions) were analyzed. The results are summarized in **Figure 2B**, and revealed that pMN1 is a low-copy-number plasmid (~6.5 ± 1.5 copies per genome equivalent), which maintains its copy number over at least 60 generations, and that its copy number is not significantly affected by conditions required for dextran synthesis.

Gene Expression of *dsrLS*

We have previously detected that *Lb. sakei* MN1 is unable to synthesize the dextran in the absence of sucrose (Nácher-Vázquez et al., 2017). This could be due not only to the lack of the substrate for the polymer synthesis, but also to the fact that *dsrLS* gene expression requires an induction mediated by the disaccharide as it has been detected in other LAB (Neubauer et al., 2003). In addition, inspection of the pMN1 DNA sequence indicated that transcription of *dsrLS* could be driven from more than one promoter located upstream of this gene. Thus, total RNA preparations were obtained from *Lb. sakei* MN1 cultures grown in medium containing either glucose or sucrose and five RT-PCR reactions were performed to generate the amplicons showed in **Figure 3A**. These amplicons contain regions of more than one gene and their corresponding intergenic regions (amplicons 1, 2, 3, and 5) or only a region of *drLS* (amplicon 4). The results revealed that four reactions generated the expected sizes 1, 2, 3, and 4 amplicons, which included regions located upstream of, or within, the *dsrLS* gene (**Figure 3B**). In addition, the

A

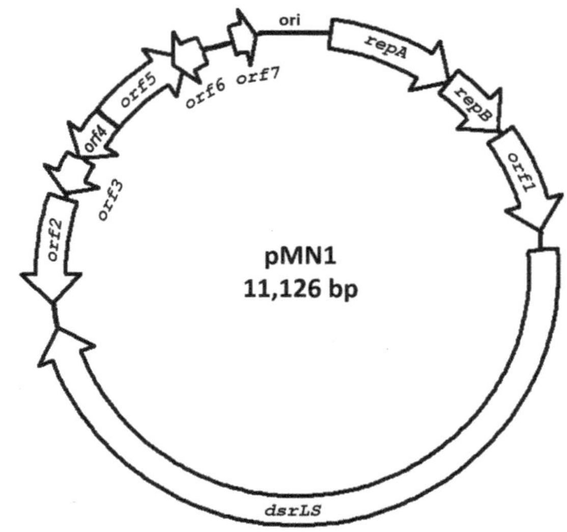

B

Nº of cell generations	pMN1 plasmid copy number	
	Glucose	Sucrose
0	6.14 ± 0.45	5.36 ± 0.23
60	8.60 ± 0.55	5.90 ± 0.41

C

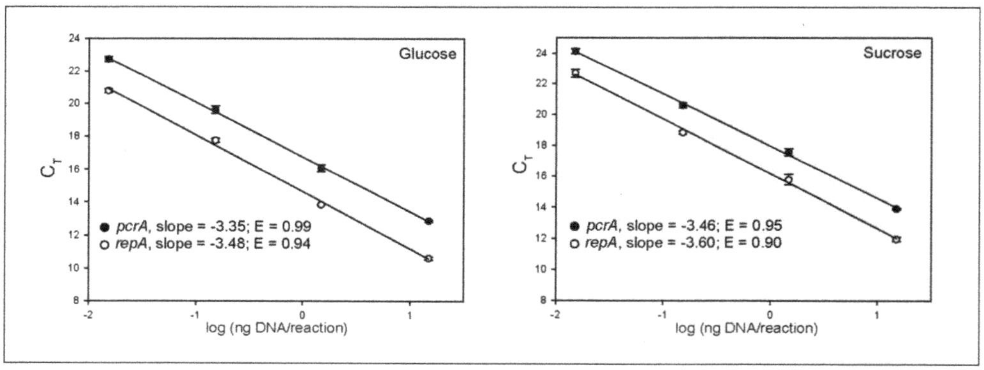

FIGURE 2 | Genetical map of plasmid pMN1 **(A)** and determination of its copy number **(B,C)**. **(A)** The genes and *orf* of pMN1 are depicted. **(B)** Results of the pMN1 copy-number determination are summarized in this table. **(C)** For each gene, determined mean Ct values were plotted against the logarithm of the four amounts of template DNA used in the qPCR assays. PCR amplification efficiencies (E values) were calculated from the slope of the curves generated by linear regression through the experimental points.

quantification of the intensity of the amplicons (results not shown) indicated that the mRNA levels were very similar in cultures grown in CDMG or CDMS media. The fifth reaction did not reveal the amplicon 5 (**Figure 3B**), which carries the 3′-end of *dsrLS* and downstream regions (**Figure 3A**). This latter result was not due to the primers used, since the expected amplicon was obtained using the plasmidic DNA preparation as substrate (**Figure 3B**). In addition, the negative result was expected due to the convergent polarity of *dsrLS* and *orf2* and to the existence of a putative bidirectional transcriptional terminator located between the 3′-ends of the two genes, which predicted the lack of their co-transcription.

Identification of Promoter Regions in pMN1

As expected, the RT-PCR analysis revealed that *repA* and *repB* were co-transcribed, but in addition showed the existence of

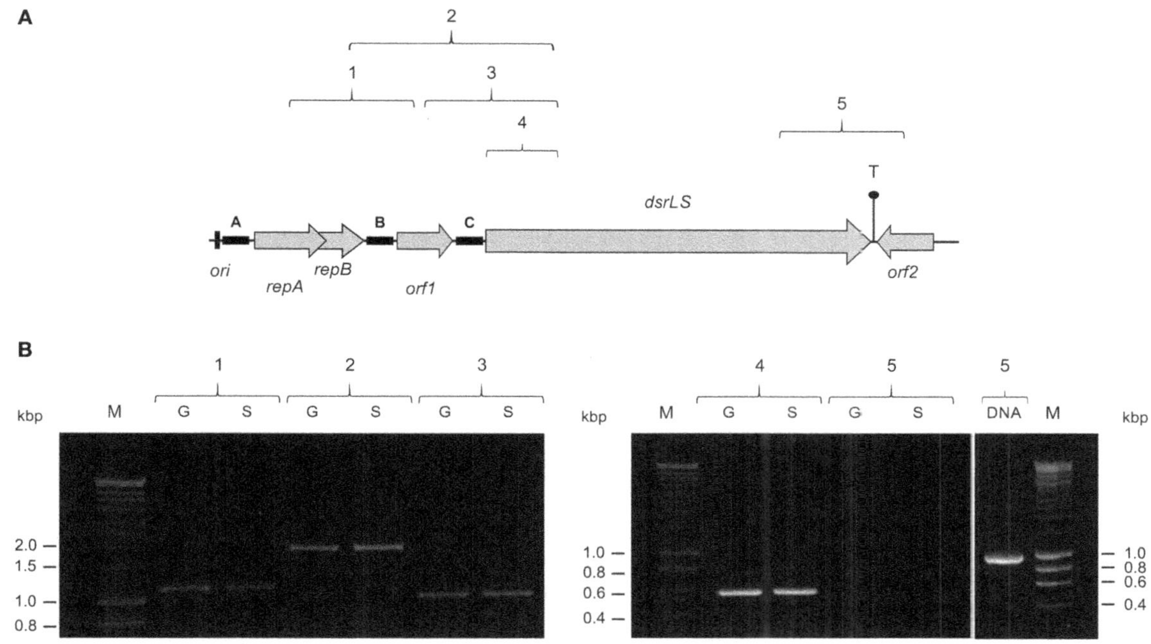

FIGURE 3 | Detection of *dsrLS* transcription by RT-PCR. **(A)** The genes and *orfs* of pMN1 are depicted. The locations of the five amplicons (1, 2, 3, 4, and 5) in pMN1 are shown. **(B)** Analysis in 0.7% agarose gel of the RT-PCR reactions performed using total RNA obtained from cultures of *Lb. sakei* MN1 grown in presence of glucose (G) or sucrose (S) and PCR reaction performed with plasmidic DNA to generate amplicon 5 (DNA). M, DNA standard marker (SmartLadder, Eurogentec).

transcripts including *dsrLS* and upstream genes. Therefore, detection of DNA regions involved in *dsrLS* expression was approached. We have previously developed the pRCR promoter probe plasmid (Figure S1) based on the pSH71 replicon, which replicates *via* a rolling circle mechanism and carries a mCherry-coding gene (*mrfp*) optimized for expression of the fluorescent protein in LAB. Moreover, we had shown functionality of this replicon concomitant with successful expression of this *mrfp* in *L. lactis* (Mohedano et al., 2015) and *Lb. sakei* MN1 (Nácher-Vázquez et al., 2017). Thus, to detect promoter regions that could drive transcription of *dsrLS* in *L. lactis* and *Lb. sakei*, A, B, and C DNA fragments, carrying intergenic regions located upstream of the *repA*, *orf1*, and *dsrLS* genes, respectively (**Figure 3**), were cloned upstream of the *mrfp* gene into the pRCR plasmid, generating the recombinant plasmids pRCR13, pRCR14, and pRCR15. Thus, these plasmids carry putative transcriptional fusions to the mCherry-coding gene (Figure S1).

The clonings were performed in *L. lactis* MG1363 and then the plasmids were transferred to *Lb. sakei* MN1. Afterwards, expression of the mCherry in the two hosts was monitored by fluorescent spectroscopy and microscopy. *L. lactis* cannot grow in media containing sucrose as the only carbon source. Therefore, expression of mCherry was monitored during growth in M17G and M17GS. The results obtained during the early and late stationary phases are shown in **Table 3** and Figure S2. As expected, MG1363 did not show fluorescence. Moreover, fluorescence was observed only in cultures of MG1363[pRCR13] and MG1363[pRCR15] and the promoter regions present in

TABLE 3 | Fluorescent detection of promoter regions in *L. lactis* by translational fusions to the *mrfp* gene.

Growth phase	[a]Initial stationary phase		[a]Late stationary phase	
Bacteria \ Growth medium	M17G	M17GS	M17G	M17GS
L. lactis MG1363	0.0 ± 0.0	0.0 ± 0.0	0.2 ± 0.1	0.10 ± 0.0
L. lactis MG1363 [pRCR13]	0.7 ± 0.5	0.8 ± 0.1	16.6 ± 0.8	12.5 ± 1.2
L. lactis MG1363 [pRCR14]	0.2 ± 0.1	0.1 ± 0.0	0.2 ± 0.0	0.1 ± 0.0
L. lactis MG1363 [pRCR15]	0.3 ± 0.2	0.9 ± 0.1	10.2 ± 0.8	10.4 ± 1.6

[a]*The specific fluorescence is depicted and it was calculated as the ratio of the detected fluorescence (10×) and the bacterial biomass estimated from the A_{600} of the culture.*

them were designated P1 and P2, respectively. Nevertheless, the levels of fluorescence in these strains were low and only significantly detected at late stationary phase. This pattern of expression had been previously observed when the *mrfp* gene was expressed in *L. lactis* MG1363 under control of a lactococcal promoter (Garcia-Cayuela et al., 2012), and it could indicate that the mCherry protein requires a long period of maturation prior to emit fluorescence in this host.

In the case of *Lb. sakei*, strains were grown in MRSG and MRSS and the presence of pRCR13 and pRCR15, and not of pCRC14, conferred fluorescence to *Lb. sakei* MN1, in both exponential and stationary phases (**Table 4** and **Figure 4**). The levels of specific florescence indicated that P1 is stronger than P2

TABLE 4 | Fluorescent detection of promoter regions in *Lb. sakei* by expression of translational fusions to the *mrfp* gene.

Growth phase	[a]Middle exponential phase		[a]Late stationary phase	
Bacteria \ Growth medium	Glucose	Sucrose	Glucose	Sucrose
Lb. sakei MN1	1.3 ± 0.6	0.9 ± 0.1	0.5 ± 0.1	0.4 ± 0.5
Lb. sakei MN1[pRCR12]	355 ± 30	351 ± 18	227 ± 14	139 ± 5.5
Lb. sakei MN1[pRCR13]	142 ± 14	76 ± 7.5	149 ± 11	68 ± 2.2
Lb. sakei MN1[pRCR14]	4.6 ± 0.4	2.0 ± 0.2	5.5 ± 0.3	1.6 ± 0.1
Lb. sakei MN1[pRCR15]	73 ± 2.6	28 ± 2.0	40 ± 0.9	8.7 ± 0.4

[a]*The specific fluorescence is depicted and it was calculated as the ratio of the detected fluorescence (10×) and the bacterial biomass estimated from the* A_{600} *of the culture.*

and that they are weaker than the pneumococcal Px promoter, which drives transcription of *mrfp* in *Lb. sakei* MN1[pRCR12] (**Table 4**) (Nácher-Vázquez et al., 2017).

In addition, at the exponential phase, the specific fluorescence in MN1[pRCR13] and MN1[pRCR15] strains was two-fold-higher in MRSG that in MRSS. These results confirmed that the sucrose present in the medium is not an inducer of the *dsrLS* gene.

DISCUSSION

The enzymes responsible for the HoPS synthesis are glycosyl hydrolases, extracellular polymerases that utilize the energy of the glucosidic bond of sucrose to link molecules of glucose. If they synthesize α-D-glucans are called glucansucrases and according to the CAZy classification (http://www.cazy.org), are members of the GH70 family. Among them, dextransucrases synthesize dextran and here we have characterized the *Lb. sakei* MN1 *dsrLS* gene of 5,304-bp, which encodes the DsrLS composed of 1,767 amino acids (aa). Analysis of these sequences with the BLAST program vs. those deposited in the NCBI databases revealed homologies with other bacterial genes and with their gene products. The highest homology was detected with the 5,094-bp *gtf1624* gene from *Lb. curvatus* TMW1624 and its product, the dextransucrase GTF1624 of 1,697 aa (Rühmkorf et al., 2013). In addition, the *dsrLS* gene and DsrLS also exhibited high homology with the *gtfkg15* gene of *Lb. sakei* Kg15 of 4,788-bp and its product GTFKg15 of 1,595 aa (Kralj et al., 2004a). An alignment of the three proteins is presented in Figure S3.

It has been demonstrated that the EPS synthesized by GTF1624 (Rühmkorf et al., 2013) and GTFKg15 (Kralj et al., 2004a) are α-(1-6)-glucans with a low percentage of substitutions at positions *O*-3, like the EPS synthesized by *Lb. sakei* MN1 (Nácher-Vázquez et al., 2015). This fact strongly supports that indeed DsrLS is the enzyme responsible for the synthesis of the MN1 dextran.

In all glucansucrases, including dextransucrases, exist: (i) a N-terminal, variable region, (ii) the catalytic domain, and (iii) a C-terminal, so-called "glucan binding" domain (van Hijum et al., 2006). All these regions were identified in DsrLS and include the following aa: (i) 49–390, (ii) 391–1,154, and (iii) 1,155–1,767. The glucansucrases are extracellular enzymes, and at the N-terminus of DsrLS it was found a sequence characteristic of the leader peptides of Gram-positive bacteria (1–48 aa), also present in GTF1624 and GTFKg15. In addition, the difference in the number of aa of the three dextransucrases is due in one hand to the absence in GTF1624 of 70 aa present in the C-terminal region of DsrLS (residues 1,541–1,610). On the other hand, DsrLS has 172 aa more than its homolog of *Lb. sakei* Kg15, being 145 of them located at the C-terminal region (1,485–1,629) and 27 aa at its N-terminal variable region (residues 64–89 and 813). Consequently, the greatest divergence of DsrLS with both GTF1624 and GTFKg15 is located at its C-terminal domain.

Currently, the 3D-structure of any entire glucansucrase has not been determined, but partial structures of: (i) the DSR-E-ΔN of *Lc. mesenteroides* NRRLB-1299 (PDB 3TTQ), (ii) the GTF-S1 of *Streptococcus mutants* (PDB 3AIE), (iii) the GTFA-ΔN of *Lb. reuteri* 121 (PDB 4AMC), and (iv) the GTF180-ΔN of *Lb. reuteri* 180 (PDB 3KLK, 4AYG, and 3HZ3) have been solved by X-ray diffraction analysis of crystals. From the crystal structure of these proteins, it has been stablished that there are five structural domains designated A, B, C, IV, and V (Leemhuis et al., 2013). The A, B, and C domains have been named following the nomenclature of the structurally homologous domains of the GH13 family of α-amilases. The domains IV and V have not homologs in GH13 and for this reason have been named with a different nomenclature (Vujicic-Zagar et al., 2010). These domains are not adjacent in the primary structure of the proteins, and are located with a "U" distribution and a pattern V, IV, B, A, C, A, B, IV, and V.

The aa sequence of DsrLS (residues 396–1,395) has an identity of 51% with that of the GTF180-ΔN of *Lb. reuteri* 180 (PDB 3HZ3) (Figure S4). Thus, it was possible to develop a model of the 3D-structure of DsrLS lacking the N- and the C-terminal regions (**Figure 5**), which predicts that the *Lb. sakei* enzyme has the same domains as its homonymous of *Lb. reuteri* (**Figure 5B**). The A domain is a barrel $(\beta/\alpha)_8$ and contains the catalytic site of the enzyme, including an amino acid triad composed of two aspartate and one glutamate residues, which are involved in the formation of a covalent glucosyl-enzyme intermediate, the key step in the transfer of D-glucosyl units. From this intermediate, the glucosyl unit is transferred to the acceptor (the growing dextran molecule) by a processive catalytic mechanism. Thus, superposition of the 3D-model of DsrLS on the co-crystal of GTF180-ΔN and sucrose indicates that D678, D789, and E716 constitute the catalytic triad of the *Lb. sakei* MN1 dextransucrase (**Figure 6**).

It has been shown that calcium is essential for GTFA-ΔN (Kralj et al., 2004b) and GTF180-ΔN (Vujicic-Zagar et al., 2010) activities. The B domain located adjacent to the A domain seems to be essential for glucansucrases activity because (i) calcium binding site includes aa from A and B domains, and (ii) some elements of the B domain contribute to stabilize the conformation of the A domain. Concerning to the C domain, although is conserved in all glucansucrases, its function is still unknown. The IV domain connects the B and V domains and seems to act as a hinge to bring the V domain close to the catalytic domain (Ito et al., 2011). The V domain is constituted

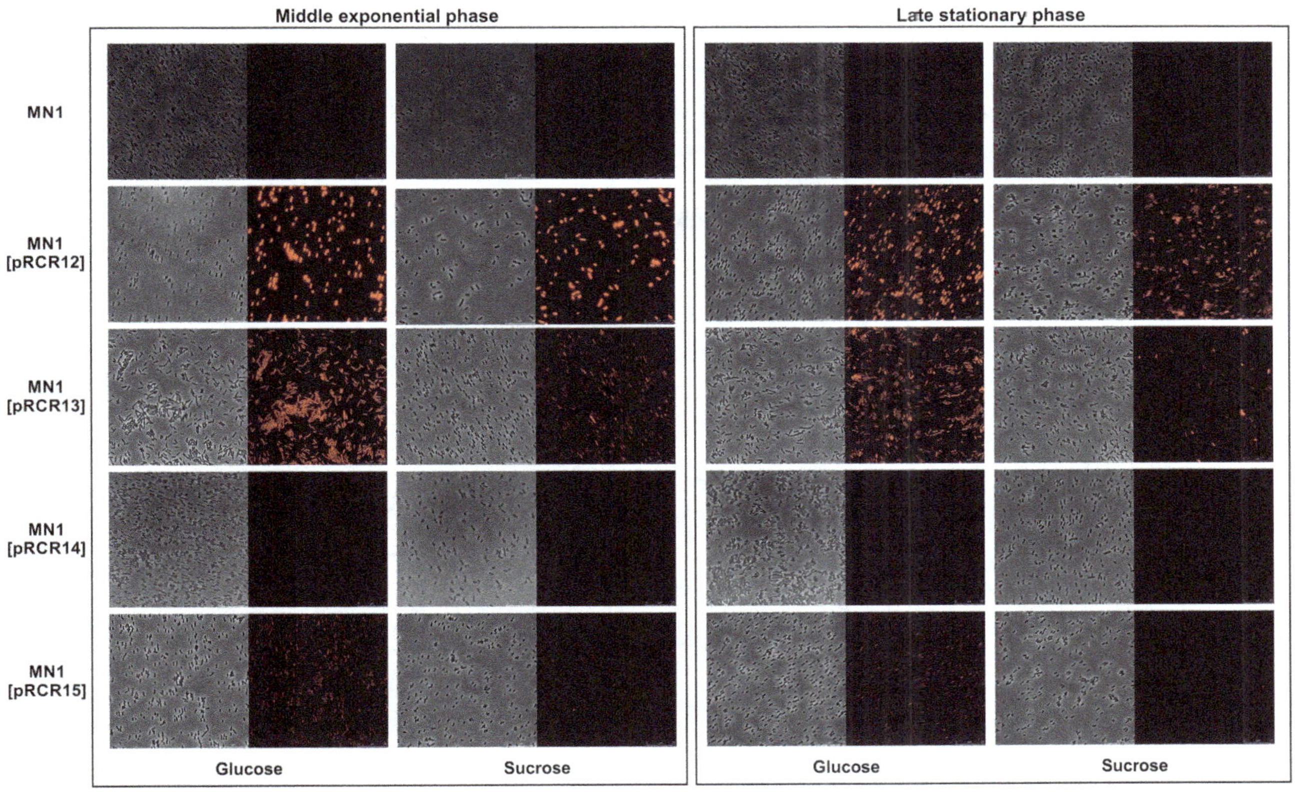

FIGURE 4 | Detection of fluorescence in *Lb. sakei* strains. Cultures of the indicated strains in MRSG (glucose) or MRSS (sucrose) were analyzed at middle exponential and late stationary phases by phase contrast **(Left)** or fluorescence **(Right)** microscopy.

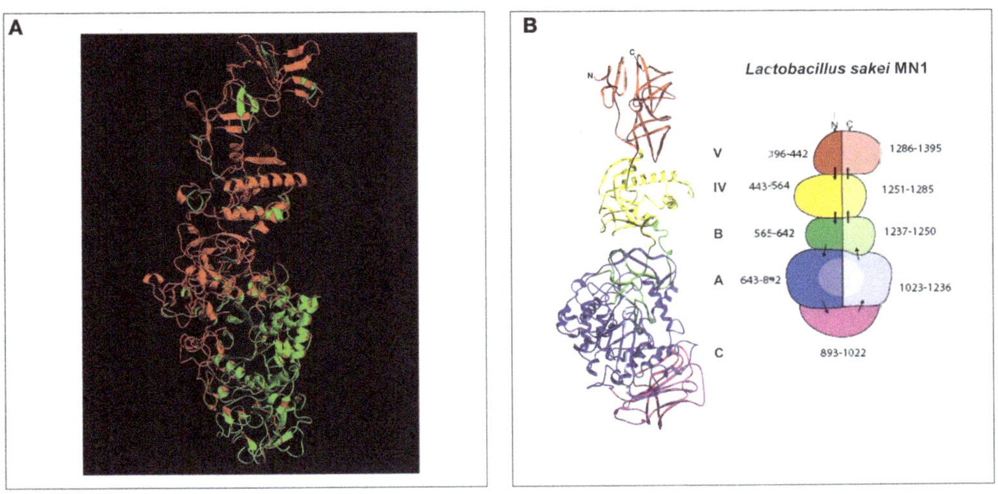

FIGURE 5 | 3D-model of the structure of DsrLS. **(A)** Superposition of the structural model of DsrLS (in red) on the crystal structure of GTF180-ΔN (in green). **(B)** The five structural domains of DsrLS as well as the numbering of the corresponding aa are shown. The color codes are: blue for A, green for B, violet for C, yellow for IV, and red for V.

by the N- and C-terminal regions, which include a series of structural modules with two or three β2/β3 units containing ~20 aa and arranged in a regularly repeating fashion, resulting in a β-solenoid fold. In some glucansucrases these modules include YG repeats (containing a tyrosine/glycine motif) (Leemhuis et al., 2013) and DsrLS carries YG in both terminal regions.

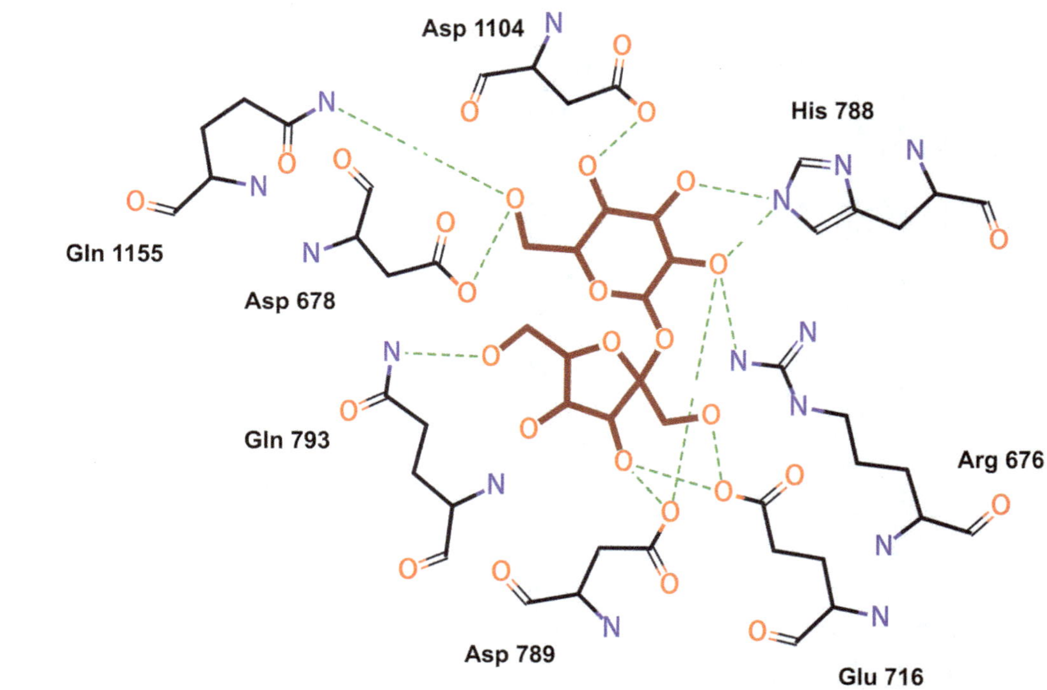

FIGURE 6 | Model of amino acids interaction with sucrose (magenta) at the catalytic center of DsrL. Model obtained by superimposing the model structure of DsrLS on the co-crystal of GTF180-ΔN and sucrose.

In *Lactobacillus,* the variable N-terminal region of glucansucrases contains 200–700 aa and mutations or deletions of this region can alter the functions of the proteins. Thus, in GTFA, deletion of this region affects the interplay between the hydrolytic and transglycosidase activities (Kralj et al., 2004b). The C-terminal domain contains ~300 aa and the function of this region is still unknown. In GTFA, its deletion diminishes affinity for sucrose (Kralj et al., 2004a). The repetitions at the C-terminal region of glucansucrases have homology with motives for binding to bacterial cell wall present in choline-binding proteins, toxins, and other bacterial surface proteins (Leemhuis et al., 2013). However, the function of the C-terminal region is currently unknown, although its implication in several functions has been proposed: (i) polymerization or glucan structure, (ii) transfer of products to the catalytic center, and (iii) anchoring of the protein to the bacterial surface.

The *Lb. sakei* DsrLS and GTFKg15 enzymes only differ in the number of aa at their C-terminal (307 and 162 residues, respectively) and N-terminal (342 and 315 residues, respectively) regions, and the *Lb. reuteri* GTF180 lacks the C-terminal region.

These differences could be related to the enzymes processivity, since the dextrans synthesized in fermentation conditions by *Lb. sakei* MN1, *Lb. sakei* Kg15 and *Lb. reuteri* 180 have molecular masses of 1.7×10^8 Da (Zarour et al., 2017), 2.7×10^7 Da, and 3.6×10^6 (Kralj et al., 2004a), respectively.

Here, we have demonstrated that the gene encoding DsrLS is carried by the 11,126 kbp pMN1 plasmid. Plasmidic localization of the genetic determinants for the production of EPS has been described previously (Wang and Lee, 1997). It has been determined, by plasmid curing, that the production of a HePS in *Lactobacillus casei* CG11 depends on the presence of a plasmid of 30 kbp (Kojic et al., 1992). Also, the *gtf* gene encoding the GTF glycosyltransferase, which synthesizes a O2-substituted (1,6)-β-D-glucan, has been identified in plasmids of *Pediococcus* and *Lactobacilli* strains (Werning et al., 2006). Concerning to dextran synthesis, it was shown that production of the polymer by two *Lactobacillus* strains isolated from meat was impaired upon curing of a 11 kbp plasmid (Ahrné et al., 1989), which could be identical or similar to pMN1. However, as far as we know, this is the first time that a plasmid carrying a gene encoding a dextransucrase has been completely sequenced.

Homology of pMN1 with other plasmids revealed that it belongs to a plasmid family whose prototype is pUCL287, which replicates *via* theta-mode. The pMN1 replicon includes two genes, *repA* and *repB,* which should encode the RepA and RepB proteins, involved in the initiation of plasmid replication and regulation of plasmid copy number. Thus, RepB could be responsible for the segregational stability of the plasmid and in fact the results obtained here indicate that low copy number pMN1 is stably inherited. In addition, the putative RelE toxin encoded by *orf4,* whose expression is probably regulated by the product of *orf3,* could be other mechanism to eliminate the bacterial population that has lost the plasmid.

We have previously shown that *Lb. sakei* MN1 utilizes very efficiently sucrose with production of dextran and without accumulation of glucose, not affecting the growth rate and resulting in a higher biomass than when the growth medium was supplemented with glucose instead of sucrose (Nácher-Vázquez et al., 2017). Thus, it is not strange that the plasmid is segregationally stable because, in addition to encode RepB, it is not a burden for the cells and rather the bacteria is beneficiated by its presence. Upstream of *repA*, several iterons were identified: (i) 4 direct repeats of 11-bp (**Figure 7A**), which according to the studies performed with pUCL287 (Benachour et al., 1997) constitute the replication origin of the plasmid and presumably the RepA binding site, and (ii) 5 direct repeats of 22-bp, which could be involved in partitioning or incompatibility processes.

The BLAST analysis of the pMN1 DNA sequences vs. those deposited in the NCBI databases revealed homologies of ORF 2, 3, 4, 5, 6, and 7 with ORF 13, 12, 11, 10, 9, and 8 of the plasmid pRV500 from *Lb. sakei* RV332 (Alpert et al., 2003), which belongs to the same plasmid family as pMN1. The major difference between pRV500 and pMN1 is that the first one carries the genes encoding a restriction-modification type I system instead of the pMN1 *dsrLS* gene and its preceding *orf1*. This last one appears to be a truncated sequence of a gene that initially encoded a specific deoxyribonuclease (R protein), belonging to a type I restriction and modification system different to that of pRV500 (**Figure 7B**).

Most lactobacilli carry more than one plasmid and both pMN1 and pRV500 have been detected in *Lb. sakei* strains isolated from meat products, although in different countries. Thus, these plasmids seem to be derived from a parental plasmid composed of a replicon and the *orfs* 2–7, which subsequently by means of transposition processes incorporated modules that allow the synthesis of the dextran or a system of restriction and modification. Possibly, the acquisition of one or other module and its fixation is due to the selective advantage that supposes for their hosts against the environmental stress or the infection by bacteriophages. Moreover, the product of the *orf5*, a site-specific integrase could be involved in the process of modules exchange.

Depending on the requirement for sucrose utilization, expression of the dextransucrases could be constitutive or inducible. To date, it has been determined that its synthesis is constitutive in *Streptococcus* (Janda and Kuramitsu, 1978; Wenham et al., 1979), while it is inducible in presence of sucrose in *Leuconostoc* (Neely and Nott, 1962; Funane et al., 1995), although the molecular mechanisms of this induction are unknown. In *Weissella*, levels of dextransucrose detected in cultures of *Weissella cibaria* and *Weissella confusa* grown in presence of different sugars indicate constitutive expression of their Dsr enzymes (Bounaix et al., 2010). Determination of levels and activity of glucansucrases in *Lactobacillus* also points to a constitutive expression. This is the case of the glucan-producing *Lb.reuteri* TMW1106 (Schwab et al., 2007), the reuteran-producing *Lb. reuteri* 121 (Kralj et al., 2004a), and the dextran-producing *Lb. reuteri* 180 and *Lactobacillus parabuchneri* 33 (Kralj et al., 2004a).

In this work we have determined by RT-PCR analysis that the expression of the *dsrLS* gene of *Lb. sakei* MN1 did not increase when sucrose is present in the culture medium. Thus, the disaccharide is not an inducing agent of the *dsrLS* expression.

The transcriptional fusions generated in this work revealed two promoter regions designated P1 and P2, which drive expression of *dsrLS*. Inspection of the pMN1 sequences cloned in pRCR13 revealed a DNA sequence TATtAT (**Figure 7A**) which only deviates a nucleotide from the canonical −10 promoter

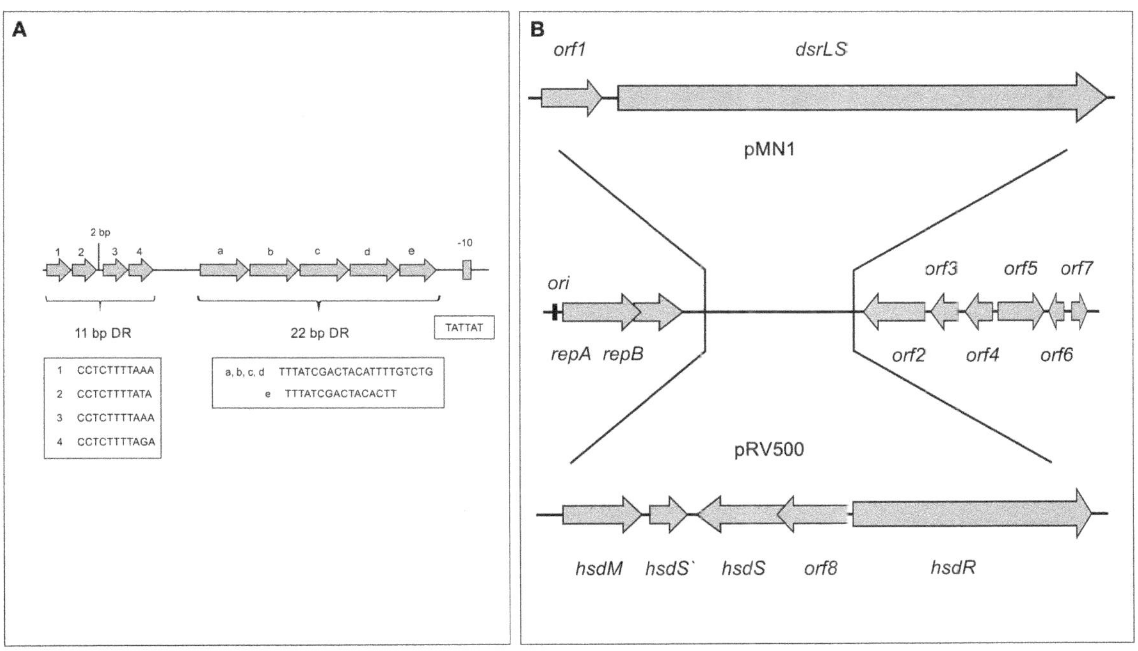

FIGURE 7 | Origin of replication of pMN1 **(A)** and model of generation of pMN1 and pRV500 from a putative parental plasmid **(B)**. In **(A)** the sequences of the iterons and of the putative promoter of the *repA-repB-orf1-dsrLS* operon are depicted. In **(B)** the genes and the *orfs* of plasmids are depicted.

sequence. This sequence is located 16 nucleotides upstream of *repA*, and could be P1. Also, within the pMN1 insert of pRCR15, a −10 extended promoter region (TGTTATtAT) with only one mismatch was observed 82 nucleotides upstream of *dsrLS*, that could correspond to P2. No −35 promoter region was detected for either P1 or P2. In addition, in late stationary phase cultures of *Lb. sakei* and not of *L. lactis* carrying pRCR15 a four-fold higher fluorescence levels was detected, when grown in medium containing only glucose (**Tables 3, 4**). Taking in consideration that *Lb. sakei* MN1 is the native background for the P2 promoter, it is feasible that a negative effector of the P2 promoter could be present or encoded by pMN1, pMN2 or the chromosome. Moreover, lower fluorescence was detected in *Lb. sakei* exponential cultures expressing mCherry from P1 or P2, when they were grown in MRSS. This fact could be due to the presence of dextran in cultures grown in MRSS, which could cause a shielding that masks fluorescence. However, analysis of individual cells by fluorescence microscopy also revealed a lower fluorescence in bacteria grown in MRSS medium (**Figure 4**), despite cells were washed repeatedly to remove the dextran prior to the analysis. In addition, difference in fluorescence levels between bacteria grown in MRSG or MRSS was not observed during exponential phase in MN1 strain when carried pRCR12 plasmid. Therefore, an alternative hypothesis is that the high synthesis of dextran (around 10 g L^{-1} in MRSS), although it does not seem to affect bacterial growth (Nácher-Vázquez et al., 2017), could finally be an energetic burden for the bacterium and the dextran itself could induce the activation of a inhibitory mechanism of its own synthesis at the transcriptional level. Another hypothesis, more feasible, is that the fructose synthesized as a consequence of the hydrolysis of sucrose, is the effector of inhibition. However, further studies will be necessary in order to elucidate the real cause of the detected effect.

An unexpected co-transcription of *dsrLS* with the *repA* and *repB* genes has been detected here. We believe that this is the first described case of co-transcription of genes involved in plasmid replication and a gene encoding a protein involved in the synthesis of EPS. However, this is not the first case in the literature of co-transcription of replication genes together with other genes, since the plasmidic *dysI* gene (encoding the immunity factor of the streptococcal bacteriocin dysgalacticin) is transcribed as part of the *copG-repB-dysI* replication-associated operon (Swe et al., 2010).

Finally, we would like to highlight the multicopy state of *dsrLS* in the pMN1 stable plasmid, its expression from two promoters not induced by sucrose, and the apparently processive DsrLS, which synthesizes a high-molecular mass dextran with antiviral and immunomodulatory activities (Nácher-Vázquez et al., 2017), as well as rheological properties (Zarour et al., 2017). Thus, all these facts support the potential of *Lb. sakei* MN1 and its dextran for multiple industrial applications including those in functional food.

AUTHOR CONTRIBUTIONS

MN-V contributed to all parts of the experimental work and wrote a draft of the manuscript. JR-M performed the plasmid characterization. MM contributed to the transcriptional gene expression analysis. GdS performed the bioinformatics analysis of pMN1 plasmid, interpreted this analysis, and revised the manuscript. RA participated in study conception and corrected the manuscript. PL participated in study conception, data interpretation, and generated the final version of the manuscript. All authors have read and approved the final manuscript.

ACKNOWLEDGMENTS

We thank Dr. Stephen W. Elson for critical reading of the manuscript. We thank Dr. Mario García Lacoba for his valuable assistance in the modeling of DsrLS.

REFERENCES

Ahrné, S., Molin, G., and StÅhl, S. (1989). Plasmids in *Lactobacillus* strains isolated from meat and meat products. *Syst. Appl. Microbiol.* 11, 320–325. doi: 10.1016/S0723-2020(89)80031-1

Alpert, C. A., Crutz-Le Coq, A. M., Malleret, C., and Zagorec, M. (2003). Characterization of a theta-type plasmid from *Lactobacillus sakei*: a potential basis for low-copy-number vectors in lactobacilli. *Appl. Environ. Microbiol.* 69, 5574–5584. doi: 10.1128/AEM.69.9.5574-5584.2003

Aman, A., Siddiqui, N. N., and Qader, S. A. U. (2012). Characterization and potential applications of high molecular weight dextran produced by *Leuconostoc mesenteroides* AA1. *Carbohydr. Polym.* 87, 910–915. doi: 10.1016/j.carbpol.2011.08.094

Amari, M., Arango, L. F., Gabriel, V., Robert, H., Morel, S., Moulis, C., et al. (2013). Characterization of a novel dextransucrase from *Weissella confusa* isolated from sourdough. *Appl. Microbiol. Biotechnol.* 97, 5413–5422. doi: 10.1007/s00253-012-4447-8

Anastasio, M., Pepe, O., Cirillo, T., Palomba, S., Blaiotta, G., and Villani, F. (2010). Selection and use of phytate-degrading LAB to improve cereal-based products by mineral solubilization during dough fermentation. *J. Food Sci.* 75, M28–M35. doi: 10.1111/j.1750-3841.2009.01402.x

Badel, S., Bernardi, T., and Michaud, P. (2011). New perspectives for lactobacilli exopolysaccharides. *Biotechnol. Adv.* 29, 54–66. doi: 10.1016/j.biotechadv.2010.08.011

Benachour, A., Frere, J., Boutibonnes, P., and Auffray, Y. (1995). Characterization and replication mode determination of the minimal replicon of *Tetragenococcus halophila* ATCC33315 plasmid pUCL287. *Biochimie* 77, 868–874. doi: 10.1016/0300-9084(95)90005-5

Benachour, A., Frere, J., Flahaut, S., Novel, G., and Auffray, Y. (1997). Molecular analysis of the replication region of the theta-replicating plasmid pUCL287 from *Tetragenococcus (Pediococcus) halophilus* ATCC33315. *Mol. Gen. Genet.* 255, 504–513. doi: 10.1007/s004380050523

Berthier, F., Zagorec, M., Champomier-Vergès, M., Ehrlich, S. D., and Morel-Deville, F. (1996). Efficient transformation of *Lactobacillus sake* by electroporation. *Microbiology* 142, 1273–1279. doi: 10.1099/13500872-142-5-1273

Bounaix, M. S., Robert, H., Gabriel, V., Morel, S., Remaud-Simeon, M., Gabriel, B., et al. (2010). Characterization of dextran-producing *Weissella* strains isolated from sourdoughs and evidence of constitutive dextransucrase expression. *FEMS Microbiol. Lett.* 311, 18–26. doi: 10.1111/j.1574-6968.2010.02067.x

Capozzi, V., Russo, P., Dueñas, M. T., López, P., and Spano, G. (2012). Lactic acid bacteria producing B-group vitamins: a great potential for functional cereals products. *Appl. Microbiol. Biotechnol.* 96, 1383–1394. doi: 10.1007/s00253-012-4440-2

Chenoll, E., Macián, M. C., Elizaquível, P., and Aznar, R. (2007). Lactic acid bacteria associated with vacuum-packed cooked meat product spoilage: population analysis by rDNA-based methods. *J. Appl. Microbiol.* 102, 498–508. doi: 10.1111/j.1365-2672.2006.03081.x

de Man, J. C., Rogosa, M., and Sharpe, M. E. (1960). A medium for the cultivation of lactobacilli. *J. Appl. Microbiol.* 23, 130–135. doi: 10.1111/j.1365-2672.1960.tb00188.x

Dimopoulou, M., Bardeau, T., Ramonet, P. Y., Miot-Certier, C., Claisse, O., Doco, T., et al. (2015). Exopolysaccharides produced by *Oenococcus oeni*: from genomic and phenotypic analysis to technological valorization. *Food Microbiol.* 53, 10–17. doi: 10.1016/j.fm.2015.07.011

Dornan, S., and Collins, M. (1987). High efficiency electroporation of *Lactococcus lactis* subps. lactis LM0230. *Lett. Appl. Microbiol.* 11, 62–64. doi: 10.1111/j.1472-765X.1990.tb01275.x

Funane, K., Yamada, M., Shiraiwa, M., Takahara, H., Yamamoto, N., Ichishima, E., et al. (1995). Aggregated form of dextransucrases from *Leuconostoc mesenteroides* NRRL B-512F and its constitutive mutant. *Biosci. Biotechnol. Biochem.* 59, 776–780. doi: 10.1271/bbb.59.776

Garcia-Cayuela, T., de Cadinanos, L. P., Mohedano, M. L., de Palencia, P. F., Boden, D., Wells, J., et al. (2012). Fluorescent protein vectors for promoter analysis in lactic acid bacteria and *Escherichia coli*. *Appl. Microbiol. Biotechnol.* 96, 171–181. doi: 10.1007/s00253-012-4087-z

Gasson, M. J. (1983). Plasmid complements of *Streptococcus lactis* NCDO 712 and other streptococci after protoplast-induced curing. *J. Bacteriol.* 154, 1–9.

Ito, K., Ito, S., Shimamura, T., Weyand, S., Kawarasaki, Y., Misaka, T., et al. (2011). Crystal structure of glucansucrase from the dental caries pathogen *Streptococcus mutans*. *J. Mol. Biol.* 408, 177–186. doi: 10.1016/j.jmb.2011.02.028

Janda, W. M., and Kuramitsu, H. K. (1978). Production of extracellular and cell-associated glucosyltransferase activity by *Streptococcus mutans* during growth on various carbon sources. *Infect. Immun.* 19, 116–122.

Kojic, M., Vujcic, M., Banina, A., Cocconcelli, P., Cerning, J., and Topisirovic, L. (1992). Analysis of exopolysaccharide production by *Lactobacillus casei* CG11, isolated from cheese. *Appl. Environ. Microbiol.* 58, 4086–4088.

Koressaar, T., and Remm, M. (2007). Enhancements and modifications of primer design program Primer3. *Bioinformatics* 23, 1289–1291. doi: 10.1093/bioinformatics/btm091

Kralj, S., Van Geel-Schutten, G. H., Dondorff, M. M. G., Kirsanovs, S., Van Der Maarel, M. J. E. C., and Dijkhuizen, L. (2004a). Glucan synthesis in the genus *Lactobacillus*: isolation and characterization of glucansucrase genes, enzymes and glucan products from six different strains. *Microbiology* 150, 3681–3690. doi: 10.1099/mic.0.27321-0

Kralj, S., Van Geel-Schutten, G. H., Van Der Maarel, M. J. E. C., and Dijkhuizen, L. (2003). Efficient screening methods for glucosyltransferase genes in *Lactobacillus* strains. *Biocatal. Biotransform.* 21, 181–187. doi: 10.1080/10242420310001618519

Kralj, S., Van Geel-Schutten, G. H., Van Der Maarel, M. J. E. C., and Dijkhuizen, L. (2004b). Biochemical and molecular characterization of *Lactobacillus reuteri* 121 reuteransucrase. *Microbiology* 150(Pt 7), 2099–2112. doi: 10.1099/mic.0.27105-0

Leemhuis, H., Pijning, T., Dobruchowska, J. M., van Leeuwen, S. S., Kralj, S., Dijkstra, B. W., et al. (2013). Glucansucrases: three-dimensional structures, reactions, mechanism, α-glucan analysis and their implications in biotechnology and food applications. *J. Biotechnol.* 163, 250–272. doi: 10.1016/j.jbiotec.2012.06.037

Macrina, F. L., Kopecko, D. J., Jones, K. R., Ayers, D. J., and McCowen, S. M. (1978). A multiple plasmid-containing *Escherichia coli* strain: convenient source of size reference plasmid molecules. *Plasmid* 1, 417–420. doi: 10.1016/0147-619X(78)90056-2

Mohedano, M. L., García-Cayuela, T., Pérez-Ramos, A., Gaiser, R., Requena, T., and López, P. (2015). Construction and validation of a mCherry protein vector for promoter analysis in *Lactobacillus acidophilus*. *J. Ind. Microbiol. Biot.* 42, 247–253. doi: 10.1007/s10295-014-1567-4

Monsan, P., Bozonnet, S., Albenne, C., Joucla, G., Willemot, R. M., and Remaud-Siméon, M. (2001). Homopolysaccharides from lactic acid bacteria. *Int. Dairy J.* 11, 675–685. doi: 10.1016/S0958-6946(01)00113-3

Nácher-Vázquez, M., Ballesteros, N., Canales, A., Rodriguez Saint-Jean, S., Perez-Prieto, S. I., Prieto, A., et al (2015). Dextrans produced by lactic acid bacteria exhibit antiviral and immunomodulatory activity against salmonid viruses. *Carbohydr. Polym.* 124, 292–301. doi: 10.1016/j.carbpol.2015.02.020

Nácher-Vázquez, M., Iturria, I., Zarour, K., Mohedano, M. L., Aznar, R., Pardo, M. A., et al. (2017). Dextran production by *Lactobacillus sakei* MN1 coincides with reduced autoagglutination, biofilm formation and epithelial cell adhesion. *Carbohydr. Polym.* 168, 22–31. doi: 10.1016/j.carbpol.2017.03.024

Naessens, M., Cerdobbel, A., Soetaert, W., and Vandamme, E. J. (2005). Review *Leuconostoc* dextransucrase and dextran: production properties and applications. *J. Chem. Tech. Biotechnol.* 80, 845–860. doi: 10.1002/jctb.1322

Neely, W. B., and Nott, J. (1962). Dextransucrase, an induced enzyme from *Leuconostoc mesenteroides*. *Biochemistry* 1, 1136–1140. doi: 10.1021/bi00912a027

Neubauer, H., Bauché, A., and Mollet, B. (2003). Molecular characterization and expression analysis of the dextransucrase DsrD of *Leuconostoc mesenteroides* Lcc4 in homologous and heterologous *Lactococcus lactis* cultures *Microbiology* 149, 973–982. doi: 10.1099/mic.0.26029-0

Pérez-Ramos, A., Nácher-Vázquez, M., Notararigo, S., López, P., and Mohedano, M. L. (2015). "Current and future applications of bacterial extracellular polysaccharides," in *Probiotics, Prebiotics, and Synbiotics*, eds V. R. Preedy and R. R. Watson (Oxford: Elsevier), 329–344.

Piret, J., Lamontagne, J., Bestman-Smith, J., Roy, S., Gourde, P., Désormeaux, A., et al. (2000). *In vitro* and *in vivo* evaluations of sodium lauryl sulfate and dextran sulfate as microbicides against herpes simplex and human immunodeficiency viruses. *J. Clin. Microbiol.* 38, 110–119.

Providenti, M. A., O'Brien, J. M., Ewing, R. J., Paterson, E. S., and Smith, M. L. (2006) The copy-number of plasmids and other genetic elements can be determined by SYBR-Green-based quantitative real-time PCR. *J. Microbiol. Methods* 65, 476–487. doi: 10.1016/j.mimet.2005.09.007

Quirasco, M., López-Munguía, A., Remaud-Simeon, M., Monsan, P., and Farrés, A. (1999). Induction and transcription studies of the dextransucrase gene in *Leuconostoc mesenteroides* NRRL B-512F. *Appl. Environ. Microbiol.* 65, 5504–5509.

Roy, A., Kucukural, A., and Zhang, Y. (2010). I-TASSER: a unified platform for automated protein structure and function prediction. *Nat. Protoc.* 5, 725–738. doi: 10.1038/nprot.2010.5

Rühmkorf, C., Bork, C., Mischnick, P., Rübsam, H., Becker, T., and Vogel, R. F. (2013). Identification of *Lactobacillus curvatus* TMW 1.624 dextransucrase and comparative characterization with *Lactobacillus reuteri* TMW 1.106 and *Lactobacillus animalis* TMW 1.971 dextransucrases. *Food Microbiol.* 34, 52–61. doi: 10.1016/j.fm.2012.11.002

Sánchez, C., Neves, A. R., Cavalheiro, J., Moreira dos Santos, M., García-Quintáns, N., López, P., et al. (2008). Contribution of citrate metabolism to the growth of *Lactococcus lactis* CRL264 at low pH. *Appl. Environ. Microbiol.* 74, 1136–1144. doi: 10.1128/AEM.01061-07

Schwab, C., Walter, J., Tannock, G. W., Vogel, R. F., and Gänzle, M. G. (2007). Sucrose utilization and impact of sucrose on glycosyltransferase expression in *Lactobacillus reuteri*. *Syst. Appl. Microbiol.* 30, 433–443. doi: 10.1016/j.syapm.2007.03.007

Swe, P. M., Heng, N. C. K., Cook, G. M., Tagg, J. R., and Jack, R. W. (2010). Identification of DysI, the immunity factor of the streptococcal bacteriocin dysgalacticin. *Appl. Environ. Microbiol.* 76, 7885–7889. doi: 10.1128/AEM.01707-10

Untergasser, A., Cutcutache, I., Koressaar, T., Ye, J., Faircloth, B. C., Remm, M., et al. (2012). Primer3—new capabilities and interfaces. *Nucleic Acids Res.* 40:e115. doi: 10.1093/nar/gks596 van Hijum, S. A. F. T., Kralj, S., Ozimek, L. K., Dijkhuizen, L., and van Geel-Schutten, I. G. H. (2006). Structure-function relationships of glucansucrase and fructansucrase enzymes from lactic acid bacteria. *Microbiol. Mol. Biol. Rev.* 70, 157–176. doi: 10.1128/MMBR.70.1.157-176.2006

Vujicic-Zagar, A., Pijning, T., Kralj, S., Lopez, C. A., Eeuwema, W., Dijkhuizen, L., et al. (2010). Crystal structure of a 117 kDa glucansucrase fragment provides insight into evolution and product specificity of GH70 enzymes. *Proc. Natl. Acad. Sci. U.S.A.* 107, 21406–21411. doi: 10.1073/pnas.1007531107

Wang, T. T., and Lee, B. H. (1997). Plasmids in *Lactobacillus*. *Crit. Rev. Biotechnol.* 17, 227–272. doi: 10.3109/07388559709146615

Wenham, D. G., Hennessey, T. D., and Cole, J. A. (1979). Regulation of lucosyl- and fructosyltransferase synthesis by continuous cultures of *Streptococcus mutans*. *J. Gen. Microbiol.* 114, 117–124. doi: 10.1099/00221287-114-1-117

Werning, M. L., Ibarburu, I., Due-as, M. T., Irastorza, A., Navas, J., and López, P. (2006). *Pediococcus parvulus* gtf gene encoding the GTF glycosyltransferase and Its application for specific PCR detection of beta-D-glucan-producing bacteria in foods and beverages. *J. Food Prot.* 69, 161–169. doi: 10.4315/0362-028X-69.1.161

Werning, M. L., Notararigo, S., Nácher, M., Fernández de Palencia, P., Aznar, R., and López, P. (2012). "Biosynthesis, purification and biotechnological use of exopolysaccharides produced by lactic acid bacteria," in *Food Additives*, ed Y. El-Samragy (Rijeka: Intech), 83–114.

Yanping, Y., Peng, Q., Guo, Y., Han, Y., Xiao, H., and Zhou, Z. (2015). Isolation and characterization of dextran produced by *Leuconostoc citreum* NM105 from manchurian sauerkraut. *Carbohydr. Polym.* 133, 365–372. doi: 10.1016/j.carbpol.2015.07.061

Zarour, K., Llamas, M. G., Prieto, A., Rúas-Madiedo, P., Due-as, M. T., de Palencia, P. F., et al. (2017). Rheology and bioactivity of high molecular weight dextrans synthesised by lactic acid bacteria. *Carbohydr. Polym.* 174, 646–657. doi: 10.1016/j.carbpol.2017.06.113

15

Comparative Genomic Analysis Reveals Organization, Function and Evolution of *ars* Genes in *Pantoea* spp.

Liying Wang[1,2]*, Jin Wang*[3] *and Chuanyong Jing*[1,2]*

[1] State Key Laboratory of Environmental Chemistry and Ecotoxicology, Research Center for Eco-Environmental Sciences, Chinese Academy of Sciences, Beijing, China, [2] College of Resources and Environment, University of Chinese Academy of Sciences, Beijing, China, [3] Department of Municipal and Environmental Engineering, School of Civil Engineering, Beijing Jiaotong University, Beijing, China

Edited by:
Manuel Espinosa,
Consejo Superior de Investigaciones Científicas (CSIC), Spain

Reviewed by:
Yunyoung Kwak,
Kyungpook National University, South Korea
Lukasz Drewniak,
University of Warsaw, Poland
Ji-Hoon Lee,
Chonbuk National University, South Korea

***Correspondence:**
Chuanyong Jing
cyjing@rcees.ac.cn

Numerous genes are involved in various strategies to resist toxic arsenic (As). However, the As resistance strategy in genus *Pantoea* is poorly understood. In this study, a comparative genome analysis of 23 *Pantoea* genomes was conducted. Two vertical genetic *arsC*-like genes without any contribution to As resistance were found to exist in the 23 *Pantoea* strains. Besides the two *arsC*-like genes, As resistance gene clusters *arsRBC* or *arsRBCH* were found in 15 *Pantoea* genomes. These *ars* clusters were found to be acquired by horizontal gene transfer (HGT) from sources related to *Franconibacter helveticus*, *Serratia marcescens*, and *Citrobacter freundii*. During the history of evolution, the *ars* clusters were acquired more than once in some species, and were lost in some strains, producing strains without As resistance capability. This study revealed the organization, distribution and the complex evolutionary history of As resistance genes in *Pantoea* spp.. The insights gained in this study improved our understanding on the As resistance strategy of *Pantoea* spp. and its roles in the biogeochemical cycling of As.

Keywords: comparative genomic, arsenic, *Pantoea* spp., arsenic resistance, *ars* genes

INTRODUCTION

Arsenic (As), one of the earliest known toxic elements, occurs naturally worldwide (Smith et al., 2002). To adapt to habitats with elevated As, microbes have evolved dynamic resistance mechanisms. The most ubiquitous and important strategy of As resistance is to reduce As(V) to As(III) and extrude it using *ars* operons with various genomic configurations in specific bacterial strains (Páezespino et al., 2009). The core genes of *ars* systems, however, are *arsR* (encoding the transcriptional repressor ArsR), *arsB* (encoding the arsenite efflux pump ArsB) and *arsC* (encoding arsenate reductase ArsC) (Xu et al., 1998). Besides this detoxification mechanism using *ars* systems, some strains possess the mechanism of As methylation-demethylation, changing inorganic As into organic forms using a distinct gene *arsM* (Qin et al., 2006; Zhao et al., 2015). Some strains are able to oxidize As(III) to As(V), which involve membrane-associated proteins, AoxAB (Levin and Tal, 2003; Ghosh et al., 2014). Some strains are able to reduce As(V) to As(III) with ArrAB as part of their respiratory processes transferring electrons to As and producing the energy for strains (Saltikov and Newman, 2003). The reported genes related to the strategy of As resistance are listed in **Table 1**.

In traditional molecular biology research, As resistance traits are revealed primarily based on the cultivation of a specific strain, and it is impossible to study the As strategy of all strains in a genus. Nevertheless, understanding of such traits in all strains of a genus is sometime more desirable. Gaining this knowledge is no longer a challenge with the explosive development of high-throughput sequencing technology. The genomic sequence of a strain contains nearly all of the genetic information. Therefore, fundamental knowledge such as the phylogenetic, the genetic traits of As resistance and its evolutionary history can be obtained through comparative genomic analysis (Arsène-Ploetze et al., 2010; Colston et al., 2014). So here we use genomic information of *Pantoea* spp. and compared these genomes to explore and predict the strategy of As resistance and their evolutionary patterns in genus *Pantoea* as an example.

Pantoea is a genus of Gram-negative, facultative anaerobic bacteria. This genus belongs to gamma *Proteobacteria*, family *Gammaproteobacteria*, and was recently separated from the genus *Enterobacter* (Gavini et al., 1989). Currently, the genus contains 26 species[1]. Members of this genus are found in various environmental matrices (Meng et al., 1995; Zhang and Birch, 1997; Rezzonico et al., 2009). In 2013, the strain *Pantoea* sp. IMH was an isolate that reported firstly as the strain having the As resistance capability within *Pantoea* species (Wu et al., 2013). Further, we sequenced the genome of *Pantoea* sp. IMH and found two *ars* clusters (*arsR1B1C1H1* and *arsR2B2C2H2*) co-contributing to its As resistance (Tian and Jing, 2014; Wang et al., 2016). However, the evolutionary history and genetic traits of As resistance in genus *Pantoea* are not fully understood.

Herein, we present the first study of the genetic traits of As resistance in *Pantoea* spp., as well as their evolutionary history. Two vertically transmitted *arsC*-like genes without any contribution to As resistance were found to exist in the 23 *Pantoea* strains. Besides these two *arsC*-like genes, As resistance gene clusters *arsRBC* or *arsRBCH* were found in 15 *Pantoea* genomes. These *ars* clusters were acquired by horizontal gene transfer (HGT) from sources related to *Franconibacter helveticus*, *Serratia marcescens*, and *Citrobacter freundii*. The insights gained in this study improve our understanding on the complex evolutionary history of As resistance genes and their roles in the biogeochemical cycling of As.

[1] http://www.bacterio.net/pantoea.html

TABLE 1 | Genes involved in arsenic resistance and transformation.

Function	Gene	Protein
Arsenate reduction	*arsR*	Transcriptional regulator ArsR
	arsB	As(III) efflux pump protein
	arsC	As(V) reductase ArsC
	arsH	Putative flavoprotein
	arsA	As(III) active ATPase
	arsD	As metallochaperone
	arsO	Monooxygenase
	arsT	Putative thioredoxin system
	arsX	Unknown function
	arsN	Acetyltransferases
Respiratory reduction	*arrA*	Large subunit of respiratory As(V) reductase
	arrB	Small subunit of respiratory As(V) reductase
Arsenite oxidation	*aioA*	Large subunit of As oxidase
	aioB	Small subunit of As oxidase
	aioD	Biosynthesis protein A with molybdenum cofactor
	aioX	Phosphonate-binding periplasmic protein
	aioS	Histidine kinase for signal transduction
	aioC	Cytochrome c
Arsenic methylation	*arsM*	As(III) S-adenosylmethyltransferase

MATERIALS AND METHODS

Phylogenetic Analysis

Phylogenetic trees of *Pantoea* species were constructed based on 100 single-copy core proteins shared by 23 *Pantoea* genomes and the genome of *Tatumella* sp. NML 06-3099 according to the following three methods: maximum likelihood (ML), neighbor joining (NJ), and Bayesian inference (BI). ML and NJ trees were computed by applying models with 1,000 bootstrap replicates and uniform rates in MEGA5 (Tamura, 2011). Multiple alignments of amino acid sequences were carried out by ClustalW, and the CONSEL program was used to select the best model of the trees (Shimodaira and Hasegawa, 2001; Thompson et al., 2002). The BI tree was generated using the MrBayes package with mixed models (Ronquist et al., 2012). The NJ tree of concatenated *arsRBC* homologs was generated according to the same method described above. MEGA5 or FigTree v.1.3.1[2] was used to illustrate the constructed trees.

Average Nucleotide Identity (ANI)

Assembled contigs were reconstituted from the RAST-generated GenBank files for 23 genomes by using the seqret function of the EMBOSS package (Rice et al., 2000). These 23 genomes were treated in the same manner to ensure that any biases were consistent across the entire dataset. JSpecies1.2.1 was used to analyze these contig sets for the ANI and tetramer usage patterns, using default parameters (Richter and Rosselló-Móra, 2009).

Comparative Genomics

All of the orthologous pairs between *Pantoea* test genomes were identified by Pan Genome Analysis Pipeline (PGAP) (Zhao et al., 2012). The common dataset of shared genes among test strains was defined as their core genome. The total set of genes with test genomes was defined as the pan genome. The set of genes in each strain not shared with other strains was defined as the unique genes. The details of the strains used are listed in Supplementary Table S1.

Construction of the Recombinant Plasmids and *Escherichia coli* Strains

A 3.86 kb *Bam*HI-*Xba*I DNA fragment containing the complete *ars1* cluster of *Pantoea stewartii* S301 (promoter region, 342 bp

[2] http://tree.bio.ed.ac.uk/software/figtree/

upstream of the start codon ATG of *arsR*, the contiguous four genes *arsR1B1C1H1* and 281 bp upstream of the start codon ATG of *arsH*) was PCR amplified with primers Ars1-F and Ars1-R (Supplementary Table S2). A 3.43 kb *Bam*HI-*Xba*I DNA fragment containing the complete *ars2* gene cluster of *P. agglomerans* Tx10 (a 280 bp region downstream of the stop codon TAA of *arsC2* and the contiguous ten genes *arsR2B2C2H2* and 328 bp downstream of the stop codon TAA of *arsH2*) was PCR amplified with primers Ars2-F and Ars2-R (Supplementary Table S2).

An 860 bp *Bam*HI-*Xba*I DNA fragment containing the complete *arsC1*-like gene of *P. stewartii* DC283 (promoter region, 221 bp upstream of the start codon ATG of *arsC1*-like gene, *arsC1*-like and 209 bp downstream of the stop codon TTA of *arsC1*-like gene) was PCR amplified with primers ArsC1-like-F and ArsC1-like-R (Supplementary Table S2). A 942 bp *Bam*HI-*Xba*I DNA fragment containing the complete *arsC2*-like gene of *P. stewartii* DC283 (promoter region, 265 bp upstream of the start codon ATG of *arsC2*-like gene, *arsC2*-like and 236 bp downstream of the stop codon TTA of *arsC2*-like gene) was PCR amplified with primers ArsC2-like-F and ArsC2-like-R (Supplementary Table S2).

The above PCR products were ligated to the *Bam*HI-*Xba*I site of plasmid pUC18, yielding plasmids pUC18-ars1, pUC18-ars2, pUC18-arsC1-like, and pUC18-arsC2-like. Then the plasmids were transferred to *E. coli* AW3110, yielding the recombinant *E. coli* AW3110-ars1, *E. coli* AW3110-ars2, *E. coli* AW3110-arsC1-like and *E. coli* AW3110-arsC2-like strains, respectively.

Strains, Plasmids, and Culture Conditions

The strains and plasmids used in this work are summarized in Supplementary Table S3. *E. coli* and *Pantoea* strains were grown in LB medium (per liter contains: 10 g tryptone, 5 g yeast, and 10 g NaCl) or LB plates (LB medium with w/v 1.5% agar) at 30°C. When appropriate, antibiotics were added at the following concentration: 100 μg/mL ampicillin. Resistance to As species was tested by plating serial dilutions of cultures of each strain onto agar plates containing filtered sodium arsenate (Na_3AsO_4).

RESULTS

Genomic Features

To date, 26 species have been reported in genus *Pantoea* and strains of nine species (*P. ananatis*, *P. agglomerans*, *P. stewartii*, *P. vagans*, *P. dispersa*, *P. septica*, *P. rodasii*, *P. rwandensis*, and *P. anthophila*) have been sequenced[3]. To study the genetic traits and phylogenetic history of As resistance in genus *Pantoea*, 23 strains were chosen, containing two to three standard strains sequenced in each species and five unidentified strains (Supplementary Table S1). A summary of features for these 23 sequenced genomes is listed in Supplementary Table S1. The G+C contents of the 23 genomes range from 53.4 to 59.1. These genomes vary in size by approximately 1.6 mega-bases in average (ranging from 4.02 to 5.68 Mb) with coding sequence (CDS) numbers ranging from 3580 to 8894, indicating a substantial strain-to-strain variation.

[3]http://www.ncbi.nlm.nih.gov/genome/?term=pantoea

Strain-Specific and Core Genes

To reveal the genomic features specific to each strain, we identified all orthologous pairs between the tested *Pantoea* genomes using PGAP. Our analysis of the total of 23 genomes revealed that a pan genome contains 48,207 putative protein-coding genes in the genus *Pantoea*. Out of these 48,207 genes, 10,896 (22.6%) were represented in the specific genomes of *Pantoea* spp., suggesting some frequency of horizontal gene acquisition from other taxa. The number of specific genes ranges from 131 to 1,285, with the smallest encoded by *P. vagans* C9-1 and the largest identified in *P. agglomerans* Tx10 (Supplementary Figure S1). The cluster of orthologous groups (COG) assignments reveal that a higher proportion of strain-specific genes in most of the strains can be assigned to the K (transcription), L (DNA replication), and M (cell wall/membrane/envelope biogenesis) categories (Supplementary Figure S2).

In contrast to the pan-genome, the core genome of *Pantoea* spp. contains 1,994 putative protein-coding genes, which represents 38.8–56.1% of the repertoire of protein coding genes of each strain, illustrating a small degree of genomic diversity in this group of bacteria (Supplementary Figure S1). The genomic analysis agrees with the fact that *Pantoea* strains are consistent in morphological and physiological appearance. Furthermore, the COG assignment results show that these core genes are in different functional categories (Supplementary Figure S3). In fact, the percentage of genes in each functional category remains rather similar (with an average divergence of 8.6%). This is consistent with an earlier report that larger prokaryotic genomes preferentially accumulate genes directly or indirectly involved in metabolism (Konstantinidis and Tiedje, 2004). These genes support a broader metabolic diversity, which, in turn, would improve the ecological success of *Pantoea* under more diverse environmental conditions.

Phylogenic Analyses

To associate the distribution of As resistance genes in *Pantoea* spp. with their phylogenetic affiliation, we constructed the phylogeny tree of the 23 *Pantoea* spp. based on 16S rRNA gene sequences using NJ methods rooted by *Tatumella* sp. NML 06-3099 (Supplementary Figure S9). This phylogenetic tree showed that the strains in the same species reported were grouped together except strain 848PVAG. At the same time, we constructed the phylogeny of the 23 genomes based on concatenation of the 100 core genes that are present as single copies in a genome using the ML method and rooted by *Tatumella* sp. NML 06-3099 (**Figure 1**). The phylogenetic trees, inferred using BI and NJ methods (Supplementary Figures S4, S5), were congruent with the ML phylogenetic tree. These trees show that some strains in different species reported were grouped together, such as, FF5 and 848PVAG, ZBG6, GB1, MP2, and Tx10. The phylogeny

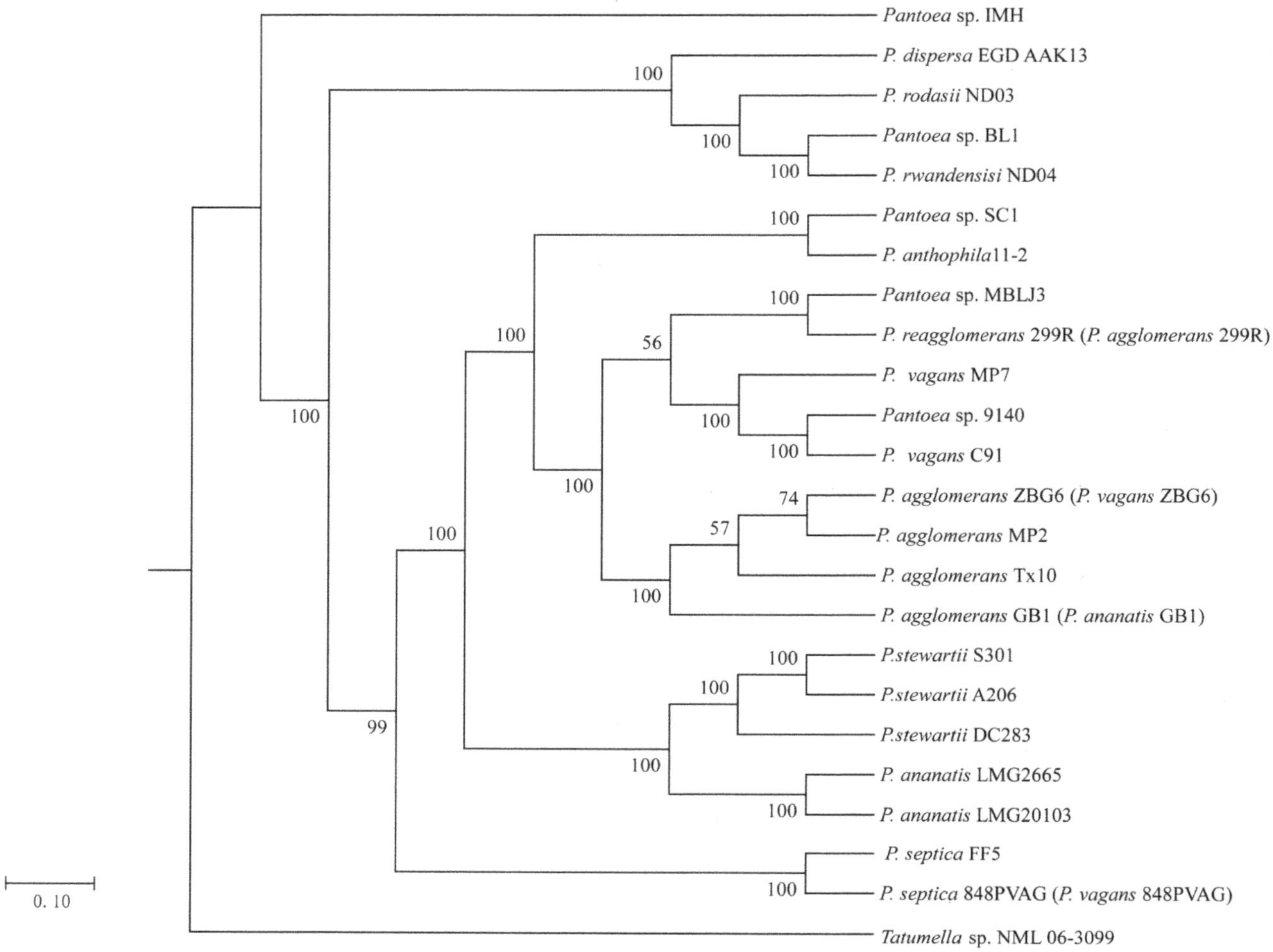

FIGURE 1 | Phylogenetic relationship of 23 *Pantoea* strains. Maximum likelihood (ML) phylogenetic tree was constructed based on 100 single-copy core proteins shared by 23 genomes and an out-group *Tatumella* sp. NML 06-3099.

based on the 100 core genes that are present as single copies in a genome showed a good correlation with that of 16S rRNA gene sequences, except for three strains ZBG6, GB1, and 299R. These results suggested that there were mistakes in the classification of *Pantoea* spp.. Further identification of the phylogenetic status of these strains was carried out as follows.

The information gained from the phylogenetic analysis provides an important depiction of the evolutionary relationship between different strains, but it does not translate directly into the overall similarity of the genomes, which is usually determined through the DNA-DNA hybridization (DDH). Herein, ANI approach was used to overcome the difficulty of conventional laboratory-based DDH in evaluating the genomic similarity of bacteria (Richter and Rosselló-Móra, 2009). The ANI results justified the conclusion of phylogenetic analysis. As shown in **Figure 2**, 23 strains were classified into 12 species based on their ANI $\geq$ 96%. For examples, LMG2665 and LMG20103 resulted in a higher ANI (99.3%), suggesting that they belong to the same species (*P. ananatis*). Strain 9140 and C91 resulted in a higher ANI (98.6%), suggesting that they belong to the same species (*P. vagans*). It was noteworthy that *Panotea* sp. IMH represented a novel species for the ANI $\geq$ 96% between IMH and other strains.

Strains MP2, Tx10, GB1, and ZBG6 which grouped together were identified as strains of *P. agglomerans*. Meanwhile, this result confirms the synonymy of *P.* FF5 and 848 PVAG (*P. vagans*), and suggests that 299R is not a member of species *P. agglomerans*. In agreement with the phylogenetic analysis, our ANI results indicate that there are mistakes in the classification of strain 299R, 848PVAG, GB1, and ZBG6. This mis-classification was also reported in other genus and generally corrected with the advance in technology (Goris et al., 2007). To associate the distribution of As-related genes with their phylogenetic affiliation, in this article below we renamed strain 299R to *P. reagglomerans* 299R (*P. agglomerans* 299R), 848PVAG to *P. septica* 848PVAG (*P. vagans* 848PVAG), GB1 to *P. agglomerans* GB1 (*P. ananatis* GB1), and ZBG6 to *P. agglomerans* ZBG6 (*P. vagans* ZBG6).

=ANI≥96%

	Pantoea ananatis LMG 2665	*Pantoea ananatis* LMG 20103	*Pantoea anthophila* 11-2	*Pantoea* sp. Sc1	*P. reagglomerans* 299R (*P. agglomerans* 299R)	*Pantoea* sp. MBLJ3	Pantoea dispersa EGD-AAK13	*Pantoea* sp. IMH	*Pantoea rodasii* ND03	*Pantoea vagans* C9-1	*Pantoea vagans* MP7	*Pantoea* sp. 9140	*Pantoea rwandensis* ND04	*Pantoea* sp. BL1	*Pantoea septica* FF5	*P. septica* 848PVAG (*P. vagans* 848PVAG)	*P. agglomerans* GB1 (*P. ananatis* GB1)	*Pantoea agglomerans* Tx10	*Pantoea agglomerans* MP2	*P. agglomerans* ZBG6 (*P. vagans* ZBG6)	*Pantoea stewartii* DC283	*Pantoea stewartii* A206	*Pantoea stewartii* S301	*Tatumella saanichensis* NML 06-3099
Pantoea ananatis LMG 2665		99.3	84.0	83.9	83.9	84.0	83.8	84.1	83.6	84.1	84.1	84.1	83.7	83.7	83.7	83.7	83.9	83.9	84.0	84.0	85.9	85.9	85.8	84.5
Pantoea ananatis LMG 20103	99.3		84.0	83.9	83.9	83.9	83.8	84.5	83.6	84.2	84.2	84.1	83.8	83.6	83.7	83.7	83.9	84.0	84.1	83.9	85.9	85.8	85.9	84.4
Pantoea anthophila 11-2	84.0	84.0		98.5	87.9	87.9	84.2	83.9	83.6	88.6	88.6	88.5	83.9	83.7	84.2	84.3	88.9	88.9	88.9	88.9	84.0	83.9	84.0	84.5
Pantoea sp. Sc1	83.9	83.9	98.5		87.9	87.9	84.2	83.6	83.7	88.5	88.5	88.5	83.8	83.7	84.1	84.2	88.8	88.8	88.8	88.8	84.0	84.0	83.9	84.4
P. reagglomerans 299R (*P. agglomerans* 299R)	83.9	83.9	87.9	87.9		99.2	84.0	83.8	83.6	90.5	90.6	90.5	83.7	83.6	83.8	83.8	90.9	90.9	90.9	90.9	83.9	83.8	83.8	84.5
Pantoea sp. MBLJ3	84.0	83.9	87.9	87.9	99.2		84.1	83.7	83.6	90.6	90.6	90.6	83.7	83.6	83.8	83.9	90.9	90.9	90.9	90.9	84.0	83.9	83.8	84.4
Pantoea dispersa EGD-AAK13	83.8	83.8	84.2	84.2	84.0	84.1		83.8	84.9	84.1	84.1	84.1	85.2	85.0	84.6	84.7	84.1	84.1	84.1	84.2	84.1	84.0	84.0	84.4
Pantoea sp. IMH	84.1	84.5	83.8	83.6	83.8	83.7	83.8		83.8	84.0	84.1	84.1	84.5	83.6	83.7	83.7	83.6	83.6	84.1	83.5	84.2	84.3	84.2	85.0
Pantoea rodasii ND03	83.6	83.6	83.6	83.7	83.6	83.6	84.9	83.8		83.7	83.7	83.7	86.8	86.5	83.7	83.7	83.6	83.6	83.7	83.6	83.8	83.7	83.7	84.7
Pantoea vagans C9-1	84.1	84.1	88.6	88.5	90.5	90.6	84.1	84.0	83.7		96.9	98.6	84.0	83.7	83.9	84.1	91.4	91.3	91.4	91.3	84.2	84.0	84.0	84.4
Pantoea vagans MP7	84.1	84.2	88.6	88.5	90.6	90.6	84.1	84.1	83.7	96.9		96.8	83.9	83.8	83.9	84.0	91.4	91.3	91.4	91.3	84.0	83.9	83.8	84.4
Pantoea sp. 9140	84.1	84.1	88.5	88.5	90.5	90.6	84.1	84.1	83.7	98.6	96.8		83.8	83.7	84.0	84.1	91.3	91.3	91.3	91.3	84.2	84.0	84.0	84.3
Pantoea rwandensis ND04	83.7	83.8	83.8	83.8	83.7	83.7	85.1	84.4	86.8	84.0	83.9	83.8		98.3	83.8	83.8	83.7	83.7	84.0	83.7	83.6	83.7	83.6	84.6
Pantoea sp. BL1	83.7	83.6	83.7	83.7	83.6	83.6	85.0	83.7	86.5	83.7	83.8	83.7	98.3		83.8	83.8	83.7	83.7	83.7	83.7	83.7	83.7	83.7	84.7
Pantoea septica FF5	83.7	83.7	84.2	84.1	83.8	83.8	84.6	83.7	83.7	84.0	84.0	84.0	83.9	83.8		96.0	83.9	83.9	83.9	83.9	83.8	83.7	83.7	84.4
P. septica 848PVAG (*P. vagans* 848PVAG)	83.7	83.7	84.3	84.2	83.8	83.9	84.7	83.7	83.7	84.1	84.0	84.1	83.8	83.8	96.0		84.1	84.1	84.1	84.1	83.8	83.7	83.7	84.4
P. agglomerans GB1 (*P. ananatis* GB1)	83.9	83.9	88.8	88.8	90.9	90.9	84.2	83.6	83.6	91.4	91.3	91.3	83.7	83.7	83.9	84.1		98.7	98.7	98.7	84.0	83.9	83.9	84.7
Pantoea agglomerans Tx10	83.9	84.0	88.8	88.8	90.9	90.9	84.1	83.6	83.6	91.3	91.3	91.3	83.7	83.7	84.0	84.1	98.7		98.7	98.7	84.0	84.0	83.9	84.6
Pantoea agglomerans MP2	84.0	84.1	88.9	88.8	90.9	90.9	84.1	84.1	83.6	91.4	91.4	91.3	84.0	83.7	83.9	84.1	98.7	98.7		98.9	84.1	84.0	83.9	84.5
P. agglomerans ZBG6 (*P. vagans* ZBG6)	84.0	83.9	88.8	88.8	90.9	90.9	84.2	83.5	83.6	91.3	91.3	91.3	83.7	83.7	83.9	84.1	98.7	98.7	98.9		84.0	84.0	83.9	84.6
Pantoea stewartii DC283	85.9	85.9	84.0	84.0	83.9	84.0	84.1	84.3	83.8	84.2	84.0	84.2	83.6	83.7	83.8	83.8	84.0	84.0	84.1	84.0		99.0	99.0	84.8
Pantoea stewartii A206	85.9	85.8	84.0	84.0	83.9	83.9	84.0	84.3	83.7	84.1	83.9	84.0	83.7	83.8	83.7	83.7	84.0	84.0	84.0	84.1	99.0		99.6	84.9
Pantoea stewartii S301	85.8	85.9	84.0	83.9	83.8	83.8	84.0	84.2	83.8	84.0	83.8	84.0	83.6	83.7	83.7	83.7	83.9	84.0	84.0	84.0	99.0	99.6		84.8
Tatumella saanichensis NML 06-3099	84.5	84.4	84.5	84.4	84.5	84.4	84.4	85.1	84.7	84.5	84.4	84.3	84.6	84.7	84.5	84.4	84.7	84.6	84.5	84.6	84.8	84.9	84.8	

FIGURE 2 | Average Nucleotide Identity (ANI) (%) based on whole genome alignments. ANI values are colored red according to historical species cutoff value (≥96%). Strains in one species are marked out the same color.

Distribution and Organization of As-Related Genes in *Pantoea* Genomes

Only As resistance genes (*ars* genes) including *arsR*, *arsB*, *arsC*, and *arsH* were detected in most *Pantoea* genomes (Supplementary Table S4 and **Figure 3**). The *arsC* gene encoding arsenate reductase is involved in the transformation of As(V) to As(III), which is then excreted by the As efflux pump ArsB encoded by the *arsB* gene. Nevertheless, *aio*, *arr*, and *arsM* were not found in *Pantoea* genomes, suggesting that cytoplasmic As(V) reduction and As(III) extrusion are the As resistance strategy used in genus *Pantoea* spp.. This mechanism benefits the bacteria itself, though it enhances the toxicity to the surrounding environment.

The *ars* genes in a genome are prone to group together as *ars* clusters (*arsRBC* and *arsRBCH*). Although comparison of the COG assignments of 23 genomes revealed that the DNA sequences between homologous genes within these *ars* clusters are conserved, some variations exist in DNA sequences, which can be divided into two sub-groups (*ars1* and *ars2*) (**Figure 3**). Unlike the two *ars* clusters in *Pantoea* sp. IMH, only one *ars* cluster, either *ars1* or *ars2*, was observed in other strains (Supplementary Table S2 and **Figure 3**). The *ars* gene clusters generally exhibited more than 80% identity within each sub-group and about 54% identity between two sub-groups. Actually, the *ars* clusters were not detected in eight strains including Sc1, BL1, 9140, DC283, MP7, C91, ND04, and FF5. Moreover, two

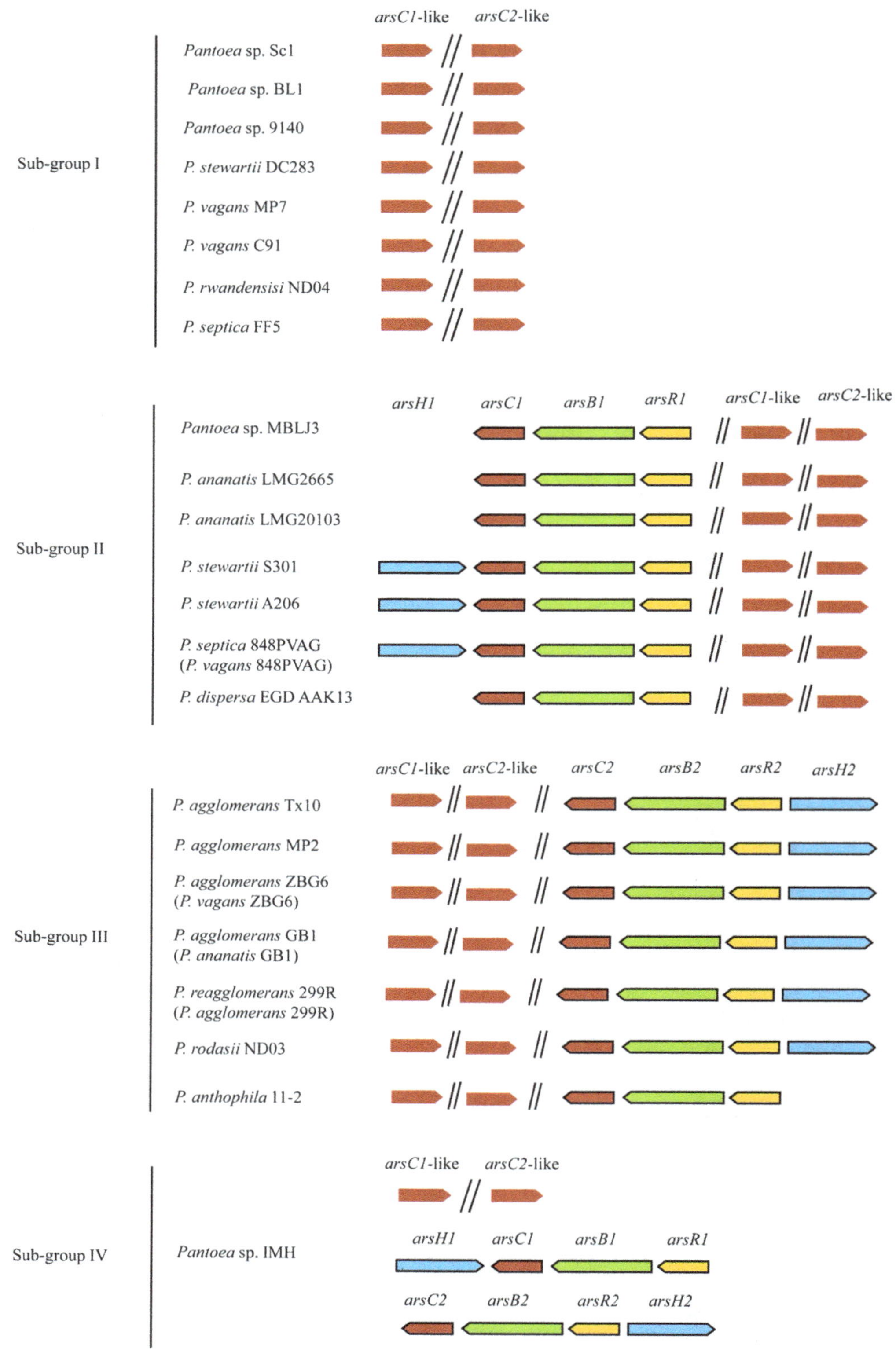

FIGURE 3 | Distribution and organization of *ars* genes and *arsC*-like genes in 23 *Pantoea* strains. *arsC*, *arsB*, *arsR*, *arsH*, and *arsC*-like genes are marked with different colors. There are only *arsC*-like genes in Sub-group I, *arsR1B1C1* or *arsR1B1C1H1* in Sub-group II, *arsR2B2C2* or *arsR2B2C2H2* in Sub-group III, and both *arsR1B1C1H1* and *arsR2B2C2H2* in Sub-group IV.

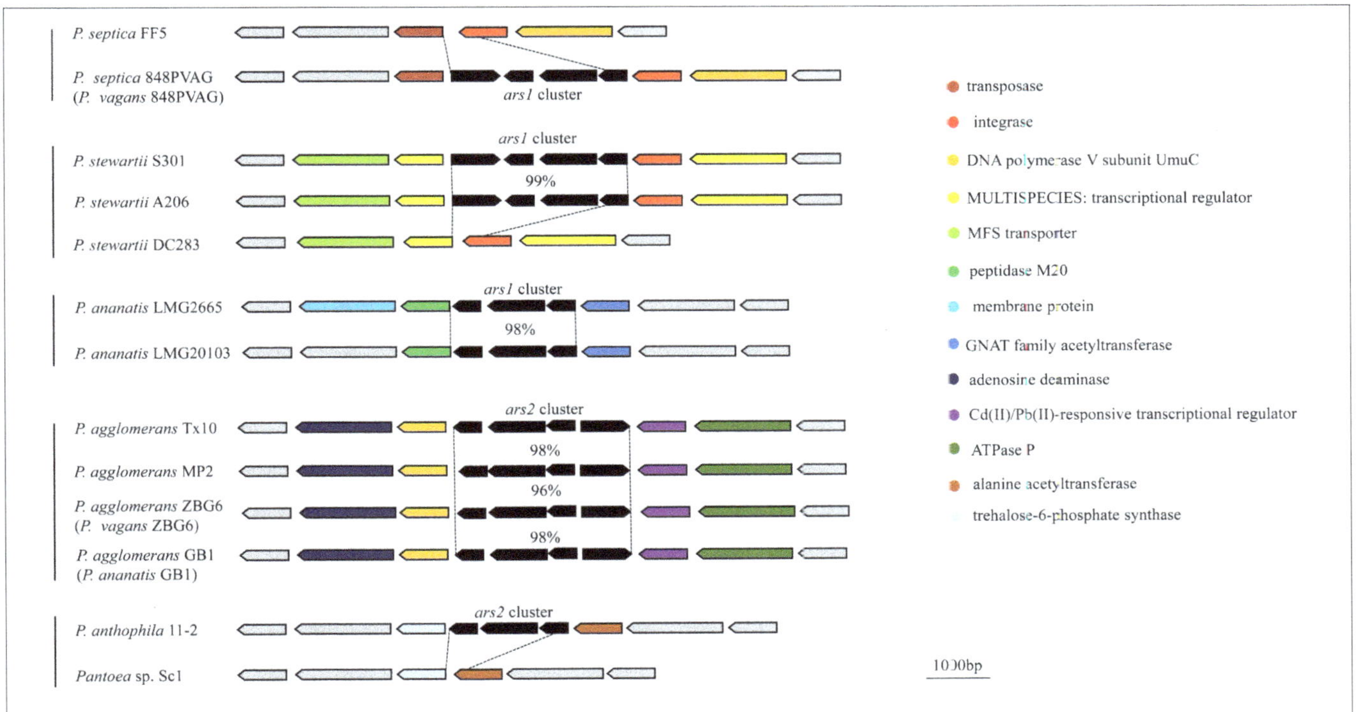

FIGURE 4 | Synteny of the chromosomal regions flanking the *ars* gene cluster among each species-group.

arsC-like genes with only 25% homology (*arsC1*-like and *arsC2*-like) were found in the 23 genomes. Based on the different *ars* genes distributions, the 23 strains were categorized into four sub-groups and discussed as follows. The overall distribution and organization of As resistance genes in 23 *Pantoea* strains are summarized in **Figure 3**.

Evolution and the Origin of *ars* Clusters

The distribution and organization of *ars* genes in *Pantoea* raise a question as to their evolution. The deviant G+C content is used as a detect method of HGT (Ochman et al., 2000; Xie et al., 2014). We detected the G+C content of *ars* clusters and their corresponding genomes. The results showed that the G+C contents of the *ars1* clusters are higher than those of the genomes in *Pantoea* strains (56.3–57.8 vs. 53.4–54.7) except *P. septica* 848PVAG (*P. vagans* 848PVAG) and *P. dispersa* EGD-AAK13; the G+C contents of the *ars2* clusters are lower than those of the genomes in *Pantoea* strains (50.6–52.4 vs. 53.7–58.8), showing variation of G+C content between clusters and the corresponding genomes. These results indicated that these *ars* clusters may be acquired in *Pantoea* strains by HGT (Supplementary Table S4 and Figure S6). To further elucidate the evolution of the *ars* gene clusters, we compared the chromosomal regions flanking the *ars* gene clusters among the 23 *Pantoea* strains and found that the genes in the upstream and downstream regions were conserved among strains of the same species (**Figure 4**). For example, the DNA polymerase V subunit UmuC gene and adenosine deaminase gene in the upstream and Cd(II)/Pb(II)-responsive transcriptional regulator gene and ATPase P gene in the downstream are conserved for the *ars* clusters in the strain Tx10, MP2, ZBG6, and GB1 within the species *P. agglomerans* strains. The same species strains share the same insertion sites, whereas the different species' strains result in different insertion sites, suggesting that *ars* clusters may be acquired more than once.

Interestingly, as shown in **Figure 4**, the flanking regions of the *ars* gene clusters in strain *P. stewartii* S301 and *P. stewartii* A206 were homologous to the corresponding regions of strain *P. stewartii* DC283; the same phenomenon was found in strain *P. septica* 848PVAG (*P. vagans* 848PVAG) and *P. septica* FF5, and strain *P. anthophila* 11-2 and *Pantoea* sp. Sc1. This result suggests that *ars* clusters may be lost in *P. stewartii* DC283, *P. septica* FF5, and *Pantoea* sp. Sc1.

To gain insights into the origin of *ars* genes clusters in *Pantoea*, a NJ phylogenetic tree was constructed based on the ArsRBC protein sequences. As shown in **Figure 5**, the strains possessing *ars1* and *ars2* clusters form separate groups. Notably, the phylogeny reveals that the *ars1* and *ars* clusters of *F. helveticus* were sister groups, and *ars2* grouped to *ars* clusters of *S. marcescens* and *C. freundii*. These results imply that the *ars1* cluster may be acquired via HGT from *F. helveticus*, and *ars2* from *S. marcescens and C. freundii* in early evolutionary history.

Two *arsC*-Like Genes in *Pantoea*

Our studies reveal that two *arsC*-like genes (*arsC1*-like and *arsC2*-like) are found in the 23 genomes with just 25% homology (**Figure 3**). Our phylogenetic analysis showed that the ArsC-like sequences formed distinct groups, which were clearly divergent from conventional arsenate reductase (**Figure 6**). It was reported

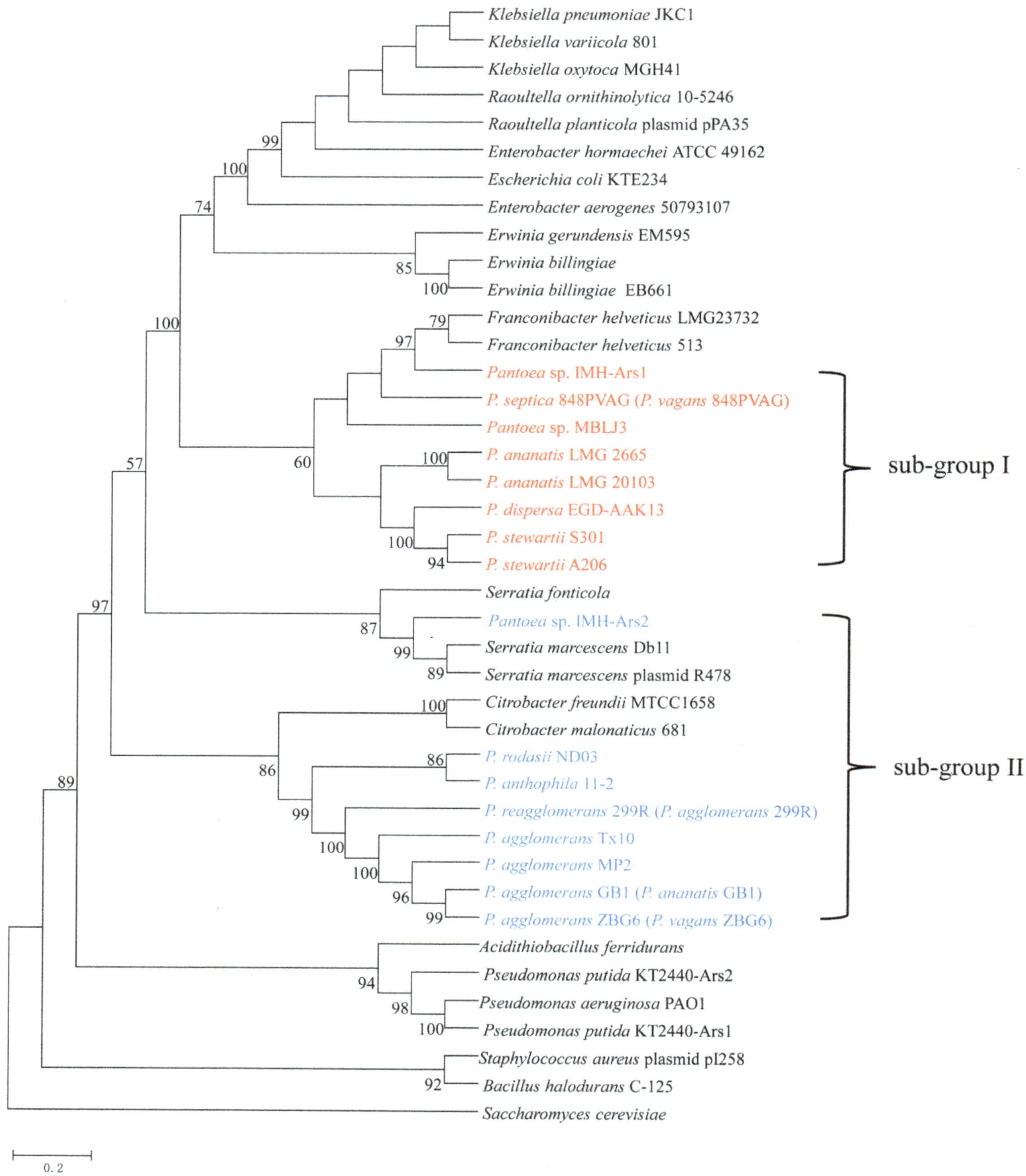

FIGURE 5 | Neighbor joining phylogenetic tree of the ArsRBC protein sequences derived from *Pantoea* spp. strains and other representative species. A total of 1,000 bootstrap replicates was made, and bootstrap values are indicated at each node. *ars1* cluster grouped together marked in red, *ars2* cluster formed in another group marked in blue.

that Cys-12, Arg-60, Arg-94, and Arg-107 were four conserved residues of the ArsC protein in the process of arsenic resistance (Gladysheva et al., 1996). Cys-12 was identified as a catalytic residue and was activated by nearby residues Arg-60, Arg-94, and Arg-107 (Martin et al., 2001). Alignment analysis of a*rsC* and *arsC*-like genes shows that Cys-12 and Arg-94 residues were conserved, but residues Arg-107 and Arg-60 in two ArsC-like proteins were not conserved (Supplementary Figure S8). These results suggest that these two *arsC*-like genes are not involved in the As resistance.

To explore the evolution of these two *arsC*-like genes in *Pantoea*, molecular phylogenetic analysis, molecular

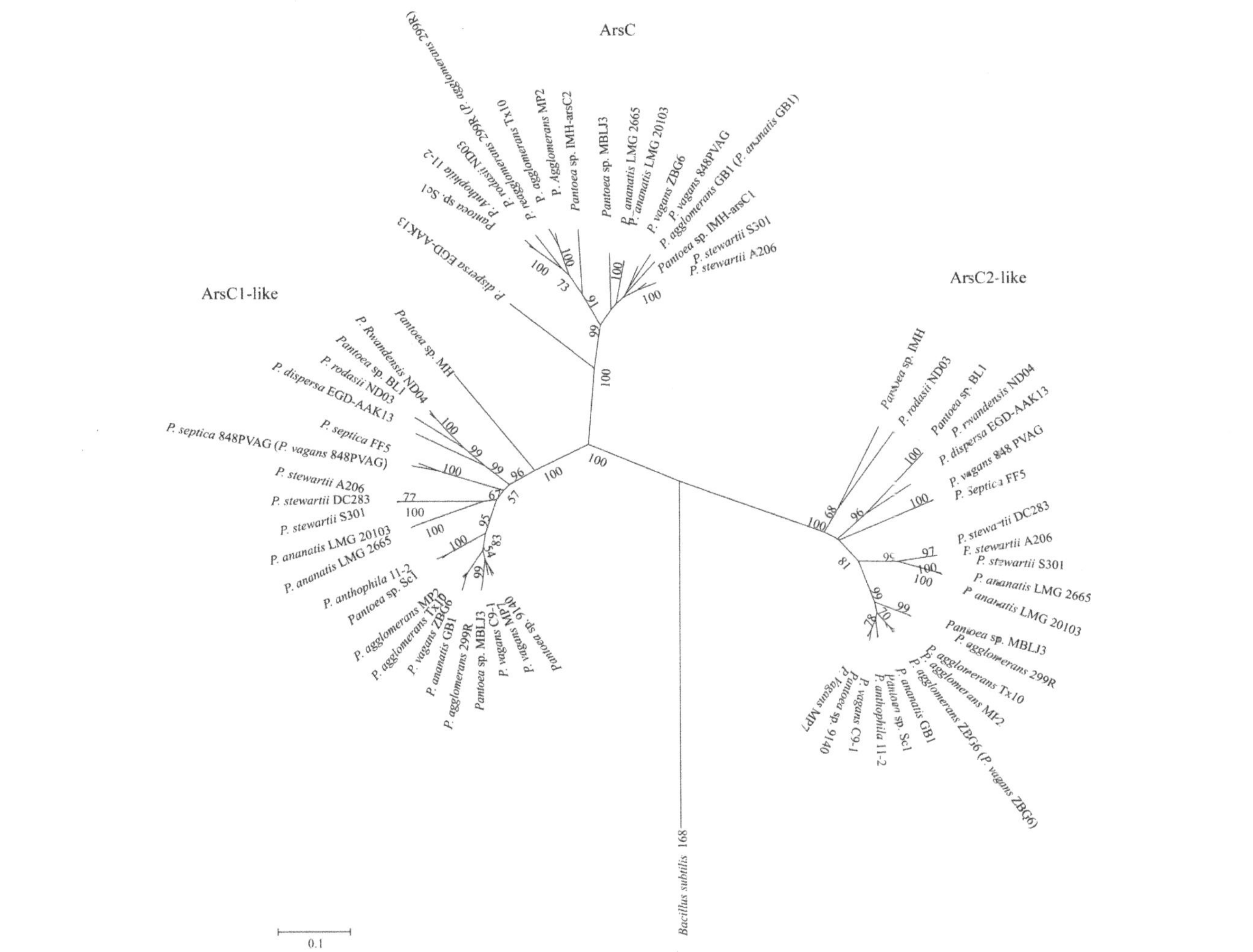

FIGURE 6 | Neighbor-joining tree based on ArsC/ArsC-like proteins. ArsC/ArsC-like sequences were derived from 23 *Pantoea* strains and the representative microorganisms. ArsC sequence of *Bacillus subtilis* 168 was used as an out group.

conservation, and linear representation analysis were used (Rice and Lampson, 1995; Nelson et al., 1999; Brochier-Armanet and Forterre, 2006; Dagan et al., 2008). The comparative analysis showed that the two *arsC*-like genes are conserved in all of the 23 (Supplementary Figure S7). Phylogenetic analysis showed that *arsC1*-like and *arsC2*-like genes were grouped together, respectively (**Figure 6**). These results suggested that the two *arsC*-like genes evolved with a possible evolutionary scenario of that there is a common ancestor. Further, we compared the flanking regions of the two *arsC*-like genes. Interestingly, two genes in the upstream (the uracil phosphoribosyl transferase genes and uracil/xanthine transporter genes) and two genes in the downstream (sulfur reduction protein DsrE and GntR family transcriptional regulator genes) are conserved for *arsC1*-like genes. The two genes in the upstream (DNA-binding response regulator genes and multidrug efflux RND transporter permease genes) and two genes in the downstream (succinyl-diaminopimelate desuccinylase genes and membrane protein genes) are also conserved for *arsC2*-like genes (Supplementary Figure S7). This observation also suggests that *arsC1*-like and *arsC2*-like genes were the vertical genetic genes in the genus *Pantoea*. Possibility, they may have been the main As resistance contributors in early times and later had evolved with deviance.

Functional Analysis of *ars* Gene and *arsC*-Like Genes

To verify that the *ars* gene clusters are the contributors to the As resistance, the *ars1* cluster with its promoter from *P. stewartii* S301, a representative strain with the *ars1* cluster, and the *ars2* cluster with its promoter from *P. agglomerans* Tx10, a representative strain with the *ars2* cluster, were PCR amplified and then ligated into vector pUC18 and further transferred to *E. coli* AW3110 (without any As resistance genes). The growth of the yielded recombinant *E. coli* strains *E. coli*-ars1 and *E. coli*-ars2, was tested in 5 mM concentration As(V). As shown in **Figure 7**, both *E. coli*-ars1 and *E. coli*-ars2 survived in 5 mM As(V), and

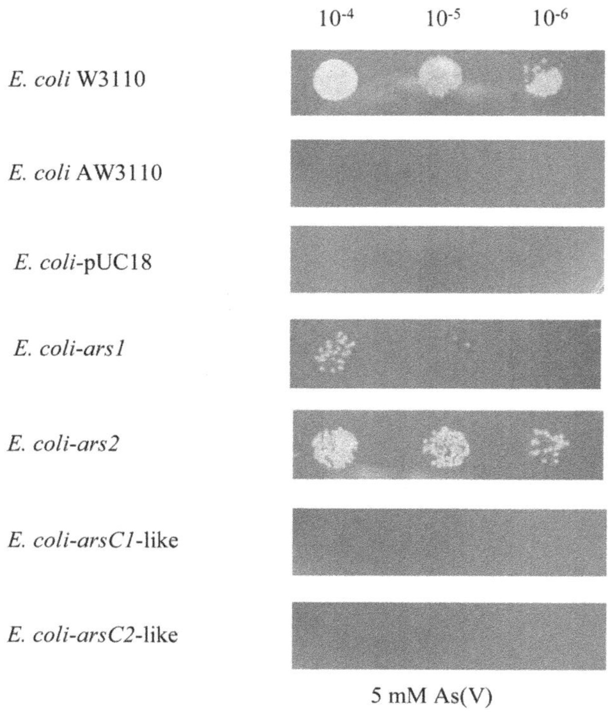

FIGURE 7 | The heterologous expression of *ars1*, *ars2*, *arsC1*-like and *arsC2*-like in *E. coli* AW3110. The pUC18-*ars1*, pUC18-*ars2*, pUC18-*arsC1*-like and pUC18-*arsC2*-like plasmids (Supplementary Table S3) were transformed, along with the empty vector pUC18 in *Escherichia coli* AW3110. *E. coli* W3110 (with *arsRBC* cluster) as a control. Serial 5 μl dilutions of the strains were then plated on LB agar medium with 5 mM As(V). Pictures were taken after 48 h at 30°C.

E. coli-arsl grew better than *E. coli*-ars2. This result suggests that both the *ars1* and *ars2* clusters enabled *E. coli* AW3110 to resist As, and *ars1* seemed to have a more effective As resistance capability than *ars2*.

To test the functions of *arsC1*-like and *arsC2*-like genes, the *arsC1*-like gene and *arsC2*-like gene with their promoters from *P. stewartii* DC283 were PCR amplified and then ligated into vector pUC18 and further transferred to *E. coli* AW3110. As shown in **Figure 7**, neither the *arsC1*-like nor *arsC2*-like gene enables *E. coli* AW3110 to resist As. In line with the alignment result, the function analysis demonstrates that these two *arsC*-like genes do not contribute to As resistance.

DISCUSSION

Pantoea is a genus with 26 members identified by DDH, a gold standard for prokaryotic species identification. However, laboratory-based DDH results may be irreproducible, and vary depending on the reannealing temperature (Gevers et al., 2005). With the rapid development in technology and decline in sequencing cost, promising new measurements such as ANI are being developed to evaluate the genomic similarity of bacteria (Richter and Rosselló-Móra, 2009). In this study, we identified the 23 *Pantoea* spp. phylogenetic status using ANI, together with phylogenetic trees based on concatenated sequences of the 100 core genes (**Figures 1, 2**). Our results showed that strain 299R, 848PVAG, GB1, and ZBG6 were misnamed. We reclassified strain 299R to *P. reagglomerans* 299R (*P. agglomerans* 299R), 848PVAG to *P. septica* 848PVAG (*P. vagans* 848PVAG), GB1 to *P. agglomerans* GB1 (*P. ananatis* GB1), and ZBG6 to *P. agglomerans* ZBG6 (*P. vagans* ZBG6). Our study provided data from genus *Pantoea* with a complex and controversial taxonomy and demonstrated the accuracy of a bioinformatics approach, such as ANI, to identify new species and to correct erroneous identifications from previous studies.

A previous study suggested that the *ars* system is a widespread As resistance mechanism (Páezespino et al., 2009). *Pantoea* sp. IMH was found to resist As by means of *ars* clusters. *arsRBC* is located on the large universal *Pantoea* plasmids of four stains including *P. agglomerans* E325, *P. agglomerans* MP2, *P. eucalyptiα*B, and *P. anthophila* Sc1 (Maayer et al., 2012). This information leads to the hypothesis that plasmids may be involved in the evolution of As resistance mechanism by *ars* genes in *Pantoea* spp. However, there are untouched questions such as what are the mechanisms of the other *Pantoea* spp. and what is the evolutionary history of genetic elements involved in the As resistance? To answer these questions, we collected the genome sequences of 23 strains in nine species in NCBI (*P. ananatis*, *P. agglomerans*, *P. stewartii*, *P. vagans*, *P. dispersa*, *P. septica*, *P. rodasii*, *P. rwandensis*, and *P. anthophila*). The sequencing results provided us with mass genomic information to detect the presence and the locations of As-related genes in *Pantoea* spp. Our study for the first time systematically analyzed the As resistance genes and revealed the As resistance traits in genus *Pantoea*. Our research provided the definitive evidence that that As resistance strategy in *Pantoea* spp. only involved the detoxification mechanism through *ars* clusters, not the respiratory reduction mechanism through *arr* clusters. This detoxification strategy was obtained by HGT. This conclusion can likely to be extended to most bacteria. We speculate that evolutionarily ancient microbes were exposed to As surroundings on ancient earth (Oremland et al., 2009). To overcome the As-induced selection pressure, microbes evolved *ars* genes in their genomes for survival by HGT. Therefore, *ars* has very early origins and represents a widespread As resistance mechanism.

Two scattered *arsC*-like genes exist in each genome of the 23 *Pantoea* strains, but they exhibited no functional As resistance. It is rare for *arsC*-like genes to show no As resistance capabilities (Butcher et al., 2000; Saltikov and Newman, 2003). Compared to functional protein ArsC, residues Arg-107 and Arg-60 of ArsC-like protein were variant (Supplementary Figure S8). We speculate that in early times, the ancestor of *Pantoea* spp. evolved the *arsC* gene to resist As, but later evolved with deviance during adaption to As-free niches, and thus retained non-functional *arsC*-like genes in some genomes.

The *ars* genes are abundant and tend to organize in typical *arsRBC* cluster structures (**Figure 3**). Apart from these operons, *arsRBCH* operons are widely observed. In genus *Pantoea*, these kinds of structures were anticipated, for these strains descended from a recent common ancestor. Our study suggests that *ars* clusters may be acquired by HGT from *F. helveticus*, *S. marcescens*, and *C. freundii* strains. This is consistent with recent literature showing that bacterial As resistance and transformation was a trait acquired via HGT, driven by adaptation to habitats containing As (Cai et al., 2009; Villegas-Torres et al., 2011). Interestingly, *ars* clusters are absent in some strains, suggesting that some microbes may have lost their As resistance genes during adaption to As-free niches. In addition, the number of As resistance genes in strains isolated from As-rich environments is much higher than in strains from other environments (Macur et al., 2004; Sutton et al., 2009). Compared to the evolutionary pattern of *ars* operons (Rosen, 1999), the evolution of As resistance genes (*ars* clusters) in *Pantoea* spp. involves a mix of HGT and loss, providing insight into the complex evolutionary history of As resistance.

AUTHOR CONTRIBUTIONS

LW and CJ conceived and designed the study. LW performed the laboratory work and data analysis. LW, JW, and CJ drafted the tables and figures, and prepared the main manuscript.

ACKNOWLEDGMENTS

We acknowledge the financial support of the National Basic Research Program of China (2015CB932003), the Strategic Priority Research Program of the Chinese Academy of Sciences (XDB14020302), and the National Natural Science Foundation of China (41373123, 41425016, 41503094, and 21321004). We thank Yongguan Zhu for the strain *E. coli* AW3110.

DATA ACCESSIBILITY

The NCBI accession numbers of 23 draft genome sequences of *Pantoea* are listed in Supplementary Table S1.

REFERENCES

Arsène-Ploetze, F., Koechler, S., Marchal, M., Coppée, J. Y., Chandler, M., Bonnefoy, V., et al. (2010). Structure, function, and evolution of the *Thiomonas* spp. genome. *PLoS Genet.* 6:e1000859. doi: 10.1371/journal.pgen.1000859

Brochier-Armanet, C., and Forterre, P. (2006). Widespread distribution of archaeal reverse gyrase in thermophilic bacteria suggests a complex history of vertical inheritance and lateral gene transfers. *Archaea* 2, 83–93.

Butcher, B. G., Deane, S. M., and Rawlings, D. E. (2000). The chromosomal arsenic resistance genes of *Thiobacillus ferrooxidans* have an unusual arrangement and confer increased arsenic and antimony resistance to *Escherichia coli*. *Appl. Environ. Microbiol.* 66, 1826–1833. doi: 10.1128/AEM.66.5.1826-1833.2000

Cai, L., Liu, G., Rensing, C., and Wang, G. (2009). Genes involved in arsenic transformation and resistance associated with different levels of arsenic-contaminated soils. *BMC Microbiol.* 9:4. doi: 10.1186/1471-2180-9-4

Colston, S. M., Fullmer, M. S., Beka, L., Lamy, B., Gogarten, J. P., and Graf, J. (2014). Bioinformatic genome comparisons for taxonomic and phylogenetic assignments using aeromonas as a test case. *MBio* 5:e2136-14. doi: 10.1128/mBio.02136-14

Dagan, T., Artzyrandrup, Y., and Martin, W. (2008). Modular networks and cumulative impact of lateral transfer in prokaryote genome evolution. *Proc. Natl. Acad. Sci. U.S.A.* 105, 10039–10044. doi: 10.1073/pnas.0800679105

Gavini, F., Mergaert, J., Bej, A., Mielcarek, C., Izard, D., Kersters, K., et al. (1989). Transfer of *Enterobacter agglomerans* (Beijerinck 1888) Ewing and Fife 1972 to *Pantoea* gen. nov. as *Pantoea agglomerans* comb. nov. and description of *Pantoea dispersa* sp. nov. *Int. J. Syst. Evol. Microbiol.* 39, 337–345.

Gevers, D., Cohan, F. M., Lawrence, J. G., Spratt, B. G., Coenye, T., Feil, E. J., et al. (2005). Opinion: re-evaluating prokaryotic species. *Nat. Rev. Microbiol.* 3, 733–739. doi: 10.1038/nrmicro1236

Ghosh, D., Bhadury, P., and Routh, J. (2014). Diversity of arsenite oxidizing bacterial communities in arsenic-rich deltaic aquifers in West Bengal, India. *Front. Microbiol.* 5:602. doi: 10.3389/fmicb.2014.00602

Gladysheva, T., Liu, J., and Rosen, B. P. (1996). His-8 lowers the pKa of the essential Cys-12 residue of the ArsC arsenate reductase of plasmid R773. *J. Biol. Chem.* 271, 33256–33260. doi: 10.1074/jbc.271.52.33256

Goris, J., Konstantinidis, K. T., Klappenbach, J. A., Coenye, T., Vandamme, P., and Tiedje, J. M. (2007). DNA-DNA hybridization values and their relationship to whole-genome sequence similarities. *Int. J. Syst. Evol. Microbiol.* 57, 81–91. doi: 10.1099/ijs.0.64483-0

Konstantinidis, K. T., and Tiedje, J. M. (2004). Trends between gene content and genome size in prokaryotic species with larger genomes. *Proc. Natl. Acad. Sci. U.S.A.* 101, 3160–3165. doi: 10.1073/pnas.0308653100

Levin, D., and Tal, A. (2003). Arsenite oxidase aox genes from a metal-resistant beta-proteobacterium. *J. Bacteriol.* 185, 135–141. doi: 10.1186/1471-2164-11-709

Maayer, P. D., Chan, W. Y., Blom, J., Venter, S. N., Duffy, B., Smits, T. H. M., et al. (2012). The large universal *Pantoea* plasmid LPP-1 plays a major role in biological and ecological diversification. *BMC Genomics* 13:625. doi: 10.1186/1471-2164-13-625

Macur, R. E., Jackson, C. R., Botero, L. M., Mcdermott, T. R., and Inskeep, W. P. (2004). Bacterial populations associated with the oxidation and reduction of arsenic in an unsaturated soil. *Environ. Sci. Technol.* 38, 104–111.

Martin, P., DeMel, S., Shi, J., Gladysheva, T., Gatti, D. L., Rosen, B. P., et al. (2001). Insights into the structure, solvation, and mechanism of ArsC arsenate reductase, a novel arsenic detoxification enzyme. *Structure* 9, 1071–1081. doi: 10.1016/S0969-2126(01)00672-4

Meng, M., Sun, W. Q., Geelhaar, L. A., Kumar, G., Patel, A. R., Payne, G. F., et al. (1995). Denitration of glycerol trinitrate by resting cells and cell extracts of *Bacillus thuringiensis/cereus* and *Enterobacter agglomerans*. *Appl. Environ. Microbiol.* 61, 2548–2553.

Nelson, K. E., Clayton, R. A., Gill, S. R., Gwinn, M. L., Dodson, R. J., Haft, D. H., et al. (1999). Evidence for lateral gene transfer between Archaea and Bacteria from genome sequence of *Thermotoga maritima*. *Nature* 399, 323–329. doi: 10.1038/20601

Ochman, H., Lawrence, J. G., and Groisman, E. A. (2000). Lateral gene transfer and the nature of bacterial innovation. *Nature* 405, 299–304. doi: 10.1038/35012500

Oremland, R. S., Saltikov, C. W., Wolfe-Simon, F., and Stolz, J. F. (2009). Arsenic in the evolution of earth and extraterrestrial ecosystems. *Geomicrobiol. J.* 26, 522–536. doi: 10.1080/01490450903102525

Páezespino, D., Tamames, J., De, L. V., and Cánovas, D. (2009). Microbial responses to environmental arsenic. *Biometals* 22, 117–130. doi: 10.1007/s10534-008-9195-y

Qin, J., Rosen, B. P., Zhang, Y., Wang, G., Franke, S., and Rensing, C. (2006). Arsenic detoxification and evolution of trimethylarsine gas by a microbial arsenite S-adenosylmethionine methyltransferase. *Proc. Natl. Acad. Sci. U.S.A.* 103, 2075–2080. doi: 10.1073/pnas.0506836103

Rezzonico, F., Smits, T. H., Montesinos, E., Frey, J. E., and Duffy, B. (2009). Genotypic comparison of *Pantoea agglomerans* plant and clinical strains. *BMC Microbiol.* 9:204. doi: 10.1186/1471-2180-9-204

Rice, P., Longden, I., and Bleasby, A. (2000). EMBOSS: the european molecular biology open software suite. *Trends Genet.* 16, 276–277. doi: 10.1016/S0168-9525(00)02024-2

Rice, S. A., and Lampson, B. C. (1995). Phylogenetic comparison of retron elements among the myxobacteria: evidence for vertical inheritance. *J. Bacteriol.* 177, 37–45. doi: 10.1128/jb.177.1.37-45.1995

Richter, M., and Rosselló-Móra, R. (2009). Shifting the genomic gold standard for the prokaryotic species definition. *Proc. Natl. Acad. Sci. U.S.A.* 106, 19126–19131. doi: 10.1073/pnas.0906412106

Ronquist, F., Teslenko, M., van, der Mark P., Ayres, D. L., Darling, A., Höhna, S., et al. (2012). MrBayes 3.2: efficient Bayesian phylogenetic inference and model choice across a large model space. *Syst. Biol.* 61, 539–542. doi: 10.1093/sysbio/sys029

Rosen, B. P. (1999). Families of arsenic transporters. *Trends Microbiol.* 7, 207–212.

Saltikov, C. W., and Newman, D. K. (2003). Genetic identification of a respiratory arsenate reductase. *Proc. Natl. Acad. Sci. U.S.A.* 100, 10983–10988. doi: 10.1073/pnas.1834303100

Shimodaira, H., and Hasegawa, M. (2001). CONSEL: for assessing the confidence of phylogenetic tree selection. *Bioinformatics* 17, 1246–1247. doi: 10.1093/bioinformatics/17.12.1246

Smith, A. H., Lopipero, P. A., Bates, M. N., and Steinmaus, C. M. (2002). Public health. Arsenic epidemiology and drinking water standards. *Science* 296, 2145–2146. doi: 10.1126/science.1072896

Sutton, N. B., van, der Kraan G. M., van, Loosdrecht M. C., Muyzer, G., Bruining, J., and Schotting, R. J. (2009). Characterization of geochemical constituents and bacterial populations associated with As mobilization in deep and shallow tube wells in Bangladesh. *Water Res.* 43, 1720–1730. doi: 10.1016/j.watres.2009.01.006

Tamura, K. (2011). MEGA5: molecular evolutionary genetics analysis using maximum likelihood, evolutionary distance, and maximum parsimony methods. *Mol. Biol. Evol.* 28, 2731–2739. doi: 10.1093/molbev/msr121

Thompson, J. D., Gibson, T. J., and Higgins, D. G. (2002). Multiple sequence alignment using ClustalW and ClustalX. *Curr. Protoc. Bioinformatics* Chater 2:Unit 2.3. doi: 10.1002/0471250953.bi0203s00

Tian, H., and Jing, C. (2014). Genome sequence of the aerobic arsenate-reducing bacterium *Pantoea* sp. strain IMH. *Genome Announc.* 2:e267-14. doi: 10.1128/genomeA.00267-14

Villegas-Torres, M. F., Bedoya-Reina, O. C., Salazar, C., Vives-Florez, M. J., and Dussan, J. (2011). Horizontal *arsC* gene transfer among microorganisms isolated from arsenic polluted soil. *Int. Biodeter. Biodegradation* 65, 147–152. doi: 10.1016/j.ibiod.2010.10.007

Wang, L., Zhuang, X., Zhuang, G., and Jing, C. (2016). Arsenic resistance strategy in *Pantoea* sp. IMH: organization, function and evolution of ars genes. *Sci. Rep.* 6:39195. doi: 10.1038/srep39195

Wu, Q., Du, J., Zhuang, G., and Jing, C. (2013). *Bacillus* sp. SXB and *Pantoea* sp. IMH, aerobic As(V)-reducing bacteria isolated from arsenic-contaminated soil. *J. Appl. Microbiol.* 114, 713–721. doi: 10.1111/jam.12093

Xie, J. B., Du, Z., Bai, L., Tian, C., Zhang, Y., Xie, J. Y., et al. (2014). Comparative genomic analysis of N2-fixing and non-N2-fixing *Paenibacillus* spp.: organization, evolution and expression of the nitrogen fixation genes. *PLoS Genet.* 10:e1004231. doi: 10.1371/journal.pgen.1004231

Xu, C., Zhou, T., Kuroda, M., and Rosen, B. P. (1998). Metalloid resistance mechanisms in prokaryotes. *J. Biochem.* 123, 16–23.

Zhang, L., and Birch, R. G. (1997). The gene for albicidin detoxification from *Pantoea dispersa* encodes an esterase and attenuates pathogenicity of *Xanthomonas albilineans* to sugarcane. *Proc. Natl. Acad. Sci. U.S.A.* 94, 9984–9989.

Zhao, C., Zhang, Y., Chan, Z., Chen, S., and Yang, S. (2015). Insights into arsenic multi-operons expression and resistance mechanisms in *Rhodopseudomonas palustris* CGA009. *Front. Microbiol.* 6:986. doi: 10.3389/fmicb.2015.00986

Zhao, Y., Wu, J., Yang, J., Sun, S., Xiao, J., and Yu, J. (2012). PGAP: pan-genomes analysis pipeline. *Bioinformatics* 28, 416–418. doi: 10.1093/bioinformatics/btr655

16

Molecular Basis of Stationary Phase Survival and Applications

*Jananee Jaishankar and Preeti Srivastava**

Department of Biochemical Engineering and Biotechnology, Indian Institute of Technology, New Delhi, India

Edited by:
Tatiana Venkova,
University of Texas Medical Branch,
United States

Reviewed by:
Grzegorz Wegrzyn,
University of Gdansk, Poland
Susana Brom,
National Autonomous University of Mexico, Mexico
Jan Nesvera,
Institute of Microbiology of the Czech Academy of Sciences, Czechia

***Correspondence:**
Preeti Srivastava
preeti@dbeb.iitd.ac.in;
preetisrivastava@hotmail.com

Stationary phase is the stage when growth ceases but cells remain metabolically active. Several physical and molecular changes take place during this stage that makes them interesting to explore. The characteristic proteins synthesized in the stationary phase are indispensable as they confer viability to the bacteria. Detailed knowledge of these proteins and the genes synthesizing them is required to understand the survival in such nutrient deprived conditions. The promoters, which drive the expression of these genes, are called stationary phase promoters. These promoters exhibit increased activity in the stationary phase and less or no activity in the exponential phase. The vectors constructed based on these promoters are ideal for large-scale protein production due to the absence of any external inducers. A number of recombinant protein production systems have been developed using these promoters. This review describes the stationary phase survival of bacteria, the promoters involved, their importance, regulation, and applications.

Keywords: stationary phase promoters, stationary phase gene expression, plasmid vectors, sigma factor, stationary phase

INTRODUCTION

The majority of the microorganisms around us in air, sea water, and soil are predominantly present in stationary phase (Gefen et al., 2014). The natural habitat of microorganisms often contains limited nutrients due to which rapid growth is usually hampered. Apart from nutrient deprivation, there are other conditions, including physical and chemical stresses, which result in unbalanced growth. All these events result in many changes at the molecular level. These molecular changes are comparable to those observed during the stationary phase of bacteria as witnessed in laboratory studies. The entry of bacteria to the stationary phase can be caused by different factors, including limitation of a specific essential nutrient, accumulation of toxic by-products, presence of stress factors such as changes in pH, temperature, osmolarity, etc. As the cell enters this phase, there is a reduction in cell size and the DNA/protein ratio is said to increase during transition to stationary phase (Nystrom, 2004). The stationary phase has received much attention due to the pattern of protein synthesis in this phase and also because of survival strategies adopted by bacteria. Numerous physiological, morphological, and gene expression changes are observed when a growing cell enters the stationary phase. These are discussed in the following sections.

PHYSIOLOGY OF THE STATIONARY PHASE

In the stationary phase, the cells become spherical and smaller with a rigid cell envelope, the cell wall is highly cross-linked, membrane fluidity reduces, and cells activate the stringent response mechanism in order to survive the calamity. The activation of this mechanism allows the bacteria to reprogram the gene expression pattern to adapt to different stresses. Two key components of the bacterial stringent response are ppGpp and pppGpp (which are explained in a later section). As a consequence, the cells divert their resources away from growth toward synthesizing amino acids so as to promote survival till nutrient conditions improve.

Figure 1 depicts the various changes observed in a cell when it enters the stationary phase. The peptidoglycan layer, being the stress-bearing component of cell, increases in thickness. It accounts for 0.7–0.8% of cell's dry weight in exponential phase cells whereas in stationary phase it increases up to 1.4–1.9% (Mengin-Lecreulx and van Heijenoort, 1985). At the subcellular level, nucleoid condensation occurs for DNA protection, the cytoplasm gets condensed with an overall decrease in protein synthesis as a consequence of stress or stationary phase (Navarro Llorens et al., 2010). At the translational level, the 70S ribosomes are converted into inactive 100S ribosome dimers by associating with ribosome modulation factor (Wada, 1998). This process, termed as ribosome hibernation, is thought to be a mechanism to fine-tune the translation process according to environmental conditions (McKay and Portnoy, 2015). Recently, 16S rRNA fragmentation at the tip of helix 6 has been shown to attenuate the activity of 30S ribosomal subunit and thereby protein synthesis (Luidalepp et al., 2016). Also, during limited nutrient availability, accumulation of truncated mRNA and deacylated tRNA occurs. The ribosomes become stuck on these mRNAs and owing to the absence of a stop codon, the ribosome is unable to get released (Pletnev et al., 2015). These mechanisms are understood to be the defense response upon starvation. As a result of the various morphological, metabolic, transcriptional, or translational alterations, the stationary phase cells become resistant to high temperature, high concentrations of H_2O_2, and very high medium osmolarity.

Cells in exponential, stationary, and long-term stationary phases have different fates (**Figure 2**). As a consequence of starvation, many bacteria including the genera *Bacillus* and *Clostridium* form resistant spores helping them withstand the harsh surrounding environment. Non-optimal growth conditions also lead to the formation of biofilm in many bacterial species. Physiologically, biofilm bacteria are similar to stationary phase bacteria. One key transition is the formation of persisters induced during stationary phase, in biofilms, and also as a consequence of a general stress response. These cells could also arise in exponential growth by the activation of ppGpp as a consequence of sub-lethal antibiotic concentration. The formation of these bacterial persisters is understood to be the reason behind relapsing infections and is a major cause of drug resistance (Harms et al., 2016).

During the late stationary phase sometimes referred to as long-term stationary phase, several remarkable adaptations take place. On continued starvation, one of the survival strategies includes bacteria entering a viable but non-culturable state (VBNC). In this state, bacteria remain metabolically active but fail to form colonies on bacteriological media. Several bacteria including *Rhodococcus biphenylivorans* (Su et al., 2015), *Escherichia coli, Agrobacterium tumefaciens, Helicobacter pylori, Lactococcus lactis,* many *Vibrio* species, and *Pseudomonas* species have been shown to enter the VBNC state (Oliver, 2005). The VBNC state poses a serious health risk as the dormant bacterial species could remain undetected in culturable conditions, though having the ability to cause infections (Navarro Llorens et al., 2010). A variety of stresses is said to lead to the manifestation of VBNC state (Pletnev et al., 2015). Prolonged starvation also results in Growth Advantage in Stationary Phase (GASP) phenotype. The GASP phenomenon is a result of mutations in the *rpoS* allele (described later) which confers a gainful ability to continue growing during starvation conditions, thus replacing the parental population (Navarro Llorens et al., 2010). These mutations allow the mutants to effectively scavenge the nutrients released by dead cells (Zambrano and Kolter, 1996). A number of Gram-positive bacteria such as *Listeria monocytogenes* (Bruno and Freitag, 2011), *Staphylococcus aureus, Enterococcus faecalis,* and *Bacillus globigii* (Finkel et al., 1997) and Gram-negative bacteria including *Campylobacter, Geobacter, Vibrio, E. coli, Pseudomonas,* etc., have been found to enter the GASP state (Chen and Chen, 2014). Gefen et al. (2014) coined the term 'constant activity stationary phase' (CASP) to describe the phenomenon of constant rate of protein synthesis observed in non-growing bacteria that have undergone over more than 60 h of starvation. On studying the protein production at this stage, they have found that both the protein synthesis machinery including ribosomes, RNA polymerases, etc., and resources such as amino acids, nucleotides, etc., remain constant at CASP. Finally, constant promoter activity was observed in this experiment for up to 10 h of starvation. Another interesting phenomenon experienced by bacterial population in stationary phase is the 'stationary phase contact-dependent inhibition' (SCDI). It requires physical contact between the evolved and original bacteria (Lemonnier et al., 2008). In this process, it was observed that the evolved strains either killed or inhibited the growth of bacteria that they were derived from. The inhibiting ability of these strains is attributed to mutations within a single gene involved in glycogen synthesis pathway: *glgC* (encoding ADP-glucose pyrophosphorylase). Astonishingly, all evolved strains overproduced glycogen which seemed to be necessary for SCDI to occur (Navarro Llorens et al., 2010).

ALTERNATIVE SIGMA FACTORS ACTIVE AT STATIONARY PHASE

A key regulator of stationary phase gene expression in *E. coli* is the transcription factor σ^S [a product of *rpoS* (*katF*) gene]. The *E. coli* genome was found to contain two genes *katE* and *katG* encoding for HPII and HPIw1-4x catalases. The expression of HPII was

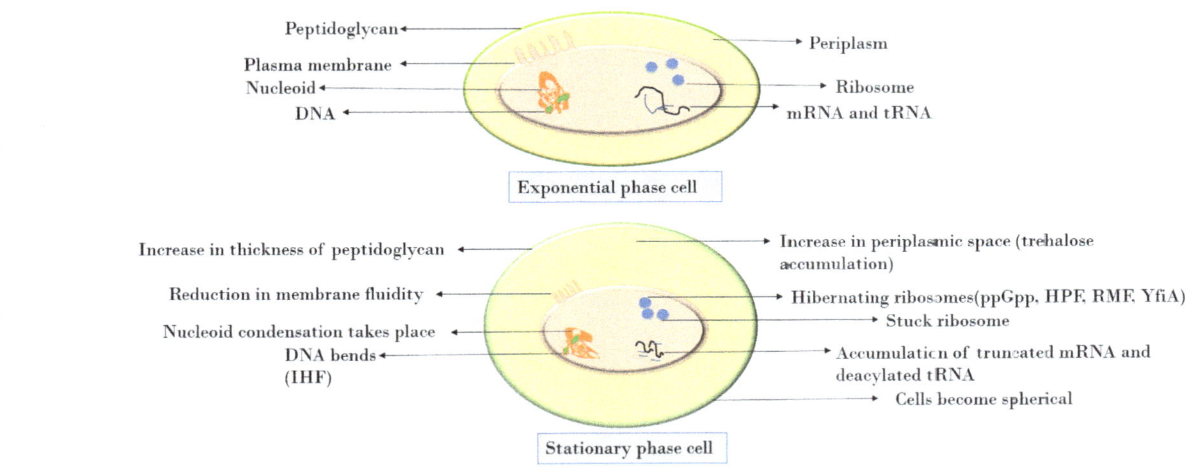

FIGURE 1 | Comparison of molecular and cellular changes in exponential vs stationary phase.

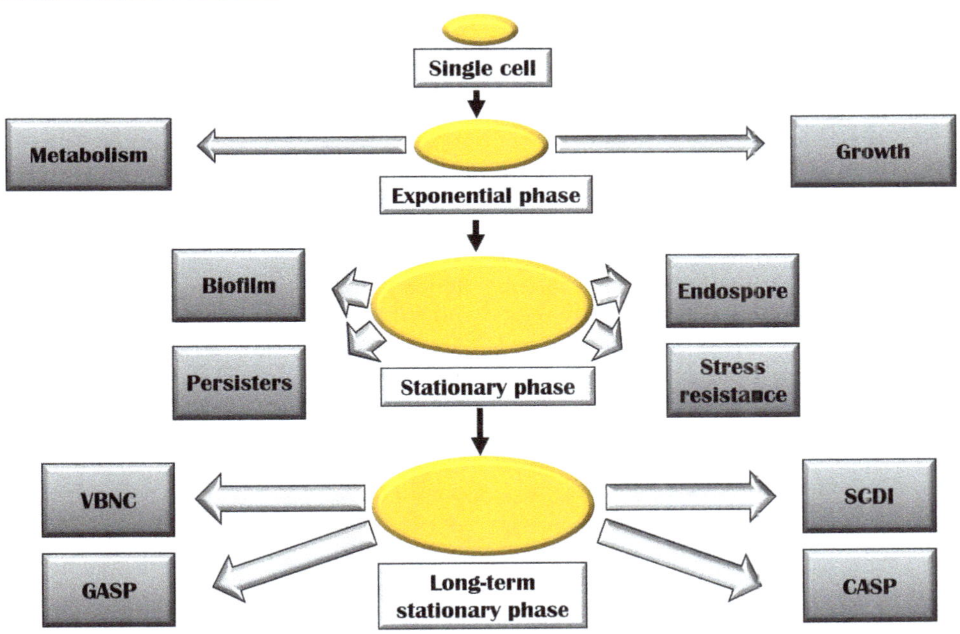

FIGURE 2 | Various bacterial adaptations at stationary and long-term stationary phase. Abbreviations are described n the text.

highest in stationary phase and has been shown to be completely dependent on *katF* gene product. The latter serves as sigma factor for RNA polymerase and therefore named as *rpoS* or σ^S or σ^{38} or stationary phase sigma factor or starvation sigma factor (Tanaka et al., 1997).

The amount of σ^S remains relatively low in the growing phase of cells but increases markedly when the cell encounters stress, starvation or enters stationary phase. The role of this protein is to aid in survival and improved resistance to stressful conditions. Induction of σ^S is observed under conditions of low pH, heat or cold shock, UV-induced DNA damage, nutrient starvation, high cell density, high osmolarity, etc. (Hengge, 2011). The σ^S-dependent genes have been attributed to morphological changes (Hengge, 2011), induction of starvation proteins (Alexander and St. John, 1994), iron uptake, carbohydrate metabolism, amino acid transport, and so on, at the onset of stationary phase (Lacour and Landini, 2004).

The *rpoS* sigma factor is selectively utilized in stationary phase. The major sigma factor rpoD (σ^{70}) is inhibited by a regulator of sigma D (Rsd). The rationale for σ^S selectivity *in vivo* is not completely understood, but it is known that many promoters can exhibit both σ^S and σ^{70} mediated expression *in vitro*. It is well known that σ^{70} is affected by changes in spacer region and consensus −10 and −35 positions, but the alternative σ^S is shown to be less affected by changes in these regions, thus making it more selective *in vivo* (Hengge, 2011). Another observation by Tanaka et al., 1995 indicates that the −35 region is not always required for stationary-phase expression (Tanaka et al., 1995). In this study, the *fic* promoter was shown to function with promoter sequences downstream from −17. Also, the promoters recognized by RpoS are found to contain curved DNA region. Hence, the absence of consensus −35 and the presence of curved DNA region imparted σ^S dependence to galP1 and galP2 promoters, whereas the presence of −35 sequence in the same promoter changed the specificity toward σ^{70} (Kolb et al., 1995). Thus, the general belief is that the σ^S promoters lack a −35 consensus sequence. However, some authors have suggested CTGCAA (Bohannon et al., 1991) or CCGACA (Wise et al., 1996) as the −35 consensus sequence. Similarly for −10, Hengge-Aronis (1993) has suggested a consensus sequence of TATACT, which was later changed to CTATACT (Espinosa-Urgel et al., 1996). More recently, a long consensus sequence KCTAYRCTTAA for −10 region has been proposed, where K could be T or G, Y could be T or C, and R could be A or G (Weber et al., 2005). Not all the stationary-phase induced genes depend on σ^S, and out of the many genes that show higher level of expression in the stationary phase, only 10% is known to be dependent on σ^S (Rava et al., 1999). Out of the genes induced in stationary phase, those that show σ^S independent behavior are *dnaK*, *groEL*, *htpG* which depend on σ^{32} (Kolter et al., 1993).

Several other alternative sigma factors have been reported. In *Salmonella typhimurium*, σ^E has been suggested to serve a complementary role in stationary phase survival. Mutants deficient in *rpoH* gene coding for σ^E have been shown to be susceptible to oxidative stress (Testerman et al., 2002).

The number of sigma factors varies from 1 in *Mycoplasma genitalium* (Dorman, 2011), 6 in *Gordonia* sp. IITR100 (Jaishankar et al., 2017), 7 in *E. coli* (Ishihama, 1997), 18 in *B. subtilis* (Gruber and Gross, 2003), 24 in *Pseudomonas aeruginosa* (Potvin et al., 2008), and 65 in *Streptomyces coelicolor* (Kim et al., 2008). **Table 1** gives a list of various sigma factors in well-known bacterial species and the types of sigma factors upregulated at stationary phase.

REGULATION OF RpoS

The RpoS is regulated at post-transcriptional level by *rpoS* mRNA secondary structure, small RNAs, Hfq, and HU proteins, ClpXP protease and RssB (phosphorylation-modulated RpoS recognition factor) (Hengge-Aronis, 2002). The *rpoS* mRNA is stimulated by regulatory factors such as Hfq (HF-1) protein and DsrA (small regulatory RNA) and repressed by H-NS (histone-like protein) and *oxyS* RNA. The 5′ UTR of *rpoS* mRNA forms a loop which represses its translation. This loop can be disrupted by non-coding RNAs such as *dsrA*, *rprA*, and *arcA* (Gaida et al., 2013). Another sRNA which positively regulates *rpoS* mRNA is *gcvB* (Jin et al., 2009).

The turnover of RpoS protein in exponential phase is very high with a half-life of 1.4 min (Lange and Hengge-Aronis, 1994). The RpoS protein is stable in stationary phase.

The levels of RpoS are also controlled by a number of other factors. These include both positive regulators such as ppGpp and polyphosphate (polyp) and negative regulators such as cAMP and UDP glucose.

The availability of ppGpp is dependent on RelA, a ppGpp synthase that is associated with ribosomes. In stationary phase, when the uncharged tRNAs accumulate due to decreased availability of amino acids, *relA* is turned on and synthesizes ppGpp. This turns on the promoters involved in amino acid biosynthesis and uptake (Barker et al., 2001). It has been shown that 6S RNA regulates *relA* gene expression, which leads to alteration in ppGpp levels in stationary phase (Cavanagh et al., 2010). The rRNA genes are turned off by ppGpp. Many stationary phase promoters (SPPs) are also regulated by 6S RNA, even in the absence of ppGpp.

In *B. subtilis*, it has been demonstrated that cells entering in stationary phase have small GTP and GDP pools. This is possibly due to conversion of GTP to (p)ppGpp or due to the lack of sufficient precursors available for nucleotide synthesis. Lopez and coworkers demonstrated that treatment of cells with decoyinine, an inhibitor of GMP synthase, can result in induction of stationary phase genes (Ratnayake-Lecamwasam et al., 2001).

The intracellular levels of certain compounds such as trehalose, glycine betaine, glycogen, and polyphosphate are high under stress conditions. Some of these compounds modulate function of the RpoS holoenzyme. For example, glutamate and trehalose modulate the holoenzyme binding to promoters. Similarly, altered promoter selectivity has been observed in *E. coli* when RpoS associates with inorganic polyphosphate. The inhibition due to PolyP is relieved by high concentrations of potassium glutamate (Shimada et al., 2004). Bacterial pheromone, Homoserine lactone (HSL), a small molecule responsible for communication between bacteria, also affects the concentration of σ^S in the cell. Mutants in the biosynthetic pathway for synthesis of HSL loose the ability to induce σ^S (Zambrano and Kolter, 1996).

EXPRESSION OF GENES IN STATIONARY PHASE

When the cells are growing, the metabolism-linked genes are highly expressed, and get turned off when the cells enter stationary phase. The stationary phase is a period of no growth, however, genes essential for survival of organisms are expressed at this stage. Around 20% of the genes of *E. coli* are found to express at higher level in the stationary phase (Rava et al., 1999). These genes are directly linked to many key events including DNA repair, glycogen production, thermotolerance, osmotolerance, etc. (Bohannon et al., 1991; Ishihama, 1997). Transcriptome profiling/expression analysis of *E. coli* in stationary phase revealed upregulation of genes which are involved in survival during osmotic stress (*ots*, *tre*, *osm*), long-term survival (e.g., *bolA*, *dps*, *cbpA*, and *glgS*), periplasmic shock (*rpoE* and *rseA*),

TABLE 1 | List of sigma factors upregulated at stationary phase in different bacteria.

Name of the organism	Sigma factors	Sigma factors upregulated at stationary phase	Reference
Escherichia coli	**7** $\sigma^{70}(\sigma^{D})$, $\sigma^{24}(\sigma^{E})$, $\sigma^{28}(\sigma^{F})$, $\sigma^{32}(\sigma^{H})$, $\sigma^{54}(\sigma^{N})$, $\sigma^{38}(\sigma^{S})$, $\sigma^{18}(\sigma^{FecI})$	$\sigma^{32}(\sigma^{H})$, $\sigma^{54}(\sigma^{N})$, $\sigma^{38}(\sigma^{S})$	Ishihama, 1997
Bacillus subtilis	**18** σ^{A}, σ^{B}, σ^{C}, σ^{D}, σ^{E}, σ^{F}, σ^{G}, σ^{H}, σ^{K}, σ^{L}, σ^{M}, σ^{V}, σ^{W}, σ^{Y}, σ^{X}, σ^{Z}, Xpf, YlacC,	σ^{B}, σ^{C}, σ^{D}, σ^{H}	Haldenwang, 1995; Gruber and Gross, 2003
Streptomyces coelicolor	**65**	σ^{B}, σ^{F}, σ^{H}, σ^{M}, σ^{N}, σ^{R}	Kim et al., 2008; Tripathi et al., 2014
Pseudomonas aeruginosa	**24**	σ^{E}, σ^{H}, σ^{S}	Potvin et al., 2008
Corynebacterium glutamicum	**7** σ^{A}, σ^{B}, σ^{C}, σ^{D}, σ^{E}, σ^{H}, σ^{M}	σ^{B}, σ^{H}, σ^{M}	Pátek and Nešvera, 2013

cold shock (*csp* genes), etc. Other genes include carbon storage regulator (*csrA*), trp repressor binding protein (*wrbA*) and universal stress protein (*uspA*) (Chang et al., 2002). Moreover, several antibiotics including lactocin B of lactic acid bacteria, alfatoxin of *Aspergillus* species are produced mainly in stationary phase (Matin, 1992).

Persister cell formation has also been attributed to genes differentially expressed in stationary phase. These cells are recalcitrant to antibiotic treatments and often are the major cause of drug resistance. Several polyamines including putrescine, spermidine, and cadaverine direct persister formation through upregulation of genes such as *rpoS*, *rmf*, *yqjD* (Tkachenko et al., 2017). This observation suggests that polyamine metabolism participates in the regulation of persister cells formation. To determine the genes upregulated at stationary phase microarray was done in *Mycobacterium smegmatis* grown under conditions of glycerol and glucose depletion. Different subset of genes were identified that were preferentially upregulated at stationary phase. The categories of genes included those involved in metabolism of sulfur, sigma factors including *sigB*, *sigE*, and *sigH*, fatty acid degradation, anaerobic respiration, etc. (Hampshire et al., 2004). Also, of key interest in this study is the presence of stationary phase operons involving many gene clusters that were significantly upregulated in stationary phase. On investigating further, the presence of other such operons were also found. The *pdh* operon of *Streptococcus mutans* is expressed only in the stationary phase. This operon was observed to be transcribed only by a subpopulation of bacteria in stationary phase and was vital for survival during long periods of sugar starvation. The *pdh* operon consists of four genes that are transcribed as an operon: *pdhD*, *pdhA*, *pdhB*, *pdhC*, which encode the components of PDH (pyruvate dehydrogenase) complex, i.e., pyruvate dehydrogenase (two subunits encoded by *pdhA* and *pdhB*), dihydrolipoyl transacetylase (*pdhC*), and dihydrolipoyl dehydrogenase (*pdhD*). The inactivation of the first gene: *pdhD* resulted in impaired survival in both batch cultures and biofilms (Busuioc et al., 2010). Similarly, phage shock protein operon (*pspABCE*) of *E. coli* was reported to be critical for survival under prolonged stationary phase at alkaline conditions. This operon was expressed strongly

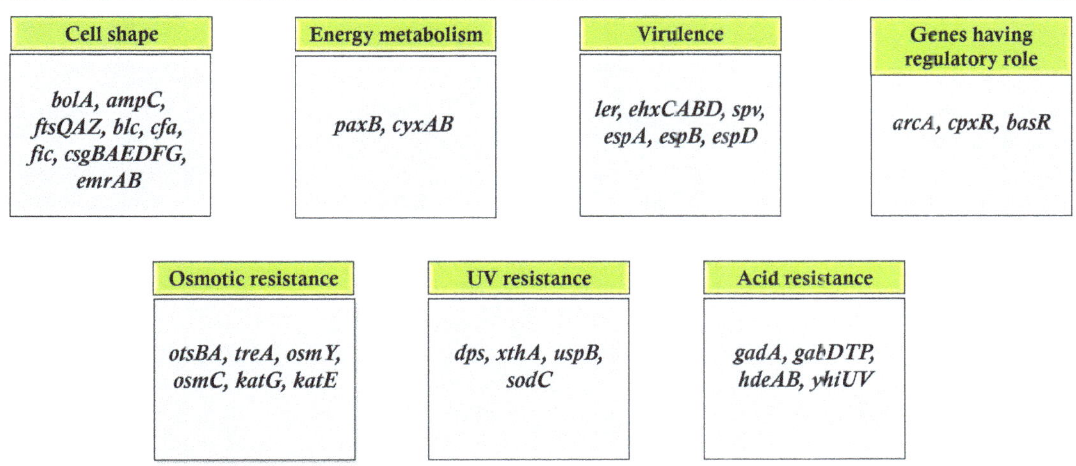

FIGURE 3 | Categories of genes transcribed in stationary phase.

TABLE 2 | Stationary phase promoters in Gram-negative bacteria.

Name of or	Name of promoter	Gene product	–10	–35	Other motifs	Length of spacer	Reference
Escherichia coli	*BolAp1*	BolA	CGGCTAGTA	CTGCAA	–	15	Aldea et al., 1989
	TreA	TreA (Osmotically inducible periplasmic trehalase)	ATGCAG	TAAGGT	–	17	Repoila and Gutierrez, 1991
	Cst-1	Cst	–	–	–	–	Tunner et al., 1992
	Fic	Fic (PABA or folate)	TATACT	–	–	–	Utsumi et al., 1993
	Hns	H-NS	TATTAT	TTGCAC	–	17	Dersch et al., 1993
	PoxB	PoxB (pyruvate oxidase)	TAAACT	–	–25: CGTCA; –60: GTTAGTG	–	Chang et al., 1994
	Slp	Slp	TATTATG	GATGAAA	–	16	Alexander and St. John, 1994
	AldB	AldB (Aldehyde dehydrogenase)	TACCCT	–	–	–	Xu and Johnson, 1995
	CsiD	CsiD (Carbon starvation inducible gene)	TATTTT	TGCGCA	-	17	Marschall et al., 1998
	OsmY (*Csi*-5)	OsmY (Periplasmic protein of unknown function)	TATATT	CGAGCG	–	15	Lange et al., 1993; Becker and Hengge-Aronis, 2004
Shigella flexneri	*GadA*	GadA (Glutamate decarboxylase)	CTACTTT	–	–	–	Waterman and Small, 2003
Vibrio anguillarum	*EmpA*	EmpA mettaloprotease	GATCCA	CCGTGCTAC		19	Croxatto et al., 2004; Denkin and Nelson, 2004

TABLE 3 | Stationary phase promoters from Gram-positive bacteria.

Organism	Promoter	Gene product	–10	–35	Length of spacer (bp)	Reference
Bacillus subtilis	*Pst*	Phosphate-specific transport	TTTACT	TTCAAA	18	Qi et al., 1997
Bacillus subtilis	*Cry3a*	Crystal proteins	TAAGCT	TTGCAA	18	Lee et al., 2010
Bacillus subtilis	*Ylb*	–	TACAAT	TTGGA	18	Yu et al., 2015
Bacillus subtilis	*SrfA* mutant	Srf operon (lipopeptide antibiotic surfactin)	TTGACT	TATAAT	–	Guan et al., 2016
Streptomyces coelicolor	*KasO* mutant	Colemycin P1	TAAAGT	TTGACA	18	Wang et al., 2013
Corynebacterium glutamicum	*Cg3141* mutant	*Cg3141 (flavohemoprotein)*	TGGGAT	TTAAGG	17	Kim et al., 2016
Gordonia sp. IITR100	Stationary phase promoter	–	AATAAT	TTAACT	22	Singh et al., 2016

under extreme stressful conditions and remained significant for survival under nutrient-limited conditions (Weiner and Model, 1994). Categories of genes that are preferentially upregulated in stationary phase is shown in **Figure 3**. Studies have demonstrated that starved cells exhibit more protective resistance to different stresses as compared to resistance induced during growing stage by non-lethal exposure of stresses (Kolter et al., 1993).

STATIONARY PHASE PROMOTERS

The genes expressed in stationary phase are controlled by promoters, which result in induction of stationary phase. The promoters, which are turned on, are called SPPs. They are recognized by RNA polymerase holoenzyme containing σ^S and therefore called RpoS.

The vast importance of SPPs had been realized way back in 1980s with the study of *mcbA* promoter and *bolA* P1 promoter of *E. coli* (Connell et al., 1987; Aldea et al., 1989). The *mcbA* promoter causes increased level of transcription initiation for Microcin B17, a DNA replication inhibitor. Promoter *mcbA*-LacZ fusion showed the induction of transcription in nitrogen, phosphate, and carbon starvation conditions. Similarly, *bolA-lacZ* fusion demonstrated an increase in expression of approximately 10- to 20-fold during transition to stationary phase. Since then many SPPs have been isolated and characterized in both Gram-positive and Gram-negative bacteria (**Tables 2, 3**). Particularly regarding *E. coli* and *B. subtilis*, the stationary-phase-specific

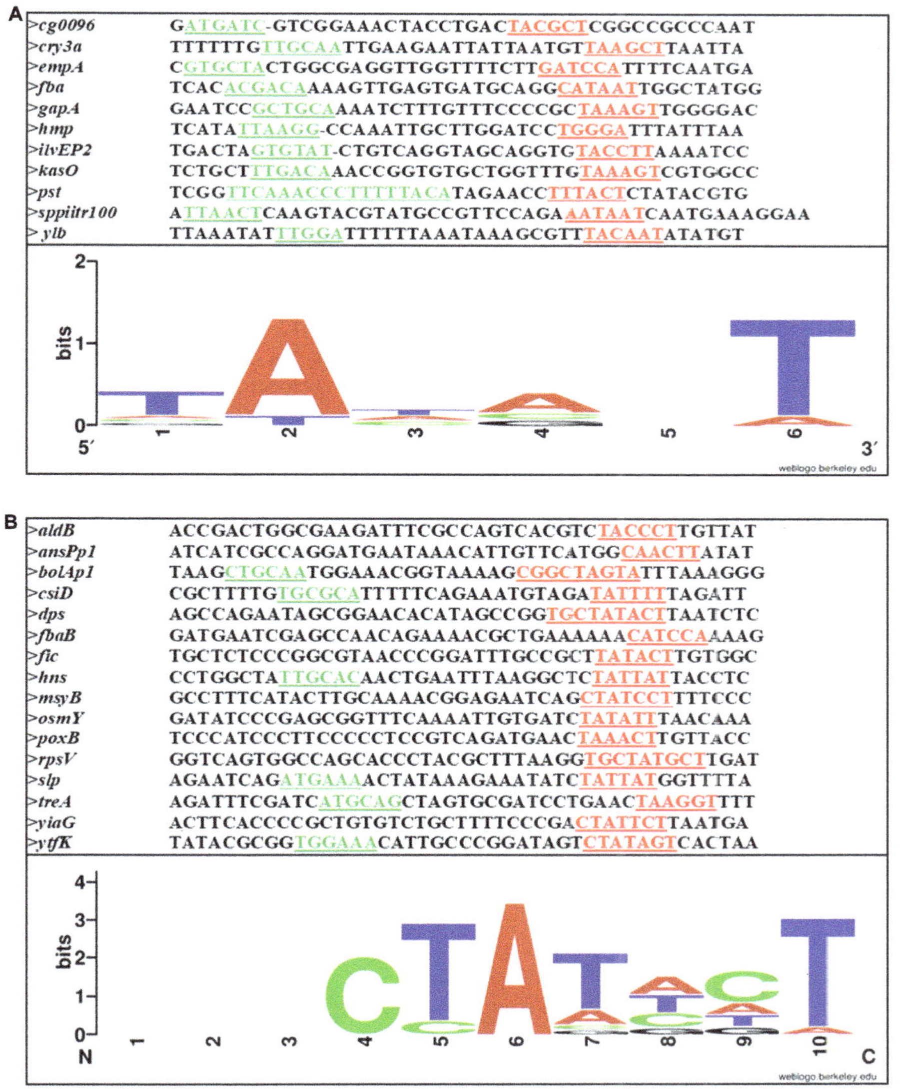

FIGURE 4 | Sequence alignment of **(A)** Gram-positive and **(B)** Gram-negative stationary phase promoters. −10 and −35 are underlined and shown in red and green, respectively. The conserved bases are shown below.

gene regulation has been intensively studied (Hengge, 2011).

On analysis of the different SPPs, our observation is that there is not much variation between this class of promoters and σ^{70} promoters. It is the sequence outside the −10 and −35 regions that distinguish between σ^{70}- and σ^{S}-dependent promoters. **Figures 4A,B** shows the −10, −35 and spacer region of few SPPs from Gram-positive and Gram-negative bacteria and the consensus sequence at the −10 region is shown as a logo designed using WebLogo software available online (Crooks et al., 2004).

Among the SPPs exist a special class of promoters known as the gearbox promoters which include *mcbAp, bolAp1, ftsQp* to name a few. This class of promoters has been studied in several Gram-negative bacteria including *E. coli*. Two different highly conserved consensus −10 and −35 sequence have been proposed by Aldea et al. (1993) for this class of promoters: CTGCAA or GTTAAGC at −35 position and CGGCAAGTA or CGTCC at −10 position. Gearbox promoter-induced gene expression seems to correlate inversely with growth rate and these promoters may or may not depend on σ^{S}.

ENERGY RESERVES CONSUMED DURING STATIONARY PHASE AND SOURCE OF NUTRIENTS FOR PROTEIN PRODUCTION

During unfavorable conditions of growth, reprogramming the cellular machinery for sustaining viability is a natural process of adaptation. Reserve polymers like glycogen and poly-β-hydroxybutyric acid that are accumulated by bacteria during growth are rapidly consumed during conditions of carbon starvation to ensure survival. In case of bacteria that do not accumulate these polymers, cellular RNA is rapidly degraded for energy generation (Matin, 1992). Among RNA, rRNA is preferentially degraded (Deutscher, 2003). Besides, 50% of ribosomes synthesized during exponential growth are degraded during entry to stationary phase (Piir et al., 2011). What is surprising is that, when in stationary phase, these ribosomes are fairly stable and so degradation occurs only in between the stages.

The yield of protein production from stationary phase systems is as high as 121% as compared to their log phase counterparts (Ou et al., 2004). This raises a very important question: What makes protein synthesis possible at stationary phase?

Balaban and coworkers devised a microfluidic device and followed the production of fluorescent proteins at stationary phase. They found that cells after entering stationary phase continue to produce proteins for several days (Gefen et al., 2014). It has been suggested that cells continue to produce proteins at stationary phase by reusing amino acids from degraded proteins. Moreover, the biosynthetic pathway of a few amino acids including serine, aspartate/asparagine, glutamine/glutamate, and alanine were shown to be active during stationary phase (Shaikh et al., 2010). In addition, it is shown that each condition resulting in starvation results in induction of specific set of proteins (Kolter et al., 1993).

DEVELOPMENT OF GENE EXPRESSION SYSTEMS USING STATIONARY PHASE PROMOTERS

A strong promoter is the key for developing efficient gene expression systems. For recombinant protein production, several bacterial hosts have been used as cell factories, with features such as easy purification, improved protein folding and secretion, high production of membrane proteins, etc. (Ferrer-Miralles and Villaverde, 2013). To develop more such expression systems in bacteria, it is necessary to ensure proper selection of a promoter that would drive the expression of genes at the right time and with maximum amount.

Promoters could be classified as constitutive or inducible, growth-stage limited, tissue specific, etc. Inducible promoters can further be classified into inducer-specific and auto-inducible promoters. Constitutive promoters are not useful for toxic proteins. Inducer-specific promoters involve the cost of inducer. Also, some chemical inducers such as Isopropyl-β-D-1-thiogalactopyranoside (IPTG) are expensive and toxic (Cao and Xian, 2011). Further, the addition of external inducers often requires growth monitoring which is vital for productivity and hence lead to difficulty in fermentation.

Auto-inducible promoters are ideal for large-scale protein production as they are induced at late log phase or stationary phase. Such promoters induce expression of the recombinant gene without any inducer and thus are economical. However, most of them have low activity (Yu et al., 2015). In *B. subtilis*, Fan and coworkers successfully identified a strong SPP Pylb by microarray approach (Yu et al., 2015). The β-galactosidase activities were observed to be up to 5000 miller units. The authors have proposed that such a promoter will be useful for protein production. A SPP-based auto-inducible gene expression system has been constructed using cry3Aa promoter. The Pcry drives the expression of crystal proteins in *B. thuringiensis*. The promoter cry3Aa was tested in *B. subtilis* and the wild type have the LacZ levels up to 1000 miller units and on mutagenesis resulted in levels up to 5200 miller units (Lee et al., 2010). Similarly, in another Gram-positive bacteria, *Gordonia* sp. IITR100, a SPP was identified and the β-galactosidase activities were up to 600 miller units (Singh et al., 2016). However, the β-galactosidase activities vary with respect to strain, copy number of plasmid, growth medium, temperature, etc., so it is difficult to assess the strength of promoter based on Miller units alone. In future, a study of such promoters based on the number of transcripts would be useful to compare the strength.

In *Corynebacterium glutamicum*, promoter of *cg3141* gene coding for flavohemoprotein was found to show higher inducibility in the stationary phase. Then, a synthetic promoter library was prepared to change the spacer and flanking regions in the promoter, to obtain a range of promoter strengths (Kim et al., 2016). At the end, one of the synthetic promoters that showed up to 20-fold higher strength compared to the original *cg3141* promoter was obtained and demonstrated for fed-batch cultivation of glutathione S-transferase in a 5L reactor. **Table 4** depicts the list of SPP-based expression vectors constructed till date. Studies like these indicate that the potential of SPPs is phenomenal. In *Streptomyces*, a high-level recombinant protein expression system has been patented (US Patent No. 7,316,914).

APPLICATIONS

SPPs have immense potential for use in many industries (**Figure 5**).

Recombinant production of toxins whose overproduction is detrimental to the growth of cells needs controlled conditions for expression. In such cases, the use of SPP is advantageous as the overproduction will not affect the growth of the host cells. Many bacteria have been used to demonstrate the utility of cell-density-dependent expression systems for heterologous protein production. Metabolic engineering of bacteria for enhanced production of industrially important chemicals has been carried out since a long time. The *fic* promoter of *E. coli* was used to express *phlD* gene at a higher titer in stationary phase, without the addition of any inducer, for the production of phloroglucinol, which has utility in pharmaceutical industry and plant tissue culture. After 20 h of cultivation in a flask with shaking, 9% of glucose supplied had converted to phloroglucinol with a productivity of 0.014g/l h (Cao and Xian, 2011). *B. subtilis* has been engineered for overproduction of aminopeptidase using a mutated P*srfA* system and has resulted in 87.89 U/ml of enzyme activity (Guan et al., 2015). Using *B. subtilis*, a *cry*-promoter-based system was developed wherein

TABLE 4 | Stationary phase promoter–based gene expression systems reported from Gram-negative and Gram-positive bacteria.

Vector	Promoter	Organism	Ori	Antibiotic resistance	Reporter gene	Reference
pTGV1	*BolAp*	*E. coli*	pBR322	Ampicillin	*bolA-LacZ*	Aldea et al., 1989
pFL1	*Fic*	*E. coli*	pMB1	Tetracycline	*LacZ*	Utsumi et al., 1993
pKS4, pKS5	*Hns*	*E. coli*	pSC101	Tetracycline	*Hns-LacZ*	Dersch et al., 1993
pNH5	*OsmY (Csi5)*	*E. coli*	pBR322	Ampicillin	*Csi5-LacZ*	Lange et al., 1993
pMC719	*Slp*	*E. coli*	pBR322	Tetracycline	*Slp-LacZ*	Alexander and St. John, 1994
pYYC128	*PoxB*	*E. coli*	pBR322	Chloramphenicol	*PoxB-LacZ*	Chang et al., 1994
pRJ4025	*AldB*	*E. coli*	pBR322	Ampicillin	*LacZ*	Xu and Johnson, 1995
pYQ23	*Pst*	*Bacillus subtilis*	pBR322	Ampicillin, Chloramphenicol	*LacZ*	Qi et al., 1997
pCM3	*CsiD*	*E. coli*	pMB1	Ampicillin	*Csi-LacZ*	Marschall et al., 1998
pDM35-EmpA	*EmpA*	*Vibrio anguillarum*	R6K	Chloramphenicol	*LacZ*	Croxatto et al., 2004; Denkin and Nelson, 2004
pGRP (many promoters)	Many	*E. coli*	pBR322	Ampicillin	green fluorescent protein (*eGFP*)	Shimada et al., 2004
pD82-aprE	*Cry3a*	*Bacillus subtilis*	–	Chloramphenicol	*AprE-LacZ*	Lee et al., 2010
pMD-ficD	*Fic*	*E. coli*	pBR322	Ampicillin	*phlD*	Cao and Xian, 2011
pDR4-K*	*KasO*	*Streptomyces coelicolor*	EBV origin	Hygromycin	*XylE-neo*	Wang et al., 2013
pBSG03	*SrfA*	*Bacillus subtilis*	pBR322	Ampicillin, Kanamycin, Neomycin	*no*	Guan et al., 2015
Pylb-bgaB-pUBC19	*Ylb*	*Bacillus subtilis*	pUBC19	Ampicillin	Beta-gal (*bgaB*)	Yu et al., 2015
pSC1	Stationary phase promoter	*Gordonia sp.* IITR100	pRC4	Kanamycin	*LacZ*	Singh et al., 2016
pCES-P_4-N_{14}-sfGFP	P_4-N_{14}	*Corynebacterium glutamicum*	pCG1	Kanamycin	*sfGFP*	Kim et al., 2016

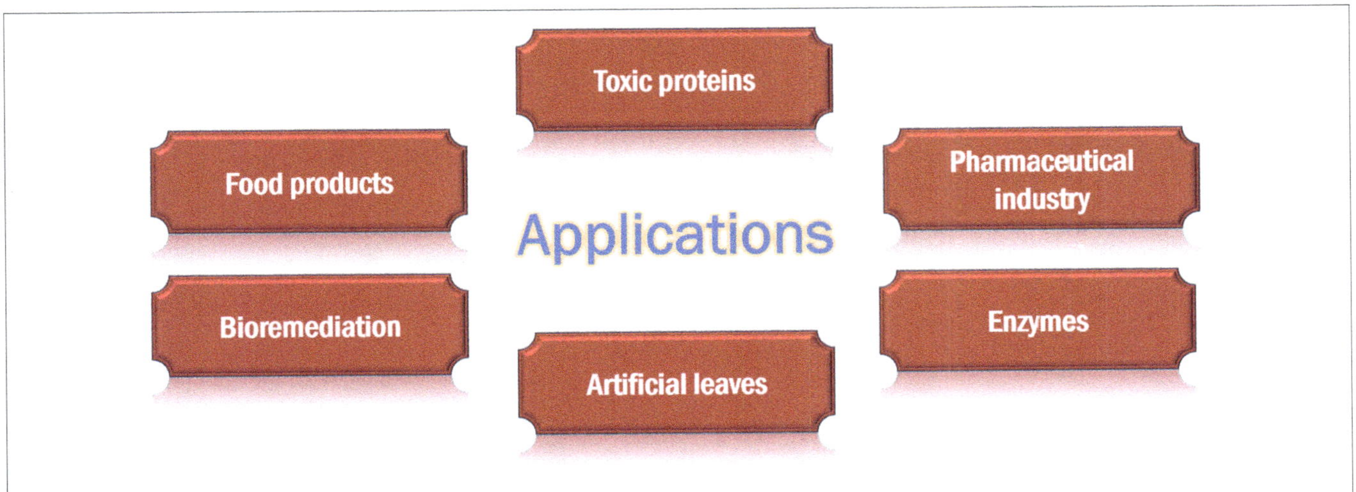

FIGURE 5 | Applications of stationary phase gene expression systems.

cellulose and alkaline protease were produced with a higher yield as compared to the wild-type cry3A promoter (Lee et al., 2010).

It is a well-known fact that the non-growing phase of lactic acid bacteria accounts for a major proportion of flavor production in lactic acid bacteria (van de Bunt et al., 2014).

Therefore, engineering bacterial cells in such a way that they are expressed at high levels, during the ripening process, by using SPPs would enhance their applicability in food industry.

In the bioremediation industry, microorganisms have routinely been employed for removing pollutants. Due to low nutrient availability in polluted sites, genetic engineering of cells resulting in higher enzymatic activities at lower growth rates have been shown to be highly efficient for bioremediation process. On studying the phenol degradation capability of two non-growing recombinant *E. coli* strains, it was found that the *groEL*-promoter-driven gene expression system caused 75% phenol degradation while the *tac*-promoter-driven expression could cause only 15% degradation of phenol (Matin, 1992). As suggested by Tunner et al. (1992), it is possible to use starvation-induced promoters for chemical waste biodegradation wherein enzymes can be induced naturally by bacteria due to the occurrence of nutrient-limited conditions in the environment. This could save the cost of induction thereby increasing the efficiency of the process.

In a very interesting experiment, *Rhodospirillum rubrum* cells grown photoheterotrophically, evolved hydrogen for about 70 h after growth ceased (Melnicki et al., 2008). Similarly, a purple non-sulfur photosynthetic bacterium, *Rhodopseudomonas palustris* under nitrogen starvation conditions, produced hydrogen gas for over 4000 h thus paving way for creation of 'artificial leaves' (Gosse et al., 2010).

CONCLUSION AND FUTURE PROSPECTS

Stationary phase survival is a means of bacterial adaptation by which bacteria survive under conditions of stress or starvation. The ugly aspect of this is that such a mechanism results in the persistence of pathogenic bacteria which can cause relapsing of infections. However, the good side is represented by the various biotechnological applications that have come up recently based on the promoters of the genes which are upregulated at stationary phase. In the present review, we have discussed not only the changes at the cellular and molecular levels at stationary phase, but also the various promoters characterized, their regulation and the gene expression systems developed. There are still many unknowns. For example, very little is known about the proteins which are involved in chromosome organization and their interaction with DNA at stationary phase. Such proteins could be important players in regulating gene expression at stationary phase. Further very few SPPs have been experimentally characterized till date. Such promoters should be highly useful for protein production as the growth and protein production phase can be uncoupled. This will pave way toward constructing improved gene expression systems for recombinant protein production.

AUTHOR CONTRIBUTIONS

JJ and PS wrote and edited the manuscript.

ACKNOWLEDGMENTS

The authors would like to thank Department of Biotechnology, Government of India for the financial support.

REFERENCES

Aldea, M., Garrido, T., Hernandez-Chico, C., Vicente, M., and Kushner, S. R. (1989). Induction of a growth-phase-dependent promoter triggers transcription of bolA, an *Escherichia coli* morphogene. *EMBO J.* 8, 3923–3931.

Aldea, M., Garrido, T., and Tormo, A. (1993). Gearbox gene expression and growth rate. *World J. Microbiol. Biotechnol.* 9, 414–420. doi: 10.1007/BF00328029

Alexander, D. M., and St. John, A. C. (1994). Characterization of the carbon starvation-inducible and stationary phase-inducible gene slp encoding an outer membrane lipoprotein in *Escherichia coli*. *Mol. Microbiol.* 11, 1059–1071. doi: 10.1111/j.1365-2958.1994.tb00383.x

Barker, M. M., Gaal, T., Josaitis, C. A., and Gourse, R. L. (2001). Mechanism of regulation of transcription initiation by ppGpp. I. Effects of ppGpp on transcription initiation *in vivo* and *in vitro*. *J. Mol. Biol.* 305, 673–688. doi: 10.1006/jmbi.2000.4327

Becker, G., and Hengge-Aronis, R. (2004). What makes an *Escherichia coli* promoter σ^S dependent? Role of the -13/-14 nucleotide promoter positions and region 2.5 of σ^S. *Mol. Microbiol.* 39, 1153–1165. doi: 10.1111/j.1365-2958.2001.02313.x

Bohannon, D. E., Connell, N., Keener, J., Tormo, A., Espinosa-Urgel, M., Zambrano, M. M., et al. (1991). Stationary-phase-inducible "gearbox" promoters: differential effects of katF mutations and role of sigma 70. *J. Bacteriol.* 173, 4482–4492. doi: 10.1128/jb.173.14.4482-4492.1991

Bruno, J. C. Jr., and Freitag, N. E. (2011). Listeria monocytogenes adapts to long-term stationary phase survival without compromising bacterial virulence. *FEMS Microbiol. Lett.* 323, 171–179. doi: 10.1111/j.1574-6968.2011.02373.x

Busuioc, M., Buttaro, B. A., and Piggot, P. J. (2010). The *pdh* operon is expressed in a subpopulation of stationary-phase bacteria and is important for survival of sugar-starved *Streptococcus mutans*. *J. Bacteriol.* 192, 4395–4402. doi: 10.1128/JB.00574-10

Cao, Y., and Xian, M. (2011). Production of phloroglucinol by *Escherichia coli* using a stationary-phase promoter. *Biotechnol. Lett.* 33, 1853–1858. doi: 10.1007/s10529-011-0638-0

Cavanagh, A. T., Chandrangsu, P., and Wassarman, K. M. (2010). 6S RNA regulation of relA alters ppGpp levels in early stationary phase. *Microbiology* 156(Pt 12), 3791–3800. doi: 10.1099/mic.0.043992-43990

Chang, D. E., Smalley, D. J., and Conway, T. (2002). Gene expression profiling of *Escherichia coli* growth transitions: an expanded stringent response model. *Mol. Microbiol.* 45, 289–306. doi: 10.1046/j.1365-2958.2002.03001.x

Chang, Y. Y., Wang, A. Y., and Cronan, J. E. Jr. (1994). Expression of *Escherichia coli* pyruvate oxidase (PoxB) depends on the sigma factor encoded by the *rpoS(katF)* gene. *Mol. Microbiol.* 11, 1019–1028. doi: 10.1111/j.1365-2958.1994.tb00380.x

Chen, H., and Chen, C. Y. (2014). Starvation induces phenotypic diversification and convergent evolution in *Vibrio vulnificus*. *PLOS ONE* 9:e88658. doi: 10.1371/journal.pone.0088658

Connell, N., Han, Z., Moreno, F., and Kolter, R. (1987). An *E. coli* promoter induced by the cessation of growth. *Mol. Microbiol.* 1, 195–201. doi: 10.1111/j.1365-2958.1987.tb00512.x

Crooks, G. E., Hon, G., Chandonia, J. M., and Brenner, S. E. (2004). WebLogo: a sequence logo generator. *Genome Res.* 14, 1188–1190. doi: 10.1101/gr.849004

Croxatto, A., Pride, J., Hardman, A., Williams, P., Camara, M., and Milton, D. L. (2004). A distinctive dual-channel quorum-sensing system operates in *Vibrio anguillarum*. *Mol. Microbiol.* 52, 1677–1689. doi: 10.1111/j.1365-2958.2004.04083.x

Denkin, S. M., and Nelson, D. R. (2004). Regulation of *Vibrio anguillarum* empA metalloprotease expression and its role in virulence. *Appl. Environ. Microbiol.* 70, 4193–4204. doi: 10.1128/AEM.70.7.4193-4204.2004

Dersch, P., Schmidt, K., and Bremer, E. (1993). Synthesis of the *Escherichia coli* K-12 nucleoid-associated DNA-binding protein H-NS is subjected to growth-phase control and autoregulation. *Mol. Microbiol.* 8, 875–889. doi: 10.1111/j.1365-2958.1993.tb01634.x

Deutscher, M. P. (2003). Degradation of stable RNA in bacteria. *J. Biol. Chem.* 278, 45041–45044. doi: 10.1074/jbc.R300031200

Dorman, C. J. (2011). Regulation of transcription by DNA supercoiling in *Mycoplasma genitalium*: global control in the smallest known self-replicating genome. *Mol. Microbiol.* 81, 302–304. doi: 10.1111/j.1365-2958.2011.07718.x

Espinosa-Urgel, M., Chamizo, C., and Tormo, A. (1996). A consensus structure for sigma S-dependent promoters. *Mol. Microbiol.* 21, 657–659. doi: 10.1111/j.1365-2958.1996.tb02573.x

Ferrer-Miralles, N., and Villaverde, A. (2013). Bacterial cell factories for recombinant protein production; expanding the catalogue. *Microb. Cell Fact.* 12:113. doi: 10.1186/1475-2859-12-113

Finkel, S. E., Zinser, E., Gupta, S., and Kolter, R. (1997). "Life and Death in stationary phase," in *Molecular Microbiology. NATO ASI Series (Series H: Cell Biology)*, Vol. 103, eds S. J. W. Busby, C. M. Thomas, and N. L. Brown (Berlin: Springer).

Gaida, S. M., Al-Hinai, M. A., Indurthi, D. C., Nicolaou, S. A., and Papoutsakis, E. T. (2013). Synthetic tolerance: three noncoding small RNAs, DsrA, ArcZ and RprA, acting supra-additively against acid stress. *Nucleic Acids Res.* 41, 8726–8737. doi: 10.1093/nar/gkt651

Gefen, O., Fridman, O., Ronin, I., and Balaban, N. Q. (2014). Direct observation of single stationary-phase bacteria reveals a surprisingly long period of constant protein production activity. *Proc. Natl. Acad. Sci. U.S.A.* 111, 556–561. doi: 10.1073/pnas.1314114111

Gosse, J. L., Engel, B. J., Hui, J. C. H., Harwood, C. S., and Flickinger, M. C. (2010). Progress toward a biomimetic leaf: 4,000 h of hydrogen production by coating-stabilized nongrowing photosynthetic *Rhodopseudomonas palustris*. *Biotechnol. Prog.* 26, 907–918. doi: 10.1002/btpr.406

Gruber, T. M., and Gross, C. A. (2003). Multiple sigma subunits and the partitioning of bacterial transcription space. *Annu. Rev. Microbiol.* 57, 441–466. doi: 10.1146/annurev.micro.57.030502.090913

Guan, C., Cui, W., Cheng, J., Zhou, L., Guo, J., Hu, X., et al. (2015). Construction and development of an auto-regulatory gene expression system in *Bacillus subtilis*. *Microb. Cell Fact.* 14, 150. doi: 10.1186/s12934-015-0341-2

Guan, C., Cui, W., Cheng, J., Zhou, L., Liu, Z., and Zhou, Z. (2016). Development of an efficient autoinducible expression system by promoter engineering in *Bacillus subtilis*. *Microb. Cell Fact.* 15, 66. doi: 10.1186/s12934-016-0464-0

Haldenwang, W. G. (1995). The sigma factors of *Bacillus subtilis*. *Microbiol. Rev.* 59, 1–30.

Hampshire, T., Soneji, S., Bacon, J., James, B. W., Hinds, J., Laing, K., et al. (2004). Stationary phase gene expression of *Mycobacterium tuberculosis* following a progressive nutrient depletion: a model for persistent organisms? *Tuberculosis* 84, 228–238. doi: 10.1016/j.tube.2003.12.010

Harms, A., Maisonneuve, E., and Gerdes, K. (2016). Mechanisms of bacterial persistence during stress and antibiotic exposure. *Science* 354:aa4268. doi: 10.1126/science.aaf4268

Hengge, R. (2011). Stationary-phase gene regulation in *Escherichia coli*. *EcoSal Plus* 4. doi: 10.1128/ecosalplus.5.6.3

Hengge-Aronis, R. (1993). Survival of hunger and stress: the role of *rpoS* in early stationary phase gene regulation in *E. coli*. *Cell* 72, 165–168. doi: 10.1016/0092-8674(93)90655-A

Hengge-Aronis, R. (2002). Signal transduction and regulatory mechanisms involved in control of the sigma(S) (RpoS) subunit of RNA polymerase. *Microbiol. Mol. Biol. Rev.* 66, 373–395. doi: 10.1128/MMBR.66.3.373-395.2002

Ishihama, A. (1997). Adaptation of gene expression in stationary phase bacteria. *Curr. Opin. Genet. Dev.* 7, 582–588. doi: 10.1016/S0959-437X(97)80003-2

Jaishankar, J., Singh, P., and Srivastava, P. (2017). Draft genome sequence of a biodesulfurizing bacterium, *Gordonia* sp. strain IITR100. *Genome Announc.* 5:e00230-17. doi: 10.1128/genomeA.00230-17

Jin, Y., Watt, R. M., Danchin, A., and Huang, J. D. (2009). Small noncoding RNA GcvB is a novel regulator of acid resistance in *Escherichia coli*. *BMC Genomics* 10:165. doi: 10.1186/1471-2164-10-165

Kim, E. S., Song, J. Y., Kim, D. W., Chater, K. F., and Lee, K. J. (2008). A possible extended family of regulators of sigma factor activity in *Streptomyces coelicolor*. *J. Bacteriol.* 190, 7559–7566. doi: 10.1128/JB.00470-08

Kim, M. J., Yim, S. S., Choi, J. W., and Jeong, K. J. (2016). Development of a potential stationary-phase specific gene expression system by engineering of SigB-dependent cg3141 promoter in *Corynebacterium glutamicum*. *Appl. Microbiol. Biotechnol.* 100, 4473–4483. doi: 10.1007/s00253-016-7297-y

Kolb, A., Kotlarz, D., Kusano, S., and Ishihama, A. (1995). Selectivity of the *Escherichia coli* RNA polymerase E sigma 38 for overlapping promoters and ability to support CRP activation. *Nucleic Acids Res.* 23, 819–826. doi: 10.1093/nar/23.5.819

Kolter, R., Siegele, D. A., and Tormo, A. (1993). The stationary phase of the bacterial life cycle. *Annu. Rev. Microbiol.* 47, 855–874. doi: 10.1146/annurev.mi.47.100193.004231

Lacour, S., and Landini, P. (2004). SigmaS-dependent gene expression at the onset of stationary phase in *Escherichia coli*: function of sigmaS-dependent genes and identification of their promoter sequences. *J. Bacteriol.* 186, 7186–7195. doi: 10.1128/JB.186.21.7186-7195.2004

Lange, R., Barth, M., and Hengge-Aronis, R. (1993). Complex transcriptional control of the sigma s-dependent stationary-phase-induced and osmotically regulated osmY (csi-5) gene suggests novel roles for Lrp, cyclic AMP (cAMP) receptor protein-cAMP complex, and integration host factor in the stationary-phase response of *Escherichia coli*. *J. Bacteriol.* 175, 7910–7917. doi: 10.1128/jb.175.24.7910-7917.1993

Lange, R., and Hengge-Aronis, R. (1994). The nlpD gene is located in an operon with rpoS on the *Escherichia coli* chromosome and encodes a novel lipoprotein with a potential function in cell wall formation. *Mol. Microbiol.* 13, 733–743. doi: 10.1111/j.1365-2958.1994.tb00466.x

Lee, S. J., Pan, J. G., Park, S. H., and Choi, S. K (2010). Development of a stationary phase-specific autoinducible expression system in *Bacillus subtilis*. *J. Biotechnol.* 149, 16–20. doi: 10.1016/j.jbiotec.2010.06.021

Lemonnier, M., Levin, B. R., Romeo, T., Garner, K., Baquero, M. R., Mercante, J., et al. (2008). The evolution of contact-dependent inhibition in non-growing populations of *Escherichia coli*. *Proc. Biol. Sci.* 275, 3–10. doi: 10.1098/rspb.2007.1234

Luidalepp, H., Berger, S., Joss, O., Tenson, T., and Polacek, N. (2016). Ribosome shut-down by 16S rRNA fragmentation in stationary-phase *Escherichia coli*. *J. Mol. Biol.* 428(10 Pt B), 2237–2247. doi: 10.1016/j.jmb.2016.01.033

Marschall, C., Labrousse, V., Kreimer, M., Weichart, D., Kolb, A., and Hengge-Aronis, R. (1998). Molecular analysis of the regulation of csiD, a carbon starvation-inducible gene in *Escherichia coli* that is exclusively dependent on sigma s and requires activation by cAMP-CRP. *J. Mol. Biol.* 276, 339–353. doi: 10.1006/jmbi.1997.1533

Matin, A. (1992). Physiology, molecular biology and applications of the bacterial starvation response. *Soc. Appl. Bacteriol. Symp. Ser.* 21, 49S–57S. doi: 10.1111/j.1365-2672.1992.tb03624.x

McKay, S. L., and Portnoy, D. A. (2015). Ribosome hibernation facilitates tolerance of stationary-phase bacteria to aminoglycosides. *Antimicrob. Agents Chemother.* 59, 6992–6999. doi: 10.1128/AAC.01532-15

Melnicki, M., Bianchi, L., Dephilippis, R., and Melis, A. (2008). Hydrogen production during stationary phase in purple photosynthetic bacteria. *Int. J. Hydrogen Energy* 33, 6525–6534. doi: 10.1016/j.ijhydene.2008.08.041

Mengin-Lecreulx, D., and van Heijenoort, J. (1985). Effect of growth conditions on peptidoglycan content and cytoplasmic steps of its biosynthesis in *Escherichia coli*. *J. Bacteriol.* 163, 208–212.

Navarro Llorens, J. M., Tormo, A., and Martinez-Garcia, E. (2010). Stationary phase in gram-negative bacteria. *FEMS Microbiol. Rev.* 34, 476–495. doi: 10.1111/j.1574-6976.2010.00213.x

Nystrom, T. (2004). Stationary-phase physiology. *Annu. Rev. Microbiol.* 58, 161–181. doi: 10.1146/annurev.micro.58.030603.123818

Oliver, J. D. (2005). The viable but nonculturable state in bacteria. *J. Microbiol.* 43, 93–100.

Ou, J., Wang, L., Ding, X., Du, J., Zhang, Y., Chen, H., et al. (2004). Stationary phase protein overproduction is a fundamental capability of *Escherichia coli. Biochem. Biophys. Res. Commun.* 314, 174–180. doi: 10.1016/j.bbrc.2003.12.077

Pátek, M., and Nešvera, J. (2013). "Promoters and plasmid vectors of *Corynebacterium glutamicum*," in *Corynebacterium glutamicum. Microbiology Monographs*, Vol. 23, eds H. Yukawa and M. Inui (Berlin: Springer), 51–88.

Piir, K., Paier, A., Liiv, A., Tenson, T., and Maivali, U. (2011). Ribosome degradation in growing bacteria. *EMBO Rep.* 12, 458–462. doi: 10.1038/embor.2011.47

Pletnev, P., Osterman, I., Sergiev, P., Bogdanov, A., and Dontsova, O. (2015). Survival guide: *Escherichia coli* in the stationary phase. *Acta Nat.* 7, 22–33.

Potvin, E., Sanschagrin, F., and Levesque, R. C. (2008). Sigma factors in *Pseudomonas aeruginosa. FEMS Microbiol. Rev.* 32, 38–55. doi: 10.1111/j.1574-6976.2007.00092.x

Qi, Y., Kobayashi, Y., and Hulett, F. M. (1997). The pst operon of *Bacillus subtilis* has a phosphate-regulated promoter and is involved in phosphate transport but not in regulation of the pho regulon. *J. Bacteriol.* 179, 2534–2539. doi: 10.1128/jb.179.8.2534-2539.1997

Ratnayake-Lecamwasam, M., Serror, P., Wong, K. W., and Sonenshein, A. L. (2001). *Bacillus subtilis* CodY represses early-stationary-phase genes by sensing GTP levels. *Genes Dev.* 15, 1093–1103. doi: 10.1101/gad.874201

Rava, P. S., Somma, L., and Steinman, H. M. (1999). Identification of a regulator that controls stationary-phase expression of catalase-peroxidase in *Caulobacter crescentus. J. Bacteriol.* 181, 6152–6159.

Repoila, F., and Gutierrez, C. (1991). Osmotic induction of the periplasmic trehalase in *Escherichia coli* K12: characterization of the treA gene promoter. *Mol. Microbiol.* 5, 747–755. doi: 10.1111/j.1365-2958.1991.tb00745.x

Shaikh, A. S., Tang, Y. J., Mukhopadhyay, A., Martin, H. G., Gin, J., Benke, P. I., et al. (2010). Study of stationary phase metabolism via isotopomer analysis of amino acids from an isolated protein. *Biotechnol. Prog.* 26, 52–56. doi: 10.1002/btpr.325

Shimada, T., Makinoshima, H., Ogawa, Y., Miki, T., Maeda, M., and Ishihama, A. (2004). Classification and strength measurement of stationary-phase promoters by use of a newly developed promoter cloning vector. *J. Bacteriol.* 186, 7112–7122. doi: 10.1128/JB.186.21.7112-7122.2004

Singh, P., Chachan, S., Singhi, D., and Srivastava, P. (2016). Isolation and molecular characterization of a stationary phase promoter useful for gene expression in *Gordonia. Gene* 591, 153–160. doi: 10.1016/j.gene.2016.07.018

Su, X., Sun, F., Wang, Y., Hashmi, M. Z., Guo, L., Ding, L., et al. (2015). Identification, characterization and molecular analysis of the viable but nonculturable *Rhodococcus biphenylivorans. Sci. Rep.* 5:18590. doi: 10.1038/srep18590

Tanaka, K., Handel, K., Loewen, P. C., and Takahashi, H. (1997). Identification and analysis of the *rpoS*-dependent promoter of *katE*, encoding catalase HPII in *Escherichia coli. Biochim. Biophys. Acta* 1352, 161–166. doi: 10.1016/S0167-4781(97)00044-4

Tanaka, K., Kusano, S., Fujita, N., Ishihama, A., and Takahashi, H. (1995). Promoter determinants for *Escherichia coli* RNA polymerase holoenzyme containing sigma 38 (the *rpoS* gene product). *Nucleic Acids Res.* 23, 827–834. doi: 10.1093/nar/23.5.827

Testerman, T. L., Vazquez-Torres, A., Xu, Y., Jones-Carson, J., Libby, S. J., and Fang, F. C. (2002). The alternative sigma factor sigmaE controls antioxidant defences required for *Salmonella* virulence and stationary-phase survival. *Mol. Microbiol.* 43, 771–782. doi: 10.1046/j.1365-2958.2002.02787.x

Tkachenko, A. G., Kashevarova, N. M., Tyuleneva, E. A., and Shumkov, M. S. (2017). Stationary-phase genes upregulated by polyamines are responsible for the formation of *Escherichia coli* persister cells tolerant to netilmicin. *FEMS Microbiol. Lett.* 364:fnx084. doi: 10.1093/femsle/fnx084

Tripathi, L., Zhang, Y., and Lin, Z. (2014). Bacterial sigma factors as targets for engineered or synthetic transcriptional control. *Front. Bioeng. Biotechnol.* 2:33. doi: 10.3389/fbioe.2014.00033

Tunner, J. R., Robertson, C. R., Schippa, S., and Matin, A. (1992). Use of glucose starvation to limit growth and induce protein production in *Escherichia coli. Biotechnol. Bioeng.* 40, 271–279. doi: 10.1002/bit.260400211

Utsumi, R., Kusafuka, S., Nakayama, T., Tanaka, K., Takayanagi, Y., Takahashi, H., et al. (1993). Stationary phase-specific expression of the fic gene in *Escherichia coli* K-12 is controlled by the rpoS gene product (sigma 38). *FEMS Microbiol. Lett.* 113, 273–278. doi: 10.1016/0378-1097(93)90217-P

van de Bunt, B., Bron, P. A., Sijtsma, L., de Vos, W. M., and Hugenholtz, J. (2014). Use of non-growing *Lactococcus lactis* cell suspensions for production of volatile metabolites with direct relevance for flavour formation during dairy fermentations. *Microb. Cell Fact.* 13:176. doi: 10.1186/s12934-014-0176-2

Wada, A. (1998). Growth phase coupled modulation of *Escherichia coli* ribosomes. *Genes Cells* 3, 203–208. doi: 10.1046/j.1365-2443.1998.00187.x

Wang, W., Li, X., Wang, J., Xiang, S., Feng, X., and Yang, K. (2013). An engineered strong promoter for streptomycetes. *Appl. Environ. Microbiol.* 79, 4484–4492. doi: 10.1128/AEM.00985-13

Waterman, S. R., and Small, P. L. (2003). Identification of the promoter regions and sigma(s)-dependent regulation of the gadA and gadBC genes associated with glutamate-dependent acid resistance in *Shigella flexneri. FEMS Microbiol. Lett.* 225, 155–160. doi: 10.1016/S0378-1097(03)00508-1

Weber, H., Polen, T., Heuveling, J., Wendisch, V. F., and Hengge, R. (2005). Genome-wide analysis of the general stress response network in *Escherichia coli*: sigmaS-dependent genes, promoters, and sigma factor selectivity. *J. Bacteriol.* 187, 1591–1603. doi: 10.1128/JB.187.5.1591-1603.2005

Weiner, L., and Model, P. (1994). Role of an *Escherichia coli* stress-response operon in stationary-phase survival. *Proc. Natl. Acad. Sci. U.S.A.* 91, 2191–2195. doi: 10.1073/pnas.91.6.2191

Wise, A., Brems, R., Ramakrishnan, V., and Villarejo, M. (1996). Sequences in the -35 region of *Escherichia coli* rpoS-dependent genes promote transcription by E sigma S. *J. Bacteriol.* 178, 2785–2793. doi: 10.1128/jb.178.10.2785-2793.1996

Xu, J., and Johnson, R. C. (1995). aldB, an RpoS-dependent gene in *Escherichia coli* encoding an aldehyde dehydrogenase that is repressed by Fis and activated by Crp. *J. Bacteriol.* 177, 3166–3175. doi: 10.1128/jb.177.11.3166-3175.1995

Yu, X., Xu, J., Liu, X., Chu, X., Wang, P., Tian, J., et al. (2015). Identification of a highly efficient stationary phase promoter in *Bacillus subtilis. Sci. Rep.* 5:18405. doi: 10.1038/srep18405

Zambrano, M. M., and Kolter, R. (1996). GASPing for life in stationary phase. *Cell* 86, 181–184. doi: 10.1016/S0092-8674(00)80089-6

17

Plasmid-Mediated Bioaugmentation for the Bioremediation of Contaminated Soils

*Carlos Garbisu[1], Olatz Garaiyurrebaso[2], Lur Epelde[1], Elisabeth Grohmann[3] and Itziar Alkorta[1]**

[1] Soil Microbial Ecology Group, Department of Conservation of Natural Resources, Neiker Tecnalia, Derio, Spain, [2] Instituto Biofisika (UPV/EHU, CSIC), Department of Biochemistry and Molecular Biology, University of the Basque Country, Bilbao, Spain, [3] Beuth University of Applied Sciences, Berlin, Germany

Edited by:
Tatiana Venkova, University of Texas Medical Branch, United States

Reviewed by:
Gloria Del Solar, Consejo Superior de Investigaciones Científicas (CSIC), Spain
Spiros Nicolas Agathos, Catholic University of Louvain, Belgium
Fabián Lorenzo, Universidad de La Laguna, Spain

***Correspondence:**
Itziar Alkorta
itzi.alkorta@ehu.eus

Bioaugmentation, or the inoculation of microorganisms (e.g., bacteria harboring the required catabolic genes) into soil to enhance the rate of contaminant degradation, has great potential for the bioremediation of soils contaminated with organic compounds. Regrettably, cell bioaugmentation frequently turns into an unsuccessful initiative, owing to the rapid decrease of bacterial viability and abundance after inoculation, as well as the limited dispersal of the inoculated bacteria in the soil matrix. Genes that encode the degradation of organic compounds are often located on plasmids and, consequently, they can be spread by horizontal gene transfer into well-established, ecologically competitive, indigenous bacterial populations. Plasmid-mediated bioaugmentation aims to stimulate the spread of contaminant degradation genes among indigenous soil bacteria by the introduction of plasmids, located in donor cells, harboring such genes. But the acquisition of plasmids by recipient cells can affect the host's fitness, a crucial aspect for the success of plasmid-mediated bioaugmentation. Besides, environmental factors (e.g., soil moisture, temperature, organic matter content) can play important roles for the transfer efficiency of catabolic plasmids, the expression of horizontally acquired genes and, finally, the contaminant degradation activity. For plasmid-mediated bioaugmentation to be reproducible, much more research is needed for a better selection of donor bacterial strains and accompanying plasmids, together with an in-depth understanding of indigenous soil bacterial populations and the environmental conditions that affect plasmid acquisition and the expression and functioning of the catabolic genes of interest.

Keywords: biodegradation, catabolic plasmid, fitness cost, horizontal gene transfer, soil pollution

INTRODUCTION

Soils play a vital role in the provision of ecosystem services and harbor one of the most complex and diverse biological communities on Earth (Barrios, 2007). Therefore, the preservation of soil quality/soil health (both terms are often used interchangeably), defined as "the capacity of soil to perform its ecosystem processes and services, while maintaining ecosystem attributes of ecological relevance" (Garbisu et al., 2011), is currently a matter of great priority. Contamination is one of the most important causes of soil degradation. Only in Europe, there are around 2.5 million potentially contaminated sites, with an annual estimated management cost of 6 billion euros (Panagos et al., 2013). Different anthropogenic activities, such as combustion of fossil fuels, incineration, mining, agricultural practices, urbanization, waste disposal, etc. have contributed to the pressing problem of soil contamination (Besser et al., 2009). Among other negative consequences, the presence of contaminants in soil can cause a negative impact on the soil biota, resulting in an altered activity, biomass and/or diversity of soil biological communities (Burges et al., 2015).

REMEDIATION OF SOIL CONTAMINANTS

Traditionally, a variety of physicochemical methods (e.g., excavation and disposal in landfills, soil washing, chemical oxidation, encapsulation, thermal treatments, incineration, vitrification, solidification, etc.) have been used for soil remediation. However, these physicochemical strategies are often expensive and, many times, reduce the concentration of soil contaminants at the expense of damaging the integrity of the soil ecosystem (Epelde et al., 2009; Gómez-Sagasti et al., 2016).

The main goal of any soil remediation technology must be not only to reduce the concentration of soil contaminants but to restore soil quality (Epelde et al., 2010; Barrutia et al., 2011; Pardo et al., 2014). A variety of soil physicochemical and biological properties (e.g., parameters that provide information on the biomass, activity and diversity of soil microbial communities) (Epelde et al., 2009; Muñoz-Leoz et al., 2013) are often used as indicators of soil quality. It has also been proposed to assess the effectiveness of remediation methods in terms of the recovery of soil ecosystem services and/or attributes of ecological relevance, such as organization, stability, redundancy, etc. (Garbisu et al., 2011; Epelde et al., 2014).

As an alternative to physicochemical treatments, several biological methods of soil remediation, included within the terms bioremediation and phytoremediation, are currently receiving much attention, mainly owing to their lower cost and environmentally friendly character (Juwarkar et al., 2014). Bioremediation, or the use of microorganisms (bacteria, fungi) to break down contaminants, takes advantage of the catabolic capacity of microorganisms to remove contaminants from soil. However, bioremediation is effective only with a limited range of contaminants and contaminant concentrations. In addition, bioremediation techniques might take too long to achieve the desired reduction in the concentration of soil contaminants (Kumavath and Deverapalli, 2013).

In relation to trace elements (a group of non-degradable contaminants of much concern due to their well-known toxicity), microorganisms can only transform them from one oxidation state or organic complex to another (Garbisu et al., 2002). Then, for the biological remediation of metal contaminated soils, metal-accumulating plants (i.e., accumulators and hyperaccumulators) offer many advantages over microbial processes, as these plants can literally extract the toxic metals from the contaminated site through a phytotechnology termed phytoextraction (Barrutia et al., 2009, 2010; Epelde et al., 2010).

Bioremediation has been successfully employed to remediate soils contaminated with organic contaminants, such as aliphatic hydrocarbons, polycyclic aromatic hydrocarbons, polychlorinated biphenyls, organic solvents and so on (Maphosa et al., 2012).

The bioremediation of organic contaminants can be approached by three different strategies: bioattenuation, biostimulation, and bioaugmentation (**Figure 1A**). *Bioattenuation* relies on natural processes to maintain the growth and degrading activity of native microbial populations, so that contaminants are biodegraded without human intervention, apart from the monitoring of contaminant dispersal and degradation rates. Instead, the term *biostimulation* refers to the adjustment of the environmental conditions (e.g., temperature, moisture, aeration, pH, redox potential) and the application of nutrients (e.g., nitrogen, phosphorus) and electron acceptors to contaminated soil, in order to enhance the growth of degrading microbial populations and, then, reduce the concentration of soil contaminants. Finally, *bioaugmentation* has been defined as the inoculation into contaminated soils of microorganisms with the ability to degrade the target contaminants (Maier, 2000; Heinaru et al., 2005). This inoculation can be performed with only one strain or, alternatively, with a consortium of microbial strains with diverse metabolic capacities. The advantage of using a consortium of different strains is that toxic intermediate products generated by one strain may be degraded by another strain (Heinaru et al., 2005). Apart from inoculating wild strains with the required degradation capacities, laboratory-constructed strains with upgraded catabolic abilities have also been considered for a more efficient bioaugmentation (Mrozik et al., 2011).

Iwamoto and Nasu (2001) and El Fantroussi and Agathos (2005) have proposed to apply bioaugmentation in those cases where biostimulation and natural attenuation are proven ineffective. In this regard, in a diesel-contaminated soil, Bento et al. (2005) found bioaugmentation to be more effective for the degradation of the light fraction (C12–C23) of petroleum hydrocarbons than biostimulation. No significant differences were detected between biostimulation and bioaugmentation in relation to the removal of the heavy fraction (C23–C40).

Bioaugmentation can be divided into two different approaches: (i) *cell bioaugmentation*, which relies on the survival and growth of the inoculated strains to perform the degradation of the target contaminants, and (ii) *genetic bioaugmentation*, based on the spread of catabolic genes, located in mobile genetic elements (MGEs), into native microbial populations.

However, despite decades of bioremediation research, the real drivers governing the degradation of organic contaminants are still poorly understood (Meckenstock et al., 2015). In order to gain insight into this question, Meckenstock et al. (2015) revisited and challenged current concepts on the controls and limitations of biodegradation, and pointed out some critical research gaps such as, for instance, the role of protozoa and bacteriophages in shaping communities of bacterial degraders and influencing contaminant degradation rates.

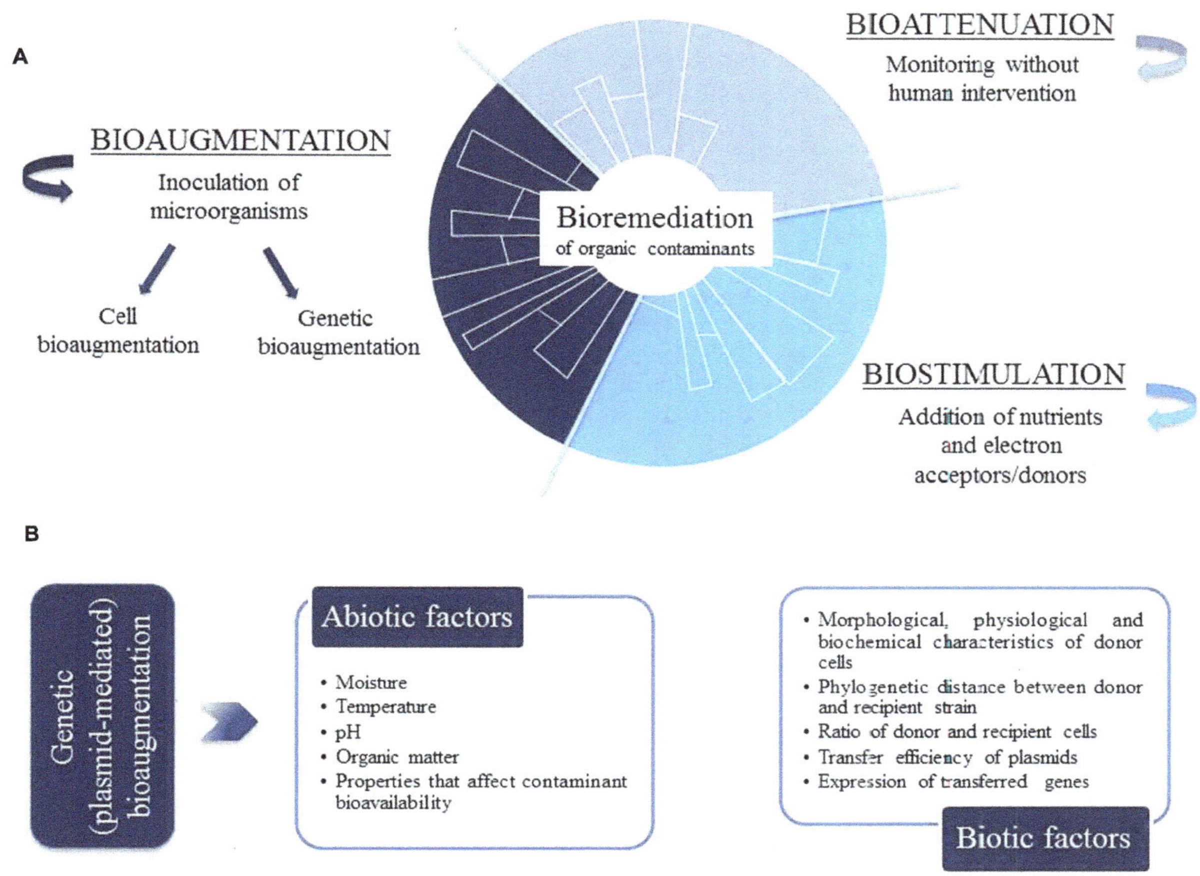

FIGURE 1 | (A) Strategies for bioremediation of organic contaminants: bioattenuation, biostimulation and bioaugmentation. **(B)** Environmental factors affecting the efficiency of plasmid-mediated bioaugmentation.

CELL BIOAUGMENTATION

Cell bioaugmentation is based on the survival and catabolic activity of inoculated microbial strains (Singh and Ward, 2004). The inoculation of bacteria harboring the necessary metabolic pathways for the degradation of the target contaminants can indeed accelerate the removal of such contaminants and, hence, reduce the time required for the intended bioremediation (Nowak and Mrozik, 2016). Inoculated microbial strains must then compete for energy and resources (e.g., nutrients and electron acceptors) with the autochthonous microbial populations already present in the soil ecosystem. The major drawbacks for the successful application of cell bioaugmentation are the (i) frequently very high mortality of the inoculated microbial strains, due to biotic or abiotic stresses, and (ii) limited dispersal of such strains throughout the soil matrix (Pepper et al., 2002; Quan et al., 2010). Many factors, including cell adhesion to soil organic matter (OM), can strongly limit the distribution of bacteria through the soil matrix. To overcome this limitation, several authors (Wang and Mulligan, 2004; Franzetti et al., 2009) have reported the use of surfactants, foams and adhesion-resistant strains.

Despite these limitations, many studies have supported the potential of cell bioaugmentation for the bioremediation of soils contaminated with organic compounds. Wang et al. (2004) reported an accelerated removal of quinoline after the inoculation of *Burkholderia pickettii*. Similarly, Mrozik et al. (2011) showed that cell bioaugmentation with *Pseudomonas* sp. JS150 significantly enhanced phenol degradation in soil, thereby reducing the possibility of formation of phenoxyl radicals (Hanscha et al., 2000). Although the number of *Pseudomonas* sp. JS150 cells decreased significantly during the first few days, the inoculated bacteria were then able to survive over the experimental period and successfully increased the rate of phenol degradation; actually, phenol biodegradation in soil bioaugmented with *Pseudomonas* sp. JS150 cells was 68 and 96 days shorter in clay and sandy soil, respectively, in comparison to non-bioaugmented soil (Mrozik et al., 2011).

GENETIC (PLASMID-MEDIATED) BIOAUGMENTATION

Genes encoding the degradation of naturally occurring or xenobiotic organic compounds are often located on MGEs, such as plasmids, integrons and transposons. By acquiring these genes through mechanisms of horizontal gene transfer (HGT), recipient bacteria may achieve the capacity to degrade those organic contaminants (Wiedenbeck and Cohan, 2011). HGT allows the exchange of genetic information among bacteria from even distantly related taxonomic groups, thereby allowing bacteria to rapidly adapt to new environmental conditions. Although mutation events can certainly contribute to bacterial adaptation, mutation rates in bacterial populations are generally low. Besides, it is currently assumed that an increased rate of mutations would result in increased death owing to deleterious effects (Martínez et al., 2009).

Out of the three mechanisms of HGT in bacteria (i.e., transformation, transduction and conjugation), conjugation is a most efficient biological process in which genetic information encoded in plasmids is transferred, from donor to recipient bacteria, by direct cell-to-cell contact (Furuya and Lowy, 2006). Bacterial conjugation is known to accelerate the dissemination of resistance to, for instance, antibiotics and heavy metals, as well as to facilitate the distribution of genes involved in the degradation of organic compounds. Nevertheless, the contribution of conjugation to HGT among soil bacteria and the factors involved in the transfer and proliferation of plasmid-containing bacteria in the soil ecosystem are yet not fully understood.

Bacterial adaptation through evolutionary time has been shaped, among other aspects, by the high plasticity of bacterial genomes, which allows bacteria to rearrange and exchange genomic sequences, thus opening the possibility to acquire beneficial traits (Sørensen et al., 2005). As a matter of fact, the loss, rearrangement and acquisition of functional genetic modules can have a vast impact on the extent and speed of the evolutionary adaptation of bacteria (Wozniak and Waldor, 2010; Bertels and Rainey, 2011). MGEs are, to a great extent, responsible for these processes of gene mobility and reorganization, both within genomes (intracellular) and between bacterial cells (intercellular).

Many of the studies on lateral dissemination of genetic material among bacteria have focused on antibiotic and metal resistance. Research on the horizontal transfer of genes associated with the degradation of organic compounds in natural environments, such as the soil ecosystem, is still insufficient to fully understand the mechanisms involved in such process (Christensen et al., 1998; Top et al., 1998; Dejonghe et al., 2000; Aspray et al., 2005; Overhage et al., 2005; Musovic et al., 2010). In any case, some plasmids, such as those implicated in the catabolic pathway of 2,4-dichlorophenoxyacetic acid (2,4-D), have been thoroughly studied (Top et al., 1998; Dejonghe et al., 2000; Newby and Pepper, 2002).

Plasmid transfer between soil bacteria has been contemplated as a promising strategy for the dissemination of catabolic functions within soil bacterial communities (Venkata Mohan et al., 2009; Mrozik and Piotrowska-Seget, 2010). As abovementioned, plasmid-encoded metabolic pathways can be transferred among bacteria, thus playing a critical role in the adaptation of bacteria to different environmental conditions (Reineke, 1998; Sayler and Ripp, 2000). Specifically, HGT has been reported to promote bacterial adaptation to the presence of organic contaminants (Top and Springael, 2003).

The underlying idea behind genetic (plasmid-mediated) bioaugmentation is to stimulate the rate of contaminant degradation by increasing, through HGT, the number and diversity of native bacteria with the capacity to metabolize the target contaminants. In this respect, it must be emphasized that numerous catabolic pathways involved in the degradation of organic contaminants have been identified in MGEs (Top et al., 2002; Jussila et al., 2007).

Genetic (plasmid-mediated) bioaugmentation is defined as a technology in which donor bacteria harboring self-transmissible catabolic plasmids are introduced into the soil matrix in order to enhance, by HGT, the potential and rate of contaminant degradation of existing bacterial populations (Top et al., 2002; Ikuma and Gunsch, 2010, 2012). Compared to cell bioaugmentation, plasmid-mediated bioaugmentation appears *a priori* a more effective strategy for the bioremediation of organic contaminants, as the bacteria that will eventually degrade the contaminants (i.e., bacteria with the recently acquired plasmids harboring the necessary catabolic genes) are expected to be adapted to live in the soil under remediation. In this manner, one of the main drawbacks for the successful application of cell bioaugmentation, i.e., the low survival of the inoculated microbial strains, appears to be overcome.

For plasmid-mediated bioaugmentation, both an appropriate selection of donor bacteria with the required plasmid and a profound knowledge of native soil bacterial populations are required to increase the probability of an efficient plasmid acquisition and the expression of the catabolic genes of interest.

Many studies on plasmid-mediated bioaugmentation have been published (**Table 1**). In a microcosm study, Halden et al. (1999) detected an enhanced degradation of 3-phenoxybenzoic acid (3-POB) as a result of the transfer of plasmids pPOB and pD30.9 from *Pseudomonas pseudoalcaligenes* POB310 (pPOB) and *Pseudomonas* sp. B13-D5 (pD30.9) and B13ST1 (pPOB) to recipient soil bacteria. Using *P. putida* as donor strain of two catabolic plasmids (pEMT1 and pJP4), Dejonghe et al. (2000) reported the degradation of 2,4-D in soil under microcosm conditions. These authors investigated the bioaugmentation potential of plasmids pEMT1 and pJP4 in two soil layers (0–30 and 30–60 cm soil depth) differing in physicochemical properties and microbial community structure, finding out a more efficient degradation of 2,4-D in the deeper soil layer where the indigenous microbial communities lacked the ability to catabolize 2,4-D. Under microcosm conditions, Inoue et al. (2012) studied the effect of bioaugmentation with *P. putida* and *Escherichia coli* cells, harboring the self-transmissible 2,4-D degradative plasmid pJP4, on the degradation of 2,4-D. These authors found that the number of *P. putida* and *E. coli* cells decreased rapidly after their inoculation in a 2,4-D contaminated soil slurry, but the degradation of this contaminant was nevertheless stimulated, most likely due to the occurrence of transconjugants resulting from the transfer of plasmid pJP4. Inoue et al. (2012) concluded that genetic bioaugmentation with *P. putida* and *E. coli* cells harboring plasmid pJP4 can stimulate the degradation of 2,4-D in soil without a substantial impact on the soil microbial community, as reflected by the values of parameters which provide information on carbon source utilization (through the use of the well-known Biolog^{TM} plates) and nitrogen transformations (nitrate reduction assay, quantification of amoA gene of ammonia-oxidizing bacteria, quantification of *nirK* and *nirS* genes of denitrifying bacteria). In sequencing batch reactors, Tsutsui et al. (2013) achieved a complete degradation of 2,4-D by plasmid (pJP4)-mediated bioaugmentation with *Cupriavidus necator* JMP134 and *E. coli* HB101 as donor strains. These authors were able to identify the emergence of 2,4-D-degrading transconjugants associated to *Achromobacter, Burkholderia, Cupriavidus* and *Pandoraea.*

Pepper et al. (2002) conducted microcosm experiments to enhance the degradation of 3-chlorobenzoate (3-CB) using plasmid pBRC60, which harbors genes for 3-CB mineralization, and *Comamonas testosteroni* as donor strain. Although they did observe degradation of 3-CB, they could not detect any transfer event of plasmid pBRC60 from *C. testosteroni* to native soil bacteria.

Miyazaki et al. (2006) isolated a plasmid (pLB1) involved in the dissemination of genes for γ-hexachlorocyclohexane (lindane) degradation. This plasmid, carrying the *linB* gene, was isolated from *Sphingobium japonicum* UT26DB and then successfully transferred, under laboratory conditions, from this strain to other α-proteobacterial strains but not to any of the β- or γ-proteobacterial strains tested.

In their study on the transfer of TOL plasmid (also designated pWW0) during bacterial conjugation *in vitro* and rhizoremediation of oil-contaminated soil *in vivo*, Jussila et al. (2007) demonstrated the successful transfer of TOL plasmid for toluene degradation from *P. putida* PaW85 to *P. oryzihabitans* 29. In rhizosphere microcosms, Mølbak et al. (2007) found that the transfer of plasmid pWW0 from *P. putida* resulted in transconjugants belonging to *Enterobacteria* and *Pseudomonas.* This well-characterized self-transmissible catabolic plasmid, pWW0, was also used by Ikuma and Gunsch (2012) to assess its potential for bioaugmentation in toluene-contaminated soil slurry.

Under laboratory conditions, horizontal transfer of plasmid pGKT2 was successfully carried out by Jung et al. (2011) from *Gordonia* sp. KTR9 to *Gordonia polyisoprenivorans, Rhodococcus jostii* RHA1 and *Nocardia* sp. TW2 strains. These transconjugants showed the ability to use hexahydro-1,3,5-trinitro-1,3,5,-triazine (RDX) as a nitrogen source.

In a contaminated field site located in Cixi, Zhejiang (China), Gao et al. (2015) achieved effective plasmid-mediated bioaugmentation for the degradation of dichlorodiphenyltrichloroethane (DDT) in soil with *E. coli TG I* (pDOD-gfp) as donor strain. In this study, the catabolic plasmid pDOD from *Sphingobacterium* sp. D-6 was conjugally transferred to soil bacteria, such as members of *Cellulomonas,* and accelerated DDT degradation. Different studies have reported the use of the GFP (green fluorescence protein) detection system to monitor plasmid transfer from donor cells to indigenous soil bacteria in soil slurries (Ikuma et al., 2012) and field contaminated soil (Gao et al., 2015).

Filonov et al. (2010) determined transfer frequencies in open soil after inoculation with genetically tagged plasmid-containing naphthalene-degrading *P. putida* KT2442 and auxotrophic donor BS394 (pNF142::Tn*Mod*-OTc) cells, finding out that plasmid pNF142 was transferred to native soil bacteria (mainly to fluorescent pseudomonads) at a frequency of 4×10^{-6}/donor cell. After bioaugmentation with *E. coli* JM109 (pDOC-gfp) strain, Zhang et al. (2012) observed that pDOC plasmid was transferred to native soil bacteria under microcosm conditions, including members of *Pseudomonas* and *Staphylococcus* which acquired the capacity to degrade chlorpyrifos (a widely used insecticide). As it is usually the case, the efficiency of this transfer, as measured by the chlorpyrifos degradation efficiency and the number of chlorpyrifos degraders, was influenced by soil type, temperature and moisture content (Zhang et al., 2012).

Finally, attention has also been paid to the use of genetically modified organisms (GMOs) for bioaugmentation. Nonetheless, the deliberate release of GMOs into the environment is subjected to regulatory constraints (Garbisu and Alkorta, 1997; Sayler and Ripp, 2000; Directive 2001/18/EC, 2001). The transfer of

TABLE 1 | Examples of plasmid-mediated bioaugmentation studies.

Contaminant	Plasmid	Strain	Reference
3-phenoxybenzoic acid	pPOB	*P. pseudoalcaligenes POB310*	Halden et al., 1999
3-chlorobenzoate	pBRC60	*C. testosteroni*	Pepper et al., 2002
2,4-dichlorophenoxyacetic acid	pJP4	*C. necator* JMP134 *E. coli* HB101	Tsutsui et al., 2013
2,4-dichlorophenoxyacetic acid	pJP4	*P. putida* KT2442 *E. coli* HB101	Inoue et al., 2012
2,4-dichlorophenoxyacetic acid	pEMT1 pJP4	*P. putida* UWC3	Dejonghe et al., 2000
2,4-dichlorophenoxyacetic acid	pJP4	*R. eutropha* JMP134 *E. coli* D11	Newby et al., 2000
γ-hexachlorocyclohexane	pLB1	*S. japonicum* UT26DB	Miyazaki et al., 2006
Chlorpyrifos	pDOC	*E. coli* JM109	Zhang et al., 2012
Dichlorodiphenyltrichloroethane	pDOD	*E. coli* TG I	Gao et al., 2015
Naphthalene	pNF142	*P. putida* BS394 *P. putida* KT2442	Filonov et al., 2010
Oil	pWW0-derivative TOL	*P. putida* PaW85	Jussila et al., 2007
Toluene	pWW0-derivative TOL	*P. putida* BBC443	Ikuma and Gunsch, 2012

catabolic genes between GMOs and wild bacterial strains might facilitate the acquisition and spread of new degradative pathways among indigenous bacterial communities. To this purpose, Massa et al. (2009) engineered the recombinant strain *P. putida* PaW340/pDH5, constructed by cloning dehalogenase genes from *Arthrobacter* sp. FG1 in *P. putida* PaW340, for the degradation of 4-chlorobenzoic acid (CBA) in soil slurry. After inoculation of this recombinant strain into soil slurry, a higher degradation of CBA was observed, compared to the slurry inoculated with pre-adapted cultures of *Arthrobacter* sp. FG1.

EFFECT OF PLASMID ACQUISITION ON HOST FITNESS

The success of plasmid-mediated bioaugmentation for the bioremediation of contaminated soil relies not only on an efficient transfer of the required plasmid from donor bacteria to soil recipient bacteria, but also on the ability of recipient cells to properly express the plasmid-harbored catabolic genes, so that the desired phenotypic changes (i.e., biodegradation of the target contaminant) can be attained. After plasmid acquisition, the capacity of recipient cells to successfully perform the desired catabolic function depends, among other factors, on their newly acquired competitive abilities and on the alteration of the host's own competitive abilities (van Rensburg et al., 2012). A thorough understanding of how plasmid acquisition can affect host fitness is fundamental to then achieve the persistence of the introduced plasmid in the recipient cells.

Plasmid acquisition can provide recipient bacteria with a large array of beneficial traits, such as catabolic potential, resistance to antibiotics and/or metals, faster growth, ability to use a wider range of compounds as energy sources, etc. (Top et al., 1998; Riley and Wertz, 2002). In many cases, plasmid-harboring hosts have been found to be competitively fitter than their plasmid-free counterparts (Dionisio et al., 2005; Starikova et al., 2013). Nonetheless, horizontally acquired genes can also function inefficiently in the genomic background of recipient cells (Chou et al., 2011; Park and Zhang, 2012). After all, horizontally acquired genes find themselves immersed in a new metabolic context and their function relies on the host's machinery. Genetic determinants often encounter the required metabolic "partners" (e.g., substrates, proteins) in the recipient cells, so that the intended changes in the host's metabolism become possible. Conversely, other times, the required metabolic partners for the proper functioning and regulation of newly acquired genes are missing in the recipient cells. Indeed, the acquisition of plasmids can negatively affect cellular networks in recipient cells and, concomitantly, trigger fitness costs as collateral damage (Bouma and Lenski, 1988; Martínez et al., 2009). Fitness (metabolic) costs derived from plasmid acquisition can be highly variable (De Gelder et al., 2007), as they can originate from a variety of factors, including: (i) *energetic costs* due to consumption of molecular building blocks and/or energy sources derived from the activity of horizontally acquired regions; (ii) *chromosomal disruption* by horizontally acquired genes, when such genes are incorporated into the chromosome; (iii) *sequestration of cellular processes and associated molecular machinery* (e.g., ribosomes) by the horizontally acquired regions; and (iv) *plasmid size*, since small plasmids can carry only a single accessory determinant but large plasmids can carry more than 10 accessory determinants as well as other genes (Shachrai et al., 2010; Baltrus, 2013; Vogwill and MacLean, 2015).

Fitness costs associated to plasmid acquisition can be offset by benefits derived from the fact that plasmids are ideal biological tools to create genetic variation within bacterial populations. A major benefit from maintaining transferable plasmids derives from the fact that, in this manner, bacterial populations can gain stability against potential environmental changes.

Bacteria with acquired genes can, on the other hand, alleviate fitness costs through compensatory evolution (San Millan et al., 2014). Thus, for instance, bacteria can minimize plasmid-related fitness costs by integrating only the desired plasmid-acquired determinants in the chromosome.

Conjugative plasmids (i) are usually large (they encode genes for the conjugation process itself and for stabilization within the host); (ii) are normally found in low copy number; (iii) appear well maintained over successive generations (Norman et al., 2009; Jung et al., 2011); and (iv) act as fundamental vehicles of HGT (Frost et al., 2005; Thomas and Nielsen, 2005).

In their laboratory study on the capacity of *Gordonia* sp. KTR9 to transfer plasmid pGKT2 and the associated RDX (hexahydro-1,3,5-trinitro-1,3,5,-triazine) degradation ability to other bacteria, Jung et al. (2011) investigated plasmid stability after HGT from *Gordonia* sp. KTR9 to *G. polyisoprenivorans*, *R. jostii* RHA1 and *Nocardia* sp. TW2, finding out a marked decrease in plasmid retention after 50 generations with *Nocardia* sp. TW2, while *G. polyisoprenivorans* and *R. jostii* RHA1 transconjugants exhibited retention of pGKT2 plasmid for 100 generations. It was speculated that this decreased stability in *Nocardia* sp. TW2 might have been caused by a larger metabolic expense incurred by the incorporation of pGKT2 in this strain, compared to the other two bacterial strains (Jung et al., 2011).

Given that positive selection cannot explain the long-term stability of costly plasmids (Hall et al., 2017), the explanation for such long-term stability remains a most challenging task, since segregational loss and the cost of plasmid carriage should drive the loss of plasmids through purifying selection (Hall et al., 2017). In this respect, two evolutionary routes to plasmid stability appear possible (Hall et al., 2017): (i) the evolution of high conjugation rates would allow plasmids to survive as infectious agents through horizontal transmission (Hall et al., 2016; Kottara et al., 2016); and (ii) compensatory evolution to ameliorate the cost of plasmid carriage can weaken purifying selection against the plasmid backbone (Harrison et al., 2015; Porse et al., 2016).

Finally, it must be taken into consideration that plasmids can be classified into incompatibility groups (incompatibility defined as the inability of plasmids sharing similar replication and partition systems to be propagated stably in the same host cell line; in other words, members of each group cannot co-reside within the same bacterial host), such as IncP, IncN, IncW, and IncF. Incompatibility groups have been independently classified in three different genera: there are 27 Inc groups in *Enterobacteriaceae*, 14 Inc groups in *Pseudomonas*, and approximately 18 Inc groups in *Staphylococcus* (Shintani et al., 2015). Plasmids classified in *E. coli* as IncP and in *Pseudomonas* as IncP-1 are a well-studied group of plasmids that can carry a variety of phenotypic markers, including antibiotic resistance, metal resistance and the ability to degrade xenobiotics. It has been reported (Popowska and Krawczyk-Balska, 2013) that a detailed analysis of IncP-1 plasmid genomes could provide useful information for the development of effective methods of soil bioremediation. After all, the evolutionary adaptation of microorganisms to the presence and utilization of organic contaminants is often due to plasmids (mainly, from IncP group) that carry genes encoding enzymes involved in the degradation of those compounds. For instance, plasmids IncP-1, IncP-7 and IncP-9 contain genes encoding enzymes required for the degradation of naphthalene, toluene, chlorobenzene, p-toluenesulfonate, 2,4-D, haloacetate and atrazine (Shintani et al., 2010a,b; Popowska and Krawczyk-Balska, 2013). Relevantly, there seems to be a distinction between (i) plasmids that harbor genes for the degradation of naturally occurring compounds and (ii) plasmids that harbor genes for the degradation of xenobiotics (Top et al., 2002): degradation of naturally occurring compounds is often encoded in IncP-2 and IncP-9 plasmids, while the degradation of xenobiotics seems to be encoded by the well-known broad host range IncP-1 plasmids. IncP-1 plasmids are very promiscuous, and this promiscuity appears to play a crucial role in the evolution of new metabolic pathways by recruiting catabolic genes or gene segments from different organisms into a suitable host (Wyndham et al., 1994; Beil et al., 1999).

Therefore, different plasmids potentially useful for plasmid-mediated bioaugmentation (with, for instance, each plasmid harboring a gene encoding a different enzyme involved in the degradation route of a specific contaminant) cannot co-reside within the same host if they belong to the same incompatibility group. Then, if we want to apply different plasmids from the same incompatibility group, each of them harboring a gene for a specific step in the contaminant degradation pathway, they must be applied in different donor cells and, for an effective biodegradation, each plasmid should be transferred to a different recipient cell, decreasing considerably the probability of successful plasmid-mediated bioaugmentation.

INFLUENCE OF ABIOTIC AND BIOTIC FACTORS ON BIOAUGMENTATION

The success of both cell and plasmid-mediated bioaugmentation greatly depends on the environmental (abiotic and biotic) conditions present in the soil to be remediated (Cho et al., 2000; Bento et al., 2005; Wolski et al., 2006). In fact, during plasmid-mediated bioaugmentation, environmental factors can play important roles in the (i) transfer efficiency of catabolic plasmids, (ii) expression of horizontally acquired genes and, finally, (iii) contaminant degradation activity (Popa et al., 2011; Ikuma and Gunsch, 2012). In particular, several abiotic factors such as soil moisture, temperature and OM content are known to affect bioaugmentation efficiency (**Figure 1B**).

Soil moisture can have an effect on plasmid transfer during plasmid-mediated bioaugmentation by affecting the contact between donor and recipient bacteria (Miller et al., 2004; Aminov, 2011). In this respect, Gao et al. (2015) evaluated the effectiveness of plasmid-mediated bioaugmentation for *p,p′*-DDT degradation at three different soil moisture conditions (40, 60, and 80%), and concluded that 60% moisture content was optimal for maximum plasmid transfer efficiency.

Temperature has been shown to affect plasmid transfer efficiency (Inoue et al., 2005; Zhang et al., 2012). For the enhancement of DDT degradation by plasmid-mediated bioaugmentation with plasmid pDOD, the optimal temperature interval for cell growth and activity of both donor and recipient soil bacteria was established at 25–30°C (Gao et al., 2015). Johnsen and Kroer (2007) found that increasing temperatures resulted in an increase in the transfer of plasmid pRO103 encoding resistance to mercury and tetracycline and partial degradation of 2,4-D.

Regarding soil OM content, in a bioaugmentation laboratory experiment, Greer and Shelton (1992) observed higher rates of mineralization of 2,4-D in soil with a low OM content, compared to soil with a high content of OM. Under laboratory conditions, Kim et al. (2008) found that *P. spadix* BD-a59 cells were able to degrade BTEX at a slower rate in soil with low OM content than in organic-rich soil. When studying the biodegradation of polychlorinated biphenyls (PCB) in soil under laboratory conditions, Haluška et al. (1995) observed that humic acids affected the survival and activity of the inoculated *Alcaligenes xylosoxidans* strain, which exhibited maximum survival rates in soil with an intermediate amount of organic carbon and the highest amount of aromatic carbon in humic acids. Highest levels of PCB degradation were found in soil with the highest content of organic carbon and an intermediate amount of aromatic carbon in humic acids (Haluška et al., 1995).

Wang et al. (2014) performed plasmid transfer experiments between soil bacteria, using a TOL-like plasmid carrying the gene encoding for catechol 2,3-dioxygenase, to study some factors (soil depth, soil type, etc.) that could affect the transfer of plasmids, finding out that these factors certainly have a considerable effect on the transfer of the TOL-like plasmid in soil. Concerning soil depth, under microcosm conditions, Wang et al. (2014) found, in general, lower frequencies of plasmid transfer at greater soil depths, a fact most likely due to the often-found gradual decrease in bacterial biomass and activity at increasing soil depths, possibly related to concomitantly decreased oxygen concentrations (Król et al., 2011). Król et al. (2011) reported that oxygen concentration can affect plasmid transfer through an oxygen-related mechanism or indirectly via its impact on cell physiology. When studying the influence of soil type (loamy sand, sandy loam, sandy clay loam, loam) on plasmid transfer, Wang et al. (2014) observed a highest frequency of plasmid transfer in loam soil, probably related to the fact that loam often contains more nutrients and humus than other soil types, and higher values of microbial biomass and metabolic activity (Djokic et al., 2013).

In the same way, the chemical nature, concentration and bioavailability of the contaminants are crucial factors influencing bioaugmentation efficiency (Davis and Madsen, 1996; Stalwood et al., 2005). Sejáková et al. (2009) reported a relationship between pentachlorophenol (PCP) concentration in soil and the number of CFU of the *C. testosteroni* CCM7530 strain used for bioaugmentation: at a PCP concentration of 100 mg kg^{-1}, the number of *C. testosteroni* CCM7530 CFUs rapidly increased over 17 days, while, at 10 mg PCP kg^{-1}, the number of CFUs initially decreased until day 7 to then increase until day 17.

In any case, the level of selective pressure required to promote conjugal plasmid transfer depends on the specific contaminant and its concentration, as well as on the specific catabolic plasmid. In soil slurry, Ikuma and Gunsch (2012) observed that environmentally relevant concentrations of toluene might not exert enough pressure for transfer of plasmid TOL from *P. putida* BBC443 to *Serratia marcescens* and *P. fluorescens* cells. In their study on the degradation of 2,4-D, DiGiovanni et al. (1996) observed that this contaminant originated the required selective pressure for conjugal transfer of the intended catabolic plasmids.

Many biotic factors can also affect the success of plasmid-mediated bioaugmentation. Some genetic differences, such as guanine-cytosine (G+C) content and phylogenetic relationship between donor and recipient strain, can negatively affect the expression of the catabolic phenotype following conjugal plasmid transfer, as described by Ikuma and Gunsch (2012). Indeed, for plasmid-mediated bioaugmentation, biological differences between donor and recipient bacterial strains such as, for example, phylogenetic distance (Popa et al., 2011) and plasmid host range (De Gelder et al., 2005; Sorek et al., 2007), can play an important role. In 2,4-D contaminated soils, Newby et al. (2000) studied the bioaugmentation efficiency of two plasmid pJP4-bearing bacteria (the natural host, *Ralstonia eutropha* JMP134, and a laboratory-generated *E. coli* strain amenable to donor counterselection, named *E. coli* D11) and concluded that the correct choice of donor strain is a factor of the utmost importance for bioaugmentation.

Ikuma and Gunsch (2012) indicated that the success of plasmid-mediated bioaugmentation is dependent on: (i) high transfer rates of the catabolic plasmid to as many indigenous bacteria as possible; and (ii) the high expression level of an active contaminant-degrading phenotype in all transconjugants following conjugal plasmid transfer. Then, prior to the bioaugmentation process itself, it is important to characterize potentially recipient soil bacterial communities, paying special attention to dominant taxonomic groups. In the last years, next generation sequencing has provided a more comprehensive analysis of indigenous soil bacterial communities (Walsh, 2000), opening the door to the identification of potential recipient bacterial populations, and therefore a more informed selection of both the donor strain and the plasmid type (Ikuma and Gunsch, 2012).

Other biotic factors, such as competition between inoculated and indigenous bacteria for carbon sources, antagonistic interactions and predation by protozoa and bacteriophages, etc. also play an essential role in bioaugmentation efficiency. The critical factor is the selection of the right bacterial strains (Thompson et al., 2005), since the inoculated strain must be able not only to degrade the target contaminant (or, in the case of plasmid-mediated bioaugmentation, to be able to effectively transfer the catabolic plasmid), but also to successfully compete with indigenous microbial populations and, in general, soil biota. On the other hand, plasmid transfer frequency has been shown to depend on the initial cell density ratio between donor and recipient cells (Pinedo and Smets, 2005; Ikuma et al., 2012).

Morphological, physiological and biochemical characteristics such as, for instance, cell size, growth rate, resource utilization ability, resistance phenotypes, biofilm formation capacity, cell motility, etc. are key traits for bacterial survival and competitiveness. Furthermore, DNA content has a marked influence on bacterial ecophysiological traits (i.e., adaptive traits to environmental changes) affecting, among other aspects, the rate of cell growth (Wickham and Lynn, 1990). Nevertheless, despite the assumption that fitness costs associated to HGT are caused by the need to maintain and replicate the extra-DNA, some studies indicate that they are predominantly due to transcription and translation processes (Bragg and Wagner, 2009; Shachrai et al., 2010).

The capacity of the host to use different carbon substrates before and after plasmid acquisition can provide an estimation of (i) its competitive ability and (ii) changes specifically associated to the plasmid transfer itself. Biolog™ plates can be employed to obtain a phenotypic fingerprint of bacterial strains in relation to their capacity to use a variety of carbon sources. Karve et al. (2016) followed phenotypic variations, using Biolog GEN III MicroPlates™, to assess functionally relevant consequences of DNA changes.

Antibiotic resistance is probably the most extensively studied bacterial competitive trait. As a consequence of the production of antibiotics by soil microbial populations (D'Costa et al., 2006), soil is thought to be the largest reservoir of antibiotic resistance genes. Owing to fitness costs associated to antibiotic resistance, when bacteria change to an antibiotic-free environment, resistance is expected to disappear (Morosini et al., 2000), according to the assumption that, in the absence of selective pressure, resistant bacteria with a lower fitness will be outcompeted by susceptible counterparts with a higher fitness. However, it seems that bacteria tend to keep the mechanisms of antibiotic resistance, in order to maintain such an advantageous trait in the face of a possible change in environmental conditions (San Millan et al., 2014). Besides, in nature, antibiotics and antibiotic resistance determinants might play a variety of roles (e.g., signaling molecules in quorum sensing and biofilm formation, production of virulence factors, host-parasite interactions) (Sengupta et al., 2013) that justify the preservation of antibiotic resistance determinants in the absence of the selective pressure.

Biofilms are known to protect bacterial cells against antimicrobials (Høiby et al., 2010), predation, oxidative stress (Geier et al., 2008), etc. Biofilms harbor spatially structured bacterial communities where plasmids can be more easily shared through HGT (Jefferson, 2004), facilitating, for instance, the dissemination of catabolic genes. Remarkably, attachment to surfaces by biofilm-associated factors is another cellular function associated to genes present in plasmids (Norman et al., 2009).

Cell motility is a critical aspect for the necessary dispersal of inoculated bacteria toward the target contaminants. Nevertheless, although highly motile bacterial cells, in their search for energy and nutrients, can disperse more easily into the surrounding environment, they also have a higher probability of encountering potential competitors (Reichenbach et al., 2007). In any case, motile bacterial populations, such as swarming bacteria, can more rapidly colonize new niches, with the associated ecological benefits (Verstraeten et al., 2008). Interestingly, there is a complex link between motility and biofilm formation because both processes appear to involve similar components at certain stages and conditions (Verstraeten et al., 2008).

(Gardin and Pauss, 2001; Gentili et al., 2006) have used different strategies of cell encapsulation and immobilization to facilitate inoculation survival, by providing a protective niche and temporary nutrition for the inoculated bacteria. Carrier materials, such as charcoal (Beck, 1991), nylon (Heitkamp and Steward, 1996), chitin, chitosan (Gentili et al., 2006; Chen et al., 2007) and zeolite (Liang et al., 2009) have been used in an attempt to maintain inoculant activity over a sufficiently long period of time after strain inoculation.

It must be taken into consideration that the influence of all these abovementioned abiotic and biotic factors has only been studied in a very limited number of bacterial strains and, in many cases, under controlled simplified environmental conditions, very different from those encountered in the natural environment. Therefore, many more in-depth studies on the impact of abiotic and biotic factors on cell and plasmid-mediated bioaugmentation are needed.

CONCLUDING REMARKS

Both cell bioaugmentation and genetic (plasmid-mediated) bioaugmentation have proven effective for the bioremediation of soils contaminated with organic compounds. However, cell bioaugmentation has an important limitation, i.e., the frequently very high mortality of the inoculated microbial strains, due to biotic or abiotic stresses. Then, *a priori*, plasmid-mediated bioaugmentation appears to have greater potential than cell bioaugmentation, since plasmids can act as gene-messenger biological tools that can transfer the required catabolic genes to indigenous bacterial populations already adapted to the soil under remediation. But for plasmid-mediated bioaugmentation to be successful and reproducible, much more research is needed for a better selection of donor bacterial strains and accompanying plasmids, together with an in-depth understanding of indigenous soil bacterial populations and the environmental conditions that affect plasmid acquisition and the expression and functioning of the catabolic genes of interest. Similarly, further research is required to better understand and then improve the ecological fitness of recipient bacterial strains in the contaminated soil.

AUTHOR CONTRIBUTIONS

CG and IA: Design of the work and the acquisition of the data, writing and revision of the content, approval of the last version and ensuring accuracy and integrity of the work. LE and OG: Acquisition of the data, writing and revision of the content, approval of the last version of the work. EG: Writing and revision of the content, approval of the last version and ensuring accuracy and integrity of the work.

ACKNOWLEDGMENTS

This work has been supported by the Spanish Ministry of Economy, Industry and Competitiveness (AGL2016-76592-R), and the Interreg SUDOE Programme (PhytoSUDOE-SOE1/P5/EO189). OG was a pre-doctoral student supported by the Basque Government and by the Fundación Biofísica Bizkaia.

REFERENCES

Aminov, R. I. (2011). Horizontal gene exchange in environmental microbiota. *Front. Microbiol.* 2:158. doi: 10.3389/fmicb.2011.0015

Aspray, T. J., Hansen, S. K., and Burns, R. G. (2005). A soil-based microbial biofilm exposed to 2,4-D: bacterial community development and establishment of conjugative plasmid pJP4. *FEMS Microbiol. Ecol.* 54, 317–327. doi: 10.1016/j.femsec.2005.04.007

Baltrus, D. A. (2013). Exploring the costs of horizontal gene transfer. *Trends Ecol. Evol.* 28, 489–495. doi: 10.1016/j.tree.2013.04.002

Barrios, E. (2007). Soil biota, ecosystem services and land productivity. *Ecol. Econ.* 64, 269–285. doi: 10.1016/j.ecolecon.2007.03.004

Barrutia, O., Epelde, L., García-Plazaola, J. I., Garbisu, C., and Becerril, J. M. (2009). Phytoextraction potential of two *Rumex acetosa* L. accessions collected from metalliferous and non-metalliferous sites: effect of fertilization. *Chemosphere* 74, 259–264. doi: 10.1016/j.chemosphere.2008.09.036

Barrutia, O., Garbisu, C., Epelde, L., Sampedro, M. C., Goicolea, M. A., and Becerril, J. M. (2011). Plant tolerance to diesel minimizes its impact on soil microbial characteristics during rhizoremediation of diesel-contaminated soils. *Sci. Total Environ.* 409, 4087–4093. doi: 10.1016/j.scitotenv.2011.06.025

Barrutia, O., Garbisu, C., Hernández-Allica, J., García-Plazaola, J. I., and Becerril, J. M. (2010). Differences in EDTA-assisted metal phytoextraction between metallicolous and non-metallicolous accessions of *Rumex acetosa* L. *Environ. Pollut.* 158, 1710–1715. doi: 10.1016/j.envpol.2009.11.027

Beck, D. P. (1991). Suitability of charcoal-amended mineral soil as carrier for *Rhizobium* inoculants. *Soil Biol. Biochem.* 23, 41–44. doi: 10.1016/0038-0717(91)90160-L

Beil, S., Timmis, K. N., and Pieper, D. H. (1999). Genetic and biochemical analyses of the tec operon suggest a route for evolution of chlorobenzene degradation genes. *J. Bacteriol.* 181, 341–346.

Bento, F. M., Camargo, F. A. O., Okeke, B. C., and Frankenberger, W. T. (2005). Comparative bioremediation of soils contaminated with diesel oil by natural attenuation, biostimulation and bioaugmentation. *Bioresour. Technol.* 96, 1049–1055. doi: 10.1016/j.biortech.2004.09.008

Bertels, F., and Rainey, P. B. (2011). Within-genome evolution of REPINs: a new family of miniature mobile DNA in bacteria. *PLOS Genet.* 7:e1002132. doi: 10.1371/journal.pgen.1002132

Besser, J. M., Brumbaugh, W. G., Allert, A. L., Poulton, B. C., and Schmitt, C. J. (2009). Ecological impacts of lead mining on Ozark streams-toxicity of sediment and pore water. *Ecotoxicol. Environ. Saf.* 72, 516–526. doi: 10.1016/j.ecoenv.2008.05.013

Bouma, J. E., and Lenski, R. E. (1988). Evolution of a bacteria/plasmid association. *Nature* 335, 351–352. doi: 10.1038/335351a0

Bragg, J. G., and Wagner, A. (2009). Protein material costs: single atoms can make an evolutionary difference. *Trends Genet.* 25, 5–8. doi: 10.1016/j.tig.2008.10.007

Burges, A., Epelde, L., and Garbisu, C. (2015). Impact of repeated single-metal and multi-metal pollution events on soil quality. *Chemosphere* 120, 8–15. doi: 10.1016/j.chemosphere.2014.05.037

Chen, Y. M., Lin, T. F., Huang, C., Lin, J. C., and Hsieh, F. M. (2007). Degradation of phenol and TCE using suspended and chitosan-bead immobilized *Pseudomonas putida*. *J. Hazard Mater.* 148, 660–670. doi: 10.1016/j.jhazmat.2007.03.030

Cho, Y. G., Rhee, S. K., and Lee, S. T. (2000). Effect of soil moisture on bioremediation of chlorophenol-contaminated soil. *Biotechnol. Lett.* 22, 915–919. doi: 10.1023/A:1005612232079

Chou, H. H., Chiu, H. C., Delaney, N. F., Segrè, D., and Marx, C. J. (2011). Diminishing returns epistasis among beneficial mutations decelerates adaptation. *Science* 332, 1190–1192. doi: 10.1126/science.1203799

Christensen, B. B., Sternberg, C., Andersen, J. B., Eberl, L., Moller, S., Givskov, M., et al. (1998). Establishment of new genetic traits in a microbial biofilm community. *Appl. Environ. Microbiol.* 64, 2247–2255.

Davis, J. W., and Madsen, S. (1996). Factors affecting the biodegradation of toluene in soil. *Chemosphere* 33, 107–130. doi: 10.1016/0045-6535(96)00152-X

D'Costa, V. M., McGrann, K. M., Hughes, D. W., and Wright, G. D. (2006). Sampling the antibiotic resistome. *Science* 311, 374–377. doi: 10.1126/science.1120800

De Gelder, L., Ponciano, J. M., Joyce, P., and Top, E. M. (2007). Stability of a promiscuous plasmid in different hosts: no guarantee for a long-term relationship. *Microbiology* 153, 452–463. doi: 10.1099/mic.0.2006/001784-0

De Gelder, L., Vandecasteele, F. P. J., Brown, C. J., Forney, L. J., and Top, E. M. (2005). Plasmid donor affects host range of promiscuous IncP-1β plasmid pB10 in an activated-sludge microbial community. *Appl. Environ. Microbiol.* 71, 5309–5317. doi: 10.1128/AEM.71.9.5309-5317.2005

Dejonghe, W., Goris, J., El Fantroussi, S., Hofte, M., De Vos, P., Verstraete, W., et al. (2000). Effect of dissemination of 2,4- dichlorophenoxyacetic acid (2,4-D) degradation plasmids on 2,4-D degradation and on bacterial community structure in two different soil horizons. *Appl. Environ. Microbiol.* 66, 3297–3304. doi: 10.1128/AEM.66.8.3297-3304.2000

DiGiovanni, G. D., Neilson, J. W., Pepper, I. L., and Sinclair, N. A. (1996). Gene transfer of *Alcaligenes eutrophus* JMP134 plasmid pJP4 to indigenous soil recipients. *Appl. Environ. Microbiol.* 62, 2521–2526.

Dionisio, F., Conceição, I. C., Marques, A. C., Fernandes, L., and Gordo, I. (2005). The evolution of a conjugative plasmid and its ability to increase bacterial fitness. *Biol. Lett.* 1, 250–252. doi: 10.1098/rsbl.2004.0275

Directive 2001/18/EC (2001). *Directive/18/EC of the European Parliament and of the Council of 12 March 2001 on the Deliberate Release into the Environment of Genetically Modified Organisms and Repealing Council Directive 90/220/EEC.* Brussels: European Commission.

Djokic, L., Narancic, T., Biocanin, M., Saljnikov, E., Casey, E., Vasiljevic, B., et al. (2013). Phenol removal from four different natural soil types by Bacillus sp. PS11. *Appl. Soil Ecol.* 70, 1–8. doi: 10.1016/j.apsoil.2013.04.002

El Fantroussi, S., and Agathos, S. N. (2005). Is bioaugmentation a feasible strategy for pollutant removal and site remediation? *Curr. Opin. Microbiol.* 8, 268–275. doi: 10.1016/j.mib.2005.04.011

Epelde, L., Becerril, J. M., Alkorta, I., and Garbisu, C. (2009). "Heavy metal phytoremediation: microbial indicators of soil health for the assessment of remediation efficiency," in *The Advances in Applied Bioremediation*, eds A. Singh, R. C. Kuhad, and O. P. Ward (Berlin: Springer), 299–313.

Epelde, L., Becerril, J. M., Alkorta, I., and Garbisu, C. (2014). Adaptive long-term monitoring of soil health in metal phytostabilization: ecological attributes and ecosystem services based on soil microbial parameters. *Int. J. Phytoremediat.* 16, 971–981. doi: 10.1080/15226514.2013.810578

Epelde, L., Becerril, J. M., Kowalchuk, G. A., Deng, Y., Zhou, J., and Garbisu, C. (2010). Impact of metal pollution and *Thlaspi caerulescens* growth on soil microbial communities. *Appl. Environ. Microbiol.* 76, 7843–7853. doi: 10.1128/AEM.01045-10

Filonov, A. E., Akhmetov, L. I., Puntus, I. F., Esikova, T. Z., Gafarov, A. B., Kosheleva, I. A., et al. (2010). Horizontal transfer of catabolic plasmids and naphthalene biodegradation in open soil. *Microbiology* 79, 184–190. doi: 10.1134/S0026261710020098

Franzetti, A., Caredda, P., Ruggeri, C., La Colla, L., Tamburini, E., Papacchini, M., et al. (2009). Potential application of surface active compounds by *Gordonia* sp. strain BS29 in soil remediation technologies. *Chemosphere* 75, 801–807. doi: 10.1016/j.chemosphere.2008.12.052

Frost, L. S., Leplae, R., Summers, A. O., and Toussaint, A. (2005). Mobile genetic elements: the agents of open source evolution. *Nat. Rev.* 3, 722–732. doi: 10.1038/nrmicro1235

Furuya, E. Y., and Lowy, F. D. (2006). Antimicrobial-resistant bacteria in the community setting. *Nat. Rev. Microbiol.* 4, 36–45. doi: 10.1038/nrmicro1325

Gao, C., Jin, X., Ren, J., Fang, H., and Yu, Y. (2015). Bioaugmentation of DDT-contaminated soil by dissemination of the catabolic plasmid pDOD. *J. Environ. Sci.* 27, 42–50. doi: 10.1016/j.jes.2014.05.045

Garbisu, C., and Alkorta, I. (1997). Utilization of genetically engineered microorganisms (GEMs) for bioremediation. *J. Chem. Technol. Biotechnol.* 74, 599–606. doi: 10.1080/07388550600842794

Garbisu, C., Alkorta, I., and Epelde, L. (2011). Assessment of soil quality using microbial properties and attributes of ecological relevance. *Appl. Soil Ecol.* 49, 1–4. doi: 10.1016/j.apsoil.2011.04.018

Garbisu, C., Hernández-Allica, J., Barrutia, O., Alkorta, I., and Becerril, J. M. (2002). Phytoremediation: a technology using green plants to remove contaminants from polluted areas. *Rev. Environ. Health* 17, 173–188. doi: 10.1515/REVEH.2002.17.3.173

Gardin, H., and Pauss, A. (2001). κ-carrageenan/gelatin gel beads for the co-immobilization of aerobic and anaerobic microbial communities degrading 2,4,6-trichlorophenol under air-limited conditions. *Appl. Microbiol. Biotechnol.* 56, 517–523. doi: 10.1007/s002530000581

Geier, H., Mostowy, S., Cangelosi, G. A., Behr, M. A., and Ford, T. E. (2008). Autoinducer-2 triggers the oxidative stress response in *Mycobacterium avium*, leading to biofilm formation. *Appl. Environ. Microbiol.* 74, 1798–1804. doi: 10.1128/AEM.02066-07

Gentili, A. R., Cubitto, M. A., Ferrero, M., and Rodriquéz, M. S. (2006). Bioremediation of crude oil polluted seawater by a hydrocarbon-degrading bacterial strain immobilized on chitin and chitosan flakes. *Int. Biodeterior. Biodegrad.* 57, 222–228. doi: 10.1016/j.ibiod.2006.02.009

Gómez-Sagasti, M. T., Epelde, L., Alkorta, I., and Garbisu, C. (2016). Reflections on soil contamination research from a biologist's point of view. *Appl. Soil Ecol.* 105, 207–210. doi: 10.1016/j.apsoil.2016.04.004

Greer, L. E., and Shelton, D. R. (1992). Effect of inoculant strain and organic matter content on kinetics of 2,4-dichloro-phenoxyacetic acid degradation in soil. *Appl. Environ. Microbiol.* 58, 1459–1465.

Halden, R. U., Tepp, S., Halden, B. G., and Dwyer, D. F. (1999). Degradation of 3-phenoxybenzoic acid in soil by *Pseudomonas pseudoalcaligenes* POB310(pPOB) and two modified *Pseudomonas* strains. *Appl. Environ. Microbiol.* 65,3354–3359.

Hall, J. P. J., Brockhurst, M. A., Dytham, C., and Harrison, E. (2017). The evolution of plasmid stability: are infectious transmission and compensatory evolution competing evolutionary trajectories. *Plasmid* 91, 90–95. doi: 10.1016/j.plasmid.2017.04.003

Hall, J. P. J., Wood, A. J., Harrison, E., and Brockhurst, M. A. (2016). Source-sink plasmid transfer dynamics maintain gene mobility in soil bacterial communities. *Proc. Natl. Acad. Sci. U.S.A.* 113, 8260–8265. doi: 10.1073/pnas.1600974113

Haluška, L., Barančiková, G., Baláž, Š, Dercová, K., Vrana, B., Paz-Weisshaar, M., et al. (1995). Degradation of PCB in different soils by inoculated *Alcaligenes xylosoxidans*. *Sci. Total Environ.* 175, 275–285. doi: 10.1016/0048-9697(95)04927-4

Hanscha, C., McKarnsb, S. C., Smith, C. J., and Doolittle, D. J. (2000). Comparative QSAR evidence for a free-radical mechanism of phenol-induced toxicity. *Chem. Biol. Interact.* 127, 61–72. doi: 10.1016/S0009-2797(00)00171-x

Harrison, E., Guymer, D., Spiers, A. J., Paterson, S., and Brockhurst, M. A. (2015). Parallel compensatory evolution stabilizes plasmids across the parasitism-mutualism continuum. *Curr. Biol.* 25, 2034–2039. doi: 10.1016/j.cub.2015.06.024

Heinaru, E., Merimaa, M., Viggor, S., Lehiste, M., Leito, I., Truu, J., et al. (2005). Biodegradation efficiency of functionally important populations selected for bioaugmentation in phenoland oil-polluted area. *FEMS Microbiol. Ecol.* 51, 363–373. doi: 10.1016/j.femsec.2004.09.009

Heitkamp, M. A., and Steward, W. P. (1996). A novel porous nylon biocarrier for immobilized bacteria. *Appl. Environ. Microbiol.* 62, 4659–4662.

Høiby, N., Bjarnsholt, T., Givskov, M., Molin, S., and Ciofu, O. (2010). Antibiotic resistance of bacterial biofilms. *Int. J. Antimicrob. Agents* 35, 322–332. doi: 10.1016/j.ijantimicag.2009.12.011

Ikuma, K., and Gunsch, C. K. (2010). Effect of carbon source addition on toluene biodegradation by an *Escherichia coli* DH5α transconjugant harboring the TOL plasmid. *Biotechnol. Bioeng.* 107, 269–277. doi: 10.1002/bit.22808

Ikuma, K., and Gunsch, C. K. (2012). Genetic bioaugmentation as an effective method for *in situ* bioremediation: functionality of catabolic plasmids following conjugal transfers. *Bioengineered* 3, 236–241. doi: 10.4161/bioe.20551

Ikuma, K., Holzem, R. M., and Gunsch, C. K. (2012). Impacts of organic carbon availability and recipient bacteria characteristics on the potential for TOL plasmid genetic bioaugmentation in soil slurries. *Chemosphere* 89, 158–163. doi: 10.1016/j.chemosphere.2012.05.086

Inoue, D., Sei, K., Soda, S., Ike, M., and Fujita, M. (2005). Potential of predominant activated sludge bacteria as recipients in conjugative plasmid transfer. *J. Biosci. Bioeng.* 100, 600–605. doi: 10.1263/jbb.100.600

Inoue, D., Yamazaki, Y., Tsutsui, T., Sei, K., Soda, S., Fujita, M., et al. (2012). Impacts of gene bioaugmentation with pJP4-harboring bacteria of 2,4-D-contaminated soil slurry on the indigenous microbial community. *Biodegradation* 23, 263–276. doi: 10.1007/s10532-011-9505-x

Iwamoto, T., and Nasu, M. (2001). Current bioremediation practice and perspective. *J. Biosci. Bioeng.* 92, 1–8. doi: 10.1016/S1389-1723(01)80190-0

Jefferson, K. K. (2004). What drives bacteria to produce a biofilm? *FEMS Microbiol. Lett.* 236, 163–173. doi: 10.1016/j.femsle.2004.06.005

Johnsen, A. R., and Kroer, N. (2007). Effects of stress and other environmental factors on horizontal plasmid transfer assessed by direct quantification of discrete transfer events. *FEMS Microbiol. Ecol.* 59, 718–728. doi: 10.1111/j.1574-6941.2006.00230.x

Jung, C. M., Crocker, F. H., Eberly, J. O., and Indest, K. J. (2011). Horizontal gene transfer (HGT) as a mechanism of disseminating RDX-degrading activity among *Actinomycete* bacteria. *J. Appl. Microbiol.* 110, 1449–1459. doi: 10.1111/j.1365-2672.2011.04995.x

Jussila, M. M., Zhao, J., Suominen, L., and Lindström, K. (2007). TOL plasmid transfer during bacterial conjugation in vitro and rhizoremediation of oil compounds in vivo. *Environ. Pollut.* 146, 510–524. doi: 10.1016/j.envpol.2006.07.012

Juwarkar, A. A., Misra, R. R., and Sharma, J. K. (2014). "Recent trends in bioremediation," in *Geomicrobiology and Biogeochemistry*, eds N. Parmar and A. Singh (Berlin: Springer), 81–100.

Karve, S. M., Tiwary, K., Selveshwari, S., and Dey, S. (2016). Environmental fluctuations do not select for increased variation or population-based resistance in *Escherichia coli. J. Biosci.* 41, 39–49. doi: 10.1007/s12038-016-9592-2

Kim, J. M., Le, N. T., Chung, B. S., Park, J. H., Bae, J. W., Madsen, E. L., et al. (2008). Influence of soil components on the biodegradation of benzene, toluene, ethylbenzene, and *o-*, *m-*, and *p-xylenes* by the newly isolated bacterium *Pseudoxanthomonas spadix* BD-a59. *Appl. Environ. Microbiol.* 74, 7313–7320. doi: 10.1128/AEM.01695-08

Kottara, A., Hall, J. P. J., Harrison, E., and Brockhurst, M. A. (2016). Multi-host environments select for host-generalist conjugative plasmids. *BMC Evol. Biol.* 16:70. doi: 10.1186/s12862-016-0642-z

Król, J. E., Nguyen, H. D., Rogers, L. M., Beyenal, H., Krone, S. M., and Top, E. M. (2011). Increased transfer of a multidrug resistance plasmid in *Escherichia coli* biofilms at the air-liquid interface. *Appl. Environ. Microbiol.* 77, 5079–5088. doi: 10.1128/AEM.00090-11

Kumavath, R. N., and Deverapalli, P. (2013). "Scientific swift in bioremediation: an overview," in *Applied Bioremediation - Active and Passive Approaches*, eds Y. B. Patil and P. Rao (Croatia: Intech), 375–388.

Liang, Y., Zhang, X., Dai, D., and Li, G. (2009). Porous biocarrier-enhanced biodegradation of crude oil contaminated soil. *Int. Biodeterior. Biodegrad.* 63, 80–87. doi: 10.1016/j.ibiod.2008.07.005

Maier, R. M. (2000). "Microorganisms and organic pollutants," in *Environmental Microbiology*, eds R. M. Maier, I. L. Pepper, and C. P. Gerba (San Diego, CA: Academic Press), 63–402.

Maphosa, F., Lieten, S. H., Dinkla, I., Stams, A. J., Smidt, H., and Fennell, D. E. (2012). Ecogenomics of microbial communities in bioremediation of chlorinated contaminated sites. *Front. Microbiol.* 3:351. doi: 10.3389/fmicb.2012.00351

Martínez, J. L., Sánchez, M. B., Martínez-Solano, L., Hernandez, A., Garmendia, L., Fajardo, A., et al. (2009). Functional role of bacterial multidrug efflux pumps in microbial natural ecosystems. *FEMS Microbiol. Rev.* 33, 430–449. doi: 10.1111/j.1574-6976.2008.00157.x

Massa, V., Infantino, A., Radice, F., Orlandi, V., Tavecchio, F., Giudici, R., et al. (2009). Efficiency of natural and engineered bacterial strains in the degradation of 4-chlorobenzoic acid in soil slurry. *Int. Biodeterior. Biodegrad.* 63, 112–125. doi: 10.1016/j.ibiod.2008.07.006

Meckenstock, R. U., Elsner, M., Griebler, C., Lueders, T., Stumpp, C., Aamand, J., et al. (2015). Biodegradation: updating the concepts of control for microbial cleanup in contaminated aquifers. *Environ. Sci. Technol.* 49, 7073–7081. doi: 10.1021/acs.est.5b00715

Miller, M. N., Stratton, G. W., and Murray, G. (2004). Effects of soil moisture and aeration on the biodegradation of pentachlorophenol contaminated soil. *Bull. Environ. Contam. Toxicol.* 72, 101–108. doi: 10.1007/s00128-003-0246-3

Miyazaki, R., Sato, Y., Ito, M., Ohtsubo, Y., Nagata, Y., and Tsuda, M. (2006). Complete nucleotide sequence of an exogenously isolated plasmid, pLB1, involved in gamma-hexachlorocyclohexane degradation. *Appl. Environ. Microbiol.* 72, 6923–6933. doi: 10.1128/AEM.01531-06

Mølbak, L., Molin, S., and Kroer, N. (2007). Root growth and exudate production define the frequency of horizontal plasmid transfer in the rhizosphere. *FEMS Microbiol. Ecol.* 59, 167–176. doi: 10.1111/j.1574-6941.2006.00229.x

Morosini, M. I., Ayala, J. A., Baquero, F., Martínez, J. L., and Blazquez, J. (2000). Biological cost of AmpC production for *Salmonella enterica* serotype *Typhimurium. Antimicrob. Agents Chemother.* 44, 3137–3143. doi: 10.1128/AAC.44.11.3137-3143.2000

Mrozik, A., Miga, S., and Piotrowska-Seget, Z. (2011). Enhancement of phenol degradation by soil bioaugmentation with *Pseudomonas* sp. JS150. *J. Appl. Microbiol.* 111, 1357–1370. doi: 10.1111/j.1365-2672.2011.05140.x

Mrozik, A., and Piotrowska-Seget, Z. (2010). Bioaugmentation as a strategy for cleaning up of soils contaminated with aromatic compounds. *Microbiol. Res.* 165, 363–375. doi: 10.1016/j.micres.2009.08.001

Muñoz-Leoz, B., Garbisu, C., Charcosset, J. Y., Sánchez-Pérez, J. M., Antigüedad, I., Ruiz, E., et al. (2013). Non-target effects of three formulated pesticides on microbially-mediated processes in a clay-loam soil. *Sci. Total Environ.* 449, 345–354. doi: 10.1016/j.scitotenv.2013.01.079

Musovic, S., Dechesne, A., Sørensen, J., and Smets, B. F. (2010). A novel assay to assess permissiveness of a soil microbial community toward receipt of mobile genetic elements. *Appl. Environ. Microbiol.* 76, 4813–4818. doi: 10.1128/AEM.02713-09

Newby, D. T., Gentry, T. J., and Pepper, I. L. (2000). Comparison of 2,4dichlorophenoxyacetic acid degradation and plasmid transfer in soil resulting from bioaugmentation with two different pJP4 donors. *Appl. Environ. Microbiol.* 66, 3399–3407. doi: 10.1128/AEM.66.8.3399-3407.2000

Newby, D. T., and Pepper, I. L. (2002). Dispersal of plasmid pJP4 in unsaturated and saturated 2,4-dichlorophenoxyacetic acid contaminated soil. *FEMS Microbiol. Ecol.* 39, 157–164. doi: 10.1111/j.1574-6941.2002.tb00917.x

Norman, A., Hansen, L. H., and Sørensen, S. J. (2009). Conjugative plasmids: vessels of the communal gene pool. *Philos. Trans. R Soc. Lond. B Biol. Sci.* 364, 2275–2289. doi: 10.1098/rstb.2009.0037

Nowak, A., and Mrozik, A. (2016). Facilitation of co-metabolic transformation and degradation of monochlorophenols by *Pseudomonas* sp. CF600 and changes in its fatty acid composition. *Water Air Soil Pollut.* 227, 83. doi: 10.1007/s11270-016-2775-5

Overhage, J., Sielker, S., Homburg, S., Parschat, K., and Fetzner, S. (2005). Identification of large linear plasmids in Arthrobacter spp. encoding the degradation of quinaldine to anthranilate. *Microbiology* 151, 491–500. doi: 10.1099/mic.0.27521-0

Panagos, P., Van Liedekerke, M., Yigini, Y., and Montanarella, L. (2013). Contaminated sites in Europe: review of the current situation based on data collected through a European network. *J. Environ. Public Health* 2013, 1–11. doi: 10.1155/2013/158764

Pardo, T., Clemente, R., Epelde, L., Garbisu, C., and Bernal, M. P. (2014). Evaluation of the phytostabilisation efficiency in a trace elements contaminated soil using soil health indicators. *J. Hazard Mater.* 268, 68–76. doi: 10.1016/j.jhazmat.2014.01.003

Park, C., and Zhang, J. (2012). High expression hampers horizontal gene transfer. *Genome Biol. Evol.* 4, 523–532. doi: 10.1093/gbe/evs030

Pepper, I. L., Gentry, T. J., Newby, D. T., Roane, T. M., and Josephson, K. L. (2002). The role of cell bioaugmentation and gene bioaugmentation in the remediation of co-contaminated soils. *Environ. Health Perspect.* 110, 943–946. doi: 10.1289/ehp.02110s6943

Pinedo, C. A., and Smets, B. F. (2005). Conjugal TOL transfer from *Pseudomonas putida* to *Pseudomonas aeruginosa*: effects of restriction proficiency, toxicant exposure, cell density ratios, and conjugation detection method on observed transfer efficiencies. *Appl. Environ. Microbiol.* 71, 51–57. doi: 10.1128/AEM.71.1.51-57.2005

Popa, O., Hazkani-Covo, E., Landan, G., Martin, W., and Dagan, T. (2011). Directed networks reveal genomic barriers and DNA repair bypasses to lateral gene transfer among prokaryotes. *Genome Res.* 21, 599–609. doi: 10.1101/gr.115592.110

Popowska, M., and Krawczyk-Balska, A. (2013). Broad-host-range IncP-1 plasmids and their resistance potential. *Front. Microbiol.* 4:44. doi: 10.3389/fmicb.2013.00044

Porse, A., Schønning, K., Munck, C., and Sommer, M. O. A. (2016). Survival and evolution of a large multidrug resistance plasmid in new clinical bacterial hosts. *Mol. Biol. Evol.* 33, 2860–2873. doi: 10.1093/molbev/msw163

Quan, X., Tang, H., Xiong, W., and Yang, Z. (2010). Bioaugmentation of aerobic sludge granules with a plasmid donor strain for enhanced degradation of 2,4-dichlorophenoxyacetic acid. *J. Hazard Mater.* 179, 1136–1142. doi: 10.1016/j.jhazmat.2010.04.00

Reichenbach, T., Mobilia, M., and Frey, E. (2007). Mobility promotes and jeopardizes biodiversity in rock-paper-scissors games. *Nature* 448, 1046–1049. doi: 10.1038/nature06095

Reineke, W. (1998). Development of hybrid strains for the mineralization of chloroaromatics by patchwork assemble. *Annu. Rev. Microbiol.* 52, 287–331. doi: 10.1146/annurev.micro.52.1.287

Riley, M. A., and Wertz, J. E. (2002). Bacteriocins: evolution, ecology and application. *Annu. Rev. Microbiol.* 56, 117–137. doi: 10.1146/annurev.micro.56.012302.161024

San Millan, A., Peña-Miller, R., Toll-Riera, M., Halbert, Z. V., McLean, A. R., Cooper, B. S., et al. (2014). Positive selection and compensatory adaptation interact to stabilize non-transmissible plasmids. *Nat. Commun.* 5, 5208. doi: 10.1038/ncomms6208

Sayler, G. S., and Ripp, S. (2000). Field applications of genetically engineered microorganisms for bioremediation processes. *Curr. Opin. Biotechnol.* 11, 286–289. doi: 10.1016/S0958-1669(00)00097-5

Sejáková, Z., Dercová, K., and Tóthová, L. (2009). Biodegradation and ecotoxicity of soil contaminated by pentachlorophenol applying bioaugmentation and addition of sorbents. *World J. Microbiol. Biotechnol.* 25, 243–252. doi: 10.1007/s11274-008-9885-1

Sengupta, S., Chattopadhyay, M. K., and Grossart, H. P. (2013). The multifaceted roles of antibiotics and antibiotic resistance in nature. *Front. Microbiol.* 4:47. doi: 10.3389/fmicb.2013.00047

Shachrai, I., Zaslaver, A., Alon, U., and Dekel, E. (2010). Cost of unneeded proteins in *E. coli* is reduced after several generations in exponential growth. *Mol. Cell* 38, 758–767. doi: 10.1016/j.molcel.2010.04.015

Shintani, M., Sanchez, Z. K., and Kimbara, K. (2015). Genomics of microbial plasmids: classification and identification based on replication and transfer systems and host taxonomy. *Front. Microbiol.* 31:242. doi: 10.3389/fmicb.2015.00242

Shintani, M., Takahashi, Y., Yamane, H., and Nojiri, H. (2010a). The behavior and significance of degradative plasmids belonging to Inc groups in *Pseudomonas* within natural environments and microcosms. *Microbes Environ.* 25, 253–265. doi: 10.1264/jsme2.ME10155

Shintani, M., Yamane, H., and Nojiri, H. (2010b). Behaviour of various hosts of the IncP-7 carbazole-degradative plasmid pCAR1 in artificial microcosms. *Biosci. Biotechnol. Biochem.* 74, 343–349. doi: 10.1271/bbb.90675

Singh, A., and Ward, O. P. (2004). "Biotechnology and bioremediation An overview," in *Biodegradation and Bioremediation*, eds A. Singh and O. P. Ward (Berlin: Springer-Verlag), 1–17.

Sorek, R., Zhu, Y., Creevey, C. J., Francino, M. P., Bork, P., and Rubin, E. M. (2007). Genome-wide experimental determination of barriers to horizontal gene transfer. *Science* 318, 1449–1452. doi: 10.1126/science.1147112

Sørensen, S. J., Bailey, M., Hansen, L. H., Kroer, N., and Wuertz, S. (2005). Studying plasmid horizontal transfer *in situ*: a critical review. *Nat. Rev. Microbiol.* 3, 700–710. doi: 10.1038/nrmicro1232

Stalwood, B., Shears, J., Williams, P. A., and Hughes, K. A. (2005). Low temperature bioremediation of oil-contaminated soil using biostimulation and bioaugmentation with a *Pseudomonas* sp. from maritime Antarctica. *J. Appl. Microbiol.* 99, 794–802. doi: 10.1111/j.1365-2672.2005.02678.x

Starikova, I., Al-Haroni, M., Werner, G., Roberts, A., Sørum, V., Nielsen, K. M., et al. (2013). Fitness costs of various mobile genetic elements in *Enterococcus faecium* and *Enterococcus faecalis*. *J. Antimicrob. Chemother.* 68, 2755–2765. doi: 10.1093/jac/dkt270

Thomas, C. M., and Nielsen, K. M. (2005). Mechanisms of, and barriers to, horizontal gene transfer between bacteria. *Nat. Rev. Microbiol.* 3, 711–721. doi: 10.1038/nrmicro1234

Thompson, I. P., Van der Gast, C. J., Ciric, L., and Singer, A. C. (2005). Bioaugmentation for bioremediation: the challenge of strain selection. *Environ. Microbiol.* 7, 909–915. doi: 10.1111/j.1462-2920.2005.00804.x

Top, E. M., and Springael, D. (2003). The role of mobile genetic elements in bacterial adaptation to xenobiotic organic compounds. *Curr. Opin. Biotechnol.* 14, 262–269. doi: 10.1016/S0958-1669(03)00066-1

Top, E. M., Springael, D., and Boon, N. (2002). Catabolic mobile genetic elements and their potential use in bioaugmentation of polluted soils and waters. *FEMS Microbiol. Ecol.* 42, 199–208. doi: 10.1111/j.1574-6941.2002.tb01009.x

Top, E. M., Van Elsas, E. D., DeSmet, S. N., and Forney, L. J. (1998). Enhancement of 2,4-dichlorophenoxyacetic acid (2,4-D) degradation in soil by dissemination of catabolic plasmids. *Antonie Van Leeuwenhoek* 73, 87–94. doi: 10.1023/A:1000663619522

Tsutsui, H., Anami, Y., Matsuda, M., Hashimoto, K., Inoue, D., Sei, K., et al. (2013). Plasmid-mediated bioaugmentation of sequencing batch reactors for enhancement of 2,4-dichlorophenoxyacetic acid removal in wastewater using plasmid pJP4. *Biodegradation* 24, 343–352. doi: 10.1007/s10532-012-9591-4

van Rensburg, E., den Haan, R., Smith, J., van Zyl, W. H., and Görgens, J. F. (2012). The metabolic burden of cellulase expression by recombinant Saccharomyces cerevisiae Y294 in aerobic batch culture. *Appl. Microbiol. Biotechnol.* 96, 197–209. doi: 10.1007/s00253-012-4037-9

Venkata Mohan, S., Falkentoft, C., Nancharaiah, Y. V., Sturm, B. S. M., Wattiau, P., Wilderer, P. A., et al. (2009). Bioaugmentation of microbial communities in laboratory and pilot scale sequencing batch biofilm reactors using the TOL plasmid. *Bioresour. Technol.* 100, 1746–1753. doi: 10.1016/j.biortech.2008.09.048

Verstraeten, N., Braeken, K., Debkumari, B., Fauvart, M., Fransaer, J., Vermant, J., et al. (2008). Living on a surface: swarming and biofilm formation. *Trends Microbiol.* 16, 496–506. doi: 10.1016/j.tim.2008.07.004

Vogwill, T., and MacLean, R. C. (2015). The genetic basis of the fitness costs of antimicrobial resistance: a meta-analysis approach. *Evol. Appl.* 8, 284–295. doi: 10.1111/eva.12202

Walsh, C. (2000). Molecular mechanisms that confer antibacterial drug resistance. *Nature* 406, 775–781. doi: 10.1038/35021219

Wang, J. L., Mao, Z. Y., Han, L. P., and Qian, Y. (2004). Bioremediation of quinoline-contaminated soil using bioaugmentation in slurry-phase reactor. *Biomed. Environ. Sci.* 17, 187–195.

Wang, S., and Mulligan, C. N. (2004). An evaluation of surfactant foam technology in remediation of contaminated soil. *Chemosphere* 57, 1079–1089. doi: 10.1016/j.chemosphere.2004.08.019

Wang, Y., Kou, S., Jiang, Q., Xu, B., Liu, X., Xiao, J., et al. (2014). Factors affecting transfer of degradative plasmids between bacteria in soils. *Appl. Soil Ecol.* 84, 254–261. doi: 10.1016/j.apsoil.2014.07.009

Wickham, S. A., and Lynn, D. H. (1990). Relation between growth rate, cell size, and DNA content in colpodean ciliates (Ciliophora: Colpodea). *Eur. J. Protistol.* 25, 345–352. doi: 10.1016/S0932-4739(11)80127-6

Wiedenbeck, J., and Cohan, F. M. (2011). Origins of bacterial diversity through horizontal genetic transfer and adaptation to new ecological niches. *FEMS Microbiol. Rev.* 35, 957–976. doi: 10.1111/j.1574-6976.2011.00292.x

Wolski, E. A., Murialdo, S. E., and Gonzales, J. F. (2006). Effect of pH and inoculum size on pentachlorophenol degradation by *Pseudomonas* sp. *Water SA* 32, 93–98.

Wozniak, R. A., and Waldor, M. K. (2010). Integrative and conjugative elements: mosaic mobile genetic elements enabling dynamic lateral gene flow. *Nat. Rev. Microbiol.* 8, 552–563. doi: 10.1038/nrmicro2382

Wyndham, R. C., Cashore, A. E., Nakatsu, C. H., and Peel, M. C. (1994). Catabolic transposons. *Biodegradation* 5, 323–357. doi: 10.1007/BF00696468

Zhang, Q., Wang, B. C., Cao, Z. Y., and Yu, Y. L. (2012). Plasmid-mediated bioaugmentation for the degradation of chlorpyrifos in soil. *J. Hazard. Mater.* 221-222, 178–184. doi: 10.1016/j.jhazmat.2012.04.024

18

Plasmid Replicons from *Pseudomonas* are Natural Chimeras of Functional, Exchangeable Modules

Leire Bardaji[1†], Maite Añorga[1†], José A. Ruiz-Masó[2], Gloria del Solar[2] and Jesús Murillo[1]*

[1] *Departamento de Producción Agraria, Escuela Técnica Superior de Ingenieros Agrónomos, Universidad Pública de Navarra, Pamplona, Spain,* [2] *Molecular Biology of Gram-Positive Bacteria, Molecular Microbiology and Infection Biology, Centro de Investigaciones Biológicas (Consejo Superior de Investigaciones Científicas), Madrid, Spain*

Edited by:
Tatiana Venkova, University of Texas Medical Branch, USA

Reviewed by:
Grzegorz Wegrzyn, University of Gdańsk, Poland
Alan Leonard, Florida Institute of Technology, USA
Jan Nesvera, Institute of Microbiology of the Czech Academy of Sciences, Czechia

***Correspondence:**
Jesús Murillo
jesus.murillo@unavarra.es

[†] *These authors have contributed equally to this work.*

Plasmids are a main factor for the evolution of bacteria through horizontal gene exchange, including the dissemination of pathogenicity genes, resistance to antibiotics and degradation of pollutants. Their capacity to duplicate is dependent on their replication determinants (replicon), which also define their bacterial host range and the inability to coexist with related replicons. We characterize a second replicon from the virulence plasmid pPsv48C, from *Pseudomonas syringae* pv. savastanoi, which appears to be a natural chimera between the gene encoding a newly described replication protein and a putative replication control region present in the widespread family of PFP virulence plasmids. We present extensive evidence of this type of chimerism in structurally similar replicons from species of *Pseudomonas*, including environmental bacteria as well as plant, animal and human pathogens. We establish that these replicons consist of two functional modules corresponding to putative control (REx-C module) and replication (REx-R module) regions. These modules are functionally separable, do not show specificity for each other, and are dynamically exchanged among replicons of four distinct plasmid families. Only the REx-C module displays strong incompatibility, which is overcome by a few nucleotide changes clustered in a stem-and-loop structure of a putative antisense RNA. Additionally, a REx-C module from pPsv48C conferred replication ability to a non-replicative chromosomal DNA region containing features associated to replicons. Thus, the organization of plasmid replicons as independent and exchangeable functional modules is likely facilitating rapid replicon evolution, fostering their diversification and survival, besides allowing the potential co-option of appropriate genes into novel replicons and the artificial construction of new replicon specificities.

Keywords: control and replication modules, chimeric replicons, gene co-option, Rep proteins, origin of replication, plasmid incompatibility, swapping of functional modules, virulence plasmids

INTRODUCTION

Plasmids are extrachromosomal elements that colonize a vast majority of bacteria and other organisms, often carrying genes that confer an adaptive advantage to the host (del Solar et al., 1998; Jackson et al., 2011; Ruiz-Masó et al., 2015). Each cell can have from none to several plasmids of diverse sizes and copy numbers. Plasmids can readily acquire large amounts of foreign DNA from different sources and are transferred between distantly related organisms, including prokaryotes and eukaryotes, which makes them a major contributor to the accessory gene pool and the most important agents in horizontal gene transfer (Halary et al., 2010; Jackson et al., 2011). Indeed, plasmids are responsible for the worldwide distribution of genes for resistance to antibiotics and other antimicrobials, rendering current strategies ineffective for the control of human, animal and plant diseases (Sundin, 2007; Jackson et al., 2011; Aviv et al., 2016; Johnson et al., 2016).

The basic replicon is the fundamental element for plasmid survival, ensuring timely duplication in coordination with cell division (Nordström, 1993; Summers, 1996; del Solar et al., 1998). Broadly, basic replicons consist of (i) a short *cis*-acting DNA sequence, the origin of replication, (ii) genes and structures involved in the control of replication and, for most plasmids, (iii) a gene coding a replication initiator (Rep) protein that recognizes the origin and promotes initiation of DNA replication. Plasmid replication is controlled by either directly repeated sequences (iterons) or by antisense RNAs, which can act alone or in coordination with a protein repressing transcription of the *rep* gene, and is tightly regulated so as to maintain the number of plasmid molecules in the cell within acceptable limits (Summers, 1996; del Solar and Espinosa, 2000). An immediate consequence of this is that plasmids sharing elements for replication or replication control cannot coexist in the same cell and are hence incompatible (Novick, 1987).

Replicons are highly diverse and can be grouped based on their general mechanism of replication, the function of their Rep proteins, their structure and genetic organization or their homology (del Solar et al., 1998; Lilly and Camps, 2015). Circular plasmids replicate by one of three general modes: rolling-circle, strand-displacement and theta-type mechanisms. According to their mode of replication initiation, the theta-type replicons have been grouped into four classes (A, B, C, and D) (Bruand et al., 1993). Class A theta replicons (e.g., R1, RK2, R6K, pSC101, pPS10, F and P) encode a Rep protein that binds to the origin and mediates melting of the duplex DNA. Class B (ColE1-like) replicons lack a *rep* gene, and melting of duplex DNA as well as synthesis of a pre-primer RNA for replication are achieved by bacterial RNA polymerase-mediated transcription. Class C (ColE2- and ColE3-like) replicons contain the smallest origins reported so far and encode a Rep primase protein that also mediates unwinding of the DNA (Itou et al., 2015). Finally, functioning of class D replicons (plasmids pAMβ1, pIP501, and pSM19035 from Gram-positive bacteria) requires transcription across the origin and participation of a Rep protein in melting of the DNA and primer processing.

The gamma proteobacterial genus *Pseudomonas* comprises very diverse species, present in all kinds of environments, including significant human, animal and plant pathogens as well as species of outstanding biotechnological interest (Ramos, 2004). *Pseudomonas syringae* is one of the most relevant plant pathogenic bacteria in the world (Mansfield et al., 2012), and many strains carry one or more highly stable plasmids, ranging from a couple of kilobases to close to 1 Mb (Murillo and Keen, 1994; Sundin, 2007; Romanchuk et al., 2014). Most plasmids from *P. syringae*, and also various from many other *Pseudomonas* species, belong to the PFP (*p*PT23A-*f*amily *p*lasmid) group (Murillo and Keen, 1994; Sesma et al., 1998, 2000; Gibbon et al., 1999; Sundin, 2007). PFPs appear to originate from a common ancestor because they share homologous RepA-PFP replicons, which are related to the ColE2 class C theta replicons (Murillo and Keen, 1994; Sesma et al., 1998; Gibbon et al., 1999; Sesma et al., 2000; Sundin, 2007). ColE2 replicons contain a *rep* gene and an upstream region coding for a small antisense RNA, which is complementary to the 5′-nontranslated region of the *rep* mRNA and negatively controls its expression posttranscriptionally (Yasueda et al., 1994). Similarly, the RepA-PFP replicons consist of the *repA* replication initiator gene, which includes the putative vegetative origin of replication (Yagura et al., 2006), preceded by a short 5′ sequence, containing diverse stem-and-loop (SaL) structures, which is probably involved in control of replication (Murillo and Keen, 1994; del Solar et al., 1998; Gibbon et al., 1999; Brantl, 2014). PFP plasmids have had tremendous evolutionary success, not only for their ubiquity across pseudomonads, but also because most *P. syringae* strains contain two to six coexisting PFP plasmids (Murillo and Keen, 1994; Sesma et al., 1998). This could be explained in part because they generally carry genes essential for the interaction with the plant host or for survival, fostering their frequent exchange among the bacterial population (Sesma et al., 2000; Vivian et al., 2001; Ma et al., 2007; Sundin, 2007; Sundin and Murillo, 2009). Additionally, their competitiveness among the bacterial plasmid pool might be enhanced by a replication machinery particularly adapted to their bacterial host. Notwithstanding, the coexistence of PFP plasmids in the same cell is difficult to explain because of their potential incompatibility (Novick, 1987; Sesma et al., 1998). In fact, PFP plasmids are generally incompatible with their cloned replicons (Murillo and Keen, 1994; Murillo et al., 1994; Sesma et al., 1998) although subcloning did not allow for the identification of the sequences responsible for this incompatibility within the replicon (Gibbon et al., 1999).

Pseudomonas syringae pv. savastanoi NCPPB 3335 contains three virulence PFP plasmids, pPsv48A (78 kb), pPsv48B (45 kb), and pPsv48C (42 kb), of which the smallest two appear to have originated by plasmid duplication and reorganization (Bardaji et al., 2011). Plasmid pPsv48C is essential for elicitation of disease symptoms in the plant host olive (M. Añorga, unpublished results), and is extremely stable (Bardaji et al., 2011). In this work, we identified a second replicon on pPsv48C, designated here as RepJ replicon, containing a putative replication control region homologous to that of the pPsv48C RepA-PFP replicon (Bardaji et al., 2011). We also show that these, and structurally similar replicons, consist of two functional modules corresponding to the putative control region (REx-C module) and the replication region (REx-R module). These modules are functionally separable, do not show specificity for each other, and are dynamically exchanged among replicons of four distinct families. Additionally, a REx-C module from pPsv48C conferred replication ability to a non-replicative *repJ* chromosomal homolog. Thus, the organization of plasmid replicons as independent and exchangeable functional modules is likely fostering their diversification and survival, besides allowing the potential co-option of appropriate genes into novel replicons and the artificial construction of new replicon specificities.

RESULTS

Definition of a Second Replicon in Plasmid pPsv48C

In an independent study (M. Añorga, unpublished data), we observed the spontaneous generation of autonomously replicating deletion derivatives of pPsv48C lacking the RepA-PFP replicon (**Figure 1B**). The smallest derivative contains five putative coding sequences (CDSs), whose annotation is not related to plasmid replication, and a 661 nt fragment that appears in two nearly identical copies in pPsv48C (Figure S1). The second copy of this fragment precedes gene *repA*, matching the typical organization of RepA-PFP replicons, and was previously shown to be essential for replication of PFP plasmids (Murillo and Keen, 1994; Sesma et al., 1998; Gibbon et al., 1999; Sundin et al., 2004).

By cloning diverse PCR fragments into the *E. coli* vectors pK184 or pKMAG (Figure S2), which do not replicate in *Pseudomonas*, we determined that a 1,375 nt fragment (coordinates 29,386–30,760 in the pPsv48C sequence, accession no. FR820587; fragment B-F, **Figure 1A** and Figure S1), comprising around half of the 661 nt repeated fragment, contained all the essential elements for autonomous replication in the plasmidless strains *P. syringae* pv. syringae B728a (**Figure 1B**) and *P. syringae* pv. savastanoi UPN912. This

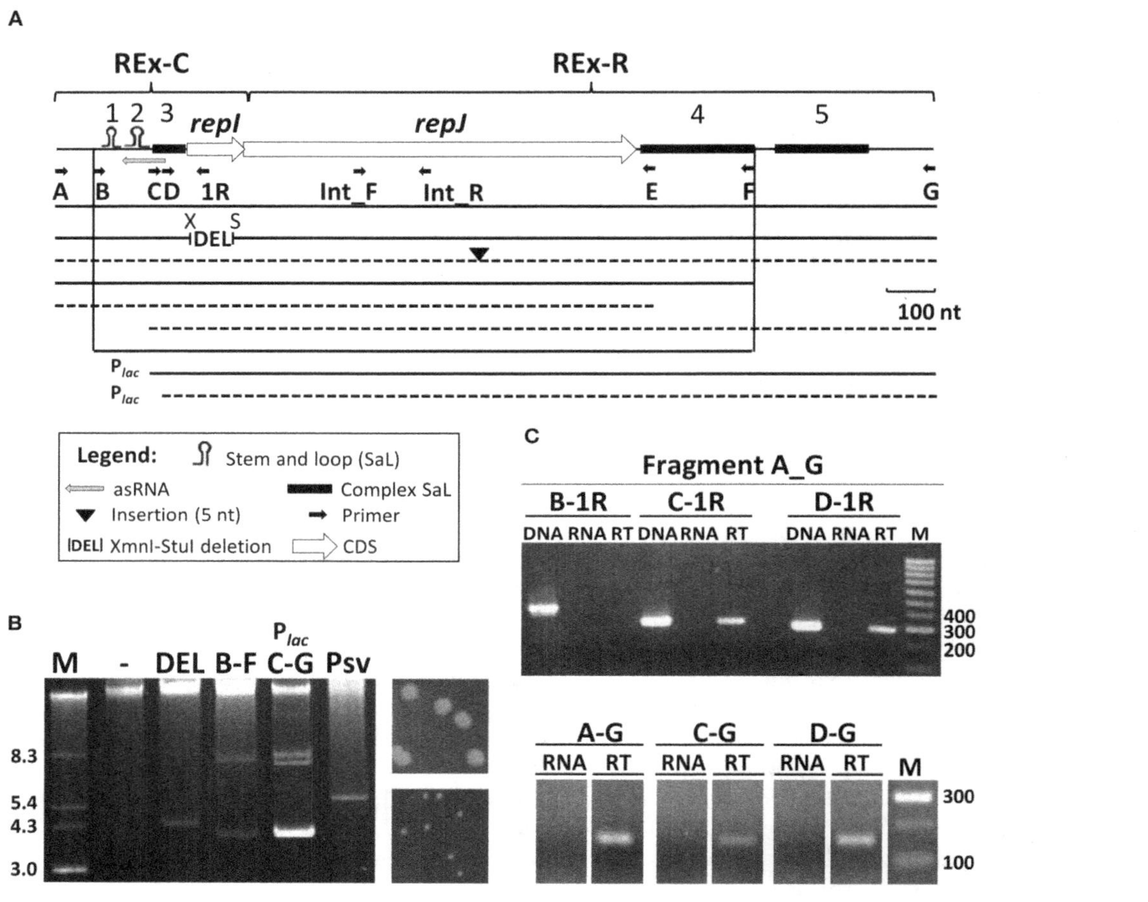

FIGURE 1 | Functional characterization of the RepJ replicon. (A) The minimal replicon consists of two modules, REx-C and REx-R, whose extent is loosely defined here; stem-and-loop structures 1 to 5 are indicated by symbols (see legend) topped with the structure number; letters under small arrows indicate the primers used for amplification of fragments and for RT-PCR, and a gray arrow, the putative antisense RNA. Lines under the map indicate the relative size of fragments cloned after a transcriptional terminator in pKMAG, or downstream of the P_{lac} promoter in pK184. Continuous lines indicate constructs able to replicate autonomously in *P. syringae* pv. syringae B728a, and dashed lines indicate non-replicative fragments. X, XmnI; S, StuI. **(B)** Plasmid profiles of B728a transformants harboring constructs from panel A (indicated by letters) able to replicate and that define the minimal replicon; multiple bands correspond to different topological forms and/or multimers with the lowest bands corresponding to monomers and the top intense band to chromosomal DNA and high-molecular-weight plasmid multimers. Lane (-) is the wild type strain B728a and Psv is a derivative of NCPPB 3335 containing a 5.6 kb spontaneous deletion derivative of pPsv48C spanning the RepJ replicon. To the right, 48 h colonies of transformants containing constructs A–G (top) or C–G (bottom). **(C)** *repJ* is transcribed from at least two promoters. Upper gel: for RT-PCR, cDNA was synthesized from UPN912 containing construct A–G (in pKMAG) using the strand-specific primer Int_R, internal to *repJ*. Subsequent PCRs were done with primer pairs indicated above the gel and shown on **(A)**. Lower gel: for RT-PCR, cDNA molecules were synthesized from UPN912 containing constructs A–G, C–G, and D–G (the last two cloned in pKMAG-C) using the strand-specific primer Int_R. Further PCR amplification of cDNA was done using primer pair Int_F-Int_R, internal to *repJ* (see panel **A**). Amplifications were done with only DNA (DNA), only RNA (RNA) or cDNA (RT); M, molecular weight marker in kb (panel **B**) or nt (panel **C**).

fragment did not contain any obvious direct repetitions reminiscent of iterons, but was rich in palindromic structures and could adopt a complex folding structure. We could distinguish two well-defined structural regions in this minimal replicating fragment; based on their conservation and functionality (see below), we have defined these regions as plasmid replicon exchangeable (REx) modules: the Rex-C module contains the putative replication control system, whereas the Rex-R module comprises the replication system.

REx-C Module

This is a 318 nt fragment (coordinates 29,386–29,703 in FR820587) that shows high identity to a fragment (coordinates 41,791-5) preceding the *repA* gene from pPsv48C and including its first two codons (**Figure 1** and Figure S1). The fragment contributes the putative start codon, the promoter(s) and the RBS for the expression of the replication initiator gene *repJ* (see below). It also contains three SaL structures, the third of which is complex, potentially folding in different ways, and gene *repI*. By analogy with replicons lacking iterons (del Solar and Espinosa, 2000; Brantl, 2014), such as ColE2, this fragment also probably codes for a small antisense RNA, with a putative promoter within SaL 3 and for which SaL 1 could function as a transcription terminator (Figure S1). Replication assays with clones spanning partial fragments of the minimal replicon showed that deletion of SaL 1 and 2 abolished autonomous replication, although they were dispensable in clones maintaining the complex SaL 3 and the strong P*lac* promoter of the vector in the same transcriptional direction as *repJ* (fragment CG in **Figure 1A**). Nevertheless, bacteria transformed with construct CG required double the time than other replicative fragments to produce visible colonies (**Figure 1B**). Additionally, clones lacking the 5′ stem of SaL 3 did not sustain autonomous replication (fragment DG, **Figure 1A** and Figure S1), even when *repJ* was cloned in the transcriptional direction of the P*lac* promoter. These results likely suggest that expression of *repJ* from the P*lac* in this clone causes a lethal runaway replication phenotype (Nordström and Wagner, 1994) or that SaL 3 is also essential for replication.

Gene *repI* (PSPSV_C0037) is short (123 nt) and appears to be translationally coupled to the replication initiator gene *repJ*, which are characteristics of leader peptide genes needed for the control of replication of certain replicons (del Solar et al., 1998; Brantl, 2014). An XmnI-StuI in-frame deletion of 87 nt, spanning most of *repI* (**Figure 1A** and Figure S1) did not have any apparent effect in the replication ability of the RepJ replicon, indicating that the product of *repI* is not essential for replication and that spacing between the SaL structures and the start of gene *repJ* is flexible.

REx-R Module

This module contains the replication initiator gene, *repJ* (PSPSV_C0038, 819 nt) and essential downstream sequences. The long, near-perfect ribosome binding site (5′-AAGGcGGTGA-3′) of *repJ* and its two first codons probably belong to the REx-C module, because they are part of a sequence highly conserved in the pPsv48C RepA-PFP replicon (Figure S1). Gene *repJ* is annotated as a putative transcriptional regulator and did not show significant homology to any domain in an InterPro search. However, the structure of 55 residues (residues 90–145) from RepJ could be modeled by Phyre2 with 81.6 % confidence, being similar to the N-terminal domain of a conserved replication initiator protein (Schumacher et al., 2014). Additionally, a construct containing a mutation causing a premature stop in *repJ* did not replicate in the plasmidless strains B728a and UPN912 (**Figure 1**). These results suggest that *repJ* codes for a replication initiator protein essential for autonomous replication.

After the *repJ* stop codon there is a *ca.* 0.5 kb fragment containing two blocks of repeated sequences that can form complex SaL structures, designated SaL 4 and 5, although only SaL 4 appears to be essential for autonomous replication (**Figure 1A**). Nevertheless, a blastn search with this fragment identifies sequences similar to SaL 5 situated 3′ of, among others, *rep* genes that are not homologous to *repJ*, such as those from plasmids pRA2 (from *P. alcaligenes* RA2), pP27494_2 (from *P. antarctica* PAMC 27494), pMBUI6 (from an uncultured bacterium), and pAOVO01 (from *Acidovorax* sp. JS42), and gene *krfA* from plasmid pTer331 (from *Colimonas fungivorans* Ter331).

The REx-C Module Contains at Least Two Active Promoters

The absolute requirement of SaL 1 and 2 for replication can be overcome when *repJ* is transcribed from the strong P*lac* promoter, suggesting a plausible role of these structures in directing *repJ* transcription. We thus examined transcription of this gene by RT-PCR in clones lacking SaL 1, 2 and 3. Using clone AG in pKMAG (**Figure 1C**, upper gel), we observed a long *repJ* transcript that extended at least to the annealing site for primer C, but not to that for primer B. This indicates transcription from a promoter situated between the annealing sites for primers A and C (Figure S1), overlapping the putative antisense RNA, and possibly involved in transcription of *repI* and *repJ*. Amplification from smaller clones showed that *repJ* was transcribed even in the absence of SaL 1, 2 and 3 (fragments CG and DG, **Figure 1C** lower gel). Since the vectors used contain a T4 transcriptional terminator upstream of the cloned fragments, this shows that there is an additional active promoter immediately upstream of *repJ*. These results indicate that the REx-C module contains at least two functional promoters for the transcription of gene *repJ* and that the failure of fragments CG and DG to replicate is not due to a lack of transcription of *repJ*, suggesting an additional role for SaL 1 and 2.

The Structure of the RepJ Replicon Is Only Partially Conserved in *Pseudomonas*

Blast comparisons revealed that a region of up to 2,785 nt containing the minimal RepJ replicon (coordinates 28,783–31,067 in FR820587) is syntenic, with very high identity, in diverse genomoespecies of the *P. syringae* group as well as, with less identity, in a few other pseudomonads (not shown). Most of the homologs are from draft genomes and it is not possible to clearly determine if they localize to the chromosome or to plasmid sequences. Sequence variation among the homologs from *P. syringae* was not distributed randomly (Figure S3): whereas the stems from SaL 1, 2, and 3 were identical in all sequences, there was a high sequence variation in the loop of SaL 2, in the 3′ end of gene *repJ* and in SaL 4 and 5, downstream of this gene. As it occurs with the *repA* gene (Gibbon et al., 1999), the nucleotide variation in *repJ* (Figure S3) leads to a higher degree of variation in the C-terminal end of the deduced product; the phylogeny of a selection of these products is shown as clade I in **Figure 2**.

Using blastp, we found RepJ homologs only within members of Pseudomonadales, mainly in species of *Pseudomonas*, and in a few *Desulfovibrio* spp. strains. An ML tree with selected sequences (**Figure 2**) grouped homologs in four well-defined clades; from this tree we can also infer that the RepJ replicon has recently moved horizontally among pathovars and species of the *P. syringae* complex and that certain strains contained two *repJ* homologs (e.g., *P syringae.* pv. fraxini CFBP5062 in clade I). The minimal RepJ replicon is moderately conserved and syntenic among homologs from clades I and II (Figure S4). Of note, the REx-C modules from the RepJ replicons from clade I were more similar among themselves than to the modules from RepA-PFP replicons, indicating that these RepJ replicons are long-time inhabitants of pseudomonads. Conversely, the REx-C module is not conserved among members in clades III and IV whereas the region downstream of *repJ* shows a degree of conservation (Figure S4); additionally these homologs show a genetic organization that is conserved among clade members, but different than that from members of clades I and II.

A 2,808 nt fragment (positions 35,079–37,886, accession no. KB644113) containing the *repJ* homolog from the chromosome of strain NCPPB 3335 (PSA3335_1080, clade III; see Figure S4) cloned in pKMAG did not generate any transformant after electroporation into *P. syringae* pv. syringae B728a or *P. syringae* pv. savastanoi UPN912, whereas in the same experiments we obtained hundreds of clones using the minimal RepJ replicon from pPsv48C. These results suggest that the chromosomal homolog from strain NCPPB 3335 is not able to sustain autonomous replication.

The REx-C Module Associates to REx-R Modules of Diverse Families and Is Exchanged among Them

Blastn comparisons showed that the REx-C module is present, with varying degrees of conservation, preceding the *rep* genes from at least four non-homologous replicon families from *Pseudomonas* (**Figure 3**; Table S1), whereas nucleotide identity is rapidly lost shortly after the CDS start codon. These families include the RepA-PFP family from *P. syringae* and other bacteria (exemplified by *repA* from pPsv48C) (Bardaji et al., 2011), the RepJ family (among others, *repJ* from pPsv48C and pA506) (Bardaji et al., 2011; Stockwell et al., 2013), and what we

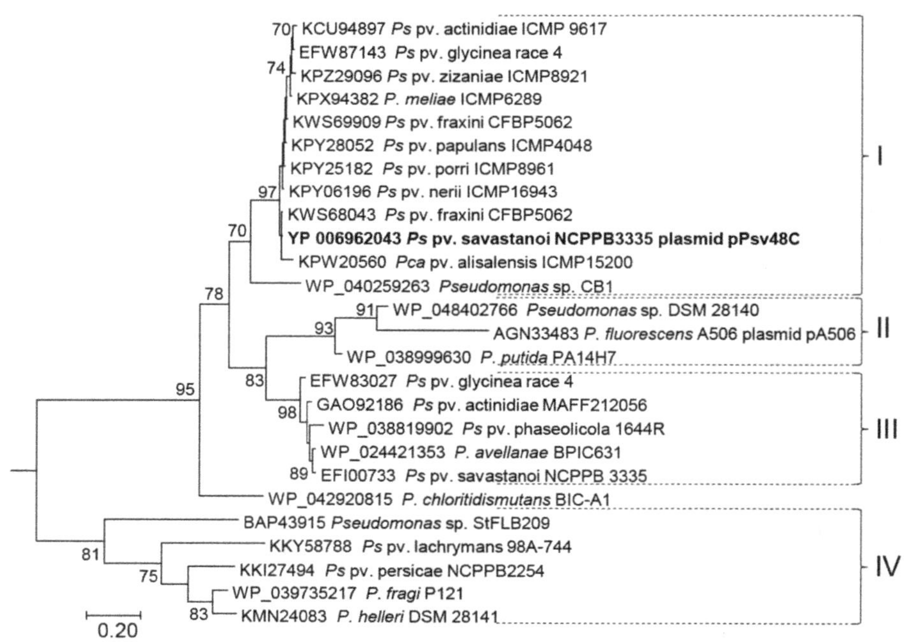

FIGURE 2 | Maximum likelihood phylogenetic tree of RepJ. Protein sequences from species of *Pseudomonas* are indicated by their accession no. followed by strain designation; the first 39 positions of the alignment were discarded to eliminate biases due to differential annotation of the start sites. The tree was rooted with sequence WP_052264087.1 from *Azotobacter chrococcum* plasmid pAcX50f and bootstrap percentages of 500 replicates higher than 70% are shown close to each node. The tree is drawn to scale, with branch lengths measured in the number of substitutions per site. Brackets followed by roman numerals indicate clades discussed in the main text. Pca, *P. cannabina*; Pco, *P. coronafaciens*; Ps, *P. syringae*.

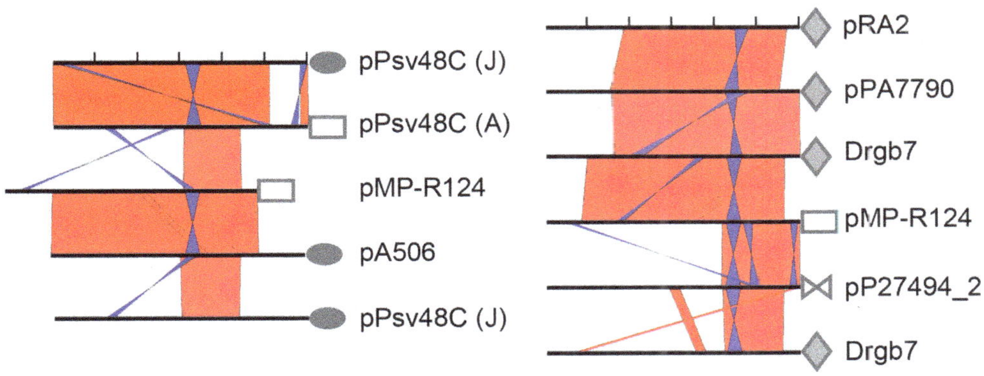

FIGURE 3 | The REx-C module is conserved among four non-homologous replicon families and subjected to recombination among them. Graphical view of NCBI blastn comparisons of the 600 nt region preceding the start codon of replication initiator protein genes (black bars, ticks every 100 nt), displayed with the ACT software; default algorithm parameters were used, except for a word size of 7. Red and blue bars indicate collinear and inverted regions of identity, respectively, with color intensity proportional to identity; only matches larger than 15 nt are shown. Symbols are followed by their corresponding plasmid name and are identical for *rep* genes of the same homology family. Sequence accession no. are pA506, NC_021361; plasmid Drgb7, KT351738; pMP-R124, NZ_CM001562; pP27494_2, CP015602; pPA7790, CP015000; pPsv48C, FR820587; pRA2, U88088. A and J refer to *repA* and *repJ*, respectively, from pPsv48C.

designated the RepA-RA2 family (named after pRA2) (Kato and Mizobuchi, 1994) and the RepA-Pa family (named here after the *P. antarctica* PAMC 27494 plasmid pP27494_2). The highest level of sequence conservation of the REx-C module among representatives of the four families is in a 175–194 nt fragment that spans SaL structures 1–3 (Figure S5). The stem sequences from SaL 1 (the putative transcriptional terminator for the antisense RNA), as well as the stretch of adenines in either side of them (Gibbon et al., 1999), are almost perfectly conserved and changes in one arm of the stems are usually compensated with complementary changes in the other arm (Figure S5) (Gibbon et al., 1999). The other structures are also well-conserved among the four replicon families; however, the number and position of palindromes, which could form SaL structures, are variable among replicons (**Figure 3**; Figure S5). Therefore, the stark conservation of the REx-C module suggests that it contains features universally essential for plasmid replication in species of *Pseudomonas*. Remarkably, we did not find REx-C sequences conserved in the control region of ColE2 replicons, which contain a replication initiator protein homologous to that from RepA-PFP replicons, or in any other plasmid outside of the genus *Pseudomonas*, indicating a diversity of REx-C modules among homologous theta replicons.

As described above, the REx-C modules accompanying genes *repA* and *repJ* from pPsv48C are nearly identical (Figure S1), although there is a gradient of conservation of this module along homologs of their corresponding gene families (**Figure 3**, Figures S3, S4) (Gibbon et al., 1999; Sesma et al., 2000; Stavrinides and Guttman, 2004). Additionally, the REx-C module preceding *repJ* from pA506 shows a very high degree of identity to that preceding *repA* from pMP-R124 (a PFP plasmid), whereas they are less similar to REx-C modules from replicons of their same family (**Figure 3**). These results show that there is frequent horizontal exchange of REx-C modules among RepA-PFP and RepJ replicons.

The results of multiple blast comparisons indicate that this exchange of REx-C modules also occurs with other replicon families. To illustrate this, we did a blast comparison against the non-redundant nucleotide collection (December 2016) of the 600 nt fragment immediately preceding the start codon for gene *repA* from the PFP plasmid pMP-R124, which comprises the REx-C module. The *rep* genes found immediately downstream of the first eight homologous sequences retrieved belong to the RepA-PFP, RepJ and RepA-RA2 families (Table S2). As before, and as an example, the REx-C module from pMP-R124 (RepA-PFP) shows a higher identity to pA506 (RepJ) than to the REx-C module from pPT14-32 (RepA-PFP) (**Figure 3**; Table S2). Incidentally, pMP-R124 and plasmids from the RepA-RA2 and RepA-Pa families also have a shorter REx-C module lacking the putative leader peptide gene sequence (**Figure 3** and Figure S5). Together, these results indicate that the REx-C modules from these four families evolved vertically within the family, but were also subjected to horizontal exchange between members of other replicon families. Likewise, they indicate that plasmids from these four families have similar mechanisms of initiation of replication and its control.

The REx-C Module Is Functionally Exchangeable among Different Replicons

Our comparative analyses of extant sequences (**Figure 3**) suggest that REx-C modules are freely exchanged among different replicons. However, previous works postulated that compatibility of coexisting RepA-PFP replicons was due to specificity between the C-terminal part of RepA and the loop sequence of SaL 2 (Figures S3, S6), which are highly variable and could coevolve for complementarity (Murillo and Keen, 1994; Gibbon et al., 1999; Stavrinides and Guttman, 2004; Ma et al., 2007). Additionally, the same pattern of variation is seen in a comparison of RepJ replicons (Figure S3). We therefore tested this putative specificity by swapping the respective SaL structures from the REx-C module (SaL fragment in **Figure 4**) and the *rep* fragments (partial *repI* plus REx-R, see **Figure 4**) from plasmids p1448A-B (RepA-PFP) and pPsv48C (RepA-PFP and RepJ) (**Figure 4** and Figure S6). These two RepA-PFP initiator proteins are 88% identical (92% similar; Table S1 and Figure S6), while they are not homologous to RepJ (Table S1). Likewise, the three replicons show a different loop in SaL 2 and 1–3 nt changes within SaL 3 (Figure S6). In spite of the differences in sequence, all the analyzed chimeras were able to sustain autonomous replication in the plasmidless strain *P. syringae* pv. syringae B728a (**Figure 4**), indicating a lack of specificity between the SaL structures from the REx-C module and the rest of the replicon.

To test whether or not sequence variations in SaL 1 (Figures S1, S6) would be specific for each Rep protein, we evaluated chimeric clones containing the SaL fragment from plasmid pPsv48A and the *rep* fragments of the RepA-PFP or RepJ replicons from pPsv48C. These two RepA-PFP proteins, from pPsv48A and pPsv48C, are 97% identical (Table S1 and Figure S6), but the pPsv48A REx-C module (and consequently the SaL fragment tested) is comparatively shorter and with several changes in SaL 1 and 2. In spite of the differences, both chimeras replicated autonomously in strain B728a (**Figure 4**), indicating that sequence variations in the loop sequence from SaL 1, and in the sequence upstream of this structure, do not significantly impact replication ability.

Finally, none of the four SaL fragments cloned in pKMAG generated any transformants when transferred to strains B728a or 1448A (the latter containing two plasmids that could supply the RepA initiator *in trans*), suggesting that the REx-C module cannot sustain replication by itself.

The REx-C Module Confers Replication Ability to a Non-replicative *repJ* Homolog

The REx-C module showed no apparent specificity with the associated Rep protein and is readily exchanged among replicons (**Figures 3, 4**), serving as a portable putative replication control region. Therefore, it is feasible that REx-C modules could move within bacterial genomes, co-opt genes with the appropriate characteristics and thereby directly create new autonomous replicons. To broadly test this, we examined the effect of the SaL structures from the RepA-PFP and RepJ replicons from pPsv48C on the replication ability of gene PSA3335_1080. This gene is located in the chromosome of *P. syringae* pv. savastanoi NCPPB 3335 and its deduced product shows 72.7% identity (81.2% similarity) with that from *repJ* (**Figure 2** and Figure S6; Table S1), but does not replicate autonomously (see above).

We therefore amplified and cloned a 1,772 nt fragment containing gene PSA3335_1080 plus the upstream 114 nt, to preserve the RBS and a similar spacing to the SaL structures as with *repJ*, and 815 nt downstream, spanning sequences homologous to SaL 5 (**Figure 4**). This fragment was ligated in the proper orientation to a partial fragment of the REx-C module containing the SaL structures preceding either *repA* or *repJ* from pPsv48C (279 nt), or preceding *repA* from pPsv48A (203 nt). The three constructions replicated in *P. syringae* pv. syringae B728a with very high efficiency (**Figure 4**). Disruption of the PSA3335_1080 reading frame by filling-in an internal restriction site abolished the replication ability of the clone containing the SaL fragment from the RepA-PFP replicon of pPsv48C. These results indicate that acquisition of a REx-C module can immediately confer autonomous replication ability to sequences containing PSA3335_1080, and that replication is dependent on the activity of this gene.

Replicon Incompatibility Associates to the REx-C Module

Their frequent exchange (**Figure 3**), and our experiments with chimeras (**Figure 4**), indicate that the REx-C modules have a low, or no specificity for their cognate REx-R module and that they probably confer a strong selective advantage. Among other possibilities, we speculated that the exchange of these modules could be a way to reduce or evade incompatibility, especially for coexisting plasmids carrying related replicons, such as the PFP group. The ColE2 replicon, related to the RepA-PFP replicons (Gibbon et al., 1999), contains two incompatibility determinants, corresponding to the control region and to the origin of replication (Tajima et al., 1988; Hiraga et al., 1994). Importantly, the RepA-PFP replicons contain sequences highly similar to the origin of replication of ColE2 replicons (Figure S7) (Yagura et al., 2006), including the primer RNA sequence (AGA), and located either within the RepA coding sequences or, for pMP-R124, situated some 300 nt after the *rep* gene stop codon.

We thus evaluated possible changes in the incompatibility behavior of chimeric replicons by transforming strain *P. syringae* pv. phaseolicola 1448A with the native RepA-PFP replicons from plasmids p1448A-B (B) and pPsv48C (C), and with their two corresponding chimeras of SaL and *rep* fragments (combinations B-C and C-B of the SaL-*rep* fragments; see **Figures 4, 5**). Strain 1448A contains naturally the native RepA-PFP plasmids p1448A-A (132 kb) and p1448A-B (52 kb). As expected, the native replicon from pPsv48C did not show any obvious incompatibility, producing a large number of transformants and coexisting with the two native plasmids from strain 1448A (**Figure 5**, lane C–C). The same results were observed with the C–B chimera (**Figure 5**, lane C–B), containing the SaL structures from plasmid C and the *rep* fragment from plasmid B. Conversely, the native replicon from plasmid B (B–B in **Figure 5**) and the B–C chimera generated about half the number of transformants than the two other replicons tested, taking double the time to reach colonies of the same size. Incompatibility mediated by the native replicon from plasmid B and the B-C chimera was evident in plasmid profile gels, where they either cointegrated with p1448A-B or induced its loss, or appeared with an apparently reduced copy number (**Figure 5**). These results indicate that strong replicon incompatibility between RepA-PFP replicons is associated to the REx-C module and, unlike what happens with ColE2 replicons (Tajima et al., 1988; Yasueda et al., 1994), not to the Rep protein or the origin of replication and that it can be overcome by only a few nucleotide changes in this module.

A

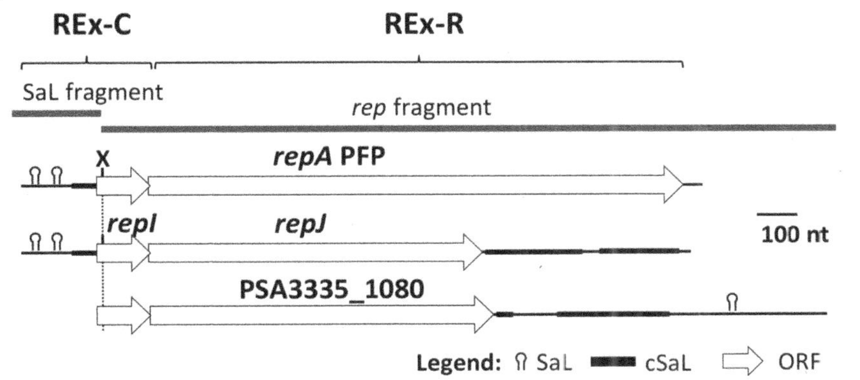

B

rep fragment / SaL fragment	pPsv48A (A)	p1448A-B (B)	RepA-pPsv48C (C)	RepJ-pPsv48C (J)	PSA3335_1080 (Ch)
pPsv48A (a)	+	nd	+	+	+
p1448A-B (b)	nd	+	+	nd	nd
RepA-pPsv48C (c)	nd	+	+	+	+
RepJ-pPsv48C (j)	nd	nd	+	+	+
PSA3335_1080 (ch)	nd	nd	nd	nd	-

C

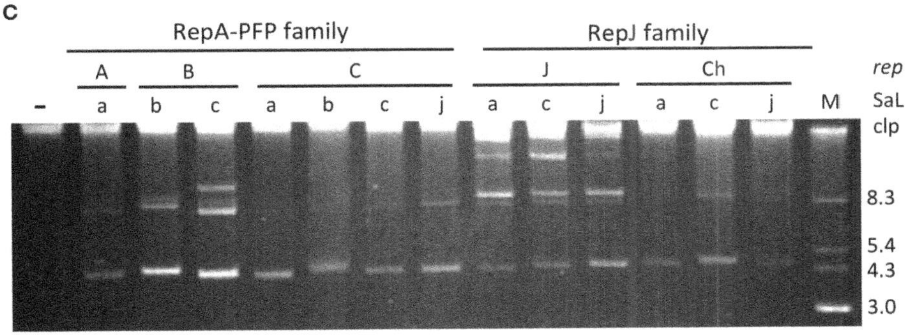

FIGURE 4 | The REx-C modules are functionally exchangeable within and among REx-R families. (A) Schema of the DNA fragments exchanged for the construction of chimeras, which were cloned in vector pKMAG immediately after the transcriptional terminator. X, XmnI restriction site. **(B)** Autonomous replication in *P. syringae* pv. syringae B728a of wild type and chimeric DNA regions: +, autonomous replication; -, no replication; nd, not determined. **(C)** Undigested plasmid profile gels of B728a harboring the different wild type and chimeric clones, with numbers indicating the combination of SaL and *rep* fragments as in **(B)**; discrete plasmid bands evidence autonomous replication, with multiple bands being different topological forms and/or multimers. Lane (-) is the wild type strain B728a, plasmidless; lane M, size markers, in kb; Clp, chromosome and high-molecular-weight plasmid multimers.

DISCUSSION

Modularity, which can be broadly defined as the degree to which a system is made up of relatively independent but interlocking parts, is ubiquitous in biology at all organization levels (Wagner et al., 2007; Kreimer et al., 2008). In this work, we demonstrate that diverse plasmid replicons from *Pseudomonas* are also modular, being composed of discrete functional units, or modules, that are readily and frequently exchanged among unrelated systems, colliding with the traditional view of minimal replicons as heritable units that evolve as a whole and that can be classified in more or less coherent incompatibility and phylogenetic groups (del Solar et al., 1998; Petersen, 2011).

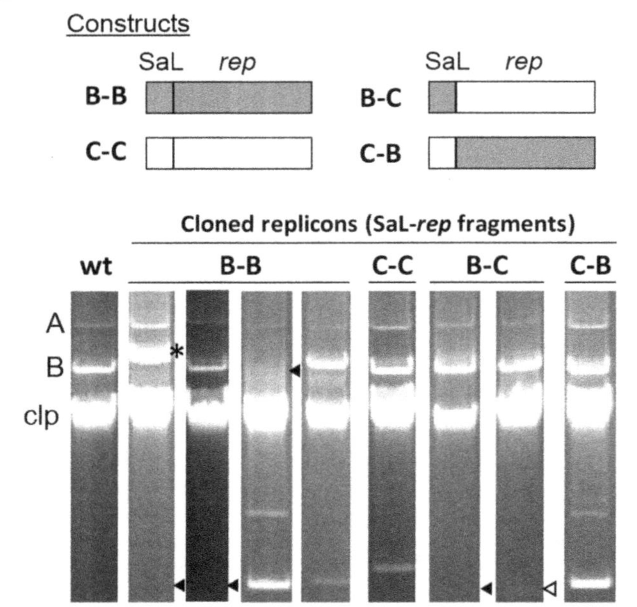

FIGURE 5 | The REx-C module, but not the REx-R module, confers plasmid incompatibility. The cartoon represents the native replicons and chimeras of the SaL fragment from the REx-C module and the *rep* fragments of plasmids p1448A-B (B) and pPsv48C (C), as indicated by the lettering, cloned in pKMAG. The bottom panels show representative examples of plasmid profiles of strain *P. syringae* pv. phaseolicola 1448A (wt) and transformants containing the native replicons and chimeras. Strain 1448A contains naturally the native RepA-PFP plasmids p1448A-A (A, 132 kb) and p1448A-B (B, 52 kb). Incompatibility is deduced from the loss of p1448A-B, or for the inability of the incoming plasmid to be maintained as an independent molecule. The asterisk indicates cointegrated plasmids; black triangles, missing plasmid band; open triangle, faint plasmid band (visible in the actual gel, but not in the picture). Clp, chromosome and linearized plasmids.

We have found extensive evidence of modularity in plasmids from many diverse species of *Pseudomonas*, including environmental species as well as plant, animal and human pathogens of paramount significance, such as *P. syringae* and *P. aeruginosa* (**Figures 2**, **3**, Table S2, and not shown). Nevertheless, it is highly likely that this concept is also applicable to plasmid replicons consisting of *rep* gene and control systems from other organisms. In particular, homologous and/or site-specific recombination has been postulated to contribute to a similar exchange of control and *rep* modules in related enterobacterial plasmids of the Incα and IncFII incompatibility groups (Kato and Mizobuchi, 1994) of the class A theta replicons, as well as different segments of ColE2 replicons (Hiraga et al., 1994); nevertheless, the concept of modularity as a general organizational model for replicons did not emerge from these studies. Based on their properties as functionally independent and exchangeable units, we defined two replicon exchangeable modules and designated them as REx-C (control) and REx-R (replication) modules, respectively (**Figure 1**).

Module REx-C consists of the replication control region, and has variable size and configuration. As defined for the RepJ replicon from pPsv48C, REx-C is a small sequence that contains three potential SaL structures, a putative leader peptide, a putative antisense RNA, and the signals for transcription and translation of the *rep* gene (Figure S1). Based on these characteristics, and by structural similarity to previously described replicons, it is highly likely that the REx-C module is involved in copy number control by the antisense RNA (del Solar et al., 1998; Brantl, 2014). Sequence comparison of extant replicons (**Figure 3**) and functional assays (**Figures 1**, **4**) indicate that this module is an autonomous but essential part of the replicon, and that it is functionally exchangeable among four non-homologous REx-R families (Table S1). The module contains a core region of 175–194 nt highly conserved among members of the four families (**Figure 3**, Figure S5) and that spans the three SaL structures essential for autonomous replication of the minimal RepJ (**Figure 1**) and the RepA-PFP replicons (Murillo and Keen, 1994; Gibbon et al., 1999). We showed that chimeric replicons containing different combinations of SaL structures and *rep* genes from the RepA-PFP and RepJ families were fully functional (**Figure 4**), demonstrating the independent functionality of the REx-C module and a lack of specificity with the REx-R module. Importantly, the *rep* fragments used for the construction of chimeras retained their corresponding peptide leader gene sequence and the RBS for the *rep* gene (**Figure 4**), suggesting that control of replication by the REx-C module does not involve the formation of secondary structures (pseudoknot) with sequences surrounding the *rep* gene. Additionally, the SaL structures also conferred replication ability to the *rep* fragment containing PSA3335_1080 (**Figure 4**), whose sequence is poorly conserved compared with the other replicons used for the construction of chimeras. Work with pA506 suggested that clones containing its REx-C module, without the *repJ* gene, were able to sustain autonomous replication (Stockwell et al., 2013), although these authors did not conclusively eliminate the possibility of ectopic integration. Conversely, evidence presented here and elsewhere (Murillo and Keen, 1994; Kwong et al., 1998; Gibbon et al., 1999; Sundin et al., 2004) shows that diverse REx-C modules were unable to replicate autonomously. Consequently, their functionality and conservation suggest that SaL structures in the REx-C module are essential for replication and replication control of a wide variety of replicons. This likely happens because the REx-C module harbors the signals required for expression of the *rep* gene, whose accessibility and proper recognition would be modulated by the interaction between the antisense RNA and the mRNA. Since the antisense RNA and its target sequence on the *rep* mRNA would be coded for in complementary strands of the same SaL region (Figure S1), they always show a perfect complementarity independently of any other downstream sequence. Therefore, the REx-C module would then act as an independent, self-contained portable unit for the control of replication.

The SaL structures are followed by a putative leader peptide that is present in diverse replicons as part of the region for the control of plasmid copy number (del Solar et al., 1998; Brantl, 2014). Previous works (Wagner et al., 1987; Blomberg et al., 1992) also showed that the level of expression of the *rep* gene is controlled by the level of translation of the leader peptide gene, and not by its product. This is compatible with the fact that an in-frame deletion of most of the putative leader peptide gene (*repI*) sequence in the RepJ replicon did not have any significant effect on replication (**Figure 1**). Likewise, lack of function for the product of this gene could justify the large sequence differences in the leader peptide genes of the RepJ and RepA-PFP replicons from pPsv48C, despite the rest of the REx-C module being highly similar (Figure S1). Remarkably, some of the REx-C modules examined here (e.g., from pMP-R124 and pRA2) lack the leader peptide gene, with the consequence that the start of the *rep* gene is in very close proximity to the end of SaL 3 (**Figure 3** and Figure S5). Nevertheless, the REx-C module is highly flexible and can accommodate different structures to ensure functionality in a diversity of replicon arrangements. For instance, functionality of the pRA2 replicon required a large region upstream of the *rep* gene, containing in this order: four potential iterons, an additional small gene (*repB*, 240 nt), and a fragment containing the conserved SaL structures shown in Figure S5 (Kwong et al., 1998).

We found four different families of REx-R modules defined by groups of homology of the corresponding Rep proteins (**Figure 3** and Table S1). The family RepA-PFP is composed of a single gene, coding for a large protein of as much as 437 amino acids homologous to many replication proteins from diverse species of Gram-negative and Gram-positive bacteria, including ColE2 replicons (Gibbon et al., 1999; Yagura et al., 2006). The RepA-PFP proteins contain highly conserved primase, PriCT and HTH domains, with the vegetative origin of replication located in the 3′ end of the *rep* gene or immediately after this gene (Figure S7) (Yagura et al., 2006). The other REx-R families contain smaller *rep* genes, with deduced products ranging from 269 to 341 amino acids and lacking homology to protein families and domains included in the InterPro database (Mitchell et al., 2015). These genes might also be accompanied by downstream sequences that can form complex folding structures and that are essential for replication, as it is the case with the RepJ replicon (**Figure 1**) and pRA2 (Kwong et al., 1998). Directed mutagenesis of the *repJ* (**Figure 1**), PSA3335_1080 and the *repA*-PFP (Gibbon et al., 1999) *rep* genes, demonstrates that they are essential for replication. Thus, the diversity of the REx-R modules found here again stresses the functional universality of the REx-C module.

The exchange of REx modules among disparate replicons is intriguing and can confer diverse evolutionary advantages, such as increasing their competitiveness by acquiring modules that are better adapted to their particular bacterial host. In this respect, the type of REx-C module characterized here is probably specific to *Pseudomonas* and its general presence in diverse replicons likely facilitates plasmid survival in these bacteria. Indeed, we were unable to find sequences homologous to the REx-C module in any other bacteria outside the genus *Pseudomonas*. Additionally, the RepA-PFP replicons analyzed here contain specific REx-C modules that are different from those associated to homologous RepA proteins, such as those from the enterobacterial ColE2 plasmids. Another likely advantage of modularity is to favor the coexistence of highly related replicons. This type of coexistence has been documented for PFP plasmids of pseudomonads (Murillo and Keen, 1994; Sundin, 2007) and *repABC*-type plasmids from rhizobia (Cevallos et al., 2002), and it also likely happens with RepJ replicons (**Figure 2**). The analysis of chimeric replicons indicates that plasmid incompatibility is conferred by the REx-C module (**Figure 5**), in agreement with previous results indicating a partial incompatibility of the corresponding module from plasmid pPT23A (Gibbon et al., 1999). Additionally, the incompatibility between plasmid p1448A-B and its cloned replicon was bypassed by swapping the REx-C module of the cloned replicon with another one differing in the sequence of the loop from SaL 2, as well as in a few other nt positions (**Figure 5** and Figure S6). Therefore, the exchange of REx-C modules could be a rapid and efficient way of reducing or eliminating incompatibility among highly related PFP plasmids, which usually carry genes conferring important adaptive advantages (Vivian et al., 2001; Sundin, 2007). Additionally, it is also likely that the coexistence of related modular replicons, requiring similar cellular resources, could help to streamline the replication process and reduce the metabolic burden of plasmids.

A further and exciting consequence derived from the modularity of replicons, is that their combination could potentially produce new replicons. In particular, the independent functionality of the REx-C module could allow it to co-opt appropriate genes that might then function as Rep initiators and result in new autonomously replicating molecules. But, would the bacterial gene pool contain genes that will function as plasmid replication initiators when associated to a REx-C module? Our experiments indicate that this is indeed the case: the non-replicative chromosomal *repJ* homolog PSA3335_1080 could replicate autonomously when preceded by a DNA fragment containing the SaL structures from the REx-C module. PSA3335_1080 is a chromosomal gene widely distributed in *P. syringae* and related species, conserving synteny with its adjacent sequences, which suggests that is a long time inhabitant of these bacteria possibly having a functional role different from replication. A priori, there could be a variety of genes that could generate new replicons when combined with an appropriate REx-C module and their identification could be challenging. For instance, manual or automatic prediction of structure or function of the RepJ protein and homologs would likely be unsuccessful because they lack obviously conserved domains typical of Rep proteins, and this might also be the case for other genes that could be recruited as replication initiators. Therefore, and as occurs with other systems (Agapakis and Silver, 2009; Lorenz et al., 2011; Melo et al., 2016), modularity of origins of replication will undoubtedly favor their evolution and adaptability, for instance by reducing incompatibility and metabolic load to the bacterial host, but could also be facilitating the generation of new plasmid replicons, with the concomitant possibility of immediate mobilization of associated chromosomal genes.

MATERIALS AND METHODS

Bacterial Strains, Plasmids, and Growth Conditions

Bacterial strains and plasmids used in this work are detailed in Table S3. *E. coli* strains DH10B and GM2929 (dcm^-, dam^-), when unmethylated DNA was needed, were used for DNA manipulations and were grown in LB at 37°C. Strains *P. syringae* pv. phaseolicola 1448A (Joardar et al., 2005), pv. syringae B728a (Feil et al., 2005), and pv. savastanoi NCPPB 3335 (Rodríguez-Palenzuela et al., 2010) and UPN912, which derives from strain NCPPB 3335 and is cured of its three native plasmids (M. Añorga, unpublished results), were propagated using King's medium B (King et al., 1954) at 25°C. When necessary, media were supplemented with 100 μg ml^{-1} ampicillin or 25 μg ml^{-1} kanamycin. We used a mixture of plasmids pME6031 (8.3 kb), pBBR1MCS-2 (5.4 kb), pKMAG-C (4.3 kb) and pBlueScript II (3.0 kb) (Table S3), purified from *E. coli* using the Illustra plasmidPrep Mini Spin kit (GE Healthcare, UK), as size markers in plasmid profile gels.

Molecular Techniques

DNA was amplified using a high fidelity enzyme (PrimeStar HS, Takara Bio Inc., Japan) and cloned using the CloneJET PCR Cloning Kit (Thermo Scientific) or the pGEM-T Easy Vector System (Promega), following the manufacturer's instructions. For plasmid profile gels, DNA was purified by alkaline lysis and separated by electrophoresis in 0.8% agarose gels with 1xTAE as described (Murillo et al., 1994). Plasmids were transferred to *P. syringae* by electroporation (Choi et al., 2006). For RT-PCR analysis, DNA-free RNA was obtained from bacterial cultures grown overnight in medium B using TriPure Isolation Reagent (Roche Diagnostics) and Ambion TURBO DNA-free Kit (Life Technologies). Concentration and purity of RNA were determined spectrophotometrically, and its integrity confirmed by electrophoresis in agarose gels. cDNA was synthesized from RNA using the ImProm-II reverse transcriptase system (Promega), following the manufacturer's recommendations. Primer Int_R, 5′-GCCGGTGCAGAGATACCC-3′, specific for the sense transcript of *repJ*, was used for the reaction: at 25°C for 5 min for primer annealing, 60 min at 42°C for reverse transcription, and 15 min at 70°C for enzyme inactivation. cDNA was amplified using primers B (5′-CGATGTAGATTCACGA ATCGCAG-3′), C (5′-CTGATTATGGCGTTCACTGC-3′), or D (5′-TGCAAGCTGTCTAAAGTGAAGC-3′) together with 1R (5′-GCTGTTGTTCAGAGAGATGACG-3′), to determine the size of the transcript, or primer pair Int_F (5′-GAGAAGTTTC TGGCCATCGAG-3′) and Int_R (5′-GCCGGTGCAGAGATAC CC-3′) to analyze the transcriptional activity of gene *repJ* (see **Figure 1**). The program used comprised 30 cycles (94°C for 30 s, 58°C for 30 s, and 72°C for 30 s) plus a final extension step of 6 min at 72°C. Control reactions included PCR amplification of pure extracted RNA, to verify the absence of contaminating DNA; amplification of purified DNA, to verify the reaction conditions, and amplification of an internal fragment of gene *gyrB*, to confirm the synthesis of cDNA.

Bioinformatics

Multiple-sequence alignments using Muscle, determination of the optimal substitution model, and Maximum likelihood phylogenetic tree construction using the JTT matrix-based model with a gamma distribution with 5 categories were done using MEGA7 (Kumar et al., 2016); confidence levels of the branching points were determined using 500 bootstraps replicates. Searches for sequence similarity in the NCBI databases were done using the BLAST algorithms (Hubbard et al., 2008) and sequences were aligned on-line using the MULTALIN program (Corpet, 1988) or the tools in the EMBL-EBI server (http://www.ebi.ac.uk/Tools/msa/). Search for protein motifs and fold recognition was done using the InterPro (Mitchell et al., 2015) (http://www.ebi.ac.uk/interpro/) and the Phyre2 (Kelley et al., 2015) web servers. Genome and nucleotide sequences were visualized and manipulated using the Artemis genome browser; when necessary, blast comparisons were visualized with ACT (Carver et al., 2008). Oligonucleotide primers were designed using Primer3plus software (Untergasser et al., 2012). Promoters were predicted using the online BPROM server (http://www.softberry.com). DNA or RNA Folding predictions were done using the Mfold web server (Zuker, 2003) using the default settings, except for a folding temperature of 25°C. Patterns of nucleotide polymorphism were calculated using DnaSP v 5.10.01 (Librado and Rozas, 2009).

Construction of Vectors

The *E. coli* pKMAG vector and the *E. coli-Pseudomonas* pKMAG-C vector were constructed to examine, respectively, the replication ability of cloned fragments and the expression of *repJ* in these fragments. To construct pKMAG, a PCR-amplified fragment from the pK184 vector (positions 71-2,064; accession no. U00800) (Jobling and Holmes, 1990), retaining the kanamycin resistant gene and the origin of replication and lacking the *lacZ* promoter, and a fragment from pME6041 (Heeb et al., 2000) (positions 3,809–4,255; accession no. AF118812), containing its polylinker and the T4 transcription terminator, were ligated together resulting in the new vector. The RepA-PFP replicon from pPsv48C (positions 41,791–1,428; accession no. FR820587) was then cloned into the unique AscI site of pKMAG, resulting into pKMAG-C. Construction details are included in Figure S2.

Replication Assays

Fragments used for the definition of a minimal RepJ replicon were amplified by PCR from strain Psv48ΔAB (Bardaji et al., 2011), a derivative of strain NCPPB 3335 containing only pPsv48C, cloned into pJET1.2 (Thermo Scientific) and then subcloned into pK184 in both directions, pKMAG and/or pKMAG-C. Primers used for amplifications were A (5′-AA AGCAGCGGATTTTGTAGG-3′), B, C, and D (described above), as forward, and E (5′-GACGCTAGGAGCCTATCC AG-3′), F (5′-TCCCTGTTTTTCCTGAAAGG-3′), and G (5′-GGTCGAACCGACCAACTG-3′), as reverse primers (see **Figure 1** and Figure S1). For the construction of chimeric replicons we first cloned in pKMAG amplicons containing the RepA-PFP and RepJ replicons, including the REx-C module in a 390–391 nt upstream fragment, from plasmids pPsv48C (coordinates 41,714–1,336, for RepA-PFP, and 23,309–31,027, for RepJ, from accession no. FR820587) and p1448A-B (coordinates 51,624–1,650 from accession no. CP000060). The SaL (coordinates 78,056–78,262, from accession no. FR820585) and the *rep* (coordinates 78,254–1,360, from FR820585) fragments from pPsv48A were amplified by PCR and cloned separately in pJET1.2 to reconstruct an XmnI site present in many RepA-PFP replicons and missing from pPsv48A, resulting in a T→C change in pos. 78,263. A double digestion XmnI-EcoRI or XmnI-XhoI of the resulting clones liberated the *rep* genes (*rep* fragment), leaving a 203–280 nt fragment attached to pKMAG that contained SaL structures 1 to 3 from the REx-C module (SaL fragment). The SaL and *rep* fragments were then separated by electrophoresis, purified and ligated in the appropriate combinations for the construction of chimeras. A 2,808 nt amplicon containing PSA3335_1080 (35,079–37,886 from accession no. KB644113), including 1,150 nt upstream and 815 nt downstream of the CDS, was cloned in pKMAG to evaluate replication of this gene in its native configuration. Additionally, an amplicon similar to the previous one but lacking the first 1,036 nt (36,115–37,886 from accession no. KB644113), and with an A→T change in pos. 36,115, was ligated in the proper orientation to the appropriate XmnI-EcoRI clones in pKMAG generated above and containing the SaL fragments from *repA* or *repJ* from pPsv48C, or from pPsv48A. For functional analyses of replication initiation protein genes, replicative constructions containing *repJ* and PSA3335_1080 were digested using a unique restriction site internal to each CDS, filled-in with Klenow enzyme (New England BioLabs Inc., UK) and subsequently religated. Enzymes used for disrupting the CDSs by changing the reading frame were SexAI, adding 5 nt to the *repJ* CDS, and EcoNI, adding 1 nt to the CDS of PSA3335_1080. For all replication experiments, at least two amplicons from two separate amplification experiments were cloned and tested; the identity and integrity of all clones was confirmed by sequencing. Replication ability was assessed by electroporation into the plasmidless strains *P. syringae* pv. syringae B728a and UPN912. Plasmids were confirmed to replicate autonomously by their ability to generate antibiotic-resistant transformants and, necessarily, by their appearance as independent bands in undigested plasmid profile gels, basically as described (Murillo and Keen, 1994; Sesma et al., 1998). Experiments were repeated at least three times, with similar results.

Incompatibility Assays

Incompatibility between native plasmids and cloned replicons was analyzed essentially as described (Nordström, 1993; Sesma et al., 1998). Briefly, constructs in pKMAG containing the native RepA-PFP replicons from plasmids pPsv48C (strain *P. syringae* pv. savastanoi NCPPB 3335) and p1448A-B (strain *P. syringae* pv. phaseolicola 1448A) and the corresponding chimeras of SaL and *rep* fragments, were isolated from *E. coli* and individually transferred to *P. syringae* pv. phaseolicola 1448A, which naturally contains the native plasmids p1448A-A (132 kb) and p1448A-B (52 kb) (Joardar et al., 2005). Resulting km^R transformants of strain 1448A were cultured overnight in liquid B medium with kanamycin at 25°C with shaking, and their plasmid content visualized by electrophoresis in plasmid profile gels to determine the possible eviction of native plasmids.

ACCESSION NUMBERS

Sequences of vectors pKMAG and pKMAG-C are deposited in GenBank under accession numbers KX714576 and KX714577.

AUTHOR CONTRIBUTIONS

LB and JM conceived the study and designed the experiments; LB and MA performed the experiments; LB, MA, JARM, GdS, and JM analyzed the data and interpreted the results; LB and JM drafted the manuscript with contributions from JARM and GdS; all authors read and approved the final manuscript.

ACKNOWLEDGMENTS

We are indebted to Pablo Llop and Theresa H. Osinga for revising the manuscript and help with English usage.

REFERENCES

Agapakis, C. M., and Silver, P. A. (2009). Synthetic biology: exploring and exploiting genetic modularity through the design of novel biological networks. *Mol. Biosyst.* 5, 704–713. doi: 10.1039/b901484e

Aviv, G., Rahav, G., and Gal-Mor, O. (2016). Horizontal transfer of the *Salmonella enterica* serovar infantis resistance and virulence plasmid pESI to the gut microbiota of warm-blooded hosts. *mBio* 7, e01395–e01316. doi: 10.1128/mBio.01395-16

Bardaji, L., Pérez-Martínez, I., Rodríguez-Moreno, L., Rodríguez-Palenzuela, P., Sundin, G. W., Ramos, C., et al. (2011). Sequence and role in virulence of the three plasmid complement of the model tumor-inducing bacterium *Pseudomonas savastanoi* pv. savastanoi NCPPB 3335. *PLoS ONE* 6:e25705. doi: 10.1371/journal.pone.0025705

Blomberg, P., Nordström, K., and Wagner, E. G. (1992). Replication control of plasmid R1: RepA synthesis is regulated by CopA RNA through inhibition of leader peptide translation. *EMBO J.* 11, 2675–2683.

Brantl, S. (2014). Plasmid replication control by antisense RNAs. *Microbiol. Spectr.* 2, PLAS-0001-2013. doi: 10.1128/microbiolspec.plas-0001-2013

Bruand, C., Le Chatelier, E., Ehrlich, S. D., and Jannière, L. (1993). A fourth class of theta-replicating plasmids: the pAM beta 1 family from gram-positive bacteria. *Proc. Natl. Acad. Sci. U.S.A.* 90, 11668–11672. doi: 10.1073/pnas.90.24.11668

Carver, T., Berriman, M., Tivey, A., Patel, C., Böhme, U., Barrell, B. G., et al. (2008). Artemis and ACT: viewing, annotating and comparing sequences stored in a relational database. *Bioinformatics* 24, 2672–2676. doi: 10.1093/bioinformatics/btn529

Cevallos, M. A., Porta, H., Izquierdo, J., Tun-Garrido, C., García-de-los-Santos, A., Dávila, G., et al. (2002). Rhizobium etli CFN42 contains at least three plasmids of the repABC family: a structural and evolutionary analysis. *Plasmid* 48, 104–116. doi: 10.1016/S0147-619X(02)00119-1

Choi, K. H., Kumar, A., and Schweizer, H. P. (2006). A 10-min method for preparation of highly electrocompetent Pseudomonas aeruginosa cells: application for DNA fragment transfer between chromosomes and plasmid transformation. *J. Microbiol. Meth.* 64, 391–397. doi: 10.1016/j.mimet.2005.06.001

Corpet, F. (1988). Multiple sequence alignment with hierarchical clustering. *Nucleic Acids Res.* 16, 10881–10890. doi: 10.1093/nar/16.22.10881

del Solar, G., and Espinosa, M. (2000). Plasmid copy number control: an ever-growing story. *Mol. Microbiol.* 37, 492–500. doi: 10.1046/j.1365-2958.2000.02005.x

del Solar, G., Giraldo, R., Ruiz-Echevarría, M. J., Espinosa, M., and Díaz-Orejas, R. (1998). Replication and control of circular bacterial plasmids. *Microbiol. Mol. Biol. Rev.* 62, 434–464.

Feil, H., Feil, W. S., Chain, P., Larimer, F., DiBartolo, G., Copeland, A., et al. (2005). Comparison of the complete genome sequences of *Pseudomonas syringae* pv. syringae B728a and pv. tomato DC3000. *Proc. Natl. Acad. Sci. U.S.A.* 102, 11064–11069. doi: 10.1073/pnas.0504930102

Gibbon, M. J., Sesma, A., Canal, A., Wood, J. R., Hidalgo, E., Brown, J., et al. (1999). Replication regions from plant-pathogenic *Pseudomonas syringae* plasmids are similar to ColE2-related replicons. *Microbiology* 145, 325–334. doi: 10.1099/13500872-145-2-325

Halary, S., Leigh, J. W., Cheaib, B., Lopez, P., and Bapteste, E. (2010). Network analyses structure genetic diversity in independent genetic worlds. *Proc. Natl. Acad. Sci. U.S.A.* 107, 127–132. doi: 10.1073/pnas.0908978107

Heeb, S., Itoh, Y., Nishijyo, T., Schnider, U., Keel, C., Wade, J., et al. (2000). Small, stable shuttle vectors based on the minimal pVS1 replicon for use in gram-negative, plant-associated bacteria. *Mol. Plant Microbe Interact.* 13, 232–237. doi: 10.1094/MPMI.2000.13.2.232

Hiraga, S., Sugiyama, T., and Itoh, T. (1994). Comparative analysis of the replicon regions of eleven ColE2-related plasmids. *J. Bacteriol.* 176, 7233–7243. doi: 10.1128/jb.176.23.7233-7243.1994

Hubbard, T. J. P., Aken, B. L., Ayling, S., Ballester, B., Beal, K., Bragin, E., et al. (2008). Ensembl 2009. *Nucleic Acids Res.* 37, D690–D697. doi: 10.1093/nar/gkn828

Itou, H., Yagura, M., Shirakihara, Y., and Itoh, T. (2015). Structural basis for replication origin unwinding by an initiator primase of plasmid ColE2-P9: DUPLEX DNA UNWINDING BY A SINGLE PROTEIN. *J. Biol. Chem.* 290, 3601–3611. doi: 10.1074/jbc.M114.595645

Jackson, R. W., Vinatzer, B., Arnold, D. L., Dorus, S., and Murillo, J. (2011). The influence of the accessory genome on bacterial pathogen evolution. *Mob. Genet. Elements* 1, 55–65. doi: 10.4161/mge.1.1.16432

Joardar, V., Lindeberg, M., Jackson, R. W., Selengut, J., Dodson, R., Brinkac, L. M., et al. (2005). Whole-genome sequence analysis of *Pseudomonas syringae* pv. phaseolicola 1448A reveals divergence among pathovars in genes involved in virulence and transposition. *J. Bacteriol.* 187, 6488–6498. doi: 10.1128/JB.187.18.6488-6498.2005

Jobling, M. G., and Holmes, R. K. (1990). Construction of vectors with the p15a replicon, kanamycin resistance, inducible lacZ alpha and pUC18 or pUC19 multiple cloning sites. *Nucleic Acids Res.* 18, 5315–5316. doi: 10.1093/nar/18.17.5315

Johnson, T. A., Stedtfeld, R. D., Wang, Q., Cole, J. R., Hashsham, S. A., Looft, T., et al. (2016). Clusters of antibiotic resistance genes enriched together stay together in swine agriculture. *mBio* 7, e02214–e02215. doi: 10.1128/mBio.02214-15

Kato, A., and Mizobuchi, K. (1994). Evolution of the replication regions of IncI alpha and IncFII plasmids by exchanging their replication control systems. *DNA Res.* 1, 201–212.

Kelley, L. A., Mezulis, S., Yates, C. M., Wass, M. N., and Sternberg, M. J. (2015). The Phyre2 web portal for protein modeling, prediction and analysis. *Nat. Protoc.* 10, 845–858. doi: 10.1038/nprot.2015.053

King, E. O., Ward, N. K., and Raney, D. E. (1954). Two simple media for the demonstration of pyocyanin and fluorescein. *J. Lab. Clin. Med.* 44, 301–307.

Kreimer, A., Borenstein, E., Gophna, U., and Ruppin, E. (2008). The evolution of modularity in bacterial metabolic networks. *Proc. Natl. Acad. Sci. U.S.A.* 105, 6976–6981. doi: 10.1073/pnas.0712149105

Kumar, S., Stecher, G., and Tamura, K. (2016). MEGA7: molecular evolutionary genetics analysis version 7.0 for bigger datasets. *Mol. Biol. Evol.* 33, 1870–1874. doi: 10.1093/molbev/msw054

Kwong, S. M., Yeo, C. C., Chuah, D., and Poh, C. L. (1998). Sequence analysis of plasmid pRA2 from *Pseudomonas alcaligenes* NCIB 9867 (P25X) reveals a novel replication region. *FEMS Microbiol. Lett.* 158, 159–165. doi: 10.1111/j.1574-6968.1998.tb12815.x

Librado, P., and Rozas, J. (2009). DnaSP v5: a software for comprehensive analysis of DNA polymorphism data. *Bioinformatics* 25, 1451–1452. doi: 10.1093/bioinformatics/btp187

Lilly, J., and Camps, M. (2015). Mechanisms of theta plasmid replication. *Microbiol. Spectr.* 3, PLAS-0029-2014. doi: 10.1128/microbiolspec.plas-0029-2014

Lorenz, D. M., Jeng, A., and Deem, M. W. (2011). The emergence of modularity in biological systems. *Phys. Life Rev.* 8, 129–160. doi: 10.1016/j.plrev.2011.02.003

Ma, Z., Smith, J. J., Zhao, Y., Jackson, R. W., Arnold, D. L., Murillo, J., et al. (2007). Phylogenetic analysis of the pPT23A plasmid family of *Pseudomonas syringae*. *Appl. Environ. Microbiol.* 73, 1287–1295. doi: 10.1128/AEM.01923-06

Mansfield, J., Genin, S., Magori, S., Citovsky, V., Sriariyanum, M., Ronald, P., et al. (2012). Top 10 plant pathogenic bacteria in molecular plant pathology. *Mol. Plant Pathol.* 13, 614–629. doi: 10.1111/j.1364-3703.2012.00804.x

Melo, D., Porto, A., Cheverud, J. M., and Marroig, G. (2016). Modularity: genes, development, and evolution. *Annu. Rev. Ecol. Evol. Syst.* 47, 463–486. doi: 10.1146/annurev-ecolsys-121415-032409

Mitchell, A., Chang, H. Y., Daugherty, L., Fraser, M., Hunter, S., Lopez, R., et al. (2015). The InterPro protein families database: the classification resource after 15 years. *Nucleic Acids Res.* 43, D213–D221. doi: 10.1093/nar/gku1243

Murillo, J., and Keen, N. T. (1994). Two native plasmids of *Pseudomonas syringae* pathovar tomato strain PT23 share a large amount of repeated DNA, including replication sequences. *Mol. Microbiol.* 12, 941–950. doi: 10.1111/j.1365-2958.1994.tb01082.x

Murillo, J., Shen, H., Gerhold, D., Sharma, A., Cooksey, D. A., and Keen, N. T. (1994). Characterization of pPT23B, the plasmid involved in syringolide production by *Pseudomonas syringae* pv. tomato PT23. *Plasmid* 31, 275–287. doi: 10.1006/plas.1994.1029

Nordström, K. (1993). "Plasmid replication and maintenance," in *Plasmids: A Practical Approach, 2nd Edn.*, ed K. G. Hardy (New York, NY: Oxford University Press), 1–38.

Nordström, K., and Wagner, E. G. (1994). Kinetic aspects of control of plasmid replication by antisense RNA. *Trends Biochem. Sci.* 19, 294–300. doi: 10.1016/0968-0004(94)90008-6

Novick, R. P. (1987). Plasmid incompatibility. *Microbiol. Rev.* 51, 381–395.

Petersen, J. (2011). Phylogeny and compatibility: plasmid classification in the genomics era. *Arch. Microbiol.* 193, 313–321. doi: 10.1007/s00203-011-0686-9

Ramos, J. L. (ed.). (2004). *Pseudomonas: Volume 1 Genomics, Life Style and Molecular Architecture*. New York, NY: Kluwer Academics-Plenum Publishers.

Rodríguez-Palenzuela, P., Matas, I. M., Murillo, J., López-Solanilla, E., Bardaji, L., Pérez-Martínez, I., et al. (2010). Annotation and overview of the *Pseudomonas savastanoi* pv. savastanoi NCPPB 3335 draft genome reveals the virulence gene complement of a tumour-inducing pathogen of woody hosts. *Environ. Microbiol.* 12, 1604–1620. doi: 10.1111/j.1462-2920.2010.02207.x

Romanchuk, A., Jones, C. D., Karkare, K., Moore, A., Smith, B. A., Jones, C., et al. (2014). Bigger is not always better: transmission and fitness burden of ~1 MB *Pseudomonas syringae* megaplasmid pMPPla107. *Plasmid* 73, 16–25. doi: 10.1016/j.plasmid.2014.04.002

Ruiz-Masó, J. A., Machón, C., Bordanaba-Ruiseco, L., Espinosa, M., Coll, M., and Del Solar, G. (2015). Plasmid rolling-circle replication. *Microbiol. Spectr.* 3, PLAS-0035-2014. doi: 10.1128/microbiolspec.plas-0035-2014

Schumacher, M. A., Tonthat, N. K., Kwong, S. M., Chinnam, N. B., Liu, M. A., Skurray, R. A., et al. (2014). Mechanism of staphylococcal multiresistance plasmid replication origin assembly by the RepA protein. *Proc. Natl. Acad. Sci. U.S.A.* 111, 9121–9126. doi: 10.1073/pnas.1406065111

Sesma, A., Sundin, G. W., and Murillo, J. (1998). Closely related replicons coexisting in the phytopathogen *Pseudomonas syringae* show a mosaic organization of the replication region and altered incompatibility behavior. *Appl. Environ. Microbiol.* 64, 3948–3953.

Sesma, A., Sundin, G. W., and Murillo, J. (2000). Phylogeny of the replication regions of pPT23A-like plasmids from *Pseudomonas syringae*. *Microbiology* 146, 2375–2384. doi: 10.1099/00221287-146-10-2375

Stavrinides, J., and Guttman, D. (2004). Nucleotide sequence and evolution of the five-plasmid complement of the phytopathogen *Pseudomonas syringae* pv. maculicola ES4326. *J. Bacteriol.* 186, 5101–5115. doi: 10.1128/JB.186.15.5101-5115.2004

Stockwell, V. O., Davis, E. W. II, Carey, A., Shaffer, B. T., Mavrodi, D. V., Hassan, K. A., et al. (2013). pA506, a conjugative plasmid of the plant epiphyte Pseudomonas fluorescens A506. *Appl. Environ. Microbiol.* 79, 5272–5282. doi: 10.1128/AEM.01354-13

Summers, D. K. (1996). *The Biology of Plasmids*. Oxford; Cambridge, MA: Blackwell Science. doi: 10.1002/9781444313741

Sundin, G. W. (2007). Genomic insights into the contribution of phytopathogenic bacterial plasmids to the evolutionary history of their hosts. *Annu. Rev. Phytopathol.* 45, 129–151. doi: 10.1146/annurev.phyto.45.062806.094317

Sundin, G. W., Mayfield, C. T., Zhao, Y., Gunasekera, T. S., Foster, G. L., and Ullrich, M. S. (2004). Complete nucleotide sequence and analysis of pPSR1 (72,601 bp), a pPT23A-family plasmid from *Pseudomonas syringae* pv. syringae A2. Mol. *Genet. Genomics* 270, 462–476. doi: 10.1007/s00438-003-0945-9

Sundin, G. W., and Murillo, J. (2009). "Gene traders: characteristics of native plasmids from plant pathogenic bacteria," in *Plant Pathogenic Bacteria: Genomics and Molecular Biology*, ed R. W. Jackson (Cambs: Caister Academic Press), 295–310.

Tajima, Y., Horii, T., and Itoh, T. (1988). Replication of ColE2 and ColE3 plasmids: two ColE2 incompatibility functions. *Mol. Gen. Genet.* 214, 451–455. doi: 10.1007/BF00330479

Untergasser, A., Cutcutache, I., Koressaar, T., Ye, J., Faircloth, B. C., Remm, M., et al. (2012). Primer3—new capabilities and interfaces. *Nucleic Acids Res.* 40:e115. doi: 10.1093/nar/gks596

Vivian, A., Murillo, J., and Jackson, R. W. (2001). The role of plasmids in phytopathogenic bacteria: mobile arsenals? *Microbiology* 147, 763–780. doi: 10.1099/00221287-147-4-763

Wagner, E. G., von Heijne, J., and Nordström, K. (1987). Control of replication of plasmid R1: translation of the 7k reading frame in the RepA mRNA leader region counteracts the interaction between CopA RNA and CopT RNA. *EMBO J.* 6, 515–522.

Wagner, G. P., Pavlicev, M., and Cheverud, J. M. (2007). The road to modularity. *Nat. Rev. Genet.* 8, 921–931. doi: 10.1038/nrg2267

Yagura, M., Nishio, S. Y., Kurozumi, H., Wang, C. F., and Itoh, T. (2006). Anatomy of the replication origin of plasmid ColE2-P9. *J. Bacteriol.* 188, 999–1010. doi: 10.1128/JB.188.3.999-1010.2006

Yasueda, H., Takechi, S., Sugiyama, T., and Itoh, T. (1994). Control of ColE2 plasmid replication: negative regulation of the expression of the plasmid-specified initiator protein, Rep, at a posttranscriptional step. *Mol. Gen. Genet.* 244, 41–48. doi: 10.1007/BF00280185

Zuker, M. (2003). Mfold web server for nucleic acid folding and hybridization prediction. *Nucleic Acids Res.* 31, 3406–3415. doi: 10.1093/nar/gkg595

Permissions

All chapters in this book were first published in TGTBTUMRBHL, by Frontiers; hereby published with permission under the Creative Commons Attribution License or equivalent. Every chapter published in this book has been scrutinized by our experts. Their significance has been extensively debated. The topics covered herein carry significant findings which will fuel the growth of the discipline. They may even be implemented as practical applications or may be referred to as a beginning point for another development.

The contributors of this book come from diverse backgrounds, making this book a truly international effort. This book will bring forth new frontiers with its revolutionizing research information and detailed analysis of the nascent developments around the world.

We would like to thank all the contributing authors for lending their expertise to make the book truly unique. They have played a crucial role in the development of this book. Without their invaluable contributions this book wouldn't have been possible. They have made vital efforts to compile up to date information on the varied aspects of this subject to make this book a valuable addition to the collection of many professionals and students.

This book was conceptualized with the vision of imparting up-to-date information and advanced data in this field. To ensure the same, a matchless editorial board was set up. Every individual on the board went through rigorous rounds of assessment to prove their worth. After which they invested a large part of their time researching and compiling the most relevant data for our readers.

The editorial board has been involved in producing this book since its inception. They have spent rigorous hours researching and exploring the diverse topics which have resulted in the successful publishing of this book. They have passed on their knowledge of decades through this book. To expedite this challenging task, the publisher supported the team at every step. A small team of assistant editors was also appointed to further simplify the editing procedure and attain best results for the readers.

Apart from the editorial board, the designing team has also invested a significant amount of their time in understanding the subject and creating the most relevant covers. They scrutinized every image to scout for the most suitable representation of the subject and create an appropriate cover for the book.

The publishing team has been an ardent support to the editorial, designing and production team. Their endless efforts to recruit the best for this project, has resulted in the accomplishment of this book. They are a veteran in the field of academics and their pool of knowledge is as vast as their experience in printing. Their expertise and guidance has proved useful at every step. Their uncompromising quality standards have made this book an exceptional effort. Their encouragement from time to time has been an inspiration for everyone.

The publisher and the editorial board hope that this book will prove to be a valuable piece of knowledge for researchers, students, practitioners and scholars across the globe.

List of Contributors

Leticia Rodríguez and Francisco García-del Portillo
Laboratorio de Patógenos Bacterianos Intracelulares, Departamento de Biotecnología Microbiana, Centro Nacional deBiotecnología-Consejo Superior de Investigaciones Científicas (CNB-CSIC), Madrid, Spain

M. Graciela Pucciarelli
Laboratorio de Patógenos Bacterianos Intracelulares, Departamento de Biotecnología Microbiana,Centro Nacional deBiotecnología-Consejo Superior de Investigaciones Científicas (CNB-CSIC), Madrid,Spain
Departamento de BiologíaMolecular, Universidad Autónoma de Madrid, Madrid, Spain
Centro de Biología Molecular Severo Ochoa-Consejo Superior de Investigaciones Científicas (CBMSO-CSIC), Madrid, Spain

Andrés Miguel-Arribas, Gayetri Ramachandran, Jorge Val-Calvo, César Gago-Córdoba,Daniel González-Álvarez, David Abia and Wilfried J. J. Meijer
Department of Virology and Microbiology, Centro de Biología Molecular "Severo Ochoa" (CSIC-UAM), Instituto de BiologíaMolecular "Eladio Viñuela" (CSIC), Autonomous University of Madrid, Madrid, Spain

Carlos Alfonso and Juan R. Luque-Ortega
Centro de Investigaciones Biológicas (CSIC), Madrid, Spain

Ling J. Wu
Centre for Bacterial Cell Biology, Institute for Cell and Molecular Biosciences, Newcastle University, Newcastle Upon Tyne,United Kingdom

Jian-An Hao
Department of Virology and Microbiology, Centro de Biología Molecular "Severo Ochoa" (CSIC-UAM), Instituto de BiologíaMolecular "Eladio Viñuela" (CSIC), Autonomous University of Madrid, Madrid, Spain
The Institute of Seawater Desalinationand Multipurpose Utilization (SOA), Tianjin, China

José Á. Ruiz-Masó, Luis M. Luengo, Inmaculada Moreno-Córdoba, Ramón Díaz-Orejas and Gloria del
Molecular Microbiology and Infection Biology Department, Centro de Investigaciones Biológicas, Consejo Superior deInvestigaciones Científicas, Madrid, Spain

Aleksandra Dydecka, Sylwia Bloch, Bozena Nejman-Falenczyk, Gracja Topka, Agnieszka Necel and Grzegorz Wegrzyn
Department of Molecular Biology, Faculty of Biology, University of Gdansk, Gdansk, Poland

Ali Rizvi, Shaili Perez, Logan W. Donaldson and Alicja
Department of Biology, YorkUniversity, Toronto, ON, Canada

Tomasz Gasior
Institute of Biochemistry and Biophysics, Polish Academy of Sciences

Brigitte Pertschy, Lisa Pöltl, Bettina Konrad, Priya Bhutada, Dina Vorkapic, Denise Atzmüller
Institute of Molecular Biosciences, University of Graz, Graz, Austria

Ellen L. Zechner
Institute of Molecular Biosciences, University of Graz, Graz, Austria
BioTechMed-Graz,Graz, Austria

Gregor Gorkiewicz
Institute of Pathology, Medical University of Graz,Graz, Austria
BioTechMed-Graz,Graz, Austria

Christoph Högenauer
Division of Gastroenterology and Hepatology, Medical University of Graz, Graz, Austria

Florian Feist
Vehicle Safety Institute, Graz University of Technology, Graz, Austria

Hanna Sprenger
Institute of Molecular Biosciences, University of Graz, Graz, Austria
Institute of Pathology, Medical University of Graz,Graz, Austria
Division of Gastroenterology and Hepatology, Medical University of Graz, Graz, Austria

Sabine Kienesberger
Institute of Molecular Biosciences, University of Graz, Graz, Austria
Institute of Pathology, Medical University of Graz,Graz, Austria
BioTechMed-Graz,Graz, Austria

María Moreno-del Álamo, Mariangela Tabone, Virginia S. Lioy and Juan C. Alonso
Department of Microbial Biotechnology, Centro Nacional de Biotecnología (CSIC), Madrid, Spain

Marko Djordjevic
Faculty of Biology, Institute of Physiology and Biochemistry, University of Belgrade, Belgrade, Serbia

Andjiela Rodic
Multidisciplinary PhD Program in Biophysics, University of Belgrade, Belgrade, Serbia

Bojana Blagojevic, Magdalena Djordjevic,
Institute of Physics Belgrade, University of Belgrade,Belgrade, Serbia

Konstantin Severinov
Waksman Institute of Microbiology, Rutgers University, Piscataway, NJ, United States
Skolkovo Institute of Science and Technology, Skolkovo, Russia

Marko Djordjevic
Institute of Physiology and Biochemistry, Faculty of Biology, University of Belgrade, Belgrade, Serbia

Magdalena Djordjevic
Institute of PhysicsBelgrade, University of Belgrade, Belgrade, Serbia

Evgeny Zdobnov
Swiss Institute of Bioinformatics and Department of Genetic Medicineand Development, University of Geneva, Geneva, Switzerland

Seema Alnajar, Bijendra Khadka and Radhey S. Gupta
Department of Biochemistry and Biomedical Sciences, McMaster University, Hamilton, ON, Canada

Ramón Díaz-Orejas and Manuel Espinosa
Centro de Investigaciones Biológicas, Consejo Superior de Investigaciones Científicas (CSIC), Madrid, Spain

Chew Chieng Yeo
Faculty ofMedicine, Biomedical Research Centre, Universiti Sultan Zainal Abidin, Kuala Terengganu, Malaysia

Soo Sum Lean
Saw Swee Hock School of Public Health, National University of Singapore, Singapore, Singapore

Chew Chieng Yeo
Faculty of Medicine,Biomedical Research Centre, Universiti Sultan Zainal Abidin, Kuala Terengganu, Malaysia

Adrian Pérez-Ramos, Alicia Prieto, Mari L. Mohedano and Paloma López
Biological Research Center (CIB), Consejo Superior de Investigaciones Científicas, Madrid, Spain

Pasquale Russo and Giuseppe Spano
Department of Agricultural, Food and Environmental Sciences, University of Foggia, Foggia, Italy

Maria L. Werning
Biological Research Center (CIB), Consejo Superior de Investigaciones Científicas, Madrid, Spain
Center of Research and Transfer of Catamarca (CITCA), Consejo Nacional de Investigaciones Científicas y Técnicas, Catamarca, Argentina

María J. Ferrándiz, Antonio J. Martín-Galiano, Jose M. Tirado-Vélez
Unidad de Genética Bacteriana, Centro Nacional de Microbiología, Instituto de Salud Carlos III, Madrid, Spain

Adela G. de la Campa
Presidencia, Consejo Superior de Investigaciones Científicas, Madrid, Spain

María T. García
Departamento de Microbiología, Facultad deCiencias Biológicas, Universidad Complutense, Madrid, Spain

Montserrat Nácher-Vázquez, José A. Ruiz-Masó, María L. Mohedano, Gloria del Solar,PalomaLópez
Department of Molecular Microbiology and Infection Biology, Biological Research Center, Spanish National ResearchCouncil (CSIC), Madrid, Spain

Rosa Aznar
Department of Food Safety and Preservation, Institute of Agrochemistry and FoodTechnology, CSIC, Paterna, Spain
Department of Microbiology and Ecology, University of Valencia, Burjassot, Spain

Liying Wang, Chuanyong Jing
State Key Laboratory of Environmental Chemistry and Ecotoxicology, Research Center for Eco-Environmental Sciences,Chinese Academy of Sciences, Beijing, China
College of Resources and Environment, University of Chinese Academy of Sciences, Beijing, China

Jin Wang
Department of Municipal and Environmental Engineering, School of Civil Engineering, Beijing Jiaotong University, Beijing, China

Elisabeth Grohmann
Beuth University of Applied Sciences, Berlin, Germany

Jananee Jaishankar and Preeti Srivastava
Department of Biochemical Engineering and Biotechnology, Indian Institute of Technology, New Delhi, India

Carlos Garbisu, Lur Epelde and Itziar Alkorta
Soil Microbial Ecology Group, Department of Conservation of Natural Resources, Neiker Tecnalia, Derio, Spain

Olatz Garaiyurrebaso
Instituto Biofisika (UPV/EHU, CSIC), Department of Biochemistry and Molecular Biology, University of the Basque Country, Bilbao,Spain

Venkova
Fox Chase Cancer Center, Research & Development Alliances, Rockledge, PA, United States

Chew Chieng Yeo
Faculty of Medicine,University Sultan Zainal Abidin Medical Campus, Kuala Terengganu, Malaysia

Manuel Espinosa
Centro de Investigaciones Biológicas,Consejo Superior de Investigaciones Científicas, Madrid, Spain

Leire Bardaji, Maite Añorga, Jesús Murillo
Departamento de Producción Agraria, Escuela Técnica Superior de Ingenieros Agrónomos, Universidad Pública de Navarra,Pamplona, Spain

José A. Ruiz-Masó and Gloria del Solar
Molecular Biology of Gram-Positive Bacteria, Molecular Microbiology and Infection Biology, Centro deInvestigaciones Biológicas (Consejo Superior de Investigaciones Científicas), Madrid, Spain

Index

www.ingramcontent.com/pod-product-compliance
Lightning Source LLC
LaVergne TN
LVHW061511220926
848485LV00017B/873
* 9 7 8 1 6 3 9 8 9 0 6 7 5 *